Numerical Linear Algebra and Applications
Second Edition

Numerical Linear Algebra and Applications
Second Edition

Biswa Nath Datta
Northern Illinois University
DeKalb, Illinois

Society for Industrial and Applied Mathematics
Philadelphia

Copyright © 2010 by the Society for Industrial and Applied Mathematics

10 9 8 7 6 5 4 3 2

All rights reserved. Printed in the United States of America. No part of this book may be reproduced, stored, or transmitted in any manner without the written permission of the publisher. For information, write to the Society for Industrial and Applied Mathematics, 3600 Market Street, 6th Floor, Philadelphia, PA 19104-2688 USA.

Trademarked names may be used in this book without the inclusion of a trademark symbol. These names are used in an editorial context only; no infringement of trademark is intended.

MATLAB is a registered trademark of The MathWorks, Inc. For MATLAB product information, please contact The MathWorks, Inc., 3 Apple Hill Drive, Natick, MA 01760-2098 USA, 508-647-7000, Fax: 508-647-7001, info@mathworks.com, *www.mathworks.com.*

Library of Congress Cataloging-in-Publication Data
Datta, Biswa Nath.
  Numerical linear algebra and applications / Biswa Nath Datta. -- 2nd ed.
      p. cm.
  Includes bibliographical references and index.
  ISBN 978-0-898716-85-6
  1. Algebras, Linear. 2. Numerical analysis. I. Title.
  QA184.2.D38 2009
  512'.5--dc22
                                                                  2009025104

 is a registered trademark.

Dedicated to my
mathematician wife

Karabi

for her love of the subject
and her endless encouragement

Contents

Preface   xvii
- 0.1   Special Features xvii
- 0.2   Additional Features and Topics for Second Edition xix
- 0.3   Intended Audience xx
- 0.4   Some Guidelines for Using this Book xx
  - 0.4.1   A First Course in Numerical Linear Algebra (Advanced Undergraduate/One Semester) xxi
  - 0.4.2   A Second Course in Numerical Linear Algebra (Beginning Graduate Course) xxi
  - 0.4.3   A One-Semester Course in Numerical Linear Algebra for Engineers xxii
- 0.5   Acknowledgments xxii

1   Linear Algebra Problems, Their Importance, and Computational Difficulties   1
- 1.1   Introduction 1
- 1.2   Fundamental Linear Algebra Problems and Their Importance 1
- 1.3   Computational Difficulties Using Theoretical Linear Algebra Techniques   3

2   A Review of Some Required Concepts from Core Linear Algebra   7
- 2.1   Introduction 7
- 2.2   Vectors 7
  - 2.2.1   Orthogonality, Subspace, and Basis 8
- 2.3   Matrices 9
  - 2.3.1   Range and Null Spaces 12
  - 2.3.2   Rank of a Matrix 12
  - 2.3.3   The Inverse of a Matrix 13
  - 2.3.4   Similar Matrices 14
- 2.4   Some Special Matrices 14
  - 2.4.1   Diagonal and Triangular Matrices 14
  - 2.4.2   Unitary and Orthogonal Matrices 15
  - 2.4.3   Symmetric and Hermitian Matrices 16
  - 2.4.4   Hessenberg Matrices (Almost Triangular) 16
- 2.5   Vector and Matrix Norms 17
  - 2.5.1   Vector Norms 17
  - 2.5.2   Matrix Norms 18

		2.5.3	Norms and Inverses	22
		2.5.4	Norm Invariant Properties of Orthogonal and Unitary Matrices	22
	2.6	Singular Value Decomposition		23
	2.7	Review and Summary		24
		2.7.1	Special Matrices	24
		2.7.2	Rank, Determinant, Inverse, and Eigenvalues	24
		2.7.3	Vector and Matrix Norms	24
	2.8	Suggestions for Further Reading		24
	Exercises on Chapter 2			25
3	**Floating Point Numbers and Errors in Computations**			**29**
	3.1	Floating Point Number Systems		29
	3.2	Rounding Errors		31
	3.3	Laws of Floating Point Arithmetic		33
	3.4	Addition of n Floating Point Numbers		36
	3.5	Multiplication of n Floating Point Numbers		37
	3.6	Inner Product Computation		38
	3.7	Error Bounds for Floating Point Matrix Operations		39
	3.8	Round-Off Errors Due to Cancellation and Recursive Computations		40
	3.9	Review and Summary		44
	3.10	Suggestions for Further Reading		45
	Exercises on Chapter 3			45
4	**Stability of Algorithms and Conditioning of Problems**			**49**
	4.1	Introduction		49
		4.1.1	Computing the Norm of a Vector	49
		4.1.2	Computing the Inner Product of Two Vectors	49
		4.1.3	Solution of an Upper Triangular System	51
		4.1.4	Solution of a Lower Triangular System	51
	4.2	Efficiency of an Algorithm		52
	4.3	Definition and Concept of Stability		52
	4.4	Conditioning of the Problem and Perturbation Analysis		57
	4.5	Conditioning of the Problem, Stability of the Algorithm, and Accuracy of the Solution		59
	4.6	Perturbation Analysis of the Linear System Problem		61
		4.6.1	Effect of Perturbation on the Right-Hand Side Vector b	61
		4.6.2	Effect of Perturbation in the Matrix A	64
		4.6.3	Effect of Perturbations in Both Matrix A and Vector b	65
	4.7	Some Properties of the Condition Number of a Matrix		67
		4.7.1	Some Well-Known Ill-Conditioned Matrices	68
		4.7.2	How Large Must the Condition Number Be for Ill-Conditioning?	69
		4.7.3	The Condition Number and Nearness to Singularity	69
		4.7.4	Examples of Ill-Conditioned Eigenvalue Problems	70
	4.8	Some Guidelines for Designing Stable Algorithms		72

Contents

	4.9	Review and Summary	72
		4.9.1 Conditioning of the Problem	72
		4.9.2 Stability of an Algorithm	73
		4.9.3 Effects of Conditioning and Stability on the Accuracy of the Solution	74
	4.10	Suggestions for Further Reading	74
	Exercises on Chapter 4	74

5 Gaussian Elimination and LU Factorization   81
  5.1 A Computational Template in Numerical Linear Algebra 81
  5.2 LU Factorization Using Gaussian Elimination 82
      5.2.1 Creating Zeros in a Vector or Matrix Using Elementary Matrix 83
      5.2.2 Triangularization Using Gaussian Elimination 84
      5.2.3 Permutation Matrices and Their Properties 96
      5.2.4 Gaussian Elimination with Partial Pivoting (GEPP) 97
      5.2.5 Gaussian Elimination with Complete Pivoting (GECP) . . . 104
      5.2.6 Summary of Gaussian Elimination and LU Factorizations . . 106
  5.3 Stability of Gaussian Elimination 106
  5.4 Summary and Table of Comparisons 109
      5.4.1 Elementary Lower Triangular Matrix 110
      5.4.2 LU Factorization 110
      5.4.3 Stability of Gaussian Elimination 110
      5.4.4 Table of Comparisons 110
  5.5 Suggestions for Further Reading 110
  Exercises on Chapter 5 111

6 Numerical Solutions of Linear Systems   117
  6.1 Introduction 117
  6.2 Basic Results on Existence and Uniqueness 118
  6.3 Some Applications Giving Rise to Linear Systems Problems 119
      6.3.1 An Electric Circuit Problem 119
      6.3.2 Analysis of a Processing Plant Consisting of Interconnected Reactors 120
      6.3.3 Linear Systems Arising from Ordinary Differential Equations: A Case Study on a Spring-Mass Problem 122
      6.3.4 Linear Systems Arising from Partial Differential Equations: A Case Study on Temperature Distribution 123
      6.3.5 Approximation of a Function by a Polynomial: Hilbert System 128
  6.4 LU Factorization Methods 129
      6.4.1 Solution of the System $Ax = b$ Using LU Factorization . . . 129
      6.4.2 Solution of $Ax = b$ Using Factorization: $MA = U$ 132
      6.4.3 Solution of $Ax = b$ without Explicit Factorization 132
      6.4.4 Solving a Linear System with Multiple Right-Hand Sides . . 134
  6.5 Scaling 135

6.6		Concluding Remarks on the Use of Gaussian Elimination for Linear Systems 136	
6.7		Inverses and Determinant 136	
	6.7.1	Avoiding Explicit Computation of the Inverses	137
	6.7.2	The Sherman–Morrison Formula for Matrix Inverse	137
	6.7.3	Computing the Inverse of an Arbitrary Nonsingular Matrix .	138
	6.7.4	Computing the Determinant of a Matrix	139
6.8		Effect of the Condition Number on Accuracy of the Computed Solution 139	
	6.8.1	Conditioning and Pivoting	140
	6.8.2	Conditioning and Scaling	141
6.9		Computing and Estimating the Condition Number	141
6.10		Componentwise Perturbations and the Errors	143
6.11		Iterative Refinement	144
6.12		Special Systems: Positive Definite, Diagonally Dominant, Hessenberg, and Tridiagonal 146	
	6.12.1	Special Linear Systems Arising from Finite Difference Methods 147	
	6.12.2	Special Linear Systems Arising from Finite Element Methods 150	
	6.12.3	Symmetric Positive Definite Systems	153
	6.12.4	Hessenberg System	158
	6.12.5	Diagonally Dominant Systems	160
	6.12.6	Tridiagonal Systems	161
6.13		Review and Summary	167
	6.13.1	Numerical Methods for Arbitrary Linear System Problems .	167
	6.13.2	Special Systems	168
	6.13.3	Inverse and Determinant	168
	6.13.4	The Condition Number and Accuracy of Solution	169
	6.13.5	Iterative Refinement	169
6.14		Suggestions for Further Reading	169
	Exercises on Chapter 6		170

7 QR Factorization, Singular Value Decomposition, and Projections   181

7.1		Introduction	181
7.2		Householder's Matrices and QR Factorization	183
	7.2.1	Definition and Basic Properties	183
	7.2.2	Householder's Method for QR Factorization	188
7.3		Complex QR Factorization	194
7.4		Givens Matrices and QR Factorization	194
	7.4.1	Definition and Basic Properties	194
	7.4.2	Givens Method for QR Factorization	199
	7.4.3	QR Factorization of a Hessenberg Matrix Using Givens Matrices 201	
7.5		Classical and Modified Gram–Schmidt Algorithms for QR Factorizations 202	
7.6		Solution of $Ax = b$ Using QR Factorization	208

	7.7	Projections Using QR Factorization 209	
		7.7.1 Orthogonal Projections and Orthonormal Bases 209	
		7.7.2 Projection of a Vector onto the Range and the Null Space of a Matrix 210	
		7.7.3 Orthonormal Bases and Orthogonal Projections onto the Range and Null Space Using QR Factorization 210	
	7.8	Singular Value Decomposition and Its Properties 211	
		7.8.1 Singular Values and Singular Vectors 212	
		7.8.2 Computation of the SVD (MATLAB Command) 213	
		7.8.3 The SVD of a Complex Matrix 213	
		7.8.4 Geometric Interpretation of the Singular Values and Singular Vectors 214	
		7.8.5 Reduced SVD 214	
		7.8.6 Sensitivity of the Singular Values 215	
		7.8.7 Norms, Condition Number, and Rank via the SVD 216	
		7.8.8 The Distance to Singularity, Rank-Deficiency, and Numerical Rank via the SVD. 217	
		7.8.9 Numerical Rank 219	
		7.8.10 Orthonormal Bases and Projections from the SVD 220	
	7.9	Some Practical Applications of the SVD 221	
	7.10	Geometric Mean and Generalized Triangular Decompositions 226	
	7.11	Review and Summary 227	
		7.11.1 QR Factorization 227	
		7.11.2 The SVD, GMD, and GTD 227	
		7.11.3 Projections 228	
	7.12	Suggestions for Further Reading 228	
	Exercises on Chapter 7 228		

8 Least-Squares Solutions to Linear Systems — 237

8.1	Introduction 237	
8.2	Geometric Interpretation of the Least-Squares Problem 238	
8.3	Existence and Uniqueness 239	
	8.3.1 Existence and Uniqueness Theorem 240	
	8.3.2 Normal Equations, Projections, and Least-Squares Solutions 241	
8.4	Some Applications of the Least-Squares Problem 242	
	8.4.1 Polynomial-Fitting to Experimental Data 242	
	8.4.2 Predicting Future Sales 245	
8.5	Pseudoinverse and the Least-Squares Problem 245	
8.6	Sensitivity of the Least-Squares Problem 246	
8.7	Computational Methods for Overdetermined Problems: Normal Equations, QR, and SVD Methods 252	
	8.7.1 The Normal Equations Method 252	
	8.7.2 QR Factorization Method 254	
	8.7.3 The SVD Method 259	
	8.7.4 Solving the Linear System Using the SVD 263	
8.8	Underdetermined Linear Systems 263	
	8.8.1 The QR Approach for the Minimum-Norm Solution 264	

		8.8.2	The SVD Approach for the Minimum-Norm Solution	265

- 8.9 Least-Squares Iterative Refinement ... 266
- 8.10 Review and Summary ... 269
  - 8.10.1 Existence and Uniqueness ... 269
  - 8.10.2 Overdetermined Problems ... 269
  - 8.10.3 The Underdetermined Problem ... 270
  - 8.10.4 Perturbation Analysis ... 270
  - 8.10.5 Iterative Refinement ... 270
  - 8.10.6 Comparison of Least-Squares Methods ... 270
- 8.11 Suggestions for Further Reading ... 271
- Exercises on Chapter 8 ... 271

9 Numerical Matrix Eigenvalue Problems   281

- 9.1 Introduction ... 281
- 9.2 Eigenvalue Problems Arising in Practical Applications ... 282
  - 9.2.1 Stability Problems for Differential and Difference Equations ... 282
  - 9.2.2 Phenomenon of Resonance ... 285
  - 9.2.3 Buckling Problem (a Boundary Value Problem) ... 286
  - 9.2.4 Simulating Transient Current for an Electric Circuit ... 288
  - 9.2.5 An Example of the Eigenvalue Problem Arising in Statistics: Principal Component Analysis ... 290
- 9.3 Localization of Eigenvalues ... 292
  - 9.3.1 The Geršgorin Disk Theorems ... 293
  - 9.3.2 Eigenvalue Bounds and Matrix Norms ... 295
- 9.4 Computing Selected Eigenvalues and Eigenvectors ... 295
  - 9.4.1 The Power Method, the Inverse Iteration, and the Rayleigh Quotient Iteration ... 296
- 9.5 Similarity Transformations and Eigenvalue Computations ... 304
  - 9.5.1 Diagonalization of a Matrix ... 305
  - 9.5.2 Numerical Instability of Nonorthogonal Diagonalization ... 306
  - 9.5.3 Reduction to Hessenberg Form via Orthogonal Similarity ... 307
  - 9.5.4 Uniqueness of Hessenberg Reduction ... 312
  - 9.5.5 Eigenvalue Computations Using the Characteristic Polynomial ... 313
- 9.6 Eigenvalue Sensitivity ... 315
  - 9.6.1 The Bauer–Fike Theorem ... 315
  - 9.6.2 Sensitivity of the Individual Eigenvalues ... 317
- 9.7 Eigenvector Sensitivity ... 319
- 9.8 The Real Schur Form and QR Iterations ... 320
  - 9.8.1 The Basic QR Iteration ... 322
  - 9.8.2 The Hessenberg QR Iteration ... 324
  - 9.8.3 Convergence of the QR Iterations and the Shift of Origin ... 325
  - 9.8.4 The Double-Shift QR Iteration ... 326
  - 9.8.5 Implicit QR Iteration ... 327
  - 9.8.6 Obtaining the Real Schur Form A ... 331
  - 9.8.7 The Real Schur Form and Invariant Subspaces ... 334

Contents   xiii

    9.9   Computing the Eigenvectors 335
          9.9.1   The Hessenberg-Inverse Iteration 335
    9.10   Review and Summary 336
          9.10.1   Applications of the Eigenvalues and Eigenvectors 336
          9.10.2   Localization of Eigenvalues 336
          9.10.3   The Power Method and the Inverse Iteration 337
          9.10.4   The Rayleigh Quotient Iteration 337
          9.10.5   Sensitivity of Eigenvalues and Eigenvectors 337
          9.10.6   Eigenvalue Computation via the Characteristic Polynomial
                     and the Jordan Canonical Form 338
          9.10.7   Hessenberg Transformation 338
          9.10.8   The QR Iteration Algorithm 338
          9.10.9   Ordering the Eigenvalues 339
          9.10.10 Computing the Eigenvectors 339
    9.11   Suggestions for Further Reading 339
    Exercises on Chapter 9 341

10 Numerical Symmetric Eigenvalue Problem and Singular Value Decomposition   **351**
    10.1   Introduction 351
    10.2   Computational Methods for the Symmetric Eigenvalue Problem 352
          10.2.1   Some Special Properties of the Symmetric Eigenvalue
                     Problem 352
          10.2.2   The Bisection Method for the Symmetric Tridiagonal
                     Matrix 354
          10.2.3   The Symmetric QR Iteration Method 357
          10.2.4   The Divide-and-Conquer Method 359
          10.2.5   The Jacobi Method 363
          10.2.6   Comparison of the Symmetric Eigenvalue Methods 364
    10.3   The Singular Value Decomposition and Its Computation 365
          10.3.1   The Relationship between the Singular Values and the
                     Eigenvalues 366
          10.3.2   Sensitivity of the Singular Values 367
          10.3.3   Computing the Variance-Covariance Matrix with SVD . . . 367
          10.3.4   Computing the Pseudoinverse with SVD 368
          10.3.5   Computing the SVD 369
          10.3.6   The Golub–Kahan–Reinsch Algorithm 370
          10.3.7   The Chan SVD Algorithm 378
    10.4   Generalized SVD 379
    10.5   Review and Summary 380
          10.5.1   The Symmetric Eigenvalue Computation 380
          10.5.2   The SVD 380
    10.6   Suggestions for Further Reading 381
    Exercises on Chapter 10 381

11 Generalized and Quadratic Eigenvalue Problems   **387**
    11.1   Introduction 387

11.2	Eigenvalue-Eigenvector Properties of Equivalent Pencils	389
11.3	Generalized Schur and Real Schur Decompositions	389
11.4	The QZ Algorithm	390
	11.4.1 Stage I: Reduction to Hessenberg Triangular Form	391
	11.4.2 Stage II: Reduction to the Generalized Real Schur Form	393
11.5	Computations of Generalized Eigenvectors	398
11.6	The Symmetric Positive Definite Generalized Eigenvalue Problem	399
	11.6.1 Eigenvalues and Eigenvectors of Symmetric Definite Pencil	399
	11.6.2 Conditioning of the Eigenvalues of the Symmetric Definite Pencil	400
	11.6.3 The QZ Method for the Symmetric Definite Pencil	400
	11.6.4 The Cholesky QR Algorithm for the Symmetric Definite Pencil	401
	11.6.5 Diagonalization of the Symmetric Definite Pencil: Simultaneous Diagonalization of A and B	402
	11.6.6 Generalized Rayleigh Quotient	404
11.7	Symmetric Definite Generalized Eigenvalue Problems Arising in Vibrations of Structures	405
	11.7.1 Vibration of a Building: A Case Study	405
	11.7.2 Forced Harmonic Vibration: Phenomenon of Resonance	406
11.8	Applications of Symmetric Positive Definite Generalized Eigenvalue Problem to Decoupling and Model Reduction	409
	11.8.1 Decoupling of a Second-Order System	410
	11.8.2 The Reduction of a Large Model	415
	11.8.3 A Case Study on the Potential Damage of a Building Due to an Earthquake	415
11.9	The Quadratic Eigenvalue Problem	418
	11.9.1 Orthogonality Relations of the Eigenvectors of Quadratic Matrix Pencil	419
	11.9.2 Applications of the Quadratic Eigenvalue Problem	421
	11.9.3 Numerical Methods for the Quadratic Eigenvalue Problem	422
11.10	Review and Summary	424
	11.10.1 Existence Results	424
	11.10.2 The QZ Algorithm	424
	11.10.3 The Generalized Symmetric Eigenvalue Problem	424
	11.10.4 The Quadratic Eigenvalue Problem	426
11.11	Suggestions for Further Reading	426
Exercises on Chapter 11		427

12 Iterative Methods for Large and Sparse Problems: An Overview   435

12.1	Introduction	435
12.2	The Jacobi, Gauss–Seidel, and SOR Methods	436
	12.2.1 Convergence of the Jacobi, Gauss–Seidel, and SOR Methods	438

Contents

12.3	Krylov Subspace Methods for Linear Systems: Lanczos, Arnoldi, GMRES, Conjugate Gradient, and QMR	445
	12.3.1 The Basic Arnoldi Method	446
	12.3.2 Solving $Ax = b$ Using the Arnoldi Method	448
	12.3.3 The GMRES Method for Solving $Ax = b$	451
	12.3.4 Solving Shifted Linear Systems Using the Arnoldi Method	454
	12.3.5 The Symmetric Lanczos Algorithm	455
	12.3.6 The Conjugate Gradient Method	456
	12.3.7 Solving Indefinite Symmetric Systems Using CG-Type Methods: MINRES and SYMMLQ.	460
	12.3.8 The Nonsymmetric Lanczos Method	462
	12.3.9 Solving Linear System $Ax = b$ Using the Lanczos Algorithm	464
	12.3.10 The Bi-conjugate Gradient Algorithm	464
	12.3.11 The QMR Algorithm	466
12.4	Preconditioners	467
	12.4.1 Classical Iterative Methods as Preconditioners	468
	12.4.2 Polynomial Preconditioners	468
	12.4.3 Incomplete LU (ILU) Factorization as a Preconditioner . .	469
	12.4.4 Preconditioning with Incomplete Cholesky Factorization .	469
	12.4.5 Numerical Experiments on Performance Comparison . . .	470
12.5	Comparison of Krylov Subspace Methods for Linear Systems	470
12.6	Eigenvalue Approximation Using Krylov Subspace Methods	473
	12.6.1 Eigenvalue Approximation Using the Arnoldi Method . . .	474
	12.6.2 Implicitly Restarted Arnoldi Method for the Nonsymmetric Eigenvalue Problem	475
	12.6.3 Computing the Eigenvalues of a Symmetric Matrix Using the Symmetric Lanczos Algorithm	477
12.7	The Bisection Method for the Tridiagonal Symmetric Positive Definite Generalized Eigenvalue Problem	478
	12.7.1 The Bisection Method for Tridiagonal A and B	479
12.8	Krylov Subspace Methods for Generalized Eigenvalue Problems . . .	479
12.9	Krylov Subspace Methods for the Quadratic Eigenvalue Problem . . .	481
12.10	The Jacobi–Davidson Method for the Quadratic Eigenvalue Problem .	482
12.11	Review and Summary	483
	12.11.1 The Classical Iterative Methods	483
	12.11.2 Krylov Subspace Methods	484
	12.11.3 Large Eigenvalue Problem	485
12.12	Suggestions for Further Reading	485
	Exercises on Chapter 12	486

13   Some Key Terms in Numerical Linear Algebra   **493**

Bibliography   **501**

Index   **523**

Special Topics Available Online at www.siam.org/books/ot116

Preface

The second edition of *Numerical Linear Algebra and Applications* not only has a different publisher but also is practically a different book in its style of presentation and coverage of topics. However, the key features of the first edition have been fully preserved and occasionally further improved in this edition.

0.1 Special Features

- *Discussions of the Computational Difficulties Using Theoretical Linear Algebra Tools*

   It is very important for students to clearly understand that some of the tools they have learned in a theoretical linear algebra course may not work in a computational setting. Some examples, such as solving linear systems using Cramer's rule or matrix inversion, finding the eigenvalues by computing the zeros of the characteristic polynomial, and computing singular values from the eigenvalues of the associated matrix product, will suffice. They will then start appreciating the beauty of numerical linear algebra early on and develop an interest in the study of numerical linear algebra. I have done this in Chapter 1, which serves as a motivating chapter for the entire book.

- *Applications to Science and Engineering*

   One of the major strengths of the first edition was that it contained a wide variety of motivating real-life examples drawn from numerous disciplines, including heat transfer, fluid dynamics, signal processing, biomedical engineering, statistics, business, control, and vibration engineering. In addition to all the applications contained in the first edition, several more, including some new SVD applications to image processing, have been added to the second edition. This feature distinguishes this book from most of the existing numerical linear algebra books.

- *A Brief Review of Theoretical Linear Algebra*

   A brief review of the basic concepts and results of theoretical linear algebra required to study the rest of the book, with a special emphasis on vector and matrix norms, has been given in Chapter 2. The importance of norm properties of orthogonal matrices, which make these matrices valuable tools for numerically reliable matrix computational algorithms, has been emphasized. In most theoretical introductory linear algebra courses, norms and norm properties are not adequately covered.

- *Early Introduction of the Basic Concepts*

  Students should be introduced early in the course to the fundamental concepts of round-off errors, efficiency, conditioning, and stability. Some basic facts should be made clear from the beginning. For example:

  - an efficient algorithm may not necessarily be a "good algorithm" (e.g., Gaussian elimination without pivoting);
  - stability is a property of the algorithm and conditioning is a property of the problem, but both have effects on the accuracy of the solution;
  - the stability of a numerical scheme depends upon the problem that is being solved using the scheme (e.g., the modified Gram–Schmidt process is stable for least-squares solution, but can perform poorly when applied to QR factorization).

  From my experience, I have seen students who, even after taking one or two major courses in numerical analysis and numerical linear algebra, do not have clear ideas about these facts. With this in mind, I have introduced these concepts early in Chapters 3 and 4 and returned to discussions of these concepts whenever a computational problem and the associated algorithm(s) have been described.

- *Presentation of Algorithms*

  Presentation of algorithms to students in the classroom is a challenging job for most instructors. Before an algorithm is presented in algorithmic form, the students should have a clear idea about the purpose of the algorithm, the tools available, and how to make good use of these tools to develop the algorithm in a stepwise fashion. This approach of algorithm presentation stimulates the creativity of the students and helps increase their interest in algorithm development and the study of numerical linear algebra. I have followed this practice throughout the whole book. Each algorithm has also been illustrated with simple examples, followed by brief discussions on efficiency, stability, accuracy, and computer implementations.

- *Discussions on Generalized Eigenvalue Problems*

  Several practical applications, especially those arising in vibration and structural engineering, give rise to generalized eigenvalue problems. A thorough treatment on theory, applications, and numerical algorithms of this problem has been given in Chapter 11. The applications include vibrations of structures, model reduction, and the effects of an earthquake on a building.

- *MATCOM Toolbox*

  A MATLAB-based toolkit, MATCOM, associated with this book contains MATLAB implementations of almost all major algorithms presented in the book. Several algorithms for the same problem have been implemented which will help students compare efficiency, stability, and accuracy of different algorithms for the same problem. Plenty of MATLAB and MATCOM exercises have been given in each chapter. Many of these problems are designed to understand why certain algorithms are better than others for the same problem.

- *Summary at the End of Each Chapter*

  Important definitions, concepts, and results have been summarized at the end of each chapter. This will help students review the material of the chapter quickly.

- *Suggestions for Further Readings*

  While it is impossible to cope with the dynamic developments of current research in numerical linear algebra, some of the latest developments as well as references to the fundamental work on topics discussed in each chapter have been included at the end of each chapter for the benefit of advanced readers. The books by Golub and Van Loan (1996), Higham (2002), Stewart (1998b, 2001a), and Björck (1996) are rich sources of references on numerical linear algebra and matrix computations.

- *Solutions and Answers to Selected Problems*

  Especially for the benefit of undergraduate students, partial solutions and answers to important problems, emphasizing those in need of proofs, have been included in an online appendix. See the book's webpage at *www.siam.org/books/ot116*.

- *List of Key Terms*

  The most common terms in numerical linear algebra have been defined and listed for a quick reference.

0.2 Additional Features and Topics for Second Edition

- *New Numbering Scheme*

  A new numbering scheme different from what was used in the first edition has been adopted in this edition. Definitions, theorems, corollaries, and examples have been consecutively numbered in each chapter, the first number of each item being the chapter number. Algorithms are numbered in order of their appearances in a chapter. Thus, Algorithm 5.1 indicates that it is the first algorithm in Chapter 5. I believe this scheme is most commonly used and will facilitate readings.

- *More Pictures and Figures*

  "A picture is worth a thousand words." Keeping this proverb in mind, I have included as many pictures and figures as possible in this edition.

- *Organization of Material*

  The material in this edition has been organized slightly differently from the first edition. Thus, the fundamental tools of numerical linear algebra, such as elementary Householder and Givens transformations, etc., have been introduced right before or along with their first application to solve a linear algebra problem. For example, elementary transformation has been introduced in the context of LU factorization using the Gaussian elimination scheme in Chapter 5; Householder and Givens transformations have been introduced in Chapter 7 right before Chapter 8, where least-squares solutions to linear systems have been discussed; reduction to Hessenberg form has been described in Chapter 9, where eigenvalue problem has been discussed, etc. In the first edition these tools were developed in an earlier chapter (Chapter 5) and their applications to solutions of linear systems, least squares, and eigenvalue problems were discussed in later chapters.

- *A Separate Chapter on Iterative Methods*

  Recognizing the importance of iterative methods in solutions of large-scale problems, a separate chapter (Chapter 12) with emphasis on development of major Krylov-subspace methods for linear systems solutions, have been included in this edition. A brief discussion on computing partial spectrum and the associated eigenvectors of large and sparse matrices, including techniques for generalized and quadratic eigenvalue problems, have also been included in this chapter.

- *Discussions on Quadratic Eigenvalue Problem*

  Quadratic eigenvalue problems arise in a wide variety of practical applications, including vibration analysis of structures, finite element model updating in structural dynamics, heat transfer, signal and acoustic studies, etc. An active research on quadratic and quadratic inverse eigenvalue problems is currently underway. Some discussions on theory, applications and computational methods of of quadratic eigenvalue problems, including the Jacobi-Davidson method, have been presented in Chapters 11 and 12 of this new edition.

- *Early Introduction of Singular Value Decomposition*

  SVD has become an indispensable tool in the treatment of major applications problems in science and engineering. It is now generally agreed that the students should be exposed to this important topic as early as possible in the course. Keeping this in mind, the SVD has already been defined in Chapter 2 of the book and a full treatment of its theory and applications have been included in Chapter 7, even before our discussions on eigenvalue problems in Chapter 9. Computational methods on SVD appear in Chapter 10 along with the treatment of the symmetric eigenvalue problem.

- *Biographical Anecdotes*

  Biographical anecdotes of several outstanding numerical linear algebraists whose noble contributions have enriched the field, and but only of those who are now deceased, have been included in the present edition.

0.3 Intended Audience

The book has been written primarily for a first course in numerical linear algebra in mathematics, computer science, and engineering, at the undergraduate and beginning graduate levels. See our Guidelines for using the book later. *Also, the book is ideal for self-study and will serve as a reference book for scientists and engineers.*

0.4 Some Guidelines for Using this Book

Far more material than can be covered in one semester course has been included, so that professors can tailor material to particular classes and easily develop syllabi. Here are some guidelines for using the book in the classroom.

0.4.1 A First Course in Numerical Linear Algebra (Advanced Undergraduate/One Semester)

- Linear Algebra Problems and Computational Difficulties Using Theoretical Tools: Chapter 1.

- Special Matrices, Vector and Matrix Norms, and SVD: Chapter 2, Sections 2.4–2.6.

- Floating Point Numbers and Errors: Chapter 3, Sections 3.1–3.3; brief discussions from Sections 3.4–3.7; and Section 3.8.

- Stability, Conditioning, and Accuracy: Chapter 4, Sections 4.1–4.5, and Sections 4.6.1, 4.7, and 4.8.

- Gaussian Elimination and LU Factorization: Chapter 5 (except possibly Section 5.2.5 on complete pivoting).

- Numerical Solutions of Linear Systems: Chapter 6, Section 6.2; some selected applications from Section 6.3; Sections 6.4, 6.5, 6.7.1, 6.7.3, 6.7.4, 6.8, 6.9 (optional), and 6.11; selected applications (giving rise to special linear systems) from Section 6.12; and Sections 6.12.3, 6.12.5, and 6.12.6.

- QR Factorization, SVD, and Projections: Chapter 7, Sections 7.2, 7.5, 7.6, 7.7, 7.8.1, 7.8.2, 7.8.4, 7.8.5, 7.8.6, 7.8.7, 7.8.9, and 7.8.10; and selected applications from Section 7.9.

- Least-Squares Solutions: Chapter 8, Sections 8.2–8.5; some basic sensitivity results from Section 8.6; and Section 8.7.

- Numerical Matrix Eigenvalue Problem: Chapter 9, selected applications from Section 9.2; and Sections 9.3, 9.4, 9.5, 9.6.1, 9.8.1, 9.8.2, 9.8.3, and 9.9.1.

Note: Some adjustments need to be made for a one-quarter course.

0.4.2 A Second Course in Numerical Linear Algebra (Beginning Graduate Course)

(It is assumed that the students have had a first course in numerical linear algebra. If that is not the case, a review of material of the first course should be done as necessary.)

- Numerical Solutions of Linear Systems: Chapters 6 and 12; a detailed perturbation analysis (Sections 4.6 and 6.10); LU factorization with complete pivoting (Section 5.2.5); Sherman–Morrison formula for matrix inverse (Section 6.7.2); condition number estimation (Section 6.9); special systems (Section 6.12); and iterative methods (Sections 12.2–12.5).

- QR Factorization: Chapter 7; a review of Householder QR factorization (Section 7.2); and Givens QR factorization and uniqueness of QR factorization (Section 7.4).

- Least-Squares Solution: Chapter 8; perturbation analysis (Section 8.6); a review of computational methods for overdetermined systems (Section 8.7); and underdetermined systems (Section 8.8); iterative refinement (Section 8.9).

- Numerical Matrix Eigenvalue Problem: Chapters 9, 10, and 11; sensitivity of individual eigenvalues and eigenvectors (Sections 9.6 and 9.7); double shift implicit QR iteration algorithm (Sections 9.8.4–9.8.6); ordering eigenvalues in real-Schur form (Section 9.8.7); symmetric eigenvalue problem (Section 10.2); and generalized and quadratic eigenvalue problems (selected sections from Chapters 11 and 12).

- SVD, Rank-Deficiency, Numerical Rank, and Possibly Some Discussions on Generalized SVD: Chapters 7 and 10, Sections 7.8.8, 7.10, 10.3, and 10.4.

- Special Topics: Chapter 14 (available online at *www.siam.org/books/ot116*); QR factorization with pivoting, updating and downdating of QR factorization, and error analyses for back substitution, forward elimination, LU factorization, and Linear Systems.

0.4.3 A One-Semester Course in Numerical Linear Algebra for Engineers

- Required Theoretical Linear Algebra Background (selected sections from Chapter 2).

- Floating Point Numbers and Errors: Sections 3.1–3.3 and Section 3.8.

- Stability, Conditioning and Accuracy: Sections 4.1–4.5, 4.6.1, and 4.7.

- Gaussian Elimination and Linear Systems: Chapters 5 and 6, Sections 5.1, 5.2.1–5.2.4, 5.3, and 6.2; selected applications from Sections 6.3 and 6.12; and Sections 6.4, 6.7.1, 6.7.3, 6.7.4, and 6.11.

- QR Factorization, SVD, and Least-Squares Solutions: Chapters 7 and 8, Sections 7.2, 7.5, 7.7, 7.8 (except 7.8.8), 8.2, 8.3, 8.4, 8.5, and 8.7.

- Numerical Eigenvalue Problems: Chapters 9, 10, 11, and 12.

    Standard Eigenvalue Problem: Selected applications from Section 9.2, and Sections 9.4, 9.5, 9.6.1, 9.8.1, and 9.8.2; and some selected methods from Section 10.2 (symmetric eigenvalue problem).

    Generalized Eigenvalue Problem: Chapter 11, Sections 11.2 and 11.3; some discussions of QZ algorithm from Section 11.4, Sections 11.5 and 11.6; and selected applications from Sections 11.7 and 11.8.

    Quadratic Eigenvalue Problem: Section 11.9

- Iterative Methods: Selected methods from Chapter 12 as needed.

0.5 Acknowledgments

Numerical linear algebra has been blessed with the publication of several excellent books in recent years. These include Golub and Van Loan (1996), Trefethen and Bau (1997), Demmel (1997), Stewart (1998b, 2001a), and Higham (2002).

I have learned my numerical linear algebra from Stewart's 1973 book. I find this book is still a rich source of knowledge. Several special topics books have also been published in

0.5. Acknowledgments

the last few years. These include Bai et al. (2000), Björck (1996), Davis (2006), Greenbaum (1997), Lawson and Hanson (1995), Parlett (1998), Saad (2003), Varga (2000, 2004), and van der Vorst (2003). These books certainly have had a great influence on the second edition of my book, and it is a pleasure to acknowledge their contributions.

In preparing the manuscript for the second edition, several colleagues and students have given their generous help. I gratefully acknowledge the contributions of James R. Bunch of the University of California, San Diego; Lothar Reichel of Kent State University; Youcef Saad of the University of Minnesota; Ronald Taylor of Wright State University; Pablo Tarazaga of Texas A & M University; Zonggang Zeng of North Eastern Illinois University; Pradip Majumadar and Abhijit Gupta of Northern Illinois University; and Marilena Mitrouli of the University of Athens, Athens, Greece. I am grateful for the feedback of several past and present students of mine who read and reread different chapters with patience. They include Mark Arnold of the University of Arkansas; Joao Carvalho of the Federal University of Rio Grande do Sul, Brazil; Daniil Sarkissian of Nizny Novgorod State University of Architecture and Civil Engineering, Russia; and Vadim Sokolov, Debasis Rooj, Resmi Sengupta, Sabyasachi Sen, Carlos Corrales, Neal Conrad, Mohan Thapa, and Osamu Chigira of Northern Illinois University.

Special thanks are due to Vadim Sokolov, who has written revised MATLAB codes for the software MATCOM.

I take this opportunity to express my gratitude to friends and colleagues whose help, constructive criticisms, and suggestions contributed to much of the success of the first edition. In addition to those mentioned above, they include Z. Bai of the University of California, Davis; M. Benbourenane of the California University of Pennsylvania; Michael Berry of the University of Tennessee; Shankar Bhattacharyya of Texas A & M University; Amit Bhaya of Universidade Federal do Rio de Janeiro, Brazil; Ake Björck of Linköping University, Sweden; Daneil Boley of the University of Minnesota; Kanti Datta of Indian Institute of Technology, Kharagpur, India; Alan Edelman of MIT; Yoopyo Hong of Northern Illinois University; Daniel Inman of Virginia State and Polytechnic University; Bart DeMoor of Katholieke University, Belgium; Mauro Leoncini of the University of Pisa, Italy; Beresford Parlett of the University of California, Berkeley; Rajnikant Patel of the University of West Ontario, Canada; Daniel Pierce of Access Analytics International; Yitshak Ram of Louisiana State University; Axel Ruhe of Chalmers University of Technology, Sweden; Gilbert Strang of MIT; and Paul Van Dooren of Catholic University of Louvain, Belgium.

The secretaries of the Mathematical Sciences Department of Northern Illinois University have done an outstanding job of typing and retyping many different versions of the book and providing timely assistance with other aspects of manuscript preparation. Thank you, Elizabeth Buck, Erika Thompson, Heather Mashare, and Jared Lash for your great help.

Eric Behr, our systems manager, supervised the typing of the manuscript and gave numerous help in the preparation of the final version of the manuscript. Thank you, Eric, for your generous help.

I am grateful to my mathematician wife, Karabi, who would never let me forget that the book must be finished, no matter how long it took and how hard it was. She also taught from several earlier versions of this book and provided me with her and her students' feedback to improve the quality and readability of the book.

The other members of my family—Bapi, Rakhi, Swati, Jayen, Shaan, Paula, Sudeep, Pragati, Sidharth, and Rahul—cheerfully accepted me as a part-time father, uncle, and grandfather during the preparation of this manuscript. My own greatest frustration was that

I was too busy to spend more quality time with my little grandsons, Jayen and Shaan, who have recently brought so much joy into our lives.

Finally, my sincere thanks are due to the SIAM editors Elizabeth Greenspan and Linda Thiel and to Development and Acquisitions editor Sara J. Murphy for their constant support, encouragement, and, above all, enormous patience with me. Thanks are also due to Production editor Lou Primus, who did an excellent job of copyediting the manuscript.

I have very sincerely tried to minimize the errors in the book. But in spite of my sincere efforts, I am sure there will be many still present in the book.

I shall remain grateful to any readers kind enough to send their comments and suggestions or to point out errors in the book. I can be reached via e-mail: *dattab@math.niu.edu*.

Chapter 1

Linear Algebra Problems, Their Importance, and Computational Difficulties

1.1 Introduction

The main objectives of this chapter are to state the fundamental linear algebra problems at the outset, make a brief mention of their importance, and point out the difficulties that one faces in computational settings when trying to solve these problems using obvious approaches.

1.2 Fundamental Linear Algebra Problems and Their Importance

The following are the fundamental linear algebra problems.

> A. *The Linear System Problem.* Given an $n \times n$ nonsingular matrix A and an n-vector b, the problem is to find an n-vector x such that $Ax = b$.

A practical variation of the problem requires solutions of several linear systems with the same matrix A on the left-hand side. That is, the problem there is to find a matrix $X = [x_1, x_2, \ldots, x_m]$ such that

$$AX = B,$$

where $B = [b_1, b_2, \ldots, b_m]$ is an $n \times m$ matrix.

Associated with linear system problems are problems of finding the inverse of a matrix, finding the rank, the determinant, the leading principal minors, an orthonormal basis for the range and the null space of A, and various projection matrices associated with A. Solutions of some of these later problems require matrix factorizations, and the problem of matrix factorizations and linear system problems are intimately related.

It is perhaps not an exaggeration to say that the linear system problem arises in almost all branches of science and engineering: applied mathematics, biology, chemistry, physics, electrical, mechanical, civil, and vibration engineering, etc.

The most common source is the numerical solution of differential equations. Many mathematical models of physical and engineering systems are systems of differential equations: ordinary and partial. A system of differential equations is normally solved numerically by discretizing the system by means of finite differences or finite element methods. The process of discretization, in general, leads to a linear system, the solution of which is an approximate solution to the differential equations (see Chapter 6 for more details).

> **B. *The Least-Squares Problem.*** Given an $m \times n$ matrix A and an m-vector b, the least-squares problem is to find an n-vector x such that the norm of the residual vector, $\|Ax - b\|_2$, is as small as possible.

Least-squares problems arise in statistical and geometric applications that require fitting a polynomial or curve to experimental data, as well as in engineering applications such as signal and image processing. See Chapter 8 for some specific applications of least-squares problems. It is worth mentioning here that methods for numerically solving least-squares problems invariably lead to solutions of linear systems problems (see again Chapter 8 for details).

> **C. *The Eigenvalue Problem.*** Given an $n \times n$ matrix A, the problem is to find n numbers λ_i and n-vectors x_i such that
> $$Ax_i = \lambda_i x_i, \quad i = 1, \ldots, n.$$

The eigenvalue problem typically arises in the explicit solution and stability analysis of a homogeneous system of first-order differential equations. The stability analysis requires only implicit knowledge of eigenvalues, whereas the explicit solution requires eigenvalues and eigenvectors explicitly.

Applications such as buckling problems, stock market analysis, and study of behavior of dynamical systems require computations of only a few eigenvalues and eigenvectors, usually the few largest or smallest ones (see Chapter 9).

In many practical instances, the matrix A is symmetric, and thus the eigenvalue problem becomes a symmetric eigenvalue problem (Chapter 10). A great number of eigenvalue problems arising in engineering applications are, however, generalized eigenvalue problems, as stated below.

> **D. *The Generalized and Quadratic Eigenvalue Problems.*** Given the $n \times n$ matrices A, B, and C, the problem is to find λ_i and x_i such that
> $$(\lambda_i^2 A + \lambda_i C + B)x_i = 0, \quad i = 1, \ldots, 2n.$$

This is known as the **quadratic eigenvalue problem**. In the special case when C is a zero matrix, the problem reduces to a **generalized eigenvalue problem**. That is, if we are given $n \times n$ matrices A and B, we must find μ and x such that
$$Ax = \mu Bx.$$
The leading equations of vibration engineering (a branch of engineering dealing with vibrations of structures, etc.) are systems of homogeneous or nonhomogeneous second-order

differential equations. A homogeneous second-order system has the form
$$A\ddot{z} + C\dot{z} + Bz = 0,$$
the solution and stability analysis of which lead to a quadratic eigenvalue problem.

Vibration problems are usually solved by setting $C = 0$. Moreover, in many practical instances, the matrices A and B are symmetric and B is positive definite. This leads to a symmetric definite generalized eigenvalue problem.

See Chapter 11 for details of some specific applications of these problems.

E. *Singular Value Decomposition Problem.* Given an $m \times n$ matrix A, the problem is to find unitary matrices U and V and a "diagonal" matrix Σ such that
$$A = U(\Sigma)V^*.$$

The above decomposition is known as the **singular value decomposition of A**. The entries of Σ are singular values. The column vectors of U and V are called the singular vectors.

Many areas of engineering such as control and systems theory, biomedical engineering, signal and image processing, and statistical applications give rise to the singular value decomposition problem. These applications typically require the rank of A, an orthonormal basis, projections, the distance of a matrix from another matrix of lower rank, etc., in the presence of certain impurities (known as noise) in the data. The singular values and singular vectors are the most numerically reliable tools to find these entities. The singular value decomposition is also the most numerically effective approach for solving the least-squares problem, especially in the rank-deficient case (see Chapters 8 and 10).

1.3 Computational Difficulties Using Theoretical Linear Algebra Techniques

In this section we would like to point out a few of the computational difficulties one might face while attempting to solve some of the above-mentioned linear algebra problems using common theoretical linear algebra methods.

- **Solving a linear system by Cramer's Rule.** Cramer's Rule, as taught at an undergraduate linear algebra course, is of significant theoretical and historical importance (for a statement of this rule, see Chapter 6). Unfortunately, it cannot be recommended as a practical computational procedure.

    Solving a 20×20 linear system, even on a fast modern-day computer, might take more than a million years to compute the solution with this rule, using the usual definition of the determinant of a matrix.

- **Computing the unique solution of a linear system by matrix inversion.** The unique solution of a nonsingular linear system can be written explicitly as $x = A^{-1}b$.

    Unfortunately, computing a solution to a linear system by explicitly computing the matrix inverse is not practical.

    The computation of the matrix inverse is about two-and-a-half times as expensive as solving the linear system problem itself using a standard elimination procedure (see Chapter 6), and often leads to more inaccuracies. Consider a trivial example: Solve $3x = 27$. An elimination procedure will give $x = 9$ and require only one division. On the other hand, solving the equation using matrix inversion will be cast as $x = (1/3) \cdot 27$, giving $x = 0.3333 \cdot 27 = 8.999$ (in four-digit arithmetic), and will require one division and one multiplication.

Note that computer time consumed by an algorithm is theoretically measured by the number of arithmetic operations needed to execute the algorithm.

- **Solving a least-squares problem by normal equations.** If the $m \times n$ matrix A has full rank, and m is greater than or equal to n, then the least-squares problem has a unique solution, and this solution is theoretically given by the solution x to the linear system
$$A^T A x = A^T b.$$
The above equations are known as the **normal equations**. Unfortunately, this procedure has some severe numerical limitations. First, in finite-precision arithmetic, during an explicit formation of $A^T A$, some vital information might be lost. Second, the normal equations are more sensitive to perturbations than the ordinary linear system $Ax = b$, and this sensitivity, in certain instances, corrupts the accuracy of the computed least-squares solution to an extent not warranted by the data. (See Chapter 8 for more details.)

- **Computing the eigenvalues of a matrix by finding the zeros of its characteristic polynomial.** The eigenvalues of a matrix A are the zeros of its characteristic polynomial.

Thus an "obvious" procedure for finding the eigenvalues would be to compute the characteristic polynomial of A and then find its zeros by a standard well-established root-finding procedure. Unfortunately, this is not a numerically viable approach. The round-off errors produced during a process for computing the characteristic polynomial will very likely produce some small perturbations in the computed coefficients. These small errors in the coefficients can affect the computed zeros very drastically in certain cases. The zeros of certain polynomials are known to be extremely sensitive to small perturbations in the coefficients. A classic example of this is the Wilkinson polynomial (see Chapter 4). Wilkinson took a polynomial of degree 20 with the distinct roots 1 through 20 and perturbed the coefficient of x^{19} by a significantly small amount. The zeros of this slightly perturbed polynomial were then computed by a well-established root-finding procedure, only to find that some zeros became totally different. Some even became complex.

- **Solving the generalized eigenvalue problem and the quadratic eigenvalue problems by matrix inversion.** The generalized eigenvalue problem in the case where B is nonsingular,
$$Ax = \mu Bx,$$
is theoretically equivalent to the ordinary eigenvalue problem
$$B^{-1} A x = \mu x.$$
However, if the nonsingular matrix B is sensitive to perturbations, then forming the matrix on the left-hand side by explicitly computing the inverse of B will lead to inaccuracies that in turn will lead to computations of inaccurate generalized eigenvalues.

Similar remarks hold for the quadratic eigenvalue problem. In major engineering applications, such as in vibration engineering, the matrix A is symmetric positive definite, and is thus nonsingular. In that case the quadratic eigenvalue problem is equivalent to the eigenvalue problem

$$Eu = \lambda u, \text{ where}$$

$$E = \begin{pmatrix} 0 & I \\ -A^{-1}B & -A^{-1}C \end{pmatrix}.$$

But numerically it is not advisable to solve the quadratic eigenvalue problem by actually computing the matrix E explicitly. If A is sensitive to small perturbations, the matrix E cannot be formed accurately, and the computed eigenvalues will then be inaccurate.

- **Finding the singular values by computing the eigenvalues of $A^T A$.** Theoretically, the singular values of A are the nonnegative square roots of the eigenvalues of $A^T A$. However, finding the singular values this way is not advisable. Again, explicit formation of the matrix product $A^T A$ might lead to the loss of significant relevant information. Consider a rather trivial example:

$$A \begin{pmatrix} 1 & 1 \\ \epsilon & 0 \\ 0 & 0 \end{pmatrix},$$

where ϵ is such that in finite-precision computation $1+\epsilon^2 = 1$. Then computationally we have $A^T A = \begin{pmatrix} 1 & 1 \\ 1 & 1 \end{pmatrix}$. The eigenvalues now are 2 and 0. So the computed singular values will now be given by $\sqrt{2}$ and 0. The exact singular values, however, are $\sqrt{2}$ and $\epsilon/\sqrt{2}$. (See Chapter 10 for details.)

Concluding Remarks

Above we have merely pointed out how certain obvious theoretical approaches to linear algebra problems might lead to computational difficulties and inaccuracies in computed results. *Numerical linear algebra deals with in-depth analysis of such difficulties, investigations of how these difficulties can be overcome in certain instances, and formulation and implementations of viable numerical algorithms for scientific and engineering use.*

Chapter 2

A Review of Some Required Concepts from Core Linear Algebra

2.1 Introduction

Although a first course in linear algebra is a prerequisite for this book, for the sake of completeness we establish in this chapter some notation and quickly review the basic definitions and concepts on matrices and vectors and then discuss in somewhat greater detail the concepts and fundamental results on vector and matrix norms and their applications. These results will be used frequently in later chapters.

2.2 Vectors

An ordered set of numbers is called a **vector**; the numbers themselves are called the **components** of the vector. A lowercase italic letter is usually used to denote a vector. A vector v having n components has the form

$$v = \begin{bmatrix} v_1 \\ v_2 \\ \vdots \\ v_n \end{bmatrix}.$$

A vector in this form is referred to as a **column vector** and its transpose is a **row vector**. The set of all n-vectors (that is, each vector having n components) will be denoted by $\mathbb{R}^{n \times 1}$ or simply $\mathbb{R}^n$. The set of all scalars will be denoted by $\mathbb{R}$. The transpose of a vector v will be denoted by v^T. Unless otherwise stated, a column vector will simply be called a vector.

Definition 2.1. *If u and v are two row vectors in $\mathbb{R}^n$, then their **sum** $u + v$ is defined by*

$$u + v = (u_1 + v_1, u_2 + v_2, \ldots, u_m + v_n)^T.$$

Definition 2.2. *If c is a scalar, then $cu = (cu_1, cu_2, \ldots, cu_n)^T$.*

*The **inner product** of two vectors u and v is the **scalar** given by*

$$u^T v = u_1 v_1 + u_2 v_2 + \cdots + u_n v_n.$$

The **length** of a vector v, denoted by $\|v\|$, is $\sqrt{v^T v}$; that is, the length of v (or Euclidean length of v) is $\sqrt{v_1^2 + v_2^2 + \cdots + v_n^2}$.

A set of vectors $\{m_1, \ldots, m_k\}$ in $\mathbb{R}^n$ is said to be **linearly dependent** if there exist scalars $c_1, \ldots, c_k$, not all zero, such that

$$c_1 m_1 + \cdots + c_k m_k = 0 \quad \text{(zero vector)}.$$

Otherwise, the set is called **linearly independent**.

Note that the **unit vectors** $\{e_i\}$ defined by

$$e_i = (0, 0, \ldots, 0,\ \underset{\underset{i\text{th component}}{\uparrow}}{1}\ , 0 \ldots, 0)^T, \qquad i = 1, \ldots, n,$$

are linearly independent.

2.2.1 Orthogonality, Subspace, and Basis

Orthogonality of two vectors. The angle θ between two vectors u and v is given by

$$\cos(\theta) = \frac{u^T v}{\|u\|\|v\|}.$$

Two vectors u and v are **orthogonal** if $\theta = 90°$, that is, $u^T v = 0$. The symbol $\perp$ is used to denote orthogonality.

Let S be a set of vectors in $\mathbb{R}^n$. Then S is called a **subspace** of $\mathbb{R}^n$ if $s_1, s_2 \in S$ implies $c_1 s_1 + c_2 s_2 \in S$, where c_1 and c_2 are any scalars. That is, S is a subspace if any linear combination of two vectors in S is also in S. Note that the space $\mathbb{R}^n$ itself is a subspace of $\mathbb{R}^n$. For every subspace there is a unique smallest positive integer r such that every vector in the subspace can be expressed as a linear combination of at most r vectors in the subspace; r is called the **dimension** of the subspace and is denoted by $\dim[S]$. Any set of r linearly independent vectors from S of $\dim[S] = r$ forms a **basis** of the subspace. A set of vectors $\{u_1, \ldots, u_n\}$ is **orthonormal** if each vector has unit length and they are pairwise orthogonal; that is, $u_i^T u_j = 0, i \neq j$, and $u_i^T u_i = 1$.

Orthogonality of two subspaces. Two subspaces S_1 and S_2 of $\mathbb{R}^n$ are said to be orthogonal if $s_1^T s_2 = 0$ for every $s_1 \in S_1$ and every $s_2 \in S_2$. Two orthogonal subspaces S_1 and S_2 will be denoted by $S_1 \perp S_2$.

2.3 Matrices

A collection of n vectors in $\mathbb{R}^n$ arranged in a rectangular array of m rows and n columns is called a **matrix**. A matrix A, therefore, has the form

$$A = \begin{pmatrix} a_{11} & b_{12} & \cdots & b_{1n} \\ a_{21} & b_{22} & \cdots & b_{2n} \\ \vdots & & & \\ a_{m1} & b_{m2} & \cdots & b_{mn} \end{pmatrix}.$$

It is denoted by $A = (a_{ij})_{m \times n}$, or simply by $A = (a_{ij})$, where it is understood that $i = 1, \ldots, m$ and $j = 1, \ldots, n$. A is said to be of order $m \times n$. The set of all $m \times n$ real matrices is denoted by $\mathbb{R}^{m \times n}$.

The set of all complex $m \times n$ matrices is denoted by $\mathbb{C}^{m \times n}$. The **complex conjugate** of a complex matrix A, denoted by $\bar{A}$, is the matrix whose every entry is the complex conjugate of the corresponding entry of A.

The transpose of the complex conjugate of A is denoted by A^*; that is,

$$A^* = (\bar{A})^T.$$

Unless otherwise specified, all matrices in this book are real matrices.

A matrix A having the same number of rows and columns is called a **square matrix**. The square matrix having ones along the main diagonal and zeros everywhere else is called the **identity matrix** and is denoted by I.

The **sum of two matrices** $A = (a_{ij})$ and $B = (b_{ij})$ in $\mathbb{R}^{m \times n}$ is a matrix of the same order as A and B and is given by

$$A + B = (a_{ij} + b_{ij}).$$

If c is a scalar, then cA is a matrix given by

$$cA = (ca_{ij}).$$

Let A be $m \times n$ and B be $n \times p$. Then their product AB is an $m \times p$ matrix given by

$$AB = \left(\sum_{k=1}^{n} a_{ik} b_{kj} \right), \quad \begin{array}{l} i = 1, \ldots, m, \\ j = 1, \ldots, p. \end{array}$$

Note that if b is a column vector, then Ab is a column vector. On the other hand, if a is a column vector and b^T is a row vector, then ab^T is a matrix, known as the **outer product** of the two vectors a and b. Thus

$$ab^T = \begin{pmatrix} a_1 \\ a_2 \\ \vdots \\ a_n \end{pmatrix} \begin{pmatrix} b_1 & b_2 & \cdots & b_m \end{pmatrix} = \begin{pmatrix} a_1 b_1 & \cdots & a_1 b_m \\ a_2 b_1 & \cdots & a_2 b_m \\ \vdots & & \vdots \\ a_n b_1 & \cdots & a_n b_m \end{pmatrix}.$$

Example 2.3.

$$a = \begin{pmatrix} 1 \\ 2 \\ 3 \end{pmatrix}, \quad b^T = (2 \ 3 \ 4).$$

$$\text{Outer product } ab^T = \begin{pmatrix} 2 & 3 & 4 \\ 4 & 6 & 8 \\ 6 & 9 & 12 \end{pmatrix} \text{ (a } matrix\text{).}$$

$$\text{Inner product } a^T b = (1 \ 2 \ 3) \begin{pmatrix} 2 \\ 3 \\ 4 \end{pmatrix} = 20 \text{ (a } scalar\text{).} \quad \blacksquare$$

The **transpose** of a matrix A of order $m \times n$, denoted by A^T, is a matrix of order $n \times m$ with rows and columns interchanged:

$$A^T = (a_{ji}), \quad \begin{matrix} i = 1, \ldots, n, \\ j = 1, \ldots, m. \end{matrix}$$

Note: The matrix product is not commutative; that is, in general

$$AB \neq BA.$$

Also, note that $(AB)^T = B^T A^T$.

An alternative way of writing the matrix product. Writing $B = (b_1, \ldots, b_p)$, where b_i is the ith column of B, the matrix product AB can be written as

$$AB = (Ab_1, \ldots, Ab_p).$$

Similarly, if a_i is the ith row of A, then

$$AB = \begin{pmatrix} a_1 B \\ a_2 B \\ \vdots \\ a_m B \end{pmatrix}.$$

Block Matrices

If two matrices A and B can be partitioned as

$$A = \begin{pmatrix} A_{11} & A_{12} \\ A_{21} & A_{22} \end{pmatrix}, \quad B = \begin{pmatrix} B_{11} & B_{12} \\ B_{21} & B_{22} \end{pmatrix},$$

then considering each block as an element of the matrix, we can perform addition, scalar multiplication, and matrix multiplication in the usual way. Thus,

$$A + B = \begin{pmatrix} A_{11} + B_{11} & A_{12} + B_{21} \\ A_{21} + B_{21} & A_{22} + B_{22} \end{pmatrix}$$

2.3. Matrices

and

$$AB = \begin{pmatrix} A_{11}B_{11} + A_{12}B_{21} & A_{11}B_{12} + A_{12}B_{22} \\ A_{21}B_{11} + A_{22}B_{21} & A_{21}B_{12} + A_{22}B_{22} \end{pmatrix},$$

assuming that the partitioning has been done *conformably* so that the corresponding matrix multiplications and additions are possible.

If $A = (A_{ij})$ and $B = (B_{ij})$ are two block matrices, then $C = AB$ is a block matrix given by

$$C = (C_{ij}) = \left(\sum_{k=1}^{n} A_{ik} B_{kj} \right),$$

where each A_{ik}, B_{kj}, and C_{ij} is a block matrix, assuming that each A_{ik} is compatible with B_{kj} for matrix multiplication.

The Determinant of a Matrix

For every square matrix A, there is a unique number associated with the matrix called the **determinant** of A, which is denoted by $\det(A)$. For a 2×2 matrix A, $\det(A) = a_{11}a_{22} - a_{12}a_{21}$; for a 3×3 matrix $A = (a_{ij})$, $\det(A) = a_{11} \cdot \det(A_{11}) - a_{12} \cdot \det(A_{12}) + a_{13} \cdot \det(A_{13})$, where A_{1i} is a 2×2 submatrix obtained by eliminating the first row and the ith column. This can be easily generalized. For an $n \times n$ matrix $A = (a_{ij})$ we have

$$\det(A) = (-1)^{i+1} a_{i1} \det(A_{i1}) + (-1)^{i+2} a_{i2} \det(A_{i2})$$
$$+ \cdots + (-1)^{i+n} a_{in} \det(A_{in}),$$

where A_{ij} is the submatrix of A of order $(n-1)$ obtained by eliminating the ith row and jth column.

Example 2.4.

$$A = \begin{pmatrix} 1 & 2 & 3 \\ 4 & 5 & 6 \\ 7 & 8 & 9 \end{pmatrix}.$$

Set $i = 1$. Then

$$\det(A) = 1 \cdot \det \begin{pmatrix} 5 & 6 \\ 8 & 9 \end{pmatrix} - 2 \cdot \det \begin{pmatrix} 4 & 6 \\ 7 & 9 \end{pmatrix} + 3 \cdot \det \begin{pmatrix} 4 & 5 \\ 7 & 8 \end{pmatrix}$$
$$= 1(-3) - 2(-6) + 3(-3) = 0. \quad \blacksquare$$

Theorem 2.5 (some determinant properties). *Let $A \in \mathbb{R}^{n \times n}$ and $B \in \mathbb{R}^{n \times n}$.*

1. $\det(A) = \det(A^T)$.

2. $\det(\alpha A) = \alpha^n \det(A)$, *where α is a scalar.*

3. $\det(AB) = \det(A) \cdot \det(B)$.

4. *If two rows or two columns of A are identical, then $\det(A) = 0$.*

5. *If C is a matrix obtained from A by interchanging two rows or two columns, then $\det(C) = -\det(A)$.*

The determinant of a block matrix. Let
$$A = \begin{pmatrix} A_{11} & A_{12} \\ 0 & A_{22} \end{pmatrix},$$
where A_{11} and A_{22} are square matrices. Then $\det(A) = \det(A_{11}) \cdot \det(A_{22})$.

The characteristic polynomial, the eigenvalues, and eigenvectors of a matrix. Let A be an $n \times n$ matrix. Then the polynomial $p_n(\lambda) = \det(\lambda I - A)$ is called the **characteristic polynomial**. The zeros of the characteristic polynomial are called the **eigenvalues** of A. Note that this is equivalent to the following: λ is an eigenvalue of A if and only if there exists a nonzero vector x such that $Ax = \lambda x$. The vector x is called a **right eigenvector** (or just an eigenvector), and the vector y satisfying $y^* A = \lambda y^*$ is called a **left eigenvector** associated with λ.

Definition 2.6. *Let $\{\lambda_i\}, i = 1, \ldots, n$, be the eigenvalues of an $n \times n$ matrix. Then the quantity $\rho(A) = \max |\lambda_i|$ is called the **spectral radius** of A.*

Theorem 2.7 (some basic eigenvalue-eigenvector properties).

1. *A and A^T have the same eigenvalues.*

2. *A matrix A is nonsingular if and only if all its eigenvalues are nonzero.*

3. *The eigenvectors corresponding to the distinct eigenvalues are linearly dependent.*

2.3.1 Range and Null Spaces

For every $m \times n$ matrix A, there are two important associated subspaces: the **range** of A, denoted by $R(A)$, and the **null space** of A, denoted by $N(A)$. They are defined as
$$R(A) = \{b \in \mathbb{R}^m \mid b = Ax \text{ for some } x \in \mathbb{R}^n\},$$
$$N(A) = \{x \in \mathbb{R}^n \mid Ax = 0\}.$$

Let S be a subspace of $\mathbb{R}^m$. Then the subspace $S^\perp$ defined by
$$S^\perp = \{y \in \mathbb{R}^m \mid y^T x = 0 \text{ for all } x \in S\}$$
is called the **orthogonal complement** of S. It can be shown (Exercise 2.4) that

(i) $N(A) = R(A^T)^\perp$;

(ii) $R(A)^\perp = N(A^T)$.

Definition 2.8. *The dimension of $N(A)$ is called the **nullity** of A and is denoted by **null**(A).*

2.3.2 Rank of a Matrix

Let A be an $m \times n$ matrix. Then the subspace spanned by the row vectors of A is called the *row space* of A. The subspace spanned by the columns of A is called the *column space* of A.

2.3. Matrices

The *rank* of a matrix A is the dimension of the column space of A. It is denoted by rank(A). A square matrix $A \in \mathbb{R}^{n \times n}$ is called **nonsingular** if rank(A) = n. Otherwise it is *singular*.

An $n \times n$ matrix $A \in \mathbb{R}^{n \times n}$ is said to have *full column rank if its columns are linearly independent*. The *full row rank* is similarly defined. A matrix A is said to have *full rank* if it has either full row rank or full column rank. If A does not have full rank, it is *rank-deficient*.

Example 2.9.
$$A = \begin{pmatrix} 1 & 2 \\ 3 & 4 \\ 5 & 6 \end{pmatrix}$$
has full rank; rank(A) = 2 (it has full column rank); null(A) = 0. ∎

Example 2.10.
$$A = \begin{pmatrix} 1 & 2 \\ 2 & 4 \\ 0 & 0 \end{pmatrix}$$
is rank-deficient; rank(A) = 1; null(A) = 1. ∎

Theorem 2.11 (some rank properties). *Let A be an $m \times n$ matrix. Then the following hold.*

1. rank(A) = rank(A^T).

2. rank(A) + null(A) = n.

3. rank(AB) $\geq$ rank(A) + rank(B) − n, *where B is $n \times p$*.

4. rank(BA) = rank(A) = rank(AC), *where B and C are nonsingular matrices of appropriate orders*.

5. rank(AB) $\leq$ min{rank(A), rank(B)}.

6. rank($A + B$) $\leq$ rank(A) + rank(B).

2.3.3 The Inverse of a Matrix

Let A be an $n \times n$ matrix. Then the matrix B such that
$$AB = BA = I,$$
where I is the $n \times n$ identity matrix, is called the **inverse** of A. The inverse of A is denoted by A^{-1}. *The inverse is unique.*

Definition 2.12. *If $A \in \mathbb{R}^{n \times n}$ has an inverse, it is called nonsingular. Otherwise, it is singular. A nonsingular matrix is also called invertible.*

Theorem 2.13 (some properties of an invertible matrix). *For a nonsingular $n \times n$ matrix A, the following properties hold.*

(i) $(A^{-1})^{-1} = A$.

(ii) $(A^T)^{-1} = (A^{-1})^T$.

(iii) $(cA)^{-1} = \frac{1}{c} A^{-1}$, where c is a nonzero constant.

(iv) $(AB)^{-1} = B^{-1} A^{-1}$.

Theorem 2.14 (characterization of nonsingularity). *For an $n \times n$ matrix A, the following are equivalent.*

(i) $\det(A) \neq 0$.

(ii) *A has linearly independent rows and columns.*

(iii) $N(A) = \{0\}$.

(iv) *The eigenvalues of A are all nonzero.*

(v) $\operatorname{rank}(A) = \operatorname{rank}(A^T) = n$.

2.3.4 Similar Matrices

Two matrices A and B are called **similar** if there exists a nonsingular matrix T such that
$$T^{-1}AT = B.$$

An important property of similar matrices. *Two similar matrices have the same eigenvalues.*

However, the converse is not true.

2.4 Some Special Matrices

2.4.1 Diagonal and Triangular Matrices

An $m \times n$ matrix $A = (a_{ij})$ is a **diagonal matrix** if $a_{ij} = 0$ for $i \neq j$. We write $A = \operatorname{diag}(a_{11}, \ldots, a_{nn})$.

For example, the matrices
$$\begin{pmatrix} 1 & 0 & 0 \\ 0 & 2 & 0 \end{pmatrix}, \begin{pmatrix} 1 & 0 \\ 0 & 4 \\ 0 & 0 \end{pmatrix}, \begin{pmatrix} 1 & 0 & 0 \\ 0 & 2 & 0 \\ 0 & 0 & 3 \end{pmatrix}$$
are all diagonal matrices.

2.4. Some Special Matrices

A **block diagonal** matrix is a diagonal matrix where each diagonal element is a square matrix. A block diagonal matrix is written in the form

$$A = \text{diag}(A_{11}, \ldots, A_{nn}),$$

where A_{ii} are square matrices.

An $m \times n$ matrix $A = (a_{ij})$ is an **upper triangular** matrix if $a_{ij} = 0$ for $i > j$.

The transpose of an upper triangular matrix is **lower triangular**; that is, $A = (a_{ij})$ is lower triangular if $a_{ij} = 0$ for $i < j$.

$$\begin{pmatrix} * & 0 & 0 & 0 \\ * & * & 0 & 0 \\ * & * & * & 0 \\ * & * & * & * \end{pmatrix} \qquad \begin{pmatrix} * & * & * & * \\ 0 & * & * & * \\ 0 & 0 & * & * \\ 0 & 0 & 0 & * \end{pmatrix}$$
$$\text{Lower Triangular} \qquad\qquad \text{Upper Triangular}$$

Theorem 2.15 (some useful properties of triangular matrices).

1. *The product of two upper (lower) triangular matrices is an upper (lower) triangular matrix. The diagonal entries of the product matrix are just the products of the diagonal entries of the individual matrices.*

2. *The inverse of a nonsingular upper (lower) triangular matrix is an upper (lower) triangular matrix. The diagonal entries of the inverse are the reciprocals of the diagonal entries of the original matrix.*

3. *The eigenvalues of a triangular matrix are its diagonal entries.*

4. *The determinant of a triangular matrix is the product of its diagonal entries. Thus, a triangular matrix is nonsingular if and only if all of its diagonal entries are nonzero.*

2.4.2 Unitary and Orthogonal Matrices

A square **complex matrix** U is **unitary** if

$$U^*U = UU^* = I,$$

where $U^* = (\overline{U})^T$; $\overline{U}$ is the complex conjugate of U.

If U is **real**, then U is **orthogonal** if

$$U^T U = UU^T = I.$$

Orthogonal matrices play a very important role in numerical matrix computations. Two very useful properties of orthogonal matrices are given by the following theorem.

Theorem 2.16.

(i) *The inverse of an orthogonal matrix O is just its transpose: $O^{-1} = O^T$.*

(ii) *The product of two orthogonal matrices is an orthogonal matrix.*

2.4.3 Symmetric and Hermitian Matrices

A square complex matrix A is *Hermitian* if $A^* = A$. A square matrix A is *symmetric* if $A^T = A$.

If A is real, then A is *real symmetric*. In this book, real symmetric matrices will be referred to as symmetric matrices.

Theorem 2.17 (eigenvalue decomposition of a symmetric matrix). *Let A be a real symmetric matrix. Then there exists an orthogonal matrix O such that*

$$O^T A O = D,$$

where $D = \mathrm{diag}(\lambda_1, \ldots, \lambda_n)$. The numbers $\lambda_1, \ldots, \lambda_n$ are the eigenvalues of A. The columns of O are the eigenvectors of A.

An important consequence of this theorem is the following.

Corollary 2.18. *The eigenvalues of a symmetric matrix are real and the eigenvectors can be chosen to be orthogonal.*

Definition 2.19. *The above decomposition is called the spectral decomposition of A.*

Theorem 2.17 and Corollary 2.18 also hold for a complex Hermitian matrix.

Theorem 2.20 (eigenvalue decomposition of a Hermitian matrix).

(i) *Let A be a complex Hermitian matrix. Then there exists a unitary matrix U such that $U^* A U = \mathrm{diag}(\lambda_1, \ldots, \lambda_n)$, where $\lambda_1, \lambda_2, \ldots, \lambda_n$ are the eigenvalues of A and are real.*

(ii) *The eigenvectors of a complex Hermitian matrix can be chosen to be unitary.*

2.4.4 Hessenberg Matrices (Almost Triangular)

A square matrix A is **upper Hessenberg** if $a_{ij} = 0$ for $i > j + 1$. The transpose of an upper Hessenberg matrix is a **lower Hessenberg matrix**; that is, a square matrix $A = (a_{ij})$ is a lower Hessenberg matrix if $a_{ij} = 0$ for $j > i + 1$. A square matrix A that is both upper and lower Hessenberg is **tridiagonal**.

$$\begin{pmatrix} \times & \times & & 0 \\ \vdots & & \ddots & \\ \times & & & \times \\ \times & \times & \cdots & \times \end{pmatrix} \quad \begin{pmatrix} \times & \cdots & \times & \times \\ \times & \cdots & \times & \times \\ & \ddots & \vdots & \vdots \\ 0 & & \times & \times \end{pmatrix} \quad \begin{pmatrix} \times & \times & & 0 \\ \times & \times & \ddots & \\ & \ddots & \ddots & \times \\ 0 & & \times & \times \end{pmatrix}$$

Lower Hessenberg Upper Hessenberg Tridiagonal

An upper Hessenberg matrix $A = (a_{ij})$ is **unreduced** if

$$a_{i,i-1} \neq 0 \quad \text{for } i = 2, 3, \ldots, n.$$

Similarly, a lower Hessenberg matrix $A = (a_{ij})$ is **unreduced** if

$$a_{i,i+1} \neq 0 \quad \text{for } i = 1, 2, \ldots, n-1.$$

2.5 Vector and Matrix Norms

Example 2.21.

$A = \begin{pmatrix} 1 & 2 & 0 \\ 2 & 3 & 4 \\ 1 & 1 & 1 \end{pmatrix}$ is an unreduced lower Hessenberg matrix.

$A = \begin{pmatrix} 1 & 1 & 1 \\ 1 & 1 & 1 \\ 0 & 2 & 3 \end{pmatrix}$ is an unreduced upper Hessenberg matrix. ∎

2.5 Vector and Matrix Norms

2.5.1 Vector Norms

A vector norm on $\mathbb{R}^n$ is a function $\|\cdot\|\colon \mathbb{R}^n \to \mathbb{R}$ that satisfies the following conditions:

1. $\|x\| > 0$ for every nonzero x. $\|x\| = 0$ if and only if x is the zero vector.
2. $\|\alpha x\| = |\alpha| \|x\|$ for all $x \in \mathbb{R}^n$ and for all scalars α.
3. $\|x + y\| \leq \|x\| + \|y\|$ for all x and $y \in \mathbb{R}^n$.

The last property is known as the **triangle inequality**.
Note:
$$\|-x\| = \|x\|,$$
$$\|x\| - \|y\| \leq \|(x - y)\|.$$

It is simple to verify that the following are vector norms.

> **Some Easily Computed Vector Norms.** Let $x = (x_1, x_2, \ldots, x_n)^T$. Then
>
> A. $\|x\|_1 = |x_1| + |x_2| + \cdots |x_n|$ (*sum norm* or *one norm*).
>
> B. $\|x\|_2 = \sqrt{x_1^2 + x_2^2 + \cdots x_n^2}$ (*Euclidean norm* or *two norm*).
>
> C. $\|x\|_\infty = \max_i |x_i|$ (*infinity norm* or *maximum norm*).

In general, if p is a real number greater than or equal to 1, then the ***p*-norm** or **Hölder norm** is defined by
$$\|x\|_p = (|x_1|^p + \cdots + |x_n|^p)^{\frac{1}{p}}.$$

Example 2.22. Let $x = (1, 1, -2)^T$. Then
$$\|x\|_1 = 4, \quad \|x\|_2 = \sqrt{1^2 + 1^2 + (-2)^2} = \sqrt{6}, \quad \|x\|_\infty = 2. \quad \blacksquare$$

An important property of the Hölder norm is the *Hölder inequality*
$$|x^T y| \leq \|x\|_p \|y\|_q,$$

where
$$\frac{1}{p} + \frac{1}{q} = 1.$$

A special case of the Hölder inequality is the *Cauchy–Schwarz* inequality

$$|x^T y| \le \|x\|_2 \|y\|_2,$$

that is,

$$\left| \sum_{j=1}^{n} x_j y_j \right| \le \sqrt{\sum_{j=1}^{n} x_j^2} \sqrt{\sum_{j=1}^{n} y_j^2}.$$

Equivalent Property of the Vector Norms

All vector norms are **equivalent** in the sense that there exist positive constants α and β such that

$$\alpha \|x\|_\mu \le \|x\|_\nu \le \beta \|x\|_\mu$$

for all x.

For the 2-, 1-, or ∞-norms, we can compute α and β easily:

$$\|x\|_2 \le \|x\|_1 \le \sqrt{n} \|x\|_2,$$
$$\|x\|_\infty \le \|x\|_2 \le \sqrt{n} \|x\|_\infty,$$
$$\|x\|_\infty \le \|x\|_1 \le n \|x\|_\infty.$$

2.5.2 Matrix Norms

Let A be an $m \times n$ matrix. Then, analogous to the vector norm, we define a matrix norm $\|A\|$ with the following properties:

1. $\|A\| > 0$; $\|A\| = 0$ if and only if A is the zero matrix.

2. $\|\alpha A\| = |\alpha| \|A\|$ for any scalar α.

3. $\|A + B\| \le \|A\| + \|B\|$.

Subordinate Matrix Norms

Given a matrix A and a vector norm $\|\cdot\|$, a nonnegative number defined by

$$\|A\|_p = \max_{x \ne 0} \frac{\|Ax\|_p}{\|x\|_p}$$

satisfies all the properties of a matrix norm. This norm is called the matrix norm **subordinate** to the vector norm or just **subordinate matrix norm**.

A very useful and frequently used property of a subordinate matrix norm (we shall sometimes call it the *p-norm of a matrix A*) is

$$\|Ax\|_p \le \|A\|_p \|x\|_p.$$

2.5. Vector and Matrix Norms

This property easily follows from the definition of p-norms. Note that

$$\|A\|_p \geq \frac{\|Ax\|_p}{\|x\|_p}$$

for any particular nonzero vector x. Multiplying both sides by $\|x\|_p$ gives the original inequality.

The two *easily computable* p-norms are

$$\|A\|_1 = \max_{1 \leq j \leq n} \sum_{i=1}^{m} |a_{ij}| \quad \text{(maximum column-sum norm)},$$

$$\|A\|_\infty = \max_{1 \leq i \leq m} \sum_{j=1}^{n} |a_{ij}| \quad \text{(maximum row-sum norm)}.$$

Example 2.23.

$$A = \begin{pmatrix} 1 & -2 \\ 3 & 4 \\ -5 & 6 \end{pmatrix}.$$

Then $\|A\|_1 = 12$ and $\|A\|_\infty = 11$. ∎

Another useful p-norm is the **spectral norm**, denoted by $\|A\|_2$:

$$\|A\|_2 = \max_{x \neq 0} \frac{\|Ax\|_2}{\|x\|_2}.$$

It can be shown that

$$\|A\|_2 = \sqrt{\text{maximum eigenvalue of } A^T A}.$$

(Note that the eigenvalues of $A^T A$ are real and nonnegative.)

Example 2.24. Let $A = \begin{pmatrix} 2 & 5 \\ 1 & 3 \end{pmatrix}$. Then the eigenvalues of $A^T A$ are 0.0257 and 38.9743, and $\|A\|_2 = \sqrt{38.9743} = 6.2429$. ∎

The Frobenius Norm

An important matrix norm compatible with the vector norm $\|x\|_2$ is the Frobenius norm:

$$\|A\|_F = \left[\sum_{j=1}^{n} \sum_{i=1}^{m} |a_{ij}|^2 \right]^{\frac{1}{2}}.$$

A matrix norm $\|\cdot\|_M$ and a vector norm $\|\cdot\|_v$ are compatible if
$$\|Ax\|_v \leq \|A\|_M \|x\|_v.$$

Example 2.25. Let $A = \begin{pmatrix} 1 & 2 \\ 3 & 4 \end{pmatrix}$. Then $\|A\|_F = \sqrt{30}$. ∎

Two important properties of the Frobenius norm.

1. For the identity matrix I, $\|I\|_F = \sqrt{n}$, whereas $\|I\|_1 = \|I\|_2 = \|I\|_\infty = 1$.

2. $\|A\|_F^2 = \text{trace}(A^T A)$, where **trace(A)** is defined as the **sum** of the diagonal entries of A; that is, if $A = (a_{ij})$, then $\text{trace}(A) = a_{11} + a_{22} + \cdots + a_{nn}$.

Consistent norm. A norm $\|\cdot\|$ is consistent if it satisfies $\|AB\| \leq \|A\|\|B\|$ whenever the product AB is defined.

The Frobenius norm and all subordinate matrix norms are consistent. The *max norm* $\|A\| = \max_{i,j} |a_{ij}|$ is not consistent.

Equivalence Property of Matrix Norms

As in the case of vector norms, the matrix norms are also related. There exist scalars α and β such that
$$\alpha \|A\|_\mu \leq \|A\|_\mu \leq \beta \|A\|_\mu.$$

In particular, the following inequalities relating various matrix norms are true and are used very frequently in practice.

Theorem 2.26. *Let A be $m \times n$.*

(1) $\frac{1}{\sqrt{n}} \|A\|_\infty \leq \|A\|_2 \leq \sqrt{m} \|A\|_\infty$.

(2) $\|A\|_2 \leq \|A\|_F \leq \sqrt{n} \|A\|_2$.

(3) $\frac{1}{\sqrt{m}} \|A\|_1 \leq \|A\|_2 \leq \sqrt{n} \|A\|_1$.

(4) $\|A\|_2 \leq \sqrt{\|A\|_1 \|A\|_\infty}$.

We prove here inequalities (1) and (2) and leave the others as exercises (Exercise 2.32).
Proof. *Proof of* (1). By definition
$$\|A\|_\infty = \max_{x \neq 0} \frac{\|Ax\|_\infty}{\|x\|_\infty}.$$

Again, from the equivalence property of the vector norms, we have
$$\|Ax\|_\infty \leq \|Ax\|_2 \text{ and } \|x\|_2 \leq \sqrt{n} \|x\|_\infty.$$

2.5. Vector and Matrix Norms

From the second inequality we get $\frac{1}{\|x\|_\infty} \leq \frac{\sqrt{n}}{\|x\|_2}$. It therefore follows that

$$\frac{\|Ax\|_\infty}{\|x\|_\infty} \leq \sqrt{n}\frac{\|Ax\|_2}{\|x\|_2}$$

or

$$\max_{x \neq 0} \frac{\|Ax\|_\infty}{\|x\|_\infty} \leq \sqrt{n} \max_{x \neq 0} \frac{\|Ax\|_2}{\|x\|_2} = \sqrt{n}\|A\|_2,$$

i.e.,

$$\frac{1}{\sqrt{n}}\|A\|_\infty \leq \|A\|_2.$$

The first part is proved. To prove the second part, we again use the definition of $\|A\|_2$ and the appropriate equivalence property of the vector norms.

$$\|A\|_2 = \max_{x \neq 0} \frac{\|Ax\|_2}{\|x\|_2}, \quad \|Ax\|_2 \leq \sqrt{m}\|Ax\|_\infty, \quad \|x\|_\infty \leq \|x\|_2.$$

Thus,

$$\frac{\|Ax\|_2}{\|x\|_2} \leq \sqrt{m}\frac{\|Ax\|_\infty}{\|x\|_\infty}.$$

So $\max_{x \neq 0} \frac{\|Ax\|_2}{\|x\|_2} \leq \sqrt{m} \max_{x \neq 0} \frac{\|Ax\|_\infty}{\|x\|_\infty}$ or $\|A\|_2 \leq \sqrt{m}\|A\|_\infty$.

The proof of (1) is now complete. □

Proof of (2). We prove (2) using a different technique. Recall that

$$\|A\|_F^2 = \text{trace}(A^T A).$$

Since $A^T A$ is symmetric, by Theorem 2.17 there exists an orthogonal matrix O such that

$$O^T(A^T A)O = D = \text{diag}(d_1, \ldots, d_n).$$

Now, the trace is invariant under similarity transformation (Exercise 2.38 (d)). We then have

$$\text{trace}(A^T A) = \text{trace}(D) = d_1 + \cdots + d_n.$$

Let $d_k = \max_i(d_i)$. Then, since $d_1, \ldots, d_n$ are also the eigenvalues of $A^T A$, we have

$$\|A\|_2^2 = d_k.$$

Thus,

$$\|A\|_F^2 = \text{trace}(A^T A) = d_1 + \cdots + d_n \geq d_k = \|A\|_2^2.$$

To prove the other part, we note that

$$\|A\|_F^2 = d_1 + \cdots + d_n \leq d_k + d_k + \cdots + d_k = nd_k.$$

That is, $\|A\|_F^2 \leq nd_k = n\|A\|_2^2$. So, $\|A\|_F \leq \sqrt{n}\|A\|_2$. □

2.5.3 Norms and Inverses

The following result plays an important role in matrix perturbation analysis (see Theorem 4.25).

In the following, $\|\ \|$ is a matrix norm for which $\|I\| = 1$.

Theorem 2.27. *Let $\|E\| < 1$. Then $(I - E)$ is nonsingular and*

$$\|(I - E)^{-1}\| \leq (1 - \|E\|)^{-1}.$$

Proof. Let $\lambda_1, \ldots, \lambda_n$ be the eigenvalues of E. It is easy to see that the eigenvalues of $I - E$ are $1 - \lambda_1, 1 - \lambda_2, \ldots, 1 - \lambda_n$.

Since $\|E\| < 1$, $|\lambda_i| < 1$ for each i (see Exercise 2.10). Thus, none of the quantities $1 - \lambda_1, 1 - \lambda_2, \ldots, 1 - \lambda_n$ is zero. This proves that $I - E$ is nonsingular. (*Note that a matrix A is nonsingular if and only if all its eigenvalues are nonzero.*)

To prove the second part, we write

$$(I - E)^{-1} = I + E + E^2 + \cdots.$$

Since $\|E\| < 1$,

$$\underset{k \to \infty}{\text{Limit}}\ E^k = 0,\ \text{because}\ \|E^k\| \leq \|E\|^k.$$

Thus, the series on the right side is convergent. Taking the norm on both sides, we have

$$\|(I - E)^{-1}\| \leq \|I\| + \|E\| + \|E^2\| \cdots = (1 - \|E\|)^{-1}\ (\text{since}\ \|I\| = 1).$$

(Note that the infinite series $1 + x + x^2 + \cdots$ converges to $\frac{1}{1-x}$ if and only if $|x| < 1$). □

Using Theorem 2.27, the following theorem can be proved (Exercise 2.37).

Theorem 2.28. *If $\|E\| < 1$, then*

$$\|(I - E)^{-1} - I\| \leq \frac{\|E\|}{1 - \|E\|}.$$

Implication of the result. If the matrix E is very small, then $1 - \|E\|$ is close to unity. Thus the above result implies that if we invert a slightly perturbed identity matrix, then the error in the inverse of the perturbed matrix does not exceed the order of the perturbation.

2.5.4 Norm Invariant Properties of Orthogonal and Unitary Matrices

We conclude the chapter by listing some very useful norm properties of orthogonal and unitary matrices that are often used in practice.

Theorem 2.29. *Let O be an orthogonal matrix. Then the following hold.*

(i) $\|O\|_2 = 1$.

(ii) $\|AO\|_2 = \|A\|_2$.

(iii) $\|AO\|_F = \|A\|_F$.

Proof. *Proof of* (i). Since A is orthogonal, by the property of the spectral norm, we have that $\|O\|_2 = \sqrt{\rho(O^T O)} = \sqrt{\rho(I)} = 1$.

Proof of (ii). $\|AO\|_2 = \sqrt{\rho(O^T A^T A O)} = \sqrt{\rho(A^T A)} = \|A\|_2$. (Note that the spectral radius remains invariant under similarity transformation (Section 2.3.4).)

Proof of (iii). $\|AO\|_F^2 = \text{trace}(O^T A^T A O) = \text{trace}(A^T A) = \|A\|_F^2$. □

Notes: (i) Theorem 2.29 is also valid for unitary matrices. That is, if U is an unitary matrix, then (i) $\|U\|_2 = 1$, (ii) $\|AU\|_2 = \|A\|_2$, and (iii) $\|AU\|_F = \|A\|_F$.

(ii) As we will see in this book that *norm invariant properties of orthogonal and unitary matrices make these matrices attractive tools for matrix computations.*

For example, if A is contaminated by an error matrix E and U is unitary, then

$$U(A + E)U^* = UAU^* + F,$$

where

$$\|F\|_2 = \|UEU^*\|_2 = \|E\|_2.$$

(iii) An immediate consequence of Theorem 2.29 and its unitary counterpart is that the *vector length is preserved by an orthogonal or unitary matrix multiplication.*

2.6 Singular Value Decomposition

Let $A \in \mathbb{R}^{m \times n}$. Then there exist orthogonal matrices $U \in \mathbb{R}^{m \times m}$ and $V \in \mathbb{R}^{n \times n}$ such that

$$A = U \Sigma V^T,$$

where $\Sigma = \text{diag}(\sigma_1, \ldots, \sigma_p) \in \mathbb{R}^{m \times n}$, $p = \min(m, n)$, and $\sigma_1 \geq \sigma_2 \geq \cdots \geq \sigma_p \geq 0$.

This decomposition is called **singular value decomposition (SVD)**.

The diagonal entries of Σ are called **singular values** and the columns of U and V are, respectively, the **left** and **right singular vectors**. The largest and smallest singular values are denoted, respectively, by $\sigma_{\max}$ and $\sigma_{\min}$. SVD is an important tool in matrix computations. Detailed discussions appear in Chapter 7 (Section 7.8) and Chapter 10.

Notes: (i) rank (A) = number of nonzero singular values.

(ii) $\|A\|_2 = \sigma_1 = \sigma_{\max}$.

(iii) $\|A^{-1}\|_2 = \frac{1}{\sigma_n}$ if A is $n \times n$ and nonsingular.

(iv) $\|A\|_F = (\sigma_1^2 + \sigma_2^2 + \cdots + \sigma_n^2)^{\frac{1}{2}}$ if A is $m \times n$, $m \geq n$.

2.7 Review and Summary

The very basic concepts that will be required for smooth reading of the rest of the book have been briefly summarized in this chapter. The most important ones are as follows.

2.7.1 Special Matrices

Diagonal, triangular, orthogonal, Hessenberg, symmetric, and Hermitian matrices have been defined and some useful properties of these matrices have been stated (Section 2.4).

2.7.2 Rank, Determinant, Inverse, and Eigenvalues

These important concepts have been defined and some useful properties have been stated (Section 2.3).

2.7.3 Vector and Matrix Norms

Some important matrix norms are *row-sum norm, column-sum norm, Frobenius norm*, and *spectral norm*.

A result on the relationship between different matrix norms is stated and proved in Theorem 2.26.

Of special importance is the norm property of orthogonal matrices. Three simple but important results have been stated and proved in Theorem 2.29. These results are (i) *the spectral norm of an orthogonal matrix is* 1, and (ii) *the spectral and the Frobenius norms remain invariant under orthogonal matrix multiplications*.

Two interesting properties relating norms and the inverse of a matrix are given in Theorems 2.27 and 2.28. These properties are useful in perturbation analysis of linear systems, as described in Chapter 4.

2.8 Suggestions for Further Reading

The material covered in this chapter can be found in any standard book on linear algebra and matrix theory. These include Bhatia (1996), Franklin (1968), Leon (2005), Lay (2003), Strang (2003, 2006), Horn and Johnson (1985), Lancaster and Tismenetsky (1985), Lancaster (1969), Meyer (2000), Noble and Daniel (1988), Ortega (1987a), Hill (1991), and Schneider and Barker (1989). There now exists a *Handbook on Linear Algebra* edited by L. Hogben (2007) that contains a wealth of information on both theoretical and numerical aspects of linear algebra.

Rich theory of linear algebra and numerically effective tools from numerical linear algebra are nowadays widely used in numerous practical applications. We will see many such applications to engineering in this book. For application to statistics, see, for example, Rao and Rao (1998), Graybill (1983), Gentle (1998), and Thisted (1988); for applications to control theory, see Datta (2003), Antoulas (2005), Petkov et al. (1991), Patel et al. (1994), Brualdi et al. (1985) and Datta et al. (1988); for applications to signal processing, see Andrews and Hunt (1988), Jain (1989), Bojanczyk (1995), and Hansen et al. (2006). For application to optimization, see Griva et al. (2009) and Nocedal and Wright (2006). For applications to vibration engineering, see Inman (2006, 2007); for applications of nonnegative matrices, see Berman and Plemmons (1994); for applications to search engine, see Berry and Browne (2005). Most linear algebra books also nowadays contain some applications.

Exercises on Chapter 2

EXERCISES ON SECTIONS 2.2 AND 2.3

2.0 Answer "True" or "False" to the following. Give reasons for your answers.

(a) The eigenvalues of an upper triangular matrix T are its diagonal entries.

(b) The eigenvalues of a real symmetric matrix are real.

(c) A matrix is nonsingular if and only if all its eigenvalues are nonzero.

(d) The eigenvalues of an orthogonal matrix are all equal to 1.

(e) An orthogonal matrix is not necessarily invertible.

(f) A real symmetric or a complex Hermitian matrix can be always transformed into a diagonal matrix by similarity transformation.

(g) Two similar matrices have the same eigenvalues.

(h) If two matrices have the same eigenvalues, they must be similar.

(i) The product of two upper (lower) triangular matrices does not need to be an upper (lower) triangular matrix.

(j) $\|I\| = 1$ for any norm.

(k) The length of a vector is preserved by an orthogonal multiplication.

(l) If $\|A\| < 1$, then $I - A$ is nonsingular.

(m) If $\|A\|_2 = 1$, then A must be orthogonal.

(n) The product of two orthogonal (unitary) matrices is an orthogonal (unitary) matrix.

2.1 Prove that

(a) a set of n linearly independent vectors in $\mathbb{R}^n$ is a basis for $\mathbb{R}^n$;

(b) the set $\{e_1, e_2, \ldots, e_n\}$ is a basis of $\mathbb{R}^n$;

(c) a set of m vectors in $\mathbb{R}^n$, where $m > n$, is linearly dependent;

(d) any two bases in a vector space $\mathbb{V}$ have the same number of vectors;

(e) $\dim(\mathbb{R}^n) = n$;

(f) $\text{span}\{v_1, \ldots, v_n\}$ is a subspace of $\mathbb{V}$, where $\text{span}\{v_1, \ldots, v_n\}$ is the set of linear combinations of the n vectors $v_1, \ldots, v_n$ from a vector space $\mathbb{V}$;

(g) $\text{span}\{v_1, \ldots, v_n\}$ is the smallest subspace of $\mathbb{V}$ containing $v_1, \ldots, v_n$.

2.2 Prove that if $S = \{s, \ldots, s_k\}$ is an orthogonal set of nonzero vectors, then S is linearly independent.

2.3 Let S be an m-dimensional subspace of $\mathbb{R}^n$. Then prove that S has an orthonormal basis.

Construct an orthonormal basis of $\mathbb{R}^3$.

2.4 Prove that (i) $N(A) = R(A^T)^\perp$, (ii) $R(A)^\perp = N(A^T)$, and (iii) null $(A) = 0$ if and only if A has linearly independent columns.

2.5 Prove the eigenvalue-eigenvector properties stated in Theorem 2.7.

2.6 Using the Gram–Schmidt process construct an orthonormal basis of $\mathbb{R}^3$.

2.7 Construct an orthonormal basis of $\mathbb{R}(A)$, where

$$A = \begin{pmatrix} 1 & 2 \\ 2 & 3 \\ 4 & 5 \end{pmatrix}.$$

2.8 Let S_1 and S_2 be two subspaces of $\mathbb{R}^n$. Then prove that
$$\dim(S_1 + S_2) = \dim(S_1) + \dim(S_2) - \dim(S_1 \cap S_2).$$

2.9 Prove Theorem 2.5 on the properties of the determinant of a matrix.

2.10 Prove that for a subordinate matrix norm $\|\cdot\|$, $|\lambda| \leq \|A\|$ for every eigenvalue of λ of A.

2.11 Let A be $m \times n$. Then A has rank 1 if and only if A can be written as $A = ab^T$, where a and b are column vectors.

2.12 Suppose a matrix A can be written as $A = LU$, where L is a lower triangular matrix with 1's along the diagonal and $U = (u_{ij})$ is an upper triangular matrix. Prove that $\det A = \prod_{i=1}^{n} u_{ii}$.

2.13 Let $A = \begin{pmatrix} A_1 & 0 \\ A_2 & A_3 \end{pmatrix}$, where A_1 and A_3 are square. Prove that $\det(A) = \det(A_1)\det(A_3)$.

EXERCISES ON SECTION 2.4

2.14 Prove the properties of a triangular matrix stated in Theorem 2.15.

2.15 Prove that the product of an upper Hessenberg matrix and an upper triangular matrix is an upper Hessenberg matrix.

2.16 Prove that a symmetric Hessenberg matrix is symmetric tridiagonal.

2.17 A square matrix $A = (a_{ij})$ is a band matrix of bandwidth $2k+1$ if $|i - j| > k$ implies that $a_{ij} = 0$. What are the bandwidths of tridiagonal and pentadiagonal matrices? Is the product of two banded matrices having the same bandwidth a banded matrix of the same bandwidth? Give reasons for your answer.

2.18 Prove Theorem 2.16.

2.19 Let A and B be two symmetric matrices.

(a) Prove that $(A + B)$ is symmetric.

(b) Prove that AB is not necessarily symmetric. Derive a condition under which AB is symmetric.

EXERCISES ON SECTION 2.5 AND 2.6

2.20 Show that $\|x\|_1$, $\|x\|_\infty$, $\|x\|_2$ (as defined in Section 2.5) are vector norms.

2.21 Show that if x and y are two vectors, then
$$\left|\|x\| - \|y\|\right| \leq \|x - y\| \leq \|x\| + \|y\|.$$

2.22 If x and y are two n-vectors, then prove that

(a) $|x^T y| \leq \|x\|_2 \|y\|_2$ (Cauchy–Schwarz inequality);

(b) $\|xy^T\|_2 \leq \|x\|_2 \|y\|_2$ (Schwarz inequality).

2.23 Let x and y be two orthogonal vectors. Then prove that
$$\|x + y\|_2^2 = \|x\|_2^2 + \|y\|_2^2.$$

2.24 Prove that for any vector x, we have
$$\|x\|_\infty \leq \|x\|_2 \leq \|x\|_1.$$

2.25 Prove that $\|A\|_1$, $\|A\|_\infty$, $\|A\|_2$ are matrix norms.

2.26 Let $A = (a_{ij})$ be $m \times n$. Define $A_\ell = \max_{ij} |a_{ij}|$. Is A_ℓ a consistent matrix norm? Give reasons for your answer.

2.27 (a) Prove that the vector length is preserved by orthogonal matrix multiplication. That is, if $x \in \mathbb{R}^n$ and $Q \in \mathbb{R}^{n \times n}$ be orthogonal, then $\|Qx\|_2 = \|x\|_2$ (isometry lemma).

(b) Is the statement in part (a) true if $\|\cdot\|_1$ and $\|\cdot\|_\infty$ are used? Give reasons. What if the Frobenius norm is used?

2.28 Prove that (i) $\|I\|_2 = 1$, and (ii) $\|I\|_F = \sqrt{n}$.

2.29 Prove that if Q and P are orthogonal matrices, then

(a) $\|QAP\|_F = \|A\|_F$;

(b) $\|QAP\|_2 = \|A\|_2$.

2.30 Prove that the spectral norm of a symmetric matrix is the same as its spectral radius.

2.31 Let $A \in \mathbb{R}^{n \times n}$ and let x, y, and z be n-vectors such that $Ax = b$ and $Ay = b + z$. Then prove that
$$\frac{\|z\|_2}{\|A\|_2} \leq \|x - y\|_2 \leq \|A^{-1}\|_2 \|z\|_2$$

(assuming that A^{-1} exists).

2.32 Prove properties (3) and (4) of Theorem 2.26.

2.33 Prove that (i) $\|A^T\|_2 = \|A\|_2$, and (ii) $\|A^T A\|_2 = \|A\|_2^2$.

2.34 Let $A = (a_1, \ldots, a_n)$, where a_j is the jth column of A. Then prove that
$$\|A\|_F^2 = \sum_{i=1}^n \|a_i\|_2^2.$$

2.35 Prove that if A and B are two matrices compatible for matrix multiplication, then

(a) $\|AB\|_F \leq \|A\|_F \|B\|_F$;

(b) $\|AB\|_F \leq \|A\|_2 \|B\|_F$.

2.36 (*Banach lemma.*) Prove that if A and $A + E$ are both nonsingular, then
$$\|(A + E)^{-1} - A^{-1}\| \leq \|E\| \|A^{-1}\| \|(A + E)^{-1}\|.$$

What is the implication of this result?

2.37 Prove Theorem 2.28.

2.38 Prove the following.

(a) $\text{trace}(AB) = \text{trace}(BA)$.

(b) $\text{trace}(AA^*) = \sum_{i=1}^m \sum_{j=1}^n |a_{ij}|^2$, where $A = (a_{ij})$ is $m \times n$.

(c) $\text{trace}(A + B) = \text{trace}(A) + \text{trace}(B)$.

(d) $\text{trace}(TAT^{-1}) = \text{trace}(A)$.

2.39 (a) Using the Jordan canonical theorem (see Theorem 9.28), prove that the matrix sequence $\{A^k\} \to 0$ if and only if $|\lambda_i| < 1$ for each eigenvalue λ_i of A.

(b) Using part (a), prove that $\{A^k\} \to 0$ if $\|A\| < 1$, where $\|\cdot\|$ is a subordinate matrix norm.

(c) Construct a 2×2 example to show that condition (b) is sufficient but not necessary.

2.40 Using SVD prove the norm properties in Exercises 2.29, 2.33, and 2.35.

Chapter 3
Floating Point Numbers and Errors in Computations

Background Material Needed

- Special matrices (Section 2.4)

- Matrix and vector norms (Section 2.5)

3.1 Floating Point Number Systems

Most scientific and engineering computations on a computer are performed using **floating point arithmetic**. Computers may have different bases, though base 2 is most common. The other commonly used bases are 10 and 16. Most hand calculators use base 10, while IBM mainframes use base 16.

A t-digit **floating point** number in base β has the form

$$x = \pm m \cdot \beta^e,$$

where m is a t-digit fraction, called the **mantissa**, and e is called the **exponent**. If the first digit of the mantissa is different from zero, then the floating point number is called **normalized**. Thus 0.3457×10^5 is a 4-digit normalized decimal floating number, whereas 0.03457×10^6 is a five-digit unnormalized decimal floating point number.

The number of digits in the mantissa is called the **precision**. On many computers, it is possible to manipulate floating point numbers so that a number can be represented with about twice the usual precision. Such a precision is called **double precision**.

Most computers nowadays conform to the IEEE floating point standard (ANSI/IEEE standard 754-1985). For single precision, IEEE standard recommends about 24 binary digits, and for double precision about 53 binary digits.

IEEE Floating Point Standard

Single Precision		
1	23	8
Sign	Mantissa	Exponent

Double Precision		
1	52	11
Sign	Mantissa	Exponent

Thus, IEEE standard for *single precision provides approximately seven decimal digits of accuracy*, since $2^{-23} \simeq 1.2 \times 10^{-7}$, and *double precision provides approximately sixteen decimal digits of accuracy*, since $2^{-52} \approx 2.2 \times 10^{-16}$.

Note: *Although computations with double precision increase accuracy, they require more computer time and storage.*

On each computer, there is an allowable range of the exponent e: L, the minimum and U, the maximum. *L and U vary from computer to computer.*

If, during computations, the computer produces a number whose exponent is too large (too small), that is, it is outside the permissible range, then we say that an **overflow (underflow)** has occurred.

Overflow is a serious problem; for most systems, the result of an overflow is $\pm\infty$. Underflow is usually considered less serious. On most computers, when an underflow occurs, the computed value is set to zero, and then computations proceed. *Unless otherwise stated, we will use only decimal arithmetic.*

Example 3.1. Examples of overflow and underflow.
1. Let $\beta = 10$, $t = 3$, $L = -3$, $U = 3$.
$$a = 0.111 \times 10^3, \; b = 0.120 \times 10^3,$$
$$c = a \times b = 0.133 \times 10^5$$
will result in an *overflow*, because the exponent 5 is too large.
2. Let $\beta = 10$, $t = 3$, $L = -2$, $U = 3$.
$$a = 0.1 \times 10^{-1},$$
$$b = 0.2 \times 10^{-1},$$
$$c = ab = 2 \times 10^{-4}$$
will result in an *underflow*. ∎

Simple mathematical computations such as finding a square root, or exponent of a number or computing factorials can give overflow. For example, consider computing
$$c = \sqrt{a^2 + b^2}.$$
If a or b is very large, then we will get an overflow while computing $a^2 + b^2$.

The IEEE standard also sets forth the results of operations with infinities and NaNs. All operations with infinities correspond to the limiting case in real analysis. Those ambiguous situations, such as $0 \cdot \infty$, result in NaNs, and all binary operations with one or two NaNs result in a NaN.

Avoiding Overflow: An Example

Overflow and underflow can *sometimes* be avoided just by organizing the computations differently. Consider, for example, the task of computing the length of an n-vector x with components, denoted by $\|x\|_2$:
$$\|x\|_2^2 = x_1^2 + x_2^2 + \cdots + x_n^2.$$

3.2. Rounding Errors

If some x_i is too big or too small, then we can get overflow or underflow with the usual way of computing $\|x\|_2$. However, if we normalize each component of the vector by dividing it by $m = \max(|x_1|, \ldots, |x_1|)$ and then form the squares and the sum, then overflow problem can be avoided. Thus, a better way to compute $\|x\|_2^2$ would be the following:

1. $m = \max(|x_1|, \ldots, |x_n|)$.
2. $y_i = x_i/m$, $i = 1, \ldots, n$.
3. $\|x\|_2 = m\sqrt{(y_1^2 + y_2^2 + \cdots + y_n^2)}$.

3.2 Rounding Errors

If a computed result of a given real number is not machine representable, then there are two ways it can be represented in the machine. Consider

$$\pm \cdot d_1 \cdots d_t d_{t+1} \cdots .$$

Then the first method, **chopping**, is the method in which the digits from d_{t+1} on are simply chopped off. The second method is **rounding**, in which the digits d_{t+1} through the rest are not only chopped off, but the digit d_t is also rounded up or down depending on whether $d_{t+1} \geq \beta/2$ or $d_{t+1} < \beta/2$.

Let fl(x) denote the floating point representation of a real number x.

Example 3.2. Rounding. Consider base 10. Let $x = 3.141596$.

$$\begin{aligned} t &= 2, & \text{fl}(x) &= 3.1, \\ t &= 3, & \text{fl}(x) &= 3.14, \\ t &= 4, & \text{fl}(x) &= 3.142. \end{aligned} \quad \blacksquare$$

We now give an expression to measure the error made in representing a real number x on the computer, and then show how this measure can be used to give bounds for errors in other floating point computations.

Definition 3.3. *Let $\hat{x}$ denote an approximation of x. Then there are two ways we can measure the error:*

$$\text{Absolute Error} = |\hat{x} - x|,$$
$$\text{Relative Error} = \frac{|\hat{x} - x|}{|x|}, \quad x \neq 0.$$

The relative error makes more sense than the absolute error. The following simple example shows this.

Example 3.4. Relative error versus absolute error. Consider

$$x_1 = 1.31, \quad \hat{x}_1 = 1.30$$
$$\text{and} \quad x_2 = 0.12, \quad \hat{x}_2 = 0.11.$$

The absolute errors in both cases are the same: $|\hat{x}_1 - x_1| = |\hat{x}_2 - x_2| = 0.01$. On the other hand, the relative error in the first case is $\frac{|\hat{x}_1 - x_1|}{|x_1|} = 0.0076335$ and the relative error in the second case is $\frac{|\hat{x}_2-x_2|}{|x_2|} = 0.0833333$. Thus, the relative errors show that $\hat{x}_1$ is closer to x_1 than $\hat{x}_2$ is to x_2, whereas the absolute errors give no indication of this at all. ∎

The relative error also gives an indication of the number of significant digits in an approximate answer. *If the relative error is about 10^{-s}, then x and $\hat{x}$ agree to about s significant digits.* We state this more specifically in the following definition.

Definition 3.5. *$\hat{x}$ is said to approximate x to s significant digits if s is the largest non-negative integer for which the relative error $\frac{|x-\hat{x}|}{|x|} < 5(10^{-s})$; that is, s is given by $s = \left[-\log\left(\frac{|x-\hat{x}|}{|x|}\right) + \frac{1}{2}\right]$.*

Thus, in the above examples, $\hat{x}_1$ and x_1 agree to *two* significant digits, while $\hat{x}_2$ and x_2 agree to about only *one* significant digit.

Round-Off Error in Representation of a Real Number

We now give an expression for the relative error in representing a real number x by its floating point representation fl(x).

Theorem 3.6. *Let* fl(x) *denote the floating point representation of a real number x. Then*

$$\frac{|\text{fl}(x) - x|}{|x|} \leq \mu = \left\{ \begin{array}{ll} \frac{1}{2}\beta^{1-t} & \text{for rounding} \\ \beta^{1-t} & \text{for chopping} \end{array} \right\}. \tag{3.1}$$

Proof. We establish the bound for rounding and leave the other part for Exercise 3.1.
Let x be written as

$$x = (\cdot d_1 d_2 \cdots d_t d_{t+1} \cdots) \times \beta^e,$$

where $d_1 \neq 0$ and $0 \leq d_i < \beta$. When we round off x we obtain one of the following floating point numbers:

$$x' = (\cdot d_1 d_2 \cdots d_t) \times \beta^e,$$
$$x'' = [(\cdot d_1 d_2 \cdots d_t) + \beta^{-t}] \times \beta^e.$$

Obviously we have $x \in (x', x'')$. Assume, without any loss of generality, that x is closer to x'. We then have

$$|x - x'| \leq \frac{1}{2}|x' - x''| = \frac{1}{2}\beta^{e-t}.$$

Thus, the relative error

$$\frac{|x - x'|}{|x|} \leq \left(\frac{1}{2}\frac{\beta^{-t}}{\cdot d_1 d_2 \cdots d_t \cdots}\right)$$

$$\leq \frac{1}{2}\frac{\beta^{-t}}{\frac{1}{\beta}}(\text{since } d_i < \beta) = \frac{1}{2}\beta^{1-t}. \quad \square$$

3.3. Laws of Floating Point Arithmetic

Example 3.7. Consider the three-digit representation of the decimal number $x = 0.2346$ ($\beta = 10$, $t = 3$). Then, if *rounding* is used, we have

$$\text{fl}(x) = 0.235,$$

$$\text{Relative Error} = 0.001705 < \frac{1}{2}10^{-2}.$$

Similarly, if *chopping* is used, we have

$$\text{fl}(x) = 0.234,$$

$$\text{Relative Error} = 0.0025575 < 10^{-2}. \quad \blacksquare$$

Definition 3.8. *The number μ in (3.1) is called the machine precision or unit round-off error. It is the smallest positive floating point number such that*

$$\text{fl}(1 + \mu) > 1.$$

The machine precision, μ, is usually between 10^{-16} and 10^{-7} (on most machines) for double and single precision, respectively. For the IBM 360 and 370, $\beta = 16$, $t = 6$, $\mu = 4.77 \times 10^{-7}$.

The machine precision is very important in scientific computations. If the particulars β, t, L, and U for a computer are not known, the following simple Fortran program can be run to estimate μ for that computer (Forsythe, Malcolm, and Moler (1977, p. 14)).

```
      REAL MEU, MEU 1
      MEU = 1.0
   10 MEU = 0.5 * MEU
      MEU 1 = MEU + 1.0
      IF (MEU 1.GT.1.0) GOTO 10
```

The above Fortran program computes an approximation of μ which differs from μ by at most a factor of 2. This approximation is quite acceptable, since an exact value of μ is not that important and is seldom needed.

The book by Forsythe, Malcolm, and Moler (1977) also contains an extensive list of L and U for various computers.

3.3 Laws of Floating Point Arithmetic

The formula

$$\frac{|\text{fl}(x) - x|}{|x|} \leq \mu = \begin{cases} \beta^{1-t} & \text{for chopping,} \\ \frac{1}{2}\beta^{1-t} & \text{for rounding} \end{cases}$$

can be written as

$$\text{fl}(x) = x(1 + \delta), \quad \text{where } |\delta| \leq \mu. \tag{3.2}$$

Assuming that the IEEE standard holds, we can easily derive the following simple **laws of floating point arithmetic**.

Theorem 3.9. *Let x and y be two floating point numbers, and let* $\text{fl}(x + y)$, $\text{fl}(x - y)$, $\text{fl}(xy)$, *and* $\text{fl}(x/y)$ *denote the computed sum, difference, product, and quotient. Then*

1. $\text{fl}(x \pm y) = (x \pm y)(1 + \delta)$, *where* $|\delta| \leq \mu$;
2. $\text{fl}(xy) = (xy)(1 + \delta)$, *where* $|\delta| \leq \mu$;
3. *if* $y \neq 0$, *then* $\text{fl}(x/y) = (x/y)(1 + \delta)$, *where* $|\delta| \leq \mu$.

On computers that do not use the IEEE standard, the following floating point law of addition might hold:

4. $\text{fl}(x + y) = x(1 + \delta_1) + y(1 + \delta_2)$, *where* $|\delta_1| \leq \mu$ *and* $|\delta_2| \leq \mu$.

Example 3.10. Simple floating point operations with rounding. Let
$$\beta = 10, \quad t = 3$$
in items 1 through 3 below.

1. $x = 0.999 \times 10^2$, $y = 0.111 \times 10^0$.
$$x + y = 100.0110 = 0.100011 \times 10^3,$$
$$\text{fl}(x + y) = 0.100 \times 10^3.$$
Thus, $\text{fl}(x + y) = (x + y)(1 + \delta)$, where
$$\delta = -1.0999 \times 10^{-4}, \quad |\delta| < \frac{1}{2}(10^{-2}).$$

2. $x = 0.999 \times 10^2$, $y = 0.111 \times 10^0$.
$$xy = 11.0889,$$
$$\text{fl}(xy) = 0.111 \times 10^2.$$
Thus, $\text{fl}(xy) = xy(1 + \delta)$, where
$$\delta = 1.00100 \times 10^{-3}, \quad |\delta| \leq \frac{1}{2}(10^{1-3}).$$

3. $x = 0.999 \times 10^2$, $y = 0.111 \times 10^0$.
$$\frac{x}{y} = 900,$$
$$\text{fl}\left(\frac{x}{y}\right) = 0.900 \times 10^3,$$
$$\delta = 0.$$

4. Let
$$\beta = 10, \quad t = 4,$$
$$x = 0.1112, \quad y = 0.2245 \times 10^5,$$
$$xy = 0.24964 \times 10^4,$$
$$\text{fl}(xy) = 0.2496 \times 10^4.$$

3.3. Laws of Floating Point Arithmetic

Thus, $|\text{fl}(xy) - xy| = 0.44$ and

$$|\delta| = 1.7625 \times 10^{-4} < \frac{1}{2} \times 10^{-3}. \quad \blacksquare$$

Computing without a Guard Digit

Theorem 3.9 and the examples following this theorem show that the relative errors in computing the sum, difference, product, and quotient in floating point arithmetic are small. However, there are computers without guard digits in which additions and subtractions may not be accurate.

A guard digit is an extra digit on the lower end of the arithmetic register whose purpose is to catch the low-order digit which would otherwise be pushed out of existence when the decimal points are aligned.

For computers with a guard digit,

$$\text{fl}(x \pm y) = (x + y)(1 + \delta), \qquad |\delta| \leq \mu.$$

However, for those without a guard digit,

$$\text{fl}(x \pm y) = x(1 + \delta_1) \pm y(1 + \delta_2),$$
$$|\delta_1| \leq \mu, \qquad |\delta_2| \leq \mu.$$

Remark. Throughout this book, we will assume that the computations have been performed with a guard digit, as they are on almost all available machines.

We shall call results 1 through 3 of Theorem 3.9 along with (3.2) the **fundamental laws of floating point arithmetic**. These fundamental laws form the basis for establishing bounds for relative errors in other floating point computations.

Example 3.11. Consider the floating point computation of $x(y + z)$:

$$\text{fl}(x(y + z)) = [x \cdot \text{fl}(y + z)](1 + \delta_1)$$
$$= x(y + z)(1 + \delta_2)(1 + \delta_1)$$
$$= x(y + z)(1 + \delta_1\delta_2 + \delta_1 + \delta_2)$$
$$\approx x(y + z)(1 + \delta_3),$$

where $\delta_3 = \delta_1 + \delta_2$; since δ_1 and δ_2 are small, their product is neglected.

We can now easily establish the bound of δ_3. Suppose $\beta = 10$, and that rounding is used. Then

$$|\delta_3| = |\delta_1 + \delta_2| \leq |\delta_1| + |\delta_2|$$
$$\leq \frac{1}{2} \times 10^{1-t} + \frac{1}{2} \times 10^{1-t}$$
$$= 10^{1-t}.$$

Thus, *the relative error due to round-off in computing* $\text{fl}(x(y + z))$ *is about* 10^{1-t} *in the worst case.* $\blacksquare$

3.4 Addition of *n* Floating Point Numbers

Consider adding n floating point numbers $x_1, x_2, \ldots, x_n$ with rounding. Define $s_2 = \text{fl}(x_1 + x_2)$. This gives

$$s_2 = \text{fl}(x_1 + x_2) = (x_1 + x_2)(1 + \delta_2), \tag{3.3}$$

where $|\delta_2| \leq \mu$. That is, $s_2 - (x_1 + x_2) = \delta_2(x_1 + x_2)$. Define $s_3, s_4, \ldots, s_n$ recursively by

$$s_{i+1} = \text{fl}(s_i + x_{i+1}), \quad i = 2, 3, \ldots, n - 1. \tag{3.4}$$

Then $s_3 = \text{fl}(s_2 + x_3) = (s_2 + x_3)(1 + \delta_3) = x_1(1 + \delta_2)(1 + \delta_3) + x_2(1 + \delta_2)(1 + \delta_3) + x_3(1 + \delta_3)$. Then

$$\begin{aligned} s_3 - (x_1 + x_2 + x_3) &= (x_1 + x_2)\delta_2 + (x_1 + x_2)(1 + \delta_2)\delta_3 + x_3\delta_3 \\ &\approx (x_1 + x_2)\delta_2 + (x_1 + x_2 + x_3)\delta_3 \end{aligned} \tag{3.5}$$

(neglecting the term $\delta_2\delta_3$, which is small, and so on). Thus, by induction we can show that

$$\begin{aligned} s_n - (x_1 + x_2 + \cdots + x_n) &\approx (x_1 + x_2)\delta_2 + (x_1 + x_2 + x_3)\delta_3 \\ &\quad + \cdots + (x_1 + x_2 + \cdots + x_n)\delta_n \end{aligned} \tag{3.6}$$

(again neglecting the terms $\delta_i\delta_j$, which are small).

Equation (3.6) can be written as

$$\begin{aligned} s_n - (x_1 + x_2 + \cdots + x_n) &\approx x_1(\delta_2 + \delta_3 + \cdots + \delta_n) \\ &\quad + x_2(\delta_2 + \cdots + \delta_n) + x_3(\delta_3 + \cdots + \delta_n) \\ &\quad + \cdots + x_n\delta_n, \end{aligned} \tag{3.7}$$

where each $|\delta_i| \leq \frac{1}{2}\beta^{1-t} = \mu$. Defining $\delta_1 = 0$, we can write the following theorem.

Theorem 3.12 (rounding error in floating point addition). *Let $x_1, x_2, \ldots, x_n$ be n floating point numbers. Then*

$$\begin{aligned} \text{fl}(x_1 + x_2 + \cdots + x_n) &- (x_1 + x_2 + \cdots + x_n) \\ &\approx x_1(\delta_1 + \delta_2 + \cdots + \delta_n) + x_2(\delta_2 + \cdots + \delta_n) + \cdots + x_n\delta_n, \end{aligned} \tag{3.8}$$

where each $|\delta_i| \leq \mu, i = 1, 2, \ldots, n$.

Remark. From the above formula we see that we should expect a smaller error in general when adding n floating point numbers in increasing order of magnitude:

$$|x_1| \leq |x_2| \leq |x_3| \leq \cdots \leq |x_n|.$$

If the numbers are arranged in increasing order of magnitude, then the larger errors will be associated with the smaller numbers (Exercise 3.6).

3.5. Multiplication of n Floating Point Numbers

Theorem 3.13. *Define the numbers n_{i1} by*

$$1 + \eta_1 = (1 + \delta_1)(1 + \delta_2) \cdots (1 + \delta_n),$$
$$1 + \eta_2 = (1 + \delta_2)(1 + \delta_3) \cdots (1 + \delta_n),$$
$$1 + \eta_3 = (1 + \delta_3) \cdots (1 + \delta_n),$$
$$\vdots$$
$$1 + \eta_{n-1} = (1 + \delta_{n-1})(1 + \delta_n),$$
$$1 + \eta_n = (1 + \delta_n).$$

Also define $\mu' = \frac{\mu}{0.9}$ and assume that $n\mu \leq 0.1$. Then

$$\mathrm{fl}(x_1 + x_2 + \cdots + x_n)$$
$$= x_1(1 + \eta_1) + x_2(1 + \eta_2) + \cdots + x_{n-1}(1 + \eta_{n-1}) + x_n(1 + \eta_n), \quad (3.9)$$

where $|\eta_1| \leq (n-1)\mu'$ and $|\eta_i| \leq (n-i+1)\mu'$, $i = 2, \ldots, n$.

Proof. See Stewart (1998b, pp. 130–132). □

3.5 Multiplication of n Floating Point Numbers

Proceeding as in the case of addition of n floating point numbers in the last section, we can show the following.

Theorem 3.14.

$$\mathrm{fl}(x_1 \times x_2 \times \cdots \times x_n) \approx (1 + \epsilon) \prod_{i=1}^{n} x_i,$$

where $\epsilon = |(1 + \delta_2)(1 + \delta_3) \cdots (1 + \delta_n) - 1|$ and $|\delta_i| \leq \mu$, $i = 1, 2, \ldots, n$.

A bound for ϵ: Assuming that $(n-1)\mu < 0.1$, it can be shown that

$$\epsilon < 1.06(n-1)\mu. \quad (3.10)$$

(This assumption is quite realistic; on most machines this assumption will hold for fairly large values of n.)

Indeed, since $|\delta_i| \leq \mu$ and $(n-1)\mu < 0.1$, we have

$$\epsilon \leq (1+\mu)^{n-1} - 1 < (n-1)\mu \left[1 + \frac{0.05}{1 - 0.05}\right] < 1.06(n-1)\mu. \quad (3.11)$$

Thus, combining Theorem 3.14 and (3.10), we can write the following.

Theorem 3.15 (rounding error in floating point multiplication). *The relative error in computing the product of n floating point numbers is at most $1.06(n-1)\mu$, assuming that $(n-1)\mu < 0.1$.*

3.6 Inner Product Computation

A frequently arising computational task in numerical linear algebra is the computation of the **inner product** of two n-vectors x and y:

$$x^T y = x_1 y_1 + x_2 y_2 + \cdots + x_n y_n, \qquad (3.12)$$

where x_i and y_i, $i = 1, \ldots, n$, are the components of x and y.

Define

$$S_1 = \mathrm{fl}(x_1 y_1), \qquad (3.13)$$
$$S_2 = \mathrm{fl}(S_1 + \mathrm{fl}(x_2 y_2)), \qquad (3.14)$$
$$\vdots$$
$$S_k = \mathrm{fl}(S_{k-1} + \mathrm{fl}(x_k y_k)), \qquad (3.15)$$
$$k = 3, 4, \ldots, n.$$

We then have, using Theorem 3.14,

$$S_1 = x_1 y_1 (1 + \delta_1), \qquad (3.16)$$
$$S_2 = [S_1 + x_2 y_2 (1 + \delta_2)](1 + \eta_2) \qquad (3.17)$$
$$\vdots$$
$$S_n = [S_{n-1} + x_n y_n (1 + \delta_n)](1 + \eta_n), \qquad (3.18)$$

where each $|\delta_i| \leq \mu$, and $|\eta_i| \leq \mu$. Substituting the values of S_1 through S_{n-1} in S_n and making some rearrangements, we can write

$$S_n = \sum_{i=1}^{n} x_i y_i (1 + \epsilon_i), \qquad (3.19)$$

where

$$\begin{aligned} 1 + \epsilon_i &= (1 + \delta_i)(1 + \eta_i)(1 + \eta_{i+1}) \cdots (1 + \eta_n) \\ &\approx 1 + \delta_i + \eta_i + \eta_{i+1} + \cdots + \eta_n \qquad (\eta_1 = 0) \end{aligned} \qquad (3.20)$$

(ignoring the products $\delta_i \eta_j$ and $\eta_j \eta_k$, which are small).

For example, when $n = 2$, it is easy to check that

$$S_2 = x_1 y_1 (1 + \epsilon_1) + x_2 y_2 (1 + \epsilon_2), \qquad (3.21)$$

where $1 + \epsilon_1 \approx 1 + \delta_1 + \eta_2$, $1 + \epsilon_2 \approx 1 + \delta_2 + \eta_2$ (neglecting the products of $\delta_1 \eta_2$ and $\delta_2 \eta_2$, which are small).

From (3.19) and (3.20), we can write the following.

Theorem 3.16.

$$\mathrm{fl}(x_1 y_1 + x_2 y_2 + \cdots + x_n y_n) = x_1 y_1 (1 + \epsilon_1) + x_2 y_2 (1 + \epsilon_2) + \cdots + x_n y_n (1 + \epsilon_n),$$

where ϵ_i are given by (3.20).

3.7. Error Bounds for Floating Point Matrix Operations

From Theorem 3.16 we have

$$|\text{fl}(x^T y) - x^T y| \leq \sum_{i=1}^{n} |x_i y_i||\epsilon_i|. \tag{3.22}$$

A bound for $|\epsilon_i|$ in terms of μ. Under the assumption that $n\mu < 0.1$, a bound for ϵ_i in terms of μ can be established.

Using this bound, we can write the following.

Theorem 3.17 (rounding error in inner product computation).

$$|\text{fl}(x^T y) - x^T y| \leq \phi(n)\mu |x|^T |y|,$$

where $|x|$ stands for the vector with components $|x_i|$ and $\phi(n)$ is a small function of n.

Remarks. (i) Note that high relative accuracy cannot be guaranteed if $|x^T y| \ll |x|^T |y|$.

(ii) The bound given in Theorem 3.17 can be improved by using extended precision or some particular implementations (see Higham (2002, pp. 63–64) for details). Typically, error becomes essentially independent of n.

3.7 Error Bounds for Floating Point Matrix Operations

Theorem 3.18. *Let $|M| = (|m_{ij}|)$. Let A and B be two floating point matrices and let c be a floating point number. Then*

1. $\text{fl}(cA) = cA + E$, $|E| \leq \mu |cA|$;

2. $\text{fl}(A + B) = (A + B) + E$, $|E| \leq \mu |A + B|$.

If A and B are two matrices compatible for matrix multiplication, then

3. $\text{fl}(AB) = AB + E$, $|E| \leq n\mu |A| |B| + O(\mu^2)$.

Proof. See Wilkinson (1965, p. 115). □

Meaning of $O(\mu^2)$

Remark. In the above expression, the notation $O(\mu^2)$ stands for a complicated expression that is bounded by $c\mu^2$, where c is a constant, depending upon the problem. The expression $O(\mu^2)$ will be used frequently in this book.

Remark. The last result shows that the matrix multiplication in floating point arithmetic can be very inaccurate, since $|A| |B|$ may be much larger than $|AB|$ itself (Exercise 3.9).

Error Bounds in Terms of Norms

Traditionally, for matrix computations the bounds for error matrices are given in terms of the norms of the matrices, rather than in terms of absolute values of the matrices as given

above. Here we rewrite the bound for error matrices for matrix multiplications using norms, for easy reference later in the book. *We must note, however, that entrywise error bounds are more meaningful than normwise errors* (see remarks in Section 4.3).
In terms of a norm, we can write

$$\text{fl}(AB) = AB + E,$$

where

$$\|E\|_p \leq \eta\mu\|A\|_p\|B\|_p + O(\mu^2), \quad p = 1, \infty, F.$$

In particular, in terms of the $\|\ \|_1$ norm, we have the following.

Theorem 3.19. $\|\text{fl}(AB) - AB\|_1 \leq n\mu\|A\|_1\|B\|_1 + O(\mu^2).$

Two Important Special Cases

A. *Matrix-vector multiplication.* If b is a vector, then from above we have

$$\|\text{fl}(Ab) - Ab\|_1 \leq n\mu\|A\|_1\|b\|_1.$$

B. *Matrix multiplication by an orthogonal matrix.* Recall that a real matrix O is called an *orthogonal matrix* if $O^T O = OO^T = I$.

Corollary 3.20. *Let $A \in R^{n \times n}$ and $Q \in R^{n \times n}$ orthogonal. Then*

$$\|\text{fl}(QA) - QA\|_F \leq n\mu\|A\|_F.$$

Implication of the above result. The result of Corollary 3.20 says that although matrix multiplication can be inaccurate in general, if one of the matrices is orthogonal, then the floating point matrix multiplication gives only a small and acceptable error. As we will see in later chapters, this result forms the basis of many numerically viable algorithms discussed in this book.

3.8 Round-Off Errors Due to Cancellation and Recursive Computations

Intuitively, it is clear that if a large number of floating point computations is done, then the accumulated error can be quite large. *However, round-off errors can be disastrous even at a single step of computation.* For example, consider the subtraction of two numbers:

$$x = 0.54617 \text{ and } y = 0.54601$$

The exact value is

$$d = x - y = 0.00016.$$

Suppose now we use four-digit arithmetic with rounding. Then we have

$$\hat{x} = 0.5462 \text{ (correct to four significant digits)},$$
$$\hat{y} = 0.5460 \text{ (correct to four significant digits)},$$
$$\hat{d} = \hat{x} - \hat{y} = 0.0002.$$

3.8. Round-Off Errors Due to Cancellation and Recursive Computations

How good is the approximation of $\hat{d}$ to d? The relative error is
$$\frac{|d - \hat{d}|}{|d|} = 0.25 \text{ (quite large!)}.$$

What happened above is the following. In four-digit arithmetic, the numbers 0.5462 and 0.5460 are of almost the same size. So, when the first one was subtracted from the second, the most significant digits got canceled and the very least significant digit was left in the answer. This phenomenon, known as **catastrophic cancellation**, occurs when two numbers of approximately the same size are subtracted.

Remark. It is to be noted that in many cases, subtraction is performed rather accurately. It is not a cause of the error—*rather it reveals the errors made in earlier computations or even those in the data associated with the subtraction.* Indeed, cancellation highlights the earlier errors.

Avoiding Cancellation

Fortunately, in many cases catastrophic cancellation can be avoided. For example, consider the case of solving the quadratic equation:
$$ax^2 + bx + c = 0, \quad a \neq 0.$$
The usual way the two roots x_1 and x_2 are computed is
$$x_1 = \frac{-b + \sqrt{b^2 - 4ac}}{2a},$$
$$x_2 = \frac{-b - \sqrt{b^2 - 4ac}}{2a}.$$

It is clear from the above that if a, b, and c are numbers such that $-b$ is about the same size as $\sqrt{b^2 - 4ac}$ (with respect to the arithmetic used), then a catastrophic cancellation will occur in computing x_2, and as a result the computed value of x_2 can be completely erroneous.

Example 3.21. Cancellation in root-finding of the quadratic. Consider solving $ax^2 + bx + c = 0$, with $a = 1$, $b = -10^5$, $c = 1$ (Forsythe, Malcolm, and Moler (1977, pp. 20–22)). Then using $\beta = 10$, $t = 8$, $L = -U = -50$, we see that
$$x_1 = \frac{10^5 + \sqrt{10^{10} - 4}}{2} = 10^5 \text{ (true answer)},$$
$$x_2 = \frac{10^5 - 10^5}{2} = 0 \text{ (completely wrong)}.$$
The true $x_2 = 0.000010000000001$ (correctly rounded to 11 significant digits). The catastrophic cancellation took place in computing x_2, since $-b$ and $(\sqrt{b^2 - 4ac})$ are of the same order. Note that in eight-digit arithmetic, $\sqrt{10^{10} - 4} = 10^5$. ∎

How Can Cancellation be Avoided in Finding Roots of the Quadratic?

Cancellation can be avoided if an equivalent pair of formulas is used:
$$x_1 = -\frac{b + \text{sign}(b)\sqrt{b^2 - 4ac}}{2a},$$
$$x_2 = \frac{c}{ax_1},$$

where sign(b) is the sign of b. Using these formulas, we easily see that
$$x_1 = 100000.00,$$
$$x_2 = \frac{1.0000000}{100000.00} = 0.000010000.$$

Remark. Cancellation may still take place during the subtraction $b^2 - 4ac$, and significant digits will be lost if $b^2 \approx 4ac$. In that case, extended precision should be used in computing $b^2 - 4ac$.

Example 3.22. For yet another example consider the problem of evaluating
$$f(x) = e^x - x - 1 \quad \text{at } x = 0.01.$$
Using five-digit arithmetic, the correct answer is 0.000050167. If $f(x)$ is evaluated directly from the expression, we have

$$f(0.01) = 1.0101 - (0.01) - 1 = 0.0001, \tag{3.23}$$

$$\text{Relative Error} = \frac{0.0001 - 0.000050167}{0.00005016}, \tag{3.24}$$

$$= 0.99 \times 10^0, \tag{3.25}$$

indicating that we cannot trust even the first significant digit.

Fortunately, cancellation can again be avoided using the convergent series for e^x:
$$e^x = 1 + x + \frac{x^2}{2} + \frac{x^3}{3!} + \cdots.$$
In this case we have
$$e^x - x - 1 = \left(1 + x + \frac{x^2}{2} + \frac{x^3}{3!} + \cdots\right) - x - 1 \tag{3.26}$$
$$= \frac{x^2}{2} + \frac{x^3}{3!} + \frac{x^4}{4!} \cdots. \tag{3.27}$$
For $x = 0.01$, this formula gives
$$\frac{(0.01)^2}{2} + \frac{(0.01)^3}{3!} + \frac{(0.01)^4}{4!} + \cdots$$
$$= 0.00005 + 0.000000166666 + 0.00000000004166 + \cdots$$
$$= 0.000050167 \quad \text{(correct up to five significant digits)}. \quad \blacksquare$$

Remark. Note that if x were negative, then use of the convergent series for e^x would not have helped. For example, to compute e^x for a negative value of x, cancellation can be avoided by using
$$e^{-x} = \frac{1}{e^x} = \frac{1}{1 + x + \frac{x^2}{2!} + \frac{x^3}{3!} + \cdots}.$$

Recursive Computations

Recursive computations are those which are performed recursively so that the computation of one step depends upon the results of the previous steps. In such cases, even if the error

3.8. Round-Off Errors Due to Cancellation and Recursive Computations

made in the first step is negligible, due to the accumulation and magnification of error at every step, the final error can be quite large, giving a completely erroneous answer.

Certain recursions propagate errors in very unhealthy fashions. Consider the following example involving recursive computations, again from Forsythe, Malcolm, and Moler (1977, pp. 16–17).

Example 3.23. Suppose we need to compute the integral

$$E_n = \int_0^1 x^n e^{x-1} dx$$

for different values of n. Integrating by parts gives

$$E_n = \int_0^1 x^n e^{x-1} dx = (x^n e^{x-1})_0^1 - \int_0^1 n x^{n-1} e^{x-1} dx$$

or

$$E_n = 1 - n E_{n-1}, \quad n = 2, 3, \ldots.$$

Thus, if E_1 is known, then for different values of n, E_n can be computed, using the above recursive formula.

Indeed, with $\beta = 10$ and $t = 6$, and starting with $E_1 = 0.367879$ as a six-digit approximation to $E_1 = 1/e$, we have from above

$$E_1 = 0.367879,$$
$$E_2 = 0.264242,$$
$$E_3 = 0.207274,$$
$$E_4 = 0.170904,$$
$$\vdots$$
$$E_9 = -0.068480 \text{ (wrong)}.$$

Although the integrand is positive throughout the interval [0, 1], *the computed value of E_9 is negative.* This phenomenon can be explained as follows.

The error in computing E_2 was -2 times the error in computing E_1, and the error in computing E_3 was -3 times the error in E_2 (therefore, the error at this step was exactly six times the error in E_1). Thus, the error in computing E_9 was $(-2)(-3)(-4)\cdots(-9) = 9!$ times the error in E_1. The error in E_1 was due to the rounding of $1/e$ using six significant digits, which is about 4.412×10^{-7}. However, this small error multiplied by $9!$ gave $9! \times 4.412 \times 10^{-7} = 0.1601$, which is quite large. ∎

Rearranging the Recurrence

Again, for this example, it turned out that we could get a much better result by simply rearranging the recursion so that the error at every step, instead of being magnified, is reduced. Indeed, if we rewrite the recursion as

$$E_{n-1} = \frac{1 - E_n}{n}, \quad n = \ldots, 3, 2,$$

then the error at each step will be reduced by a factor of $1/n$. Thus, starting with a large value of n (say, $n = 20$) and working backward, we will see that E_9 will be accurate to full six-digit precision.

To obtain a starting value, we note that

$$E_n = \int_0^1 x^n e^{n-1} dx \leq \int_0^1 x^n dx = \frac{1}{n+1}.$$

With $n = 20$, $E_{20} \leq \frac{1}{21}$. Let's take $E_{20} = 0$. Then, starting with $E_{20} = 0$, it can be shown (Forsythe, Malcolm, and Moler (1977, p. 17)) that $E_9 = 0.0916123$, which is correct to full six-digit precision.

The reason for obtaining this accuracy was that the error in E_{20} was at most $\frac{1}{21}$; this error was multiplied by $\frac{1}{20}$ in computing E_{19}, giving an error of at most $\frac{1}{20} \cdot \frac{1}{21} = 0.0024$ in the computation of E_{19}, and so on.

3.9 Review and Summary

The concepts of floating point numbers and rounding errors have been introduced and discussed in this chapter.

1. *Floating point numbers.* A normalized floating point number has the form

    $$x = \pm r \beta^e,$$

    where e is called **exponent**, r is the **significant**, and β is the **base** of the number system. The floating point number system is characterized by four parameters: β, the **base**; t, the **precision**; and L, U, the **lower** and **upper limits of the exponent**.

2. *Errors.* The error(s) in a computation is measured either by *absolute error* or *relative error*.

    Relative errors make more sense than absolute errors.

    The relative error gives an indication of the number of significant digits in an approximate answer.

    The relative error in representing a real number x by its floating point representation fl(x) is bounded by a number μ, called the *machine precision* (Theorem 3.6).

3. *Laws of floating point arithmetic.*

    $$\text{fl}(x \odot y) = (x \odot y)(1 + \delta),$$

    where $\odot$ indicates any of the four basic arithmetic operations $+$, $-$, $\times$, or $\div$, and $|\delta| \leq \mu$.

4. *Addition, multiplication, and inner product computations.* The results of addition and multiplication of n floating point numbers and inner product computation are given in Theorems 3.12, 3.15, and 3.17, respectively.

    - While adding n floating point numbers, it is advisable that they be added in increasing order of magnitude.

5. *Floating point matrix multiplications.* The entrywise and normalized error bounds for matrix multiplication of two floating point matrices are given in Theorems 3.18 and 3.19, respectively.

   - Matrix multiplication in floating point arithmetic can be very inaccurate, unless one of the matrices is orthogonal (or unitary, if complex).
   - The high accuracy in a matrix product computation involving an orthogonal matrix (Corollary 3.20) makes the use of orthogonal matrices in matrix computations very attractive.

6. *Round-off errors due to cancellation and recursive computation.* Subtractive cancellation or catastrophic cancellation (as it is commonly called) is a phenomenon in which a number of significant digits in a computation gets cancelled due to subtraction of two almost equal numbers. In most cases, however, subtractions are done exactly. *Catastrophic cancellation signals some errors made in previous steps.* In fact, it brings this error in prominence. Recursive computations are those which are performed recursively so that the computation of one step depends upon the results of the previous steps.

   These have been discussed in some detail in Section 3.8.

   Examples have been given to show how these errors come up in many basic computations. *An encouraging message here is that in several instances, computations can be reorganized so that cancellation can be avoided, and the error in recursive computations can be diminished at each step of computation.*

3.10 Suggestions for Further Reading

For details of IEEE standard, see the monograph *An American National Standard: IEEE Standard for Binary Floating-Point Arithmetic* (IEEE, 1985), *IEEE, Standard for Radix-Independent Floating-Point Arithmetic* (IEEE, 1987), and *Numerical Computing with IEEE Floating Point Arithmetic* (Overton, 2001).

For results on error bounds for basic floating point matrix operations, the classic books by James H. Wilkinson (1963, 1995) are extremely useful and valuable resources. The most recent authoritative book on error analysis is the one by Higham (2002). *Every researcher of numerical analysis must have a copy of this book.*

Discussion on basic floating point operations and rounding errors due to cancellations and recursive computations are given nowadays in many numerical analysis textbooks.

Exercises on Chapter 3

3.1 (a) Prove the expression

$$\frac{|\text{fl}(x) - x|}{|x|} \leq \mu = \begin{cases} \frac{1}{2}\beta^{1-t} & \text{for rounding,} \\ \beta^{1-t} & \text{for chopping.} \end{cases}$$

(b) Show that (a) can be written in the form

$$\text{fl}(x) = x(1 + \delta), \ |\delta| \leq \mu.$$

3.2 Let x be a floating point number and let k be a positive integer. Then prove that
$$\mathrm{fl}\left(\frac{x^k}{k!}\right) = \frac{x^k}{k!}(1 + e_k),$$
where
$$|e_k| \leq 2k\mu + O(\mu^2).$$

3.3 Construct examples to show that the distributive law for floating point addition and multiplication does not hold. What can you say about the commutativity and associativity for these operations? Give reasons for your answers.

3.4 Let $x_1, x_2, \ldots, x_n$ be the n floating point numbers. Define
$$s_2 = \mathrm{fl}(x_1 + x_2), \quad s_k = \mathrm{fl}(s_{k-1} + x_k), \ k = 3, \ldots, n.$$

(a) Then from Theorem 3.12 show that
$$\mathrm{fl}(x_1 + x_2 + \cdots + x_n) = x_1(1 + \eta_1) + x_2(1 + \eta_2) + \cdots + x_n(1 + \eta_n).$$

(b) Give a bound for each $\eta_i, i = 1, 2, \ldots, n$.

3.5 (a) Give a proof of Theorem 3.14.

(b) Prove Theorem 3.15 by first establishing the result (3.11).

3.6 (a) Construct an example to show that, when adding a list of floating point numbers, the rounding error will generally be less if the numbers are added in order of increasing magnitude.

(b) Find another example to show that this is not always necessarily true.

3.7 Using Theorem 3.17, show that high relative accuracy is obtained in computing $x^T x$.

3.8 Show that

(a) $\mathrm{fl}(cA) = cA + E, \ |E| \leq \mu |cA|$;

(b) $\mathrm{fl}(A + B) = (A + B) + E, \ |E| \leq \mu(|A| + |B|)$;

(c) $\mathrm{fl}(AB) = AB + E, \ |E| \leq n\mu |A|\,|B| + O(\mu^2)$.

(Consult Wilkinson (1965, p. 115)).

3.9 Construct a simple example to show that the matrix multiplication in floating point arithmetic need not be accurate.

3.10 Prove that if Q is orthogonal, then
$$\mathrm{fl}(QA) = Q(A + E), \text{ where } \|E\|_2 \leq n^2 \mu \|A\|_2 + O(\mu^2).$$

3.11 Let $y_1, \ldots, y_n$ be n column vectors defined recursively:
$$y_{i+1} = Ay_i, \quad i = 1, 2, \ldots, n-1.$$
Let $\hat{y}_i = \mathrm{fl}(y_i)$. Find a bound for the relative error in computing each $y_i, i = 1, \ldots, n$.

3.12 Let $\beta = 10$, $t = 4$. Compute
$$\text{fl}(A^T A),$$
where
$$A = \begin{pmatrix} 1 & 1 \\ 10^{-4} & 0 \\ 0 & 10^{-4} \end{pmatrix}.$$

Repeat your computation with $t = 9$. Compare the results.

3.13 Show how to arrange computation in each of the following, so that the loss of significant digits can be avoided. Do one numerical example in each case to support your answer.

(a) $e^x - x - 1$ for negative values of x.

(b) $\sqrt{x^4 + 1} - x^2$ for large values of x.

(c) $\dfrac{1}{x} - \dfrac{1}{x+1}$ for large values of x.

(d) $x - \sin x$ for values of x near zero.

(e) $1 - \cos x$ for values of x near zero.

(f) $\dfrac{e^x - 1}{x}$ for $|x| \ll 1$.

(g) $\dfrac{(1 - \cos x)}{x^2}$ for small x.

3.14 What are the relative and absolute errors in approximating

(a) π by $\dfrac{22}{7}$?

(b) $\dfrac{1}{3}$ by 0.333?

(c) $\dfrac{1}{6}$ by 0.166?

How many significant digits are there in each computation?

3.15 Let $\beta = 10$, $t = 4$. Consider computing
$$a = \left(\dfrac{1}{6} - 0.1666\right) / 0.1666.$$

How many correct digits of the exact answer will you get?

3.16 Consider evaluating
$$e = \sqrt{a^2 + b^2}.$$

How can the computation be organized so that overflow in computing $a^2 + b^2$ for large values of a or b can be avoided?

3.17 What answers will you get if you compute the following numbers on your calculator or computer?

(a) $\sqrt{10^8 - 1}$.

(b) $\sqrt{10^{-20} - 1}$.

(c) $10^{16} - 50$.

Compute the absolute and relative errors in each case.

3.18 What problem do you foresee in solving the quadratic equations

(a) $x^2 - 10^6 x + 1 = 0$,

(b) $10^{-10} x^2 - 10^{10} x + 10^{10} = 0$

using the well-known formula

$$x = \frac{-b \pm \sqrt{b^2 - 4ac}}{2a}?$$

What remedy do you suggest? Now solve the equations using your suggested remedy, with $t = 4$.

3.19 Show that the integral

$$y_i = \int_0^1 \frac{x^i}{x+5} dx$$

can be computed by using the recursion formula:

$$y_i = \frac{1}{i} - 5 y_{i-1}.$$

Compute $y_1, y_2, \ldots, y_{10}$ using this formula, taking

$$y_0 = \ln(x+5)|_{x=0}^1 = \ln 6 - \ln 5 = \ln(1.2).$$

What abnormalities do you observe in this computations? Explain what happened.

Now rearrange the recursion so that the values of y_i can be computed more accurately.

Chapter 4
Stability of Algorithms and Conditioning of Problems

Background Material Needed

- Vector and matrix norms (Section 2.5)

4.1 Introduction

In this chapter, we introduce the basic concepts of *algorithm*, two of its important properties, *efficiency* and *stability*, and an important property of the problem, called *conditioning*.

We begin with a definition of algorithm and state some basic algorithms for matrix computations in this section itself.

Definition 4.1. An *algorithm* *is an ordered set of operations, logical and arithmetic, which when applied to a computational problem defined by a given set of data, called the **input data**, produces a solution to the problem. A solution comprises a set of data called the **output data**.*

In this book, for the sake of convenience and simplicity, we will very often describe algorithms by means of **pseudocodes** which can be translated into computer codes easily. Describing algorithms by pseudocodes has been made popular by Stewart (1973). Here are some examples.

4.1.1 Computing the Norm of a Vector

Given $x = (x_1, \ldots, x_n)^T$, compute $\|x\|_2$.

4.1.2 Computing the Inner Product of Two Vectors

Given two n-vectors x and y, $x = (x_1, x_2, \ldots, x_n)^T$ and $y = (y_1, y_2, \ldots, y_n)^T$, compute the inner product $x^T y = x_1 y_1 + x_2 y_2 + \cdots + x_n y_n$.

ALGORITHM 4.1. Computing the Norm of a Vector.

Input: $n, x_1, \ldots, x_n$.
Output: $s = \|x\|_2$.

Step 1. Compute $\mathbf{r} = \max(|x_1|, \ldots, |x_n|)$.

Step 2. Compute $y_i = x_i/\mathbf{r}$, $i = 1, \ldots, n$

Step 3. Compute $s = \|x\|_2 = \mathbf{r}\sqrt{(y_1^2 + \cdots + y_n^2)}$.

Pseudocodes

$\mathbf{r} = \max(|x_1|, \ldots, |x_n|)$
$s = 0$
For $i = 1$ to n do
$\quad y_i = x_i/\mathbf{r}; \quad s = s + y_i^2$
$s = \mathbf{r}(s)^{1/2}$
End

An Algorithmic Note

In order to avoid overflow, each entry of x was normalized before using the norm formula

$$\|x\|_2 = \sqrt{x_1^2 + \cdots + x_n^2}.$$

ALGORITHM 4.2. Computing the Inner Product of Two Vectors.

Input: A positive integer n and two sets of numbers $\{x_i\}_{i=1}^n$ and $\{y_i\}_{i=1}^n$.
Output: Sum = Inner product $x^T y$

Step 1. Compute the partial products: $s_i = x_i y_i$, $i = 1, \ldots, n$.

Step 2. Add the partial products: Sum $= \sum_{i=1}^{n} s_i$.

Pseudocodes

Sum $= 0$
For $i = 1, \ldots, n$ do
$\quad$ Sum $=$ Sum $+ x_i y_i$
End

4.1. Introduction

4.1.3 Solution of an Upper Triangular System

Consider the system

$$Ty = b,$$

where $T = (t_{ij})$ is a nonsingular upper triangular matrix and $y = (y_1, y_2, \ldots, y_n)^T$. Specifically,

$$t_{11}y_1 + t_{12}y_2 + \cdots + t_{1n}y_n = b_1,$$
$$t_{22}y_2 + \cdots + t_{2n}y_n = b_2,$$
$$t_{33}y_3 + \cdots + t_{3n}y_n = b_3,$$
$$\vdots$$
$$t_{n-1,n-1}y_{n-1} + t_{n-1,n}y_n = b_{n-1},$$
$$t_{nn}y_n = b_n,$$

where each $t_{ii} \neq 0$ for $i = 1, 2, \ldots, n$.

The last equation is solved first to obtain y_n; then this value is inserted into the next to last equation to obtain y_{n-1}, and so on. This process is known as **back substitution**. The algorithm can easily be written down.

ALGORITHM 4.3. Back Substitution Method for Upper Triangular System.

Input: An $n \times n$ upper triangular matrix $T = (t_{ij})$ and an n-vector b.
Output: The vector $y = (y_1, \ldots, y_n)^T$, such that $Ty = b$.

Step 1. Compute $y_n = \dfrac{b_n}{t_{nn}}$.

Step 2. Compute y_{n-1} through y_1 successively:
For $i = n - 1, \ldots, 2, 1$ do

$$y_i = \frac{1}{t_{ii}} \left(b_i - \sum_{j=i+1}^{n} t_{ij} y_j \right).$$

End

4.1.4 Solution of a Lower Triangular System

A lower triangular system can be solved in an analogous manner. The process is known as the **forward elimination method**. Let $L = (l_{ij})$ and $b = (b_1, b_2, \ldots, b_n)^T$.

> **ALGORITHM 4.4. The Forward Elimination Method for Lower Triangular System.**
>
> **Input:** A $n \times n$ lower triangular matrix $L = (l_{ij})$ and an n-vector b.
> **Output:** An n-vector $y = (y_1, y_2, \ldots, y_n)^T$ such that $Ly = b$.
>
> **Step 1.** Compute $y_1 = \dfrac{b_1}{l_{11}}$.
>
> **Step 2.** For $i = 2, 3, \ldots, n$ do
> $$y_i = \frac{1}{l_{ii}} \left(b_i - \sum_{j=1}^{i-1} l_{ij} y_j \right)$$
> End

MATCOM Notes: Algorithms 4.3 and 4.4 have been implemented in MATCOM programs, BACKSUB and FORELIM, respectively.

4.2 Efficiency of an Algorithm

Two most desirable properties of an algorithm are *efficiency* and *stability*.

The efficiency of an algorithm is measured by the amount of computer time consumed in its implementation. A *theoretical and crude measure of efficiency* is the number of floating point operations (**flops**) needed to implement the algorithm.

Definition 4.2. *A flop is a basic floating point operation:* $+, -, *, or /$.

Flop-count for Algorithm 4.3 and Algorithm 4.4 substitution. Each of these algorithms requires n^2 flops.

The big O notation. An algorithm will be called an $O(n^p)$ algorithm if the dominant term in the operations count of the algorithm is a multiple of n^p. Thus, the solution of a triangular system is an $O(n^2)$ algorithm.

Notation for overwriting and interchange. We will use the notation
$$a \equiv b$$
to denote that "**b** **overwrites** **a**." Similarly, if two computed quantities a and b are interchanged, they will be written symbolically
$$a \leftrightarrow b.$$

4.3 Definition and Concept of Stability

The examples on catastrophic cancellations and recursive computations in the last chapter (Section 3.8) had one thing in common: *the inaccuracy of the computed result in each*

4.3. Definition and Concept of Stability

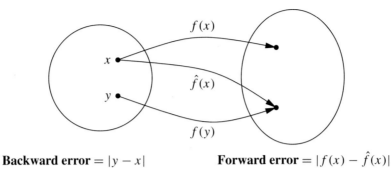

Backward error $= |y - x|$ **Forward error** $= |f(x) - \hat{f}(x)|$

Figure 4.1. *Backward error vs. forward error.*

case was entirely due to the algorithm used, because as soon as the algorithm was changed or rearranged and applied to the problem with the same data, the computed result became satisfactory. Thus, we are talking about two different types of algorithms for a given problem. The algorithms of the first type are examples of *unstable algorithms*, while the ones of the second type—giving satisfactory results—are *stable algorithms*.

There are two types of stable algorithms: *backward stable* and *forward stable*.

In this context, we first define **forward error** and **backward error**. Let $\hat{f}(x)$ be the computed approximate value of $f(x)$ with an input data x. Then we have the following.

- *Forward error* $= |f(x) - \hat{f}(x)|$.

  On the other hand, backward errors relate the errors to the data of the problem rather than to the problem's solution.

- *Backward error.* Here we ask for what value of the input data y does $f(y) = \hat{f}(x)$? Backward error $= |y - x|$.

See Figure 4.1.

Example 4.3. Backward vs. forward errors. Suppose we would like to estimate $f(x) = e^x$ at $x = 1$. Consider the truncated series

$$\hat{f}(x) = 1 + x + \frac{x^2}{2} + \frac{x^3}{3!}.$$

Then

$$f(1) = 2.7183, \quad \hat{f}(1) = 2.6667$$

Forward error $= 2.7183 - 2.6667 = 0.0516$.

To find backward error we must find y such that $f(y) = \hat{f}(1)$. For e^x, $y = \log(\hat{f}(x))$. At $x = 1$ we have $y = \log(\hat{f}(1)) = 0.9808$.

Backward Error: $|y - x| = 0.0192$.

Verify: $e^y = e^{0.9809} = 2.6667 = \hat{f}(1) = 2.6667$. ∎

Example 4.4. Forward error bound for the inner product. Let x and y be two n-vectors. Then the error bound in Theorem 3.17, $|\text{fl}(x^T y) - x^T y| \leq \phi(n)\mu |x|^T |y|$, is the *forward error bound* for the inner product of x and y.

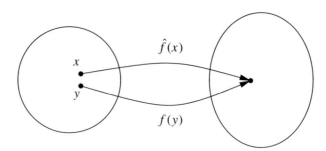

Figure 4.2. *Backward stable algorithm.*

This result shows that high accuracy in inner product computation cannot be guaranteed unless $|x^T y| \ll |x|^T |y|$; on the other hand, if $y = x$, then the result will be accurate. ∎

Example 4.5. Forward error bound for matrix multiplication. The error bound in Theorem 3.19, $\text{fl}(AB) = AB + E$, where $\|E\|_1 \leq n\mu \|A\|_1 \|B\|_1 + O(\mu^2)$, gives the *forward error* bound for matrix multiplication. ∎

Backward and Forward Stability

Definition 4.6. *An algorithm is called **backward stable** if for any x it produces a value $\hat{f}(x)$ with a small backward error. In other words, an algorithm is backward-stable if it produces an exact solution to a nearby problem. That is, an algorithm is backward stable if $\hat{f}(x) = f(y)$, for some y close to x.*

Remark. The forward stability of the algorithm is defined in a similar way. *In this book, by "stability" we will imply backward stability.* Thus, an algorithm will be called stable if it is backward stable.

Note that an algorithm can be forward stable without being backward stable; that is, a small error in $\hat{f}(x)$ may or may not correspond to a small perturbation of the data.

The process of analyzing the backward errors in a numerical computation is called the *backward error analysis*. Backward error analysis, introduced in the literature by J. H. Wilkinson,[1] is nowadays widely used in matrix computations; using this analysis, the stability (or instability) of many algorithms in numerical linear algebra has been established in recent years.

Example 4.7. Backward stability of arithmetic operations of two floating point numbers. Consider computing the sum of two floating point numbers x and y. We have seen

---

[1] James H. Wilkinson, a British mathematician, is well known for his pioneering work on backward error analysis for matrix computations. He was affiliated with the National Physical Laboratory in Britain, and held visiting appointments at Argonne National Laboratory, Stanford University, etc. Wilkinson died an untimely death in 1986. A fellowship in his name has since been established at Argonne National Laboratory. Wilkinson's book *The Algebraic Eigenvalue Problem* is an extremely important and very useful book for any numerical analyst.

4.3. Definition and Concept of Stability

before (Theorem 3.9) that

$$\text{fl}(x + y) = (x + y)(1 + \delta)$$
$$= x(1 + \delta) + y(1 + \delta) = x' + y'.$$

Thus, the computed sum of two floating point numbers x and y is the exact sum of another two floating point numbers x' and y'. Since $|\delta| \leq \mu$, both x' and y' are close to x and y, respectively. Thus we conclude that *the operation of adding two floating point numbers is backward stable*. Similar statements, of course, hold for other arithmetic operations of two floating point numbers. ∎

Example 4.8. Backward stability of addition of n. Recall from Chapter 3 (Theorem 3.13) that $\text{fl}(x_1 + x_2 + \cdots + x_n) = x_1(1 + \eta_1) + x_2(1 + \eta_2) + \cdots + x_n(1 + \eta_n)$, where each η_i is small. Thus, the *computed sum of n floating point numbers is the exact sum of n perturbed numbers with small perturbations*. ∎

Example 4.9. Backward stability and instability of the inner and outer products. *The inner product of two vectors x and y is backward stable.* Theorem 3.16 shows that the computed inner product is the exact inner product of a perturbed set of data: $x_1, x_2, \ldots, x_n$ and $y_1(1 + \epsilon_1), \ldots y_n(1 + \epsilon_n)$, where each perturbation is small.

The outer product of the vectors x and y is, however, not backward stable (Exercise 4.1(b)). ∎

Examples of Backward Stability and Instability of Linear Systems Solvers

Definition 4.10. *An algorithm for solving $Ax = b$ will be called **backward stable** if the computed solution $\hat{x}$ is such that*

$$(A + E)\hat{x} = b + \delta b$$

with E and δb small.

How Do We Measure Smallness?

The "smallness" of a matrix or a vector is measured either by looking into its entries or by computing its norm.

Normwise vs. Entrywise Errors

While measuring errors in computations using norms is traditional in matrix computations, componentwise measure of errors is becoming increasingly important. It really does make more sense.

An $n \times n$ matrix A has n^2 entries, but the norm of A is a single number. Thus the smallness or largeness of the norm of an error matrix E does not truly reflect the smallness or largeness of the individual entries of E. For example, if $E = (10, 0.00001, 1)^T$, then $\|E\| = 10.0499$. Thus the small entry 0.00001 was not reflected in the norm measure.

Example 4.11. A stable algorithm: Solution of an upper triangular system by back substitution. Consider Algorithm 4.3 (the back substitution method). It can be shown (see Chapter 14, available online at *www.siam.org/books/ot116*) that the computed solution $\hat{x}$, obtained by this algorithm, satisfies

$$(T + E)\hat{x} = b,$$

where the entries of the error matrix E are quite small. In fact, if $E = (e_{ij})$ and $T = (t_{ij})$, then

$$|e_{ij}| \leq n\mu|t_{ij}| + O(\mu^2),$$

showing that the error can be even smaller than the error made in rounding the entries of T. Thus, *the back substitution process for solving an upper triangular system is stable.* ∎

Example 4.12. An unstable algorithm: Gaussian elimination without pivoting. Consider solving the 2×2 system using the standard elimination method, called Gaussian elimination: $Ax = b$, where $A = \begin{pmatrix} 10^{-10} & 1 \\ 1 & 2 \end{pmatrix}$, $b = \begin{pmatrix} 1 \\ 3 \end{pmatrix}$. That is,

$$10^{-10}x_1 + x_2 = 1,$$
$$x_1 + 2x_2 = 3.$$

Eliminating x_2 from the second equation, we obtain

$$10^{-10}x_1 + x_2 = 1,$$
$$(2 - 10^{10})x_2 = 3 - 10^{10}.$$

In computer arithmetic, we will have

$$10^{-10}x_1 + x_2 = 1,$$
$$-10^{-10}x_2 = -10^{10},$$

giving $x_2 = 1$, $x_1 = 0$, whereas the exact solution is $x_1 = x_2 = 1$.

Thus, *the above process is clearly unstable.* The readers are asked to verify for themselves that the computed solution $\hat{x} = (1, 0)^T$ is the exact solution of the system $(A + E)\hat{x} = b$, where E is large. ∎

If an algorithm is stable for a given matrix A, then one would like to see that the algorithm is stable for every matrix A in a given class. Thus, we may give a formal definition of stability as follows.

Definition 4.13. *An algorithm is stable for a class of matrices C if for every matrix A in C, the computed solution by the algorithm is the exact solution of a nearby problem.*

Thus, for the linear system problem

$$Ax = b,$$

an algorithm is stable for a class of matrices C if for every $A \in C$ and for each b, it produces a computed solution $\hat{x}$ that satisfies

$$(A + E)\hat{x} = \delta = b + \delta b$$

for some E and δb, where $(A + E)$ is close to A and $b + \delta b$ is close to b.

4.4 Conditioning of the Problem and Perturbation Analysis

From the preceding discussion we should not form the opinion that if a stable algorithm is used to solve a problem, then the computed solution will be accurate. A property of the problem, called **conditioning**, also contributes to the accuracy or inaccuracy of the computed result.

The conditioning of a problem is a property of the problem itself. It is concerned with how the solution of the problem will change if the input data contains some impurities. This concern arises from the fact that in practical applications very often the data come from some experimental observations where the measurements can be subjected to disturbances (or "noise") in the data. There are other sources of error also, for example, *round-off* errors and *discretization errors*. Thus, when a numerical analyst has a problem in hand to solve, he or she must frequently solve the problem not with the original data, but with data that has been perturbed. *The question naturally arises: What effects do these perturbations have on the solution?*

A theoretical study done by numerical analysts to investigate these effects, which is independent of the particular algorithm used to solve the problem, is called **perturbation analysis**. This study helps one detect whether a given problem is "bad" or "good" in the sense of whether small perturbations in the data will create a large or small change in the solution.

When the result of a perturbation analysis is combined with that of backward error analysis of a particular algorithm, an error bound in the computed solution by the algorithm can be obtained.

Definition 4.14. *A problem (with respect to a given set of data) is called an ill-conditioned or badly conditioned problem if a small relative perturbation in data can cause a large relative error in the computed solution, regardless of the method of solution. Otherwise, it is called well-conditioned; that is, a problem is well-conditioned if all small perturbations in data produce only small relative errors in the solution.*

Let x and y denote the original and the slightly perturbed data, and let $f(x)$ and $f(y)$ be the respective solutions. Then we have the following.

Well-conditioned problem. If y is close to x, then $f(y)$ is close to $f(x)$.
Ill-conditioned problem. Even if y is close to x, then $f(y)$ can depart from $f(x)$ drastically.

Numerical analysts attempt to assign a number to each problem, called the *condition number*, to determine if the problem is ill-conditioned or well-conditioned. Formally, the relative condition number or simply the condition number can be defined as follows.

Graphically, these concepts are illustrated in Figure 4.3.

Definition 4.15. *The condition number of the problem f with respect to the data x is defined as*

$$\frac{\text{Relative error in the solution}}{\text{Relative perturbation in the data}} = \frac{|f(x) - f(y)|}{|f(x)|} \bigg/ \left|\frac{x-y}{x}\right|. \quad (4.1)$$

If $f : \mathbb{R}^n \to \mathbb{R}^m$ and x and $y \in \mathbb{R}^n$, then the condition number is formally defined as

$$\limsup_{\epsilon \to 0} \left\{ \left(\frac{\|f(y) - f(x)\|}{\|f(x)\|}\right) \bigg/ \left(\frac{\|x-y\|}{\|x\|}\right) \bigg| \, \|x-y\| \leq \epsilon \right\}.$$

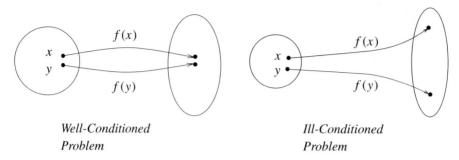

Figure 4.3. *Well-conditioned (left) and ill-conditioned (right) problems.*

A problem is *ill-conditioned* if the condition number is $\gg 1$ ($\gg$ stands for much greater than), for example, 10^{10}, 10^{15}, etc.

Example 4.16. Condition number of a function. If $f(x)$ is a differentiable function of one variable, then it is easy to see (Exercise 4.10) that for small perturbations, the condition number of $f(x)$, denoted by $c(x)$, is given by

$$c(x) = \frac{|x||f'(x)|}{|f(x)|}. \tag{4.2}$$

As an example, let $f(x) = e^x$. Then $c(x) = |x|$, which is large for large values of x. That means *a small relative error in x can produce a large relative error in e^x, so this problem is ill-conditioned when x is large.*

If f or x is a vector, then the condition number can be defined in the same way using norms instead of the absolute values.

Thus, the condition number of a function of several variables (or a vector) can be defined by replacing $f'(x)$ by its gradient. In this case $c(x) = \frac{\|x\|\|\nabla f\|}{\|f(x)\|}$, where ∇f is the gradient.

For example, if $x = (x_1, x_2)^T$ is a vector and the problem is to obtain the scalar $f(x) = x_1 - x_2$, then $\nabla f = (1, -1)$, and the condition number $c(x)$ of f (with respect to the infinite norm) is given by

$$c(x) = \frac{\|x\|_\infty \|\nabla f\|_\infty}{\|f(x)\|_\infty} = \frac{2\max\{|x_1|, |x_2|\}}{|x_1 - x_2|}, \tag{4.3}$$

which shows that the problem is *ill-conditioned if $x_1 \simeq x_2$.* ∎

Example 4.17. Condition number of the matrix-vector product. Suppose $A \in \mathbb{R}$ and x is an n-vector. Then it can be shown (Exercise 4.12) that the condition number κ of Ax (with respect to perturbations of x) is given by

$$\kappa = \|A\| \frac{\|x\|}{\|Ax\|}, \tag{4.4}$$

where the matrix norm is the subordinate matrix norm.

If A is square and nonsingular, then

$$\kappa \leq \|A\|\|A^{-1}\|. \qquad \blacksquare \tag{4.5}$$

4.5. Conditioning of the Problem, Stability, and Accuracy

Example 4.18. Condition number of the polynomial roots. The problem of finding roots of a polynomial can be highly ill-conditioned. We first illustrate this with a simple quadratic polynomial. Consider solving the quadratic equation
$$f(x) = x^2 - 2x + 1 = 0.$$
The roots are $x = 1, 1$. Now perturb the coefficient 2 by 0.00001. The computed roots of the perturbed polynomial $\hat{f}(x) = x^2 - 2.00001x + 1$ are $x_1 = 1.0032$ and $x_2 = 0.9968$. The relative errors in x_1 and x_2 are 0.0032; on the other hand, the relative error in the data is 5×10^{-6}. Thus, a *small perturbation in the data changed the roots substantially.* ∎

The Wilkinson Polynomial

The above example involved multiple roots. Multiple roots or roots close to each other invariably make the root-finding problem ill-conditioned; however, the *problem can be ill-conditioned even when the roots are very well separated.* Consider the following well-known example by Wilkinson:
$$p(x) = (x-1)(x-2) \cdots (x-20)$$
$$= x^{20} - 210x^{19} + \cdots .$$

The roots of $p(x)$ are $1, 2, \ldots, 20$. Now perturb the coefficient of x^{19} from -210 to $-210 - 2^{-23}$, leaving other coefficients unchanged. This change amounts to approximately 1.12×10^{-7}, which is small. Several roots of the perturbed polynomial, carefully computed by Wilkinson, were found to be very different from the original roots. For example, the roots $x = 16$ and $x = 17$ became approximately equal to $16.73 \pm 2.81i$. This change can be easily explained by computing the condition numbers of the individual roots. It can be shown (Exercise 4.23(a)) that the condition number of the root $x = x_j$ with respect to the perturbation of the single coefficient a_i is

$$\text{cond}_j = \frac{|a_i x_j^{i-1}|}{|p'(x_j)|}.$$

Using this definition it is easy to verify that cond_{16} and cond_{17} are both of order $O(10^{10})$, which are quite large.

Note: The definition of conditioning is data-dependent. Thus, a problem which is ill-conditioned for *one* set of data could be well-conditioned for *another* set.

Root-finding and eigenvalue computation. The above examples teach us a very useful lesson: *it is not a good idea to compute the eigenvalues of a matrix by explicitly finding the coefficients of the characteristic polynomial,* since the round-off errors in computations will invariably put some small perturbations in the computed coefficients of the characteristic polynomial, and these small perturbations in the coefficients may cause large changes in the zeros. The eigenvalues will then be computed inaccurately.

4.5 Conditioning of the Problem, Stability of the Algorithm, and Accuracy of the Solution

As stated in the previous section, the conditioning of a problem is a property of the problem itself and has nothing to do with the algorithm used to solve the problem. To a user, of course, the accuracy of the computed solution is of primary importance. However, the accuracy of a computed solution by a given algorithm is directly connected with both the stability of the algorithm and the conditioning of the problem. *If the problem is ill-conditioned, no matter how stable the algorithm is, the accuracy of the computed solution cannot be guaranteed.*

In general, if a backward stable algorithm is applied to a problem with the condition number κ, then the accuracy of the solution depends upon κ. *If it is small, the results will be accurate; but if it is large, the accuracy cannot be guaranteed, the accuracy will depend upon the condition number.*

Also, it is to be kept in mind that a method may be stable for one problem but unstable for another. For example, the modified Gram–Schmidt (MGS) method is stable for least-squares problem (see Chapter 8), but can be unstable for finding an orthonormal basis of a matrix (Chapter 7).

Conditioning, Stability, and Accuracy

Note that the definition of backward stability does not say that the computed solution $\hat{x}$ by a backward stable algorithm will be close to the exact solution of the original problem. However, when a stable algorithm is applied to a well-conditioned problem, the computed solution should be near the exact solution. Also, if a "stable" algorithm is applied to an ill-conditioned problem, it should not introduce more error than what the data warrants.

Stable Algorithm + Well-Conditioned Problem
  $\equiv$ Accurate Solution (the computed solution is near
     the exact solution).

Stable Algorithm + Ill-Conditioned Problem $\equiv$ Accuracy not guaranteed.

An illustration: Suppose that an algorithm to solve the computational problem f defined by the input x produces the function $\hat{f}$ as an approximation of f. Let y be close to x. Then the behavior of a stable algorithm in two cases—when the problem is well-conditioned and ill-conditioned—is illustrated in Figure 4.4 (Stewart (1998b, p. 133)).

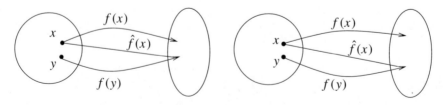

Figure 4.4. *Performance of a backward stable algorithm with well-conditioned problem (left) and ill-conditioned problem (right).*

4.6. Perturbation Analysis of the Linear System Problem

Why is the answer inaccurate? Based on our discussions here, we can state some of the following reasons why the answer might be inaccurate (see Higham (2002, p. 31) for details).

- The problem might be ill-conditioned.
- The algorithm might be unstable.
- The test examples may be too special.
- The algorithm, though successful, might have failed in the particular circumstances.

4.6 Perturbation Analysis of the Linear System Problem

Consider the following linear system:
$$x_1 + 2x_2 = 3,$$
$$2x_1 + 3.999x_2 = 5.999.$$

The exact solution is $\boxed{x_1 = x_2 = 1.}$ Now make a small perturbation on the right-hand side, obtaining the system
$$x_1 + 2x_2 = 3,$$
$$2x_1 + 3.999x_2 = 6.$$

The solution of the perturbed system, obtained by Gaussian elimination with partial pivoting (considered to be a stable method in practice), is $\boxed{x_1 = 3, \ x_2 = 0.}$

Thus, a very small change on the right-hand side changed the solution altogether.

In this section we study the effect of small perturbations of the input data A and b on the computed solution x of the system $Ax = b$; that is, *sensitivity* of linear system solutions.

Since in the linear system problem $Ax = b$ the input data are A and b, there could be impurities either in b or in A or in both. We will therefore consider the effect of perturbations on the solution x in each of these cases separately. We will see that in all of these cases, a number called the *condition number* of the matrix A plays an important role.

4.6.1 Effect of Perturbation on the Right-Hand Side Vector b

We assume here that there are impurities in b but that matrix A is exact.

$$
\begin{array}{ll}
A \to A & \text{(unchanged)} \\
b \to b + \delta b & (\delta b = \text{perturbation in the vector } b) \\
x \to x + \delta x & (\delta x = \text{change in the solution } x)
\end{array}
$$

Theorem 4.19 (right perturbation theorem). If δb and δx are, respectively, the perturbations of b and x in the linear system $Ax = b$, A is nonsingular, and $b \neq 0$, then

$$\frac{\|\delta b\|}{\|A\|\|A^{-1}\|} \leq \frac{\|\delta x\|}{\|x\|} \leq \|A\|\|A^{-1}\|\frac{\|\delta b\|}{\|b\|}.$$

Proof. Since

$$Ax = b$$

and

$$A(x + \delta x) = b + \delta b,$$

we have

$$A\delta x = \delta b.$$

That is,

$$\delta x = A^{-1}\delta b.$$

Taking a subordinate matrix-vector norm we get

$$\|\delta x\| \leq \|A^{-1}\| \|\delta b\|. \tag{4.6}$$

Again, taking the same norm on both sides of $Ax = b$, we get $\|Ax\| = \|b\|$ or

$$\|b\| = \|Ax\| \leq \|A\| \|x\|. \tag{4.7}$$

Combining (4.6) and (4.7), we have

$$\frac{\|\delta x\|}{\|x\|} \leq \|A\| \|A^{-1}\|\frac{\|\delta b\|}{\|b\|}. \tag{4.8}$$

On the other hand, $A\delta x = \delta b$ gives

$$\|\delta x\| \geq \frac{\|\delta b\|}{\|A\|}. \tag{4.9}$$

Also, from $Ax = b$, we have

$$\frac{1}{\|x\|} \geq \frac{1}{\|A^{-1}\|\|b\|}. \tag{4.10}$$

Combining (4.9) and (4.10), we have

$$\frac{\|\delta x\|}{\|x\|} \geq \frac{\|\delta b\|}{\|A\|\|A^{-1}\|\|b\|}.$$

The other part can be similarly proved. $\square$

Definition 4.20. *The number $\|A\| \|A^{-1}\|$ is called the condition number of A and is denoted by* Cond(A).

Interpretation of Theorem 4.19

Theorem 4.19 says that a relative change in the solution can be as large as Cond(A) multiplied by the relative change in the vector b. Thus, *if the condition number is not too large, then a*

4.6. Perturbation Analysis of the Linear System Problem

small perturbation in the vector b will have very little effect on the solution. On the other hand, if the condition number is large, then even a small perturbation in b might change the solution drastically.

Remark. In view of Theorem 4.19, what happened with the above example can be easily explained. Note that for this example $\text{Cond}(A) = O(10^4)$.

Example 4.21. An ill-conditioned linear system problem.

$$A = \begin{pmatrix} 1 & 2 & 1 \\ 2 & 4.0001 & 2.002 \\ 1 & 2.002 & 2.004 \end{pmatrix}, \quad b = \begin{pmatrix} 4 \\ 8.0021 \\ 5.006 \end{pmatrix}.$$

The exact solution is

$$x = \begin{pmatrix} 1 \\ 1 \\ 1 \end{pmatrix}.$$

Change b to $b' =$

$$\begin{pmatrix} 4 \\ 8.0020 \\ 5.0061 \end{pmatrix}.$$

Relative perturbation:

$$\frac{\|b' - b\|}{\|b\|} = \frac{\|\delta b\|}{\|b\|} = 1.3795 \times 10^{-5} \ (small).$$

If we solve the system $Ax' = b'$, we get

$$x' = x + \delta x = \begin{pmatrix} 3.0850 \\ -0.0436 \\ 1.0022 \end{pmatrix}$$

(x' is completely different from x).

Relative error in the solution: $\frac{\|\delta x\|}{\|x\|} = 1.3461$. It is easily verified that the inequality in Theorem 4.19 is satisfied: $\text{Cond}(A) \cdot \frac{\|\delta b\|}{\|b\|} = 4.4418 > 1.3461 = \frac{\|\delta x\|}{\|x\|}$.
The predicted change was, however, overly estimated. ∎

Example 4.22. A well-conditioned problem.

$$A = \begin{pmatrix} 1 & 2 \\ 3 & 4 \end{pmatrix}, \quad b = \begin{pmatrix} 3 \\ 7 \end{pmatrix}.$$

The exact solution is $x = \begin{pmatrix} 1 \\ 1 \end{pmatrix}$. Let $b' = b + \delta b = \begin{pmatrix} 3.0001 \\ 7.0001 \end{pmatrix}$.

The relative change in b is $\frac{\|b'-b\|}{\|b\|} = 1.875 \times 10^{-5}$ (small). Note that $\text{Cond}(A) = 14.9330$ (small). *Thus a drastic change in the solution x is not expected.* In fact x' satisfying $Ax' = b'$ is

$$x' = \begin{pmatrix} 0.9999 \\ 1.0001 \end{pmatrix} \approx x = \begin{pmatrix} 1 \\ 1 \end{pmatrix}. \qquad Note: \ \frac{\|\delta x\|}{\|x\|} = 10^{-5} \ (small). \quad ∎$$

4.6.2 Effect of Perturbation in the Matrix A

Here we assume that there are impurities only in A, and as a result we have $A + \Delta A$ in hand, but b is exact.

$$
\begin{array}{ll}
A \to A + \Delta A & (\Delta A = \text{perturbation in the matrix } A) \\
b \to b & (\text{unchanged}) \\
x \to x + \delta x & (\delta x = \text{change in the solution } x)
\end{array}
$$

Theorem 4.23 (left perturbation theorem). Assume A is nonsingular and $b \neq 0$. Suppose that ΔA and δx are, respectively, the perturbations of A and x in the linear system $Ax = b$. Furthermore, assume that ΔA is such that $\|\Delta A\| < \frac{1}{\|A^{-1}\|}$. Then

$$\frac{\|\delta x\|}{\|x\|} \leq \mathrm{Cond}(A) \frac{\|\Delta A\|}{\|A\|} \bigg/ \left(1 - \mathrm{Cond}(A)\frac{\|\Delta A\|}{\|A\|}\right).$$

Proof. We have

$$(A + \Delta A)(x + \delta x) = b$$

or

$$(A + \Delta A)x + (A + \Delta A)\delta x = b. \tag{4.11}$$

Since $Ax = b$, we have from (4.11)

$$(A + \Delta A)\delta x = -\Delta A x \tag{4.12}$$

or

$$\delta x = -A^{-1}\Delta A(x + \delta x). \tag{4.13}$$

Taking the norm on both sides, we have

$$\|\delta x\| \leq \|A^{-1}\| \, \|\Delta A\| \cdot (\|x\| + \|\delta x\|) \tag{4.14}$$
$$= \frac{\|A^{-1}\| \, \|A\| \, \|\Delta A\|}{\|A\|}(\|x\| + \|\delta x\|),$$

that is,

$$\left(1 - \frac{\|A\|^{-1}\|A\|\|\Delta A\|}{\|A\|}\right)\|\delta x\| \leq \frac{\|A\|\|A^{-1}\|\|\Delta A\|}{\|A\|}\|x\|. \tag{4.15}$$

Since $\|A^{-1}\| \, \|\Delta A\| < 1$, the expression in parentheses on the left-hand side is positive. We can thus divide both sides of the inequality by this number without changing the inequality.

4.6. Perturbation Analysis of the Linear System Problem

After this, if we also divide by $\|x\|$, we obtain

$$\frac{\|\delta x\|}{\|x\|} \leq \frac{\|A\|\|A^{-1}\|\frac{\|\Delta A\|}{\|A\|}}{\left(1 - \|A\|\|A^{-1}\|\frac{\|\Delta A\|}{\|A\|}\right)} = \text{Cond}(A)\frac{\|\Delta A\|}{\|A\|} \bigg/ \left(1 - \text{Cond}(A)\frac{\|\Delta A\|}{\|A\|}\right),$$
(4.16)

which proves the theorem. □

Remarks. Because of the assumption that $\|\Delta A\| < \frac{1}{\|A^{-1}\|}$ (which is quite reasonable to assume), the denominator on the right-hand side of the inequality in Theorem 4.23 is less than one. *Thus even if $\frac{\|\Delta A\|}{\|A\|}$ is small, then there could be a drastic change in the solution if* Cond(A) *is large.*

Example 4.24. Consider Example 4.21 once more. Change $a_{23} = 2.002$ to 2.0021; keep b fixed. Thus

$$\Delta A = -10^{-3} \begin{pmatrix} 0 & 0 & 0 \\ 0 & 0 & 0.1000 \\ 0 & 0 & 0 \end{pmatrix} \quad \text{(small)}.$$

Now solve the system $(A + \Delta A)x' = b$:

$$x' = \begin{pmatrix} 3.0852 \\ -0.0437 \\ 1.0011 \end{pmatrix}, \quad \delta x = x' - x = \begin{pmatrix} 2.0852 \\ -1.0437 \\ 0.0021 \end{pmatrix},$$

$$\text{Relative error} = \frac{\|\delta x\|}{\|x\|} = 1.3463 \quad \text{(quite large)}.$$

Note that $\text{Cond}(A) = O(10^5)$. ∎

4.6.3 Effect of Perturbations in both matrix A and vector b

Finally, we assume now that both the input data A and b have impurities. As a result we have the system with $A + \Delta A$ as the matrix and $b + \delta b$ as the right-hand-side vector.

$$\begin{aligned} A &\to A + \Delta A & (\Delta A &= \text{perturbation in } A) \\ b &\to b + \delta b & (\delta b &= \text{perturbation in the vector } b) \\ x &\to x + \delta x & (\delta x &= \text{change in the solution}) \end{aligned}$$

Theorem 4.25 (general perturbation theorem). *Assume that A is nonsingular, $b \neq 0$, and $\|\Delta A\| < \frac{1}{\|A^{-1}\|}$. Then*

$$\frac{\|\delta x\|}{\|x\|} \leq \left(\frac{\text{Cond}(A)}{1 - \text{Cond}(A) \cdot \frac{\|\Delta A\|}{\|A\|}}\right) \left(\frac{\|\Delta A\|}{\|A\|} + \frac{\|\delta b\|}{\|b\|}\right).$$

Proof. Subtracting
$$Ax = b$$
from
$$(A + \Delta A)(x + \delta x) = b + \delta b$$
we have
$$(A + \Delta A)(x + \delta x) - Ax = \delta b$$
or
$$(A + \Delta A)(x + \delta x) - (A + \Delta A)x + (A + \Delta A)x - Ax = \delta b$$
or
$$(A + \Delta A)(\delta x) + \Delta A x = \delta b$$
or
$$A(I - A^{-1}(-\Delta A))\delta x = \delta b - \Delta A x. \tag{4.17}$$

Let $A^{-1}(-\Delta A) = F$. Then
$$\|F\| = \|A^{-1}(-\Delta A)\| \leq \|A^{-1}\| \, \|\Delta A\| < 1 \quad \text{(by assumption)}.$$

Since $\|F\| < 1$, $I - F$ is invertible (see Theorem 2.27), and then from (4.17) we have
$$\delta x = (I - F)^{-1} A^{-1} (\delta b - \Delta A x).$$

Again, using Theorem 2.27, we can write
$$\|(I - F)^{-1}\| \leq \frac{1}{1 - \|F\|}. \tag{4.18}$$

Thus,
$$\|\delta x\| \leq \frac{\|A^{-1}\|}{1 - \|F\|} (\|\delta b\| + \|\Delta A\| \, \|x\|)$$

or
$$\frac{\|\delta x\|}{\|x\|} \leq \frac{\|A^{-1}\|}{(1 - \|F\|)} \cdot \left(\frac{\|\delta b\|}{\|x\|} + \|\Delta A\| \right) \tag{4.19}$$

$$\leq \frac{\|A^{-1}\|}{(1 - \|F\|)} \left(\frac{\|\delta b\| \, \|A\|}{\|b\|} + \|\Delta A\| \right). \quad \text{Note: } \frac{1}{\|x\|} \leq \frac{\|A\|}{\|b\|}.$$

That is,
$$\frac{\|\delta x\|}{\|x\|} \leq \frac{\|A^{-1}\| \, \|A\|}{(1 - \|F\|)} \left(\frac{\|\delta b\|}{\|b\|} + \frac{\|\Delta A\|}{\|A\|} \right). \tag{4.20}$$

Again
$$\|F\| = \|A^{-1}(-\Delta A)\| \leq \|A^{-1}\| \, \|\Delta A\| = \frac{\|A^{-1}\| \, \|A\|}{\|A\|} \cdot \|\Delta A\|. \tag{4.21}$$

Since $\|F\| \leq 1$, we can write from (4.20) and (4.21)

$$\frac{\|\delta x\|}{\|x\|} \leq \left(\frac{\|A^{-1}\| \|A\|}{1 - \left(\frac{\|A^{-1}\| \|A\|}{\|A\|}\right) \cdot \|\Delta A\|} \right) \left(\frac{\|\delta b\|}{\|b\|} + \frac{\|\Delta A\|}{\|A\|} \right)$$
$$= \left(\frac{\text{Cond}(A)}{1 - \frac{\text{Cond}(A)}{\|A\|} \cdot \|\Delta A\|} \right) \left(\frac{\|\delta b\|}{\|b\|} + \frac{\|\Delta A\|}{\|A\|} \right). \quad \square \quad (4.22)$$

Remarks. We again see from (4.22) that even if the relative perturbations $\frac{\|\delta_b\|}{\|b\|}$ and $\frac{\|\Delta A\|}{\|A\|}$ are small, there might be a drastic change in the solution if Cond(A) is large. Thus, Cond(A) plays a crucial role in the sensitivity of the solution.

Notation for Condition Numbers

Unless otherwise stated, when we write Cond(A), we mean $\text{Cond}_2(A)$, that is, the condition number with respect to the 2-norm. The condition number of a matrix A with respect to a subordinate p norm ($p = 1, 2, \infty$) will be denoted by $\text{Cond}_p(A)$. $\text{Cond}_F(A)$ will stand for condition number with respect to Frobenius norm.

4.7 Some Properties of the Condition Number of a Matrix

The following are some important (but easy to prove) properties of the condition number of a matrix (Exercises 4.16, 4.17, and 4.20).

(I) $\text{Cond}_p(A) \geq 1$ for any p-norm.

(II) $\text{Cond}(\alpha A) = \text{Cond}(A)$, where α is a nonzero scalar, for any given norm.

(III) $\text{Cond}_2(A) = 1$ if and only if A is a nonzero scalar multiple of an orthogonal matrix, i.e., $A^T A = \alpha I$, where $\alpha \neq 0$.

(Note that this property of an orthogonal matrix A makes the matrix very attractive for its use in numerical computations).

(IV) $\text{Cond}_2(A^T A) = (\text{Cond}_2(A))^2$.

(V) $\text{Cond}_2(A) = \text{Cond}_2(A^T)$; $\text{Cond}_1(A) = \text{Cond}_\infty(A^T)$.

(VI) For any given norm, $\text{Cond}(AB) \leq \text{Cond}(A)\text{Cond}(B)$ if A and B are compatible for matrix multiplication.

(VII) $\text{Cond}_2(A) = \frac{\sigma_{\max}}{\sigma_{\min}}$, where $\sigma_{\max}$ and $\sigma_{\min}$ are, respectively, the largest and smallest singular values of A.

We now formally define the ill-conditioning and well-conditioning in terms of the condition number.

Definition 4.26. *The system $Ax = b$ is ill-conditioned if* $\text{Cond}(A)$ *is large (e.g., 10^5, 10^8, 10^{10}, etc.). Otherwise, it is well-conditioned.*

A convention. Unless otherwise stated, by $\text{Cond}(A)$ we will mean $\text{Cond}_2(A)$.

Remarks. Though the condition number, as defined above, is norm-dependent, the condition numbers with respect to two different norms are related (see Golub and Van Loan (1996, p. 26)). (For example, it can be shown that if A is an $n \times n$ matrix, then $\frac{1}{n} \leq \frac{\text{Cond}_2(A)}{\text{Cond}_\infty(A)} \leq n$.) In general, *if a matrix is well-conditioned or ill-conditioned with respect to one norm, it is also ill-conditioned or well-conditioned with respect to some other norms.*

Example 4.27. (a) Consider
$$A = \begin{pmatrix} 1 & 0.9999 \\ 0.9999 & 1 \end{pmatrix}; \text{ then } A^{-1} = 10^3 \begin{pmatrix} 5.0003 & -4.99997 \\ -4.9997 & 5.0003 \end{pmatrix}.$$

1. The condition numbers with respect to the ∞-norm and the 1-norm are
$$\|A\|_\infty = \|A\|_1 = 1.9999, \quad \|A^{-1}\|_\infty = \|A^{-1}\|_1 = 10^4,$$
$$\text{Cond}_\infty(A) = \text{Cond}_1(A) = 1.9999 \times 10^4.$$

2. The condition number with respect to the 2-norm is
$$\|A\|_2 = \sqrt{\rho(A)} = 1.9999, \quad \|A^{-1}\|_2 = \sqrt{\rho(A^{-1})} = 10^4,$$
$$\text{Cond}_2(A) = 1.9999 \times 10^4.$$

3. $\text{Cond}_F(A) = 1.9999 \times 10^4$. ∎

Remark. For the above example, it turned out that the condition number with respect to any norm is the same. This is, however, not always the case, but in general they are closely related. (See below for the condition number of the Hilbert matrix with respect to different norms.)

4.7.1 Some Well-Known Ill-Conditioned Matrices

1. *The Hilbert matrix.*
$$A = \begin{pmatrix} 1 & \frac{1}{2} & \frac{1}{3} & \cdots & \frac{1}{n} \\ \frac{1}{2} & \frac{1}{3} & \frac{1}{4} & \cdots & \frac{1}{n+1} \\ \vdots & & & & \vdots \\ \frac{1}{n} & \frac{1}{n+1} & \cdots & \cdots & \frac{1}{2n-1} \end{pmatrix}.$$

For $n = 10$, $\text{Cond}_2(A) = 1.6025 \times 10^{13}$; $\text{Cond}_\infty(A) = 3.5353 \times 10^{13}$; $\text{Cond}_1(A) = 3.5353 \times 10^{13}$.

2. *The Pei matrix.* $A = (a_{ij})$ with $a_{ii} = \alpha$, $a_{ij} = 1$ for $i \neq j$. The matrix becomes ill-conditioned when α is close to 1 or $n - 1$, for example, when $\alpha = 0.9999$ and $n = 5$, $\text{Cond}(A) = 5 \times 10^4$.

3. *Vandermonde matrix.* $A = (a_{ij})$, where $a_{ij} = v_i^{n-j}$; $v_i = i$th component of an n-vector v. For $n = 5$, $v = (1, 2, 3, 4, 5)^T$, $\text{Cond}(A) = 2.6170 \times 10^4$. This matrix arises in several practical applications, including polynomial interpolation.

4.7.2 How Large Must the Condition Number Be for Ill-Conditioning?

A frequently asked question is, how large must $\text{Cond}(A)$ be before the system $Ax = b$ is considered ill-conditioned? We will use Theorem 4.25 to answer the question.

Suppose for simplicity that

$$\frac{\|\Delta A\|}{\|A\|} = \frac{\|\delta b\|}{\|b\|} = 10^{-d}.$$

Then, from Theorem 4.25, it follows that $\frac{\|\delta x\|}{\|x\|}$ is *approximately* less than or equal to $2 \times \text{Cond}(A) \times 10^{-d}$.

This says that if the data have a relative error of 10^{-d} and if the relative error in the solution has to be guaranteed to be less than or equal to 10^{-t}, then $\text{Cond}(A)$ has to be less than or equal to $\frac{1}{2} \times 10^{d-t}$. Thus, *whether a system is ill-conditioned or well-conditioned depends on* (i) *the accuracy of the data, and* (ii) *how much error in the solution can be tolerated.*

For example, suppose that the data have a relative error of about 10^{-5} and an accuracy of about 10^{-3} is sought. Then $\text{Cond}(A) \leq \frac{1}{2} \times 10^2 = 50$. On the other hand, if an accuracy of about 10^{-2} is sought, then $\text{Cond}(A) \leq \frac{1}{2} \times 10^3 = 500$. Thus, in the first case the system will be well-conditioned if $\text{Cond}(A)$ is less than or equal to 50, while in the second case the system will be well-conditioned if $\text{Cond}(A)$ is less than or equal to 500.

Estimating Accuracy from the Condition Number

In general, if the data are approximately accurate and if $\text{Cond}(A) = 10^s$, then there will be only about $t - s$ significant digit accuracy in the computed solution when the solution is computed in t-digit arithmetic.

For better understanding of conditioning, stability, and accuracy, we refer the readers to the paper of Bunch (1987). For discussions on "*strong and weak stability*" (not discussed in this book), see Bunch (1987) and Bunch et al. (1989).

4.7.3 The Condition Number and Nearness to Singularity

The condition number also gives an indication when a matrix A is computationally close to a singular matrix: *if* $\text{Cond}(A)$ *is large, A is close to singular.*

This measure of nearness to singularity is a more accurate measure than the determinant of A. For example, consider the well-known $n \times n$ upper triangular matrix $A = (a_{ij})$ with $a_{ii} = 1$, and $a_{ij} = -1$ if $j > i$. The matrix has the determinant equal to 1; however, it is nearly singular for large n. Note that $\text{Cond}_\infty(A) = n2^{n-1}$. Similarly, *the smallness of the determinant of a matrix does not necessarily mean that A is close to a singular matrix.* For example, consider $A = \text{diag}(0.1, 0.1, \ldots, 0.1)$ of order 1000. $\det(A) = 10^{-1000}$, which is a small number. However, A is considered to be perfectly nonsingular, because $\text{Cond}_2(A) = 1$.

4.7.4 Examples of Ill-Conditioned Eigenvalue Problems

Perturbation analysis of the eigenvalue problem will be discussed in Chapter 9. The conditioning of the eigenvalues and eigenvectors will be introduced there. Here we just present a few examples of the well-known ill-conditioned eigenvalue problems.

Example 4.28. Consider the 10×10 matrix

$$A = \begin{pmatrix} 1 & 1 & & & & \\ & 1 & 1 & & 0 & \\ & & \ddots & \ddots & & \\ & 0 & & \ddots & 1 & \\ & & & & & 1 \end{pmatrix}.$$

The eigenvalues of A are all 1. Now perturb the (10,1) coefficient of A by a small quantity $\epsilon = 10^{-10}$. Then the eigenvalues of the perturbed matrix computed using the MATLAB function **eig** (that uses a numerically effective eigenvalue-computation algorithm) were found to be

$$0$$
$$1.0184 + 0.0980i,$$
$$0.9506 + 0.0876i,$$
$$1.0764 + 0.0632i,$$
$$0.9051 + 0.0350i,$$
$$1.0999 + 0.00i,$$
$$1.0764 - 0.0632i,$$
$$0.9051 - 0.0350i,$$
$$1.0184 - 0.0980i,$$
$$0.9506 - 0.0876i.$$

(Note the change in the eigenvalues.) ∎

Example 4.29. The Wilkinson-Bidiagonal matrix. Again, it should not be thought that an eigenvalue problem can be ill-conditioned only when the eigenvalues are multiple or are close to each other. An eigenvalue problem with well-separated eigenvalues can be very

4.7. Some Properties of the Condition Number of a Matrix

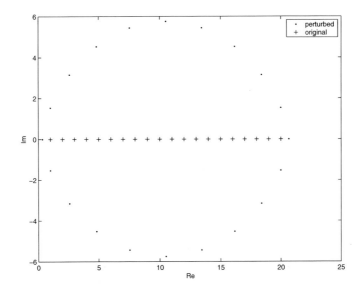

Figure 4.5. *The eigenvalues of a slightly perturbed Wilkinson matrix.*

ill-conditioned too. Consider the 20×20 triangular matrix (known as the Wilkinson-bidiagonal matrix):

$$A = \begin{pmatrix} 20 & 20 & & & & \\ & 19 & 20 & & \text{\Large 0} & \\ & & \ddots & \ddots & & \\ & \text{\Large 0} & & \ddots & 20 & \\ & & & & & 1 \end{pmatrix}.$$

The eigenvalues of A are $1, 2, \ldots, 20$. Now perturb the (20,1) entry of A by $\epsilon = 10^{-10}$. If the eigenvalues of this slightly perturbed matrix are computed using MATLAB function **eig**, it will be seen that some of them will change drastically; they will even become complex, as shown in Figure 4.5. Again, this can be explained by using the definition of condition number of an individual eigenvalue given in Chapter 9. ∎

Example 4.30 (Wilkinson (1965, p. 92)).

$$A = \begin{pmatrix} n & (n-1) & (n-2) & \cdots & 3 & 2 & 1 \\ (n-1) & (n-1) & (n-2) & \cdots & 3 & 2 & 1 \\ 0 & (n-2) & (n-2) & \ddots & & \vdots & \vdots \\ \vdots & & \ddots & \ddots & \ddots & \vdots & \vdots \\ \vdots & & & & \ddots & 2 & \vdots \\ \vdots & & & & 2 & 2 & 1 \\ 0 & \cdots & \cdots & \cdots & 0 & 1 & 1 \end{pmatrix}.$$

As n increases, the smallest eigenvalues become progressively ill-conditioned. For example, when $n = 12$, the condition numbers of the first few eigenvalues are of order unity, while those of the last three are of order 10^7. ∎

4.8 Some Guidelines for Designing Stable Algorithms

Following Higham (2002, pp. 26–27), and based on our discussions in this chapter, we state a few helpful guidelines for designing a stable algorithms. However, note that *"there is no simple recipe for designing stable algorithms"* (Higham (2002), pp. 26–27).

- Avoid catastrophic cancellations if possible.

- Avoid unnecessary overflow and underflow.

- In transforming the problem to another mathematically equivalent problem, use only well-conditioned transformation, such as orthogonal matrix multiplicability.

- If a numerical scheme appears to be unstable, look for different formulations which are mathematically, but not numerically, equivalent (see the use of modified Gram–Schmidt processes versus classical Gram–Schmidt processes in solving least-squares problems in Chapter 8).

- Arrange your computational scheme (if possible) in such a way that the intermediate quantities are much smaller than the final answer.

- Update the solution by using only a small correction; that is, update as new solution = old solution + small correction if the *correction can be computed with sufficient figures*.

4.9 Review and Summary

In this chapter we have introduced two of the most important concepts in numerical linear algebra, namely, the *conditioning* of the problem and *stability* of the algorithm, and have discussed how they affect the *accuracy* of the solution.

4.9.1 Conditioning of the Problem

The conditioning of the problem is a property of the problem. A problem is said to be *ill-conditioned* if a small change in the data can cause a large change in the solution; otherwise it is *well-conditioned*.

Examples of ill-conditioned problems:

- Wilkinson's polynomial of degree 20 for the root-finding problem

- Wilkinson's bidiagonal matrix for the eigenvalue problem

- The Hilbert matrix for the algebraic linear system problem

4.9. Review and Summary

The conditioning of a problems is data-dependent. A problem can be ill-conditioned with respect to one set of data, while it may be quite well-conditioned with respect to another set.

Ill-conditioning or well-conditioning of a matrix problem is generally measured by means of a number called the *condition number*.

In the linear system problem $Ax = b$, the input data are A and b. There may exist impurities either in A or in b, or in both.

We have presented perturbation analyses in all three cases. The results are contained in Theorems 4.19, 4.23, and 4.25. Theorem 4.25 is the most general theorem.

In all three cases, it turns out that

$$\text{Cond}(A) = \|A\| \, \|A^{-1}\|$$

is the deciding factor. If this number is large, then a small perturbation in the input data can cause a large relative error in the computed solution. In this case, the system is called an *ill-conditioned system*, otherwise it is *well-conditioned*. The matrix A having a large condition number is called an *ill-conditioned matrix*.

Some important properties of the condition number of a matrix have been listed (Section 4.7).

The condition number, of course, has a noticeable effect on the accuracy of the solution.

A frequently asked question is, *How large does* $\text{Cond}(A)$ *have to be before the system* $Ax = b$ *is considered to be ill-conditioned?*

The answer depends upon the accuracy of the input data and the level of tolerance of the error in the solution.

In general, if the data are approximately accurate and if $\text{Cond}(A) = 10^s$, *then there are about* $(t - s)$ *significant digits of accuracy in the solution if it is computed in t-digit arithmetic.*

4.9.2 Stability of an Algorithm

An algorithm is said to be a *backward stable algorithm* if it computes the exact solution of a nearby problem. Some examples of stable algorithms (as we will see later in the book) are

- Backward substitution and forward elimination for triangular systems

- Gaussian elimination with complete pivoting for linear systems

- QR factorization using Householder and Givens transformations

- QR iteration algorithm for eigenvalue computations, etc.

The *Gaussian elimination algorithm without row changes is unstable for arbitrary matrices. It is stable for special matrices such as strictly diagonally dominant, Hessenberg, and symmetric positive definite. Gaussian elimination with partial pivoting is stable in practice.*

4.9.3 Effects of Conditioning and Stability on the Accuracy of the Solution

The conditioning of the problem and the stability of the algorithm both have effects on accuracy of the solution computed by the algorithm.

- Stable Algorithm + Well-Conditioned Problem ≡ Accurate Solution
- Stable Algorithm + Ill-Conditioned Problem ≡ Accuracy not guaranteed

4.10 Suggestions for Further Reading

The basic concepts and results of stability and conditioning can be found in most numerical linear algebra books (e.g., Golub and Van Loan (1996), Stewart (1973), Trefethen and Bau (1997)). The two most authoritative books on these topics are the classical book by Wilkinson (1965) and the most recent one by Higham (2002). Stewart's recent books (1998b, 2001a) also give a good amount of coverage of these topics. A book devoted entirely to the perturbation analysis is by Stewart and Sun (1990). An advanced book containing a fair amount of matrix perturbation results is by Bhatia (1996). See also Bhatia (2007). For a condensed review of material of this chapter, see the article of Byers and Datta (2007). Some classical well-known papers on conditioning and stability include DeJong (1977) and Rice (1966). Some earlier papers of Stewart (1977a, 1977b, 1978, 1979, 1991, 1993a) contain a wealth of information on perturbation analysis of various numerical linear algebra problems. For concepts and results on weak and strong stability, see Bunch (1987) and Bunch et al. (1989).

Exercises on Chapter 4

4.0 Answer "True" or "False" to the following. Give reasons for your answers.

(a) If a backward stable algorithm is applied to a computational problem, the solution will be accurate.

(b) A backward stable algorithm produces a good approximation to an exact solution.

(c) Well-conditioning is a good property of an algorithm.

(d) Cancellation is always bad.

(e) If the zeros of a polynomial are all distinct, then they must be well-conditioned.

(f) An efficient algorithm is necessarily a stable algorithm.

(g) Backward errors relate the errors to the data of the problem.

(h) A backward stable algorithm applied to a well-conditioned problem produces an accurate solution.

(i) Stability analysis of an algorithm is performed by means of perturbation analysis.

(j) A symmetric matrix must be well-conditioned.

(k) If the determinant of a matrix A is small, then it must be close to a singular matrix.

(l) One must perform a large amount of computations to obtain a large round-off error.

(m) If a matrix A is ill-conditioned, then its smallest singular value is very small.

4.1 (a) Show that the floating point computations of the sum, product, and division of two numbers are backward stable.

(b) Show that the floating computation of the inner product of two vectors is backward stable and that, on the other hand, the outer product is not.

4.2 Are the following floating point computations backward stable? Give reasons for your answer in each case.

(a) $\text{fl}(x+1)$.

(b) $\text{fl}(x(y+z))$.

(c) $\text{fl}(x_1 + x_2 + \cdots + x_n)$.

(d) $\text{fl}(x_1 x_2 \cdots x_n)$.

(e) $\text{fl}(x^T y/c)$, where x and y are vectors and c is a scalar.

(f) $\text{fl}\left(\sqrt{x_1^2 + x_2^2 + \cdots + x_n^2}\right)$.

4.3 Identify which of the roots of the following polynomials are ill-conditioned and give reasons for your answer.

(a) $x^3 - 3x^2 + 3x - 1$.

(b) $(x-1)^3(x-2)$.

(c) $(x-1)(x-0.9999)(x-2)$.

4.4 Work out the flop-counts for the following simple matrix operations.

(i) Multiplication of matrices A and B of orders $n \times m$ and $m \times p$, respectively.

(ii) Multiplication of a matrix A of order $m \times n$ by a vector b.

(iii) Multiplication of a column vector u by a row vector v.

(iv) Computation of $\|u\|_2$.

(v) Multiplication of row vector u by a column vector v.

(vi) Computation of the matrix $A = \frac{uv^T}{u^T v}$, where u and v are m column vectors.

(vii) Computation of the matrix $B = A - uv^T$, where A and B are two $n \times n$ matrices and u and v are two column vectors.

4.5 Develop an algorithm to compute the following matrix products. Your algorithm should take advantage of the special structure of the matrices in each case. Give flop-count and show storage requirement in each case.

(a) A and B are both lower triangular matrices.

(b) A is arbitrary and B is lower triangular.

(c) A and B are both tridiagonal.

(d) A is arbitrary and B is upper Hessenberg.

(e) $(I + xy^T)A$, where x and y are vectors.

(f) A is upper Hessenberg and B is upper triangular.

4.6 A square matrix $A = (a_{ij})$ is said to be a **band matrix** of bandwidth $2k + 1$ if
$$a_{ij} = 0 \quad \text{whenever } |i - j| > k.$$
Develop an algorithm to compute the product $C = AB$, where A is arbitrary and B is a band matrix of bandwidth 3, taking advantage of the structure of the matrix B. Overwrite A with AB and give flop-count.

4.7 Let A and B be two symmetric matrices of the same order. Develop an algorithm to compute $C = A + B$, taking advantage of symmetry for each matrix. Your algorithm should overwrite B with C. What is the flop-count?

4.8 Let a_r and b_r denote, respectively, the rth columns of the matrices A and B. Then develop an algorithm to compute the product AB from the formula
$$AB = \sum_{i=1}^{n} a_i b_i^T.$$
Give flop-count and storage requirement of the algorithm.

4.9 Consider the matrix
$$A = \begin{pmatrix} 12 & 11 & 10 & \cdots & & & 3 & 2 & 1 \\ 11 & 11 & 10 & \cdots & & & 3 & 2 & 1 \\ 0 & 10 & 10 & & \ddots & & & \vdots & \vdots \\ \vdots & & \ddots & \ddots & \ddots & & \vdots & \vdots \\ \vdots & & & \ddots & \ddots & & 2 & \vdots \\ \vdots & & & & & 2 & 2 & 1 \\ 0 & \cdots & & \cdots & & & 0 & 1 & 1 \end{pmatrix}.$$

Find the eigenvalues of this matrix using MATLAB command **eig**. Now perturb the (1,12) element to 10^{-9} and compute the eigenvalues of this perturbed matrix. What conclusion do you make about the conditioning of the eigenvalues?

4.10 If $f(x)$ is a real-valued differentiable function of a real variable x, then prove that $\frac{|f'(x)||x|}{|f(x)|}$ is the condition number of $f(x)$ at x.

4.11 (a) Show that if $f(x) = \log x$, then the condition number, $c(x) = \left| \frac{1}{\log x} \right|$.

(b) Using the above result (or otherwise), show that $\log x$ is ill-conditioned near $x = 1$.

4.12 Show that the condition number κ for the product Ax (with respect to the perturbation of x) is $\kappa = \|A\| \frac{\|x\|}{\|Ax\|}$.

4.13 Show, by computing the condition number, that the problem of computing $\sqrt{x}$ for $x > 0$ is a well-conditioned problem.

Exercises on Chapter 4

4.14 Work out a bound for the relative error when a backward stable algorithm is applied to a problem with the condition number κ.

4.15 Let A be nonsingular and ΔA be such that $\frac{\|\Delta A\|}{\|A\|} <$ Cond(A). Then prove that $A + \Delta A$ is nonsingular.

4.16 (a) How are Cond$_2(A)$ and Cond$_2(A^{-1})$ related?

(b) Show that
- (i) Cond$(A) \geq 1$ for a norm $\|\cdot\|$ such that $\|I\| \geq 1$;
- (ii) Cond$_2(A^T A) = ($Cond$_2(A))^2$;
- (iii) Cond$(cA) =$ Cond(A) for any given norm.

4.17 (a) Let A be an orthogonal matrix. Then show that Cond$_2(A) = 1$.

(b) Show that Cond$_2(A) = 1$ if and only if A is a scalar multiple of an orthogonal matrix.

4.18 Let $U = (u_{ij})$ be a nonsingular upper triangular matrix. Then show that
$$\text{Cond}_\infty(U) \geq \frac{\max |u_{ii}|}{\min |u_{ii}|}.$$
Hence construct a simple example of an ill-conditioned nondiagonal symmetric positive definite matrix.

4.19 Let $A = LDL^T$ be a symmetric positive definite matrix, where L is unit lower triangular. Let $D = \text{diag}(d_{ii})$. Then show
$$\text{Cond}_2(A) \geq \frac{\max(d_{ii})}{\min(d_{ii})}.$$
Hence construct an example for an ill-conditioned nondiagonal symmetric positive definite matrix.

4.20 Prove that for a given norm, Cond$(AB) \leq$ Cond$(A) \cdot$ Cond(B).

4.21 (a) Find for what values of a the matrix $A = \begin{pmatrix} 1 & a \\ a & 1 \end{pmatrix}$ is ill-conditioned?

(b) What is the condition number of A?

4.22 Give an example to show that a stable algorithm applied to an ill-conditioned problem can produce an inaccurate solution.

4.23 (a) Let a_i be the ith coefficient of a polynomial $p(x)$ and let δa_i and δx_j denote small perturbations of a_i and the jth root x_j. Then show that the condition number of root x_j with respect to perturbations of the single coefficient a_i is
$$\frac{|\delta x_j|}{|x_j|} \bigg/ \frac{|\delta a_i|}{|a_i|} = \frac{|a_i x_j^{i-1}|}{|p'(x_j)|}.$$

(b) *Using MATLAB functions **poly** and **polyder***, compute the condition numbers of the roots $x = i$, $i = 1, 2, \ldots, 20$, of the Wilkinson polynomial
$$p(x) = (x-1)(x-2)\ldots(x-20) = x^{20} - 210x^{19} + \cdots$$
with respect to perturbation of the coefficient x^{19} from -210 to $-210 + 2^{-23}$. Present your results in tabular form and write your conclusion on the ill-conditioning of the roots of the Wilkinson polynomial. Explain why certain roots are more ill-conditioned than others.

MATLAB and MATCOM Programs and Problems on Chapter 4

> **MATLAB RESOURCES**
>
> - Free on-line **MATLAB Tutorial** is available from the website: *www.math.ufl.edu/help/matlab-tutorial*.
>
> - Appendix B of this book, available online at *www.siam.org/books/ot116*, also shows how to use basic MATLAB commands and write simple MATLAB programs.
>
> - See also MATLAB guide books by Higham and Higham (2005), Davis and Sigmon (2005), and Chapman (2009).
>
> - *A Practical Introduction to MATLAB* by Mark S. Gockenbach is available from *http://www.math.mtu.edu/ ~msgocken*.

M4.1 Using the MATLAB function **rand**, create a 5×5 random matrix and then print out the following outputs:
A(2,:), A(:,1), A(:,5),
A(1, 1: 2 : 5), A([1, 5]), A(4: -1: 1, 5: -1: 1).

M4.2 Using the function **for**, write a MATLAB program to find the *inner product* and *outer product* of two n-vectors u and v.

$[s] = \text{inpro}(u,v)$
$[A] = \text{outpro}(u,v)$

Test your program by creating two different vectors u and v using rand (4,1).

M4.3 Learn how to use the following MATLAB commands to create special matrices:

compan	Companion matrix
diag	Diagonal matrices or the diagonals of a matrix
ones	Matrix with all entries equal to one
zeros	Zero matrix
rand	Random matrix
wilkinson	Wilkinson's eigenvalue test matrix
hankel	Hankel matrix
toeplitz	Toeplitz matrix
hilb	Hilbert matrix
triu	Extract the upper triangular part of a matrix
tril	Extract the lower triangular part of a matrix
vander	Vandermonde matrix
rand(n)	Matrix with random entries, chosen from a normal distribution with mean zero, variance one, and standard deviation one.

M4.4 Learn how to use the following MATLAB functions for *basic matrix computations* (you will learn about the algorithms of these functions later in this book):

a\b	Linear equation solution of $Ax = b$.
inv	Matrix inverse
det	Determinant
cond	Condition number
eig	Eigenvalues and eigenvectors
norm	Various matrix and vector norms
poly	Characteristic polynomial
polyval	The value of a polynomial at a given number
plot	Plotting various functions
rank	Rank of a matrix
lu	LU factorization
qr	QR factorization
svd	Singular value decomposition

M4.5 Write MATLAB programs to create the following well-known matrices:

(a) $[A] = \mathbf{wilk}(n)$ to create the Wilkinson bidiagonal matrix $A = (a_{ij})$ of order n:

$$a_{ii} = n - i + 1, \; i = 1, 2, \ldots, n,$$
$$a_{i-1,i} = n, \; i = 2, 3, \ldots, n,$$
$$a_{ij} = 0, \; \text{otherwise}.$$

(b) $[A] = \mathbf{Pei}(n, \alpha)$ to create the Pei matrix $A = (a_{ij})$ of order n:

$$a_{ii} = \alpha \geq 0,$$
$$a_{ij} = 1 \text{ for } i \neq j.$$

(c) Print the condition numbers of the Wilkinson matrix with $n = 10, 20, 50$, and 100, using the MATLAB function **cond**.

(d) Fix $n = 20$, and then perform an experiment to demonstrate the fact that the Pei matrix becomes more ill-conditioned as $\alpha \to 1$.

M4.6 Using "help" commands for **clock** and **etime**, learn how to measure timing for an algorithm.

M4.7 Using MATLAB functions **for, size, zero**, write a MATLAB program to find the product of two upper triangular matrices A and B of order $m \times n$ and $n \times p$, respectively. Test your program using

$A = \text{triu(rand (4,3))},$
$B = \text{triu(rand (3,3))}.$

M4.8 *The purpose of this exercise is to test that the Hilbert matrix is ill-conditioned with respect to solving the linear system problem.*

(i) Create $A = \text{hilb}(10)$. Perturb the (10,1) entry of A by 10^{-5}. Call the perturbed matrix B. Let $b = \text{rand}(10, 1)$. Compute $x = A\backslash b$, $y = B\backslash b$. Compute $\|x - y\|$ and $\frac{\|x-y\|}{\|x\|}$. What conclusion can you draw from this?

(ii) Compute the condition numbers of both A and B: $\operatorname{Cond}(A)$, $\operatorname{Cond}(B)$.

(iii) Compute the condition number of A using the MATLAB command $\operatorname{Cond}(A)$ and then use it to compute the theoretical upper bound given in Theorem 4.23. Compare this bound with the actual relative error.

M4.9 Perform the respective experiments stated in Section 4.7 on Examples 4.28–4.30 to show that the eigenvalue problems for these matrices are ill-conditioned.

M4.10 (a) Write a MATLAB program to construct the $n \times n$ lower triangular matrix $A = (a_{ij})$ as follows:
$$a_{ij} = 1 \quad \text{if} \quad i = j,$$
$$a_{ij} = -1 \quad \text{if} \quad i > j,$$
$$a_{ij} = 0 \quad \text{if} \quad i < j.$$

(b) Perform an experiment to show that the solution of $Ax = b$ with A as above and the vector b created such that $b = Ax$, where $x = (1, 1, \ldots, 1)^T$, becomes more and more inaccurate as n increases due to the increasing ill-conditioning of A. Let $\hat{x}$ denote the computed solution.

Present your results in the following form:

n	$\operatorname{Cond}(A)$	$\hat{x} = A \backslash b$	Relative error $\dfrac{\|x - \hat{x}\|_2}{\|x\|_2}$	Residual norm $\dfrac{\|b - A\hat{x}\|_2}{\|b\|_2}$
10				
20				
30				
40				
50				

M4.11 Using MATLAB function **vander**(v), where $v = $ rand $(20, 1)$, create a 20×20 Vandermonde matrix A. Now take $x = $ ones $(20,1)$ and $b = A * x$. Now compute $y = A \backslash b$. Compare y with x by computing $y - x$ and $\|y - x\|$. What conclusions can you draw?

M4.12 (*Higham's Gallery of Test Matrices.*)

Learn how to use Higham's Gallery of Test Matrices in MATLAB (type *help gallery* for a complete list).

M4.13 (*Computing the sample variance (Higham (2002, pp. 11–12)).*)

Consider computing *sample variance* of n numbers $x_1, \ldots, x_n$ defined by
$$S_n^2 = \frac{1}{n-1} \sum_{i=1}^{n} (x_i - \bar{x})^2,$$
$$\text{where } \bar{x} = \frac{1}{n} \sum_{i=1}^{n} x_i.$$

Describe various mathematically equivalent ways of computing this quantity and discuss their different numerical stability properties.

Chapter 5
Gaussian Elimination and LU Factorization

Background Material Needed

- Vector and matrix norms and their properties (Section 2.5)
- Special matrices (Section 2.4)
- Concepts of errors, floating point operations, and stability (Sections 3.2, 4.2, and 4.3)

5.1 A Computational Template in Numerical Linear Algebra

Most computational algorithms to be presented in this book have a common basic structure that can be described in the following three steps:

Step 1. The problem is first transformed into an "easier-to-solve" problem by transforming the associated matrices to "condensed" forms with special structures.

Step 2. The transformed problem is then solved by exploiting the special structures of these condensed forms.

Step 3. Finally, the solution of the original problem is recovered from the solution of the transformed problem. Sometimes the solution of the transformed problem is the solution of the original problem.

Some typical **condensed** forms used in the following computations are shown in Figures 5.2–5.5:

- The system of linear equation $Ax = b$ is solved by transforming A into an upper triangular matrix (Gaussian elimination), followed by solving two triangular systems: upper and lower (Chapter 6).

- The eigenvalues of a matrix A are computed by transforming A first to an upper Hessenberg matrix H, followed by reducing H further to a real Schur matrix iteratively (Chapter 9).

Triangular: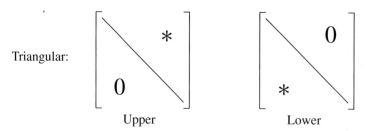

Figure 5.1. *Upper and lower triangular matrices.*

Hessenberg: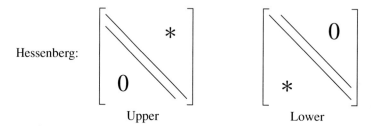

Figure 5.2. *Upper and lower Hessenberg matrices.*

- The singular values of A are computed by transforming A first into a bidiagonal matrix followed by further reduction of the bidiagonal matrix to a diagonal matrix (Chapters 7 and 10).

5.2 LU Factorization Using Gaussian Elimination

The tools of Gaussian elimination are **elementary matrices**.

Definition 5.1. *An elementary lower triangular matrix of order n of type k is a matrix of the form*

$$M_k = \begin{pmatrix} 1 & 0 & \cdots & & \cdots & & 0 & 0 & 0 \\ 0 & 1 & 0 & \cdots & & \cdots & & 0 & 0 \\ 0 & 0 & 1 & 0 & & & & & 0 \\ \vdots & \vdots & \ddots & \ddots & \ddots & & & & \vdots \\ \vdots & \vdots & & 0 & 1 & \ddots & & & \vdots \\ 0 & \vdots & & & m_{k+1,k} & \ddots & 0 & & \vdots \\ 0 & 0 & & & \vdots & \ddots & \ddots & & 0 \\ 0 & 0 & 0 & \cdots & m_{n,k} & \cdots & & 0 & 1 \end{pmatrix}. \quad (5.1)$$

5.2. LU Factorization Using Gaussian Elimination

Tridiagonal: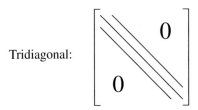

Figure 5.3. *Tridiagonal matrix.*

Bidiagonal: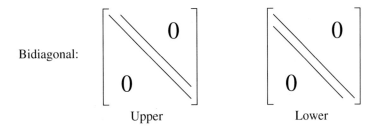

Upper   Lower

Figure 5.4. *Upper and lower bidiagonal matrices.*

Thus, *it is an identity matrix except possibly for a few nonzero elements below the diagonal of the kth column.* The matrix M_k can be written in the form (Exercise 5.1(a))

$$M_k = I + m_k e_k^T,$$

where I is the identity matrix of order n, $m_k = (0, 0, \ldots, 0, m_{k+1,k}, \ldots, m_{n,k})^T$, and e_k is the kth unit vector, that is, $e_k^T = (0, 0, \ldots, 0, 1, 0, \ldots, 0)$, where "1" is at the kth entry.

5.2.1 Creating Zeros in a Vector or Matrix Using Elementary Matrix

Lemma 5.2. *Let*

$$a_1 = \begin{pmatrix} a_{11} \\ a_{21} \\ \vdots \\ a_{n1} \end{pmatrix}.$$

Then the elementary matrix

$$M_1 = \begin{pmatrix} 1 & 0 & 0 & \cdots & \cdots & 0 \\ \dfrac{-a_{21}}{a_{11}} & 1 & 0 & \cdots & \cdots & 0 \\ \dfrac{-a_{31}}{a_{11}} & 0 & 1 & 0 & \cdots & 0 \\ \vdots & \vdots & \ddots & \ddots & & \vdots \\ \dfrac{-a_{n1}}{a_{11}} & 0 \cdots & 0 & 0 & & 1 \end{pmatrix} \quad (5.2)$$

Real Schur: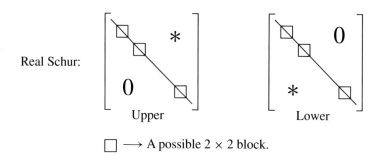

☐ ⟶ A possible 2 × 2 block.

Figure 5.5. *Real Schur matrix.*

is such that

$$M_1 a_1 = \begin{pmatrix} a_{11} \\ 0 \\ 0 \\ \vdots \\ 0 \end{pmatrix}. \qquad (5.3)$$

We leave the proof to the reader (see Exercise 5.1(c)).
The elements $m_{i1} = -\frac{a_{i1}}{a_{11}}, i = 2, \ldots, n$, are called **multipliers**.

Example 5.3. Let

$$a = \begin{pmatrix} 2 \\ 5 \\ 1 \\ 2 \end{pmatrix}.$$

Then

$$M_1 = \begin{pmatrix} 1 & 0 & 0 & 0 \\ -\frac{5}{2} & 1 & 0 & 0 \\ -\frac{1}{2} & 0 & 1 & 0 \\ -1 & 0 & 0 & 1 \end{pmatrix}$$

is such that

$$M_1 a = \begin{pmatrix} 2 \\ 0 \\ 0 \\ 0 \end{pmatrix}. \quad \blacksquare$$

5.2.2 Triangularization Using Gaussian Elimination

The elementary matrices can be conveniently used in triangularizing a matrix. The process is called *Gaussian elimination, after the famous German mathematician and astronomer, Karl Friedrich Gauss.*[2]

---

[2]Karl Friedrich Gauss (1777–1855) was noted for the development of many classical mathematical theories, and for his calculation of the orbits of the asteroids Ceres and Pallas. Gauss is still regarded as one of the greatest mathematicians the world has ever produced.

5.2. LU Factorization Using Gaussian Elimination

> The idea is to triangularize the matrix A to an upper triangular matrix U by successively premultiplying A with a series of elementary matrices (which are unit lower triangular).

Thus, given an $n \times n$ matrix A, Gaussian elimination process consists of finding the elementary matrices $M_1, \ldots, M_{n-1}$ such that

- $A^{(1)} = M_1 A$ has zeros in the first column below the (1,1) entry;

- $A^{(2)} = M_2 A^{(1)}$ has zeros in the second column below (2,2) entry;

- $A^{(n-1)} = M_{n-1} A^{(n-2)}$ has zeros in the $(n-1)$th column below the $(n-1, n-1)$ entry.

The final matrix $A^{(n-1)}$ is upper triangular. The key observation is that each of the matrices $A^{(k)}$ is the result of the premultiplication of $A^{(k-1)}$ by an elementary matrix.

Figure 5.6 is an illustrative diagram showing case $n = 4$.

Step 1. $\quad A \xrightarrow{M_1} M_1 A = \begin{pmatrix} a_{11} & a_{12} & a_{13} & a_{14} \\ 0 & a_{22}^{(1)} & a_{23}^{(1)} & a_{24}^{(1)} \\ 0 & a_{32}^{(1)} & a_{33}^{(1)} & a_{34}^{(1)} \\ 0 & a_{42}^{(1)} & a_{43}^{(1)} & a_{44}^{(1)} \end{pmatrix} = A^{(1)}.$

Step 2. $\quad A^{(1)} \xrightarrow{M_2} M_2 A^{(1)} = \begin{pmatrix} a_{11} & a_{12} & a_{13} & a_{14} \\ 0 & a_{22}^{(1)} & a_{23}^{(1)} & a_{24}^{(1)} \\ 0 & 0 & a_{33}^{(2)} & a_{34}^{(2)} \\ 0 & 0 & a_{43}^{(2)} & a_{44}^{(2)} \end{pmatrix} = A^{(2)}.$

Step 3. $\quad A^{(2)} \xrightarrow{M_3} M_3 A^{(2)} = \begin{pmatrix} a_{11} & a_{12} & a_{13} & a_{14} \\ 0 & a_{22}^{(1)} & a_{23}^{(1)} & a_{24}^{(1)} \\ 0 & 0 & a_{33}^{(2)} & a_{34}^{(2)} \\ 0 & 0 & 0 & a_{44}^{(3)} \end{pmatrix} = A^{(3)}.$

Figure 5.6. *Illustration of Gaussian elimination.*

Notes: (i) The matrix $A^{(1)}$ in Step 1 can be formed as

- Find the elementary matrix M_1 such that $M_1 \begin{pmatrix} a_{11} \\ a_{12} \\ a_{13} \\ a_{14} \end{pmatrix} = \begin{pmatrix} a_{11} \\ 0 \\ 0 \\ 0 \end{pmatrix}.$

- Update: $A^{(1)} = M_1 A.$

(ii) The matrix $A^{(2)}$ in Step 2 can be formed in two smaller steps as follows:

- Find the elementary matrix $\hat{M}_2$ such that $\hat{M}_2 \begin{pmatrix} a_{22}^{(1)} \\ a_{32}^{(1)} \\ a_{42}^{(1)} \end{pmatrix} = \begin{pmatrix} a_{22}^{(2)} \\ 0 \\ 0 \end{pmatrix}.$

- Form $M_2 = \begin{pmatrix} 1 & 0 \\ 0 & \hat{M}_2 \end{pmatrix}.$

- Update: $A^{(2)} = M_2 A^{(1)}.$

And so on.

(iii) In practice, *neither the matrices M_k nor the products $M_k A^{(k-1)}$ need to be explicitly formed*, as shown below with a 4 × 4 *numerical example*.

Example 5.4. Consider
$$A = \begin{pmatrix} 1 & 2 & 3 & 4 \\ 5 & 6 & 7 & 8 \\ 1 & 1 & 3 & 3 \\ 2 & 1 & 1 & 1 \end{pmatrix}.$$

Step 1 (*eliminate the entries of the first column of A below the diagonal*). Multiply the first row by $-5, -1, -2$, and add, respectively, to the second through fourth rows. At the end of this step, we have
$$A^{(1)} = \begin{pmatrix} 1 & 2 & 3 & 4 \\ 0 & -4 & -8 & -12 \\ 0 & -1 & 0 & -1 \\ 0 & -3 & -5 & -7 \end{pmatrix}.$$

Note that in terms of the matrix multiplication, we have
$$A^{(1)} = \underbrace{\begin{pmatrix} 1 & 0 & 0 & 0 \\ -5 & 1 & 0 & 0 \\ -1 & 0 & 1 & 0 \\ -2 & 0 & 0 & 1 \end{pmatrix}}_{M_1} \underbrace{\begin{pmatrix} 1 & 2 & 3 & 4 \\ 5 & 6 & 7 & 8 \\ 1 & 1 & 3 & 3 \\ 2 & 1 & 1 & 1 \end{pmatrix}}_{A} = M_1 A.$$

Step 2 (*eliminate the entries of the second column of $A^{(1)}$ below the diagonal*). Multiply the second row of $A^{(1)}$ by $-\frac{1}{4}$ and $-\frac{3}{4}$ and add, respectively, to the third and fourth rows. At the end of this step, we have
$$A^{(2)} = \begin{pmatrix} 1 & 2 & 3 & 4 \\ 0 & -4 & -8 & -12 \\ 0 & 0 & 2 & 2 \\ 0 & 0 & 1 & 2 \end{pmatrix}.$$

5.2. LU Factorization Using Gaussian Elimination

Again, in terms of matrix multiplication, we have

$$A^{(2)} = \begin{pmatrix} 1 & 0 & 0 & 0 \\ 0 & 1 & 0 & 0 \\ 0 & -\frac{1}{4} & 1 & 0 \\ 0 & -\frac{3}{4} & 0 & 1 \end{pmatrix} \begin{pmatrix} 1 & 2 & 3 & 4 \\ 0 & -4 & -8 & -12 \\ 0 & -1 & 0 & -1 \\ 0 & -3 & -5 & -7 \end{pmatrix} = M_2 A^{(1)}.$$

Step 3 (*eliminate the entries of the third column of $A^{(2)}$ below the diagonal*). Multiply the third row of $A^{(2)}$ by $-\frac{1}{2}$ and add it to the fourth, giving

$$A^{(3)} = \begin{pmatrix} 1 & 2 & 3 & 4 \\ 0 & -4 & -8 & -12 \\ 0 & 0 & 2 & 2 \\ 0 & 0 & 0 & 1 \end{pmatrix} = \text{Upper triangular}.$$

Again,

$$A^{(3)} = \begin{pmatrix} 1 & 0 & 0 & 0 \\ 0 & 1 & 0 & 0 \\ 0 & 0 & 1 & 0 \\ 0 & 0 & -\frac{1}{2} & 1 \end{pmatrix} \begin{pmatrix} 1 & 2 & 3 & 4 \\ 0 & -4 & -8 & -12 \\ 0 & 0 & 2 & 2 \\ 0 & 0 & 1 & 2 \end{pmatrix} = M_3 A^{(2)}. \quad \blacksquare$$

Remark. Note that to form $A^{(k)}$ from $A^{(k-1)}$, $k = 1, 2, 3$, neither the matrices M_k nor the products $M_k A^{(k-1)}$ need to be performed explicitly.

The general process is now quite clear.

Starting with A, the process constructs successively the matrices $A^{(1)}, A^{(2)}, \ldots, A^{(n-1)}$ such that $A^{(1)}$ has zeros on the first column below the diagonal, $A^{(2)}$ has zeros on the second column below its diagonal, and so on. The final matrix $A^{(n-1)}$ is an upper triangular matrix. *The key observation is that each of these matrices is a result of premultiplication of the previous one by an elementary lower triangular matrix.*

General process. There are $(n-1)$ steps. Let $A^{(k)} = (a_{ij}^{(k)})$, $k \geq 1$.
Step 1 (*eliminate the entries of the first column of A below the diagonal*). Multiply the entries of the first row of A by the numbers

$$m_{i1} = -\frac{a_{i1}}{a_{11}}, \quad i = 2, \ldots, n,$$

and add them, respectively, to those of the second through nth rows. We have a new matrix $A^{(1)}$,

$$A^{(1)} = \begin{pmatrix} a_{11} & a_{12} & \cdots & \cdots & a_{1n} \\ 0 & a_{22}^{(1)} & \cdots & \cdots & a_{2n}^{(1)} \\ 0 & \vdots & \ddots & & \vdots \\ \vdots & & & \ddots & \vdots \\ 0 & a_{n2}^{(1)} & \cdots & \cdots & a_{nn}^{(1)} \end{pmatrix},$$

which can be written as

$$A^{(1)} = \begin{pmatrix} 1 & 0 & 0 & \cdots & 0 \\ m_{21} & 1 & 0 & \cdots & 0 \\ \vdots & & \ddots & & \\ m_{n1} & 0 & \cdots & 0 & 1 \end{pmatrix} \begin{pmatrix} a_{11} & \cdots & a_{1n} \\ a_{21} & \cdots & a_{22} \\ \vdots & & \\ a_{n1} & \cdots & a_{nn} \end{pmatrix} = M_1 A.$$

Step 2 (*eliminate the entries of the second column of $A^{(1)}$ below the diagonal*). Multiply the entries of second row of $A^{(1)}$ by the numbers

$$m_{i2} = -\frac{a_{i2}^{(1)}}{a_{22}^{(1)}}, \quad i = 3, \ldots, n,$$

and add them, respectively, to those of third through nth rows.

We now have a new matrix $A^{(2)}$,

$$A^{(2)} = \begin{pmatrix} a_{11} & a_{12} & a_{13} & \cdots & a_{1n} \\ 0 & a_{22}^{(1)} & a_{23}^{(1)} & \cdots & a_{2n}^{(1)} \\ 0 & 0 & a_{32}^{(2)} & \cdots & a_{3n}^{(2)} \\ \vdots & \vdots & & \ddots & \vdots \\ 0 & 0 & a_{n3}^{(2)} & \cdots & a_{nn}^{(2)} \end{pmatrix},$$

which can be written as

$$A^{(2)} = \begin{pmatrix} 1 & & & & \\ 0 & 1 & & & 0 \\ 0 & m_{32} & \ddots & & \\ \vdots & \vdots & & \ddots & \\ 0 & m_{n2} & \cdots & \cdots & 1 \end{pmatrix} \begin{pmatrix} a_{11} & a_{12} & \cdots & a_{1n} \\ 0 & a_{22}^{(1)} & \cdots & a_{2n}^{(1)} \\ 0 & a_{32}^{(1)} & \cdots & a_{3n}^{(1)} \\ \vdots & \vdots & & \vdots \\ 0 & a_{n2}^{(1)} & \cdots & a_{nn}^{(1)} \end{pmatrix} = M_2 A^{(1)}.$$

The process is fairly general. The general kth step can now easily be written down.

Step k ($k > 1$) (*eliminate the entries below the diagonal of the kth column of $A^{(k-1)}$*). Multiply the kth row of $A^{(k-1)}$ by the numbers

$$m_{ik} = -\frac{a_{ik}^{(k-1)}}{a_{kk}^{(k-1)}}, \quad i = k+1, \ldots, n,$$

and add, respectively, to the $(k + 1)$th through nth rows. This will yield a matrix $A^{(k)}$ given by

$$A^{(k)} = M_k A^{(k-1)},$$

where M_k is defined by (5.1).

Step $n - 1$. At the end of the $(n - 1)$th step, the matrix $A^{(n-1)}$ is upper triangular:

$$A^{(n-1)} = \begin{pmatrix} a_{11} & a_{12} & \cdots & \cdots & \cdots & a_{1n} \\ 0 & a_{22}^{(1)} & \cdots & \cdots & \cdots & a_{2n}^{(1)} \\ 0 & 0 & a_{33}^{(2)} & & & a_{3n}^{(2)} \\ 0 & & & \ddots & & \vdots \\ \vdots & & & & \ddots & \vdots \\ 0 & 0 & 0 & \cdots & 0 & a_{nn}^{(n-1)} \end{pmatrix}.$$

5.2. LU Factorization Using Gaussian Elimination

Similar to the other steps, we have (*in terms of matrix multiplications*)

$$A^{(n-1)} = M_{n-1} A^{(n-2)}.$$

The LU Factorization of a Matrix from Gaussian Elimination

The process we just described yields a factorization of the matrix: $A = LU$, where L is *unit lower triangular* and U is *upper triangular*, as shown below. This factorization is known as *LU factorization of A*. *LU factorization of a matrix is an important matrix factorization useful for solving a linear system and computing the determinant and the inverse of a matrix (see Chapter 6).* Figure 5.7 illustrates LU factorization.

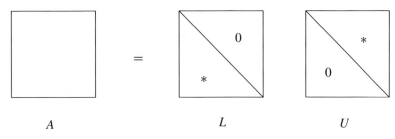

Figure 5.7. *LU factorization of a matrix obtaining L and $\tilde{U}$.*

First, observe that the final matrix $A^{(n-1)}$ is an upper triangular matrix. So, we can take this matrix as our U matrix.

Thus $U = A^{(n-1)} = M_{n-1} A^{(n-2)}$. Again, $A^{(n-2)} = M_{n-2} A^{(n-3)}$.
So, $U = M_{n-1} M_{n-2} A^{(n-3)}$. Continuing this way, we can write

$$U = M_{n-1} M_{n-2} \ldots M_2 M_1 A.$$

Now set $L_1 = M_{n-1} M_{n-2} \ldots M_2 M_1$. Then $U = L_1 A$.

Since each of the elementary lower triangular matrices is a unit lower triangular matrix (*a lower triangular matrix with* 1's *along the diagonal*), it follows that L_1 is invertible and L_1^{-1} is also **unit lower triangular.** (Note that *the product of unit lower triangular matrices is a unit lower triangular matrix, and so is the inverse*). Now set $L = L_1^{-1}$. Then $A = LU$.

LU Factorization of a Matrix from Gaussian Elimination

$$A = LU$$

- $L = (M_{n-1} M_{n-2} \ldots M_2 M_1)^{-1}$ (Unit Lower Triangular).

- $U = A^{n-1}$ (Upper Triangular).

Definition 5.5. *The entries $a_{11}, a_{22}^{(1)}, \ldots, a_{n-1,n-1}^{(n-2)}$ are called pivots, and the above process of obtaining LU factorization is known as Gaussian elimination without row interchanges. It is commonly known as Gaussian elimination without pivoting. The numbers m_{ik} are called multipliers.*

Obtaining *L* without Matrix Inversion

We will now show that the matrix L can be formed without explicitly computing any matrix product and without any matrix inversion.

$$L = L_1^{-1} = M_1^{-1} M_2^{-1} \ldots M_{n-1}^{-1}.$$

First, we observe (Exercise 5.1(b)) that

$$M_i^{-1} = I - m e_i^T, \quad i = 1, 2, \ldots, n-1,$$

where $m = (0, 0, \ldots, 0, m_{i+1,i} \ldots, m_{n,i})^T$ and e_i is the ith unit vector.

This simply means that the M_i^{-1} is just the matrix M_i except that the entries on the ith column below the diagonal are just the negatives of the corresponding entries of M_i.

Thus, $M_i^{-1} M_{i+1}^{-1}$ is a unit lower triangular matrix with the nonzero entries below the diagonal only on the columns i and $(i+1)$, which are the negatives of the corresponding entries of M_i and M_{i+1}. For example,

$$M_1^{-1} M_2^{-1} = \begin{pmatrix} 1 & 0 & 0 & \ldots & 0 \\ -m_{21} & 1 & & \ldots & 0 \\ -m_{31} & -m_{32} & \ddots & & \vdots \\ \vdots & \vdots & 0 & \ddots & \\ -m_{n1} & -m_{n2} & \ldots & \ldots & 1 \end{pmatrix}.$$

This implies that

$$L = M_1^{-1} M_2^{-1} \ldots M_{n-1}^{-1} = \begin{pmatrix} 1 & 0 & \ldots & & \ldots & 0 \\ -m_{21} & 1 & 0 & & & 0 \\ -m_{31} & -m_{32} & 1 & \ddots & & 0 \\ & & & \ddots & \ddots & \vdots \\ \vdots & \vdots & \ddots & & & 0 \\ -m_{n1} & -m_{n2} & \ldots & & -m_{n,n-1} & 1 \end{pmatrix}.$$

Thus to form L do the following:

- Save the multipliers at each step.

- Insert the negative of the multipliers of Step 1 in the first column of the identity matrix below the diagonal, the negatives of the multipliers of Step 2 in the second column below the diagonal, and so on.

Example 5.6.

$$A = \begin{pmatrix} 2 & 2 & 3 \\ 4 & 5 & 6 \\ 1 & 2 & 4 \end{pmatrix}.$$

Step 1 (*eliminate the entries on the first column of A below the diagonal*). Multiply the first row of A by -2 and $-\frac{1}{2}$ and add, respectively, to the second and third rows.

$$A^{(1)} = \begin{pmatrix} 2 & 2 & 3 \\ 0 & 1 & 0 \\ 0 & 1 & \frac{5}{2} \end{pmatrix}; \quad m_{21} = -2, \; m_{31} = -\frac{1}{2}.$$

5.2. LU Factorization Using Gaussian Elimination

Step 2 (*eliminate the entries on the second column of $A^{(1)}$ below the diagonal*). Multiply the second row of $A^{(1)}$ by -1 and add it to the third row.

$$A^{(2)} = \begin{pmatrix} 2 & 2 & 3 \\ 0 & 1 & 0 \\ 0 & 0 & \frac{5}{2} \end{pmatrix}; \quad m_{32} = -1.$$

So,

$$U = A^{(2)} = \begin{pmatrix} 2 & 2 & 3 \\ 0 & 1 & 0 \\ 0 & 0 & \frac{5}{2} \end{pmatrix},$$

$$L = L_1^{-1} = \begin{pmatrix} 1 & 0 & 0 \\ -m_{21} & 1 & 0 \\ -m_{31} & -m_{32} & 1 \end{pmatrix} = \begin{pmatrix} 1 & 0 & 0 \\ 2 & 1 & 0 \\ \frac{1}{2} & 1 & 1 \end{pmatrix}.$$

(Note that neither L_1 nor its inverse needs to be computed explicitly.) ■

Existence and Uniqueness of LU Factorization

Note that for an LU factorization to exist, the pivots must be different from zero. Thus, LU factorization may not exist even for a very simple matrix. Take $A = \begin{pmatrix} 0 & 1 \\ 1 & 0 \end{pmatrix}$. The pivot $a_{11}^{(0)}$ is zero. So, the *Gaussian elimination scheme cannot be carried out*.

The following theorem gives conditions on the existence and uniqueness of LU factorization.

Definition 5.7. *The kth leading principal minor of a matrix, A, denoted by A_k, is defined to be the $k \times k$ leading principal submatrix consisting of the first k rows and the first k columns.*

Theorem 5.8 (existence and uniqueness of LU factorization).

(i) *An $n \times n$ matrix A has an LU factorization if A_k, $k = 1, \ldots, n-1$, are nonsingular.*

(ii) *If the LU factorization exists and A is nonsingular, then this factorization is unique.*

Proof. *Existence:* From the derivation of Gaussian elimination scheme, it follows that the process can break down only if any of the pivots $a_{11}, a_{22}^{(1)}, \ldots, a_{n-1,n-1}^{(n-2)}$ is zero. Again, it can be shown (Exercise 5.3) that

$$\det(A_k) = a_{11} a_{22}^{(1)} \ldots a_{kk}^{(k-1)}, \quad k = 1, \ldots, n-1 \text{ (note that } \det(A_1) = a_{11}).$$

This means that if the first $(n-1)$ leading principal minors are nonsingular, then Gaussian elimination scheme does not fail, and we always have an LU factorization of A in this case, as shown by the above discussion.

Uniqueness: The uniqueness will be proved by *contradiction*. Suppose there are two different LU factorizations of A: $A = L_1 U_1 = L_2 U_2$. Then, we must show that $L_1 = L_2$ and $U_1 = U_2$.

Because A is nonsingular, the matrices L_1, L_2, U_1, and U_2 are all nonsingular. Thus it follows from the above two factorizations of A that $L_2^{-1}L_1 = U_2 U_1^{-1}$.

Now $L_2 L_1^{-1}$ is a unit lower triangular matrix and $U_2 U_1^{-1}$ is an upper triangular matrix, and the only way they can be equal is that both of these are the identity. Thus $L_1 = L_2$ and $U_1 = U_2$. □

Remark. Note that in the above theorem, if the diagonal entries of L are not specified, then the factorization is *not* unique. (Do an example to verify this.)

A Storage Scheme for a Practical LU Factorization

Example 5.6 shows that for a practical Gaussian elimination scheme, the kth step consists of the following.

- Forming the multipliers $m_{ik} = -\dfrac{a_{ik}^{(k-1)}}{a_{kk}^{(k-1)}}, i = k+1, \ldots, n$.

- Updating the entries of the submatrix: $A(k+1:n, k+1:n)$ (the submatrix consisting of the rows $(k+1)$ through n and columns $(k+1)$ through n).

  The following storage scheme thus can be used:

- The multipliers are stored below the main diagonal of A. These multipliers then can be used to form L.

- The entries of the upper triangular matrix U are stored in the upper half of A including the diagonal.

ALGORITHM 5.1. LU Factorization using Gaussian Elimination without Pivoting (GEWP).

Input: An $n \times n$ matrix A.
Outputs: (i) An upper triangular matrix U, and (ii) the multipliers m_{ij} needed to form the unit lower triangular matrix L such that $A = LU$.
Storage: The upper triangular part of U is stored over the upper triangular part of A including the diagonal. The multipliers needed to compute L are stored in the lower triangular part of A below the diagonal.

For $k = 1, 2, \ldots, (n-1)$ do

1. (*Form the multipliers*):
$$a_{ik} \equiv m_{ik} = -\frac{a_{ik}}{a_{kk}} \quad (i = k+1, k+2, \ldots, n).$$

2. (*Update the entries* of $A(k+1:n, k+1:n)$):
$$a_{ij} \equiv a_{ij} + m_{ik}a_{kj} (i = k+1, \ldots, n; j = k+1, n).$$

End

5.2. LU Factorization Using Gaussian Elimination

With this storage scheme, the upper triangular matrix $A^{(n-1)}$ at the end of the $(n-1)$th step will look like this:

$$A \equiv A^{(n-1)} = \begin{pmatrix} a_{11} & a_{12} & \cdots & \cdots & a_{1n} \\ m_{21} & a_{22}^{(1)} & \cdots & \cdots & a_{2n}^{(1)} \\ \vdots & \ddots & \ddots & & \vdots \\ \vdots & & \ddots & \ddots & \vdots \\ m_{n1} & \cdots & \cdots & m_{n,n-1} & a_{nn}^{(n-1)} \end{pmatrix}$$

Remark. The algorithm does not give the matrix L explicitly; however, it can be formed out of the multipliers saved at each step, as shown earlier (see the expression for L).

Example 5.9. Consider Example 5.6 again.

k = 1: Multipliers: $m_{21} = -2, \; m_{31} = -\frac{1}{2},$

Updated A: $A \equiv \begin{pmatrix} 2 & 2 & 3 \\ 0 & 1 & 0 \\ 0 & 1 & \frac{5}{2} \end{pmatrix}.$

k = 2: Multiplier: $m_{32} = -1,$

Updated A: $A \equiv U = \begin{pmatrix} 2 & 2 & 3 \\ \boxed{0} & 1 & 0 \\ \boxed{0} & \boxed{0} & \frac{5}{2} \end{pmatrix}.$

Form L and U: $L = \begin{pmatrix} 1 & 0 & 0 \\ 2 & 1 & 0 \\ \frac{1}{2} & 1 & 1 \end{pmatrix}, \quad U = \begin{pmatrix} 2 & 2 & 3 \\ 0 & 1 & 0 \\ 0 & 0 & \frac{5}{2} \end{pmatrix}.$ ∎

Note: In practical computation, the boxes of the output matrix A will hold the multipliers.

Flop-count. Algorithm 5.1 requires roughly $\frac{2n^3}{3}$ *flops*. This can be seen as follows:

Step 1. We compute $(n-1)$ multipliers and update $(n-1)^2$ entries of A. Each multiplier requires one flop and updating each entry requires two flops. Thus, Step 1 requires $2(n-1)^2 + (n-1)$ flops.

Step 2. Computing $(n-2)$ multipliers and updating $(n-2)^2$ entries require $2(n-2)^2 + (n-2)$ flops.

In general, Step k requires $2(n-k)^2 + (n-k)$ flops. Since there are $(n-1)$ steps, we have

$$\begin{aligned} \text{Total flops} &= \sum_{k=1}^{n-1} 2(n-k)^2 + \sum_{k=1}^{n-1}(n-k) \\ &= 2\frac{n(n-1)(2n-1)}{6} + \frac{n(n-1)}{2} \simeq \left[\frac{2n^3}{3} + O(n^2)\right]. \end{aligned}$$

> Recall that
> - $1^2 + 2^2 + \cdots + r^2 = \frac{r(r+1)(2r+1)}{6}$,
> - $1 + 2 + \cdots + r = \frac{r(r+1)}{2}$.

Gaussian Elimination for a Rectangular Matrix

The above-described Gaussian elimination process for an $n \times n$ matrix A can be easily extended to an $m \times n$ matrix to compute its LU factorization, when it exists. The process is identical. However, the number of required steps in this case is $k = \min\{m-1, n\}$. We illustrate this with an example.

MATCOM Note: Algorithm 5.1 has been implemented in the MATCOM program **LUGSEL**.

Example 5.10. Let

$$A = \begin{pmatrix} 1 & 2 \\ 3 & 4 \\ 5 & 6 \end{pmatrix}, \quad m = 3, \ n = 2.$$

$k = \min(2, 2) = 2$.

$k = 1$. (*Eliminate the entries in the first column of A below the diagonal.*) The multipliers are $m_{21} = -3$, $m_{31} = -5$.

$$\text{Update: } a_{22} \equiv a_{22}^{(1)} = a_{22} + m_{21}a_{12} = -2,$$
$$a_{32} \equiv a_{32}^{(1)} = a_{32} + m_{31}a_{12} = -4,$$
$$A \equiv A^{(1)} = \begin{pmatrix} 1 & 2 \\ 0 & -2 \\ 0 & -4 \end{pmatrix}.$$

$k = 2$. (*Eliminate the entries in the second column of $A^{(1)}$ below the diagonal.*) The multiplier is $m_{32} = -2$, $a_{32} \equiv a_{32}^{(2)} = 0$.

$$\text{Update: } A \equiv A^{(2)} = \begin{pmatrix} 1 & 2 \\ 0 & -2 \\ 0 & 0 \end{pmatrix}.$$

$$\text{So, } U = \begin{pmatrix} 1 & 2 \\ 0 & -2 \\ 0 & 0 \end{pmatrix}.$$

Note that U in this case is a 3×2 upper triangular matrix.
Form

$$L = \begin{pmatrix} 1 & 0 & 0 \\ -m_{21} & 1 & 0 \\ -m_{31} & -m_{32} & 1 \end{pmatrix} = \begin{pmatrix} 1 & 0 & 0 \\ 3 & 1 & 0 \\ 5 & 2 & 1 \end{pmatrix}.$$

5.2. LU Factorization Using Gaussian Elimination

Verify that

$$LU = \begin{pmatrix} 1 & 0 & 0 \\ 3 & 1 & 0 \\ 5 & 2 & 1 \end{pmatrix} \begin{pmatrix} 1 & 2 \\ 0 & -2 \\ 0 & 0 \end{pmatrix} = \begin{pmatrix} 1 & 2 \\ 3 & 4 \\ 5 & 6 \end{pmatrix} = A. \quad \blacksquare$$

Flop-count. For an $m \times n$ matrix, the Gaussian elimination process requires $mn^2 - \frac{n^3}{3}$ flops (Exercise 5.16(a)).

Difficulties of Gaussian Elimination without Pivoting

As we have seen before, Gaussian elimination without pivoting fails if any of the pivots is zero. However, it is worse yet if any pivot becomes close to zero: *in this case the method can be carried to completion, but the obtained results may be totally wrong.*

Consider the following celebrated example from Forsythe and Moler (1967, p. 34):

Let Gaussian elimination without pivoting be applied to

$$A = \begin{pmatrix} 0.0001 & 1 \\ 1 & 1 \end{pmatrix},$$

and use three-digit arithmetic. *There is only one step.* We have just one multiplier: $m_{21} = \frac{-1}{10^{-4}} = -10^4$.

$$U = A^{(1)} = \begin{pmatrix} 0.0001 & 1 \\ 0 & 1 - 10^4 \end{pmatrix} \equiv \begin{pmatrix} 0.0001 & 1 \\ 0 & -10^4 \end{pmatrix}, \text{ and } L = \begin{pmatrix} 1 & 0 \\ 10^4 & 1 \end{pmatrix}.$$

The product of the computed L and U is $LU = \begin{pmatrix} 0.0001 & 1 \\ 1 & 0 \end{pmatrix}$, which is different from A. *Who is to blame?*

Note that the pivot $a_{11}^{(1)} = 0.0001$ is very close to zero (in three-digit arithmetic). *This small pivot gave a large multiplier.* The large multiplier, when used to update the entries, eliminated the smaller entries (e.g., $(1 - 10^4)$ became -10^4).

Fortunately, *we can avoid this small pivot just by row interchanges.* Consider the matrix with the first and second rows interchanged, giving

$$A' = \begin{pmatrix} 1 & 1 \\ 0.0001 & 1 \end{pmatrix}.$$

Gaussian elimination applied to A' now gives

$$U = A^{(1)} = \begin{pmatrix} 1 & 1 \\ 0 & 1 \end{pmatrix}, \quad L = \begin{pmatrix} 1 & 0 \\ 0.0001 & 1 \end{pmatrix}.$$

Note that the pivot in this case is $a_{11}^{(1)} = 1$. The product

$$LU = \begin{pmatrix} 1 & 1 \\ 0.0001 & 1.0001 \end{pmatrix} = A'.$$

Remark. It is true that with the interchange above, we now obtained an LU factorization of the matrix A', a permuted version of the matrix A, and not of the original matrix. However, as we will see in Chapter 6, this will suffice our purpose for solving a linear system of equations.

5.2.3 Permutation Matrices and Their Properties

A nonzero square matrix P is called a **permutation matrix** if there is exactly one nonzero entry in each row and column which is one and the rest are all zero. Thus, if $(\alpha_1, \ldots, \alpha_n)$ is a permutation of $(1, 2, \ldots, n)$, then the associated permutation matrix P is given by

$$P = \begin{pmatrix} e_{\alpha_1}^T \\ \vdots \\ e_{\alpha_n}^T \end{pmatrix},$$

where e_i^T is the ith row of the $n \times n$ identity matrix I. Similarly,

$$P = (e_{\alpha_1}, e_{\alpha_2}, \ldots, e_{\alpha_n}),$$

where e_i is the ith column of I, is a permutation matrix.

Example 5.11.

$$P_1 = \begin{pmatrix} 0 & 1 & 0 \\ 0 & 0 & 1 \\ 1 & 0 & 0 \end{pmatrix}, \quad P_2 = \begin{pmatrix} 1 & 0 & 0 \\ 0 & 1 & 0 \\ 0 & 0 & 1 \end{pmatrix}, \quad P_3 = \begin{pmatrix} 1 & 0 & 0 \\ 0 & 0 & 1 \\ 0 & 1 & 0 \end{pmatrix}$$

are all permutation matrices. ∎

Effects of Premultiplication and Postmultiplication by a Permutation Matrix.

If

$$P_1 = \begin{pmatrix} e_{\alpha_1^T} \\ \vdots \\ e_{\alpha_n^T} \end{pmatrix},$$

then

$$P_1 A = \begin{pmatrix} \alpha_1 \text{th row of } A \\ \alpha_2 \text{th row of } A \\ \vdots \\ \alpha_n \text{th row of } A \end{pmatrix}.$$

Similarly, if $P_2 = (e_{\alpha_1} e_{\alpha_1} \cdots e_{\alpha_n})$, where e_{α_i} is the ith column of A, then $AP_2 = (\alpha_1$th column of A, α_2th column of A, $\ldots$, α_nth column of A).

Thus, the effect of premultiplication of A by a permutation matrix is a permutation of the associated rows of A, and that of postmultiplication is the permutation of the associated columns.

Example 5.12.

1. $A = \begin{pmatrix} a_{11} & a_{12} & a_{13} \\ a_{21} & a_{22} & a_{23} \\ a_{13} & a_{23} & a_{33} \end{pmatrix}$, $P_1 = \begin{pmatrix} 0 & 1 & 0 \\ 0 & 0 & 1 \\ 1 & 0 & 0 \end{pmatrix} = \begin{pmatrix} e_2^T \\ e_3^T \\ e_1^T \end{pmatrix}$;

$$P_1 A = \begin{pmatrix} a_{21} & a_{22} & a_{23} \\ a_{31} & a_{32} & a_{33} \\ a_{11} & a_{12} & a_{13} \end{pmatrix} = \begin{pmatrix} \text{2nd row of } A \\ \text{3rd row of } A \\ \text{1st row of } A \end{pmatrix}.$$

5.2. LU Factorization Using Gaussian Elimination

2. $P_1 = \begin{pmatrix} 0 & 1 & 0 \\ 0 & 0 & 1 \\ 1 & 0 & 0 \end{pmatrix} = (e_3, e_1, e_2);$

$AP_1 = \begin{pmatrix} a_{13} & a_{11} & a_{12} \\ a_{23} & a_{21} & a_{22} \\ a_{33} & a_{31} & a_{32} \end{pmatrix} =$ (3rd column of A, 1st column of A, 2nd column of A). ∎

An important property of a permutation matrix P is that *it is orthogonal*; that is, $PP^T = I$. As a consequence of this, we have the following.

- The inverse of a permutation matrix P is its transpose, and it is also a permutation matrix.

- The product of two permutation matrices is a permutation matrix, and therefore is orthogonal.

5.2.4 Gaussian Elimination with Partial Pivoting (GEPP)

As the above example suggests, disaster in Gaussian elimination without pivoting can perhaps be avoided by identifying a "good pivot" (*a pivot as large as possible in magnitude*) at each step, before the process of elimination is applied. The good pivot may be located among the entries in a column or among all the entries in a submatrix of the current matrix. In the former case, since the search is only partial, the method is called **partial pivoting**; in the latter case, the method is called **complete pivoting**. *It is important to note that the purpose of pivoting is to prevent large growth in the reduced matrices, which can wipe out original data.* One way to do this is to keep multipliers less than or equal to one in magnitude, and this is exactly what is accomplished by pivoting. However, *large multipliers do not necessarily mean instability* (see our discussion of Gaussian elimination without pivoting for symmetric positive definite matrices in Chapter 6). We first describe Gaussian elimination with partial pivoting (GEPP).

The process consists of $(n - 1)$ steps.

The process is just a slight modification of Gaussian elimination in the following sense:

At each step do the following:

- Identify the pivot as the largest entry (in magnitude) among all the entries in the pivot column.

- Interchange the appropriate rows to bring the pivot entry to the diagonal position of the current matrix.

- Perform Gaussian elimination to the row-permuted matrix.

The process is illustrated with a 4×4 example in the following. *For this example, we assume that rows* 3, 4, *and* 4 *are pivot rows in* Steps 1, 2, and 3, respectively.

Step 1.

$$A \to \begin{pmatrix} a_{11} & a_{12} & a_{13} & a_{14} \\ a_{21} & a_{22} & a_{23} & a_{24} \\ \boxed{a_{31}} & a_{32} & a_{33} & a_{34} \\ a_{41} & a_{42} & a_{43} & a_{44} \end{pmatrix} \to \begin{pmatrix} \boxed{a_{31}} & a_{32} & a_{33} & a_{34} \\ a_{21} & a_{22} & a_{23} & a_{24} \\ a_{11} & a_{12} & a_{13} & a_{14} \\ a_{41} & a_{42} & a_{43} & a_{44} \end{pmatrix}$$

Pivot Identification   *Row Interchange (1st and 3rd)*

$$\to \begin{pmatrix} a_{31} & a_{32} & a_{33} & a_{34} \\ 0 & a_{22}^{(1)} & a_{23}^{(1)} & a_{24}^{(1)} \\ 0 & a_{12}^{(1)} & a_{13}^{(1)} & a_{14}^{(1)} \\ 0 & a_{42}^{(1)} & a_{43}^{(1)} & a_{44}^{(1)} \end{pmatrix}.$$

Gaussian Elimination

Step 2.

$$A^{(1)} \begin{pmatrix} a_{31} & a_{32} & a_{33} & a_{34} \\ 0 & a_{22}^{(1)} & a_{23}^{(1)} & a_{24}^{(1)} \\ 0 & a_{12}^{(1)} & a_{13}^{(1)} & a_{14}^{(1)} \\ 0 & \boxed{a_{42}^{(1)}} & a_{43}^{(1)} & a_{44}^{(1)} \end{pmatrix} \to \begin{pmatrix} a_{31} & a_{32} & a_{33} & a_{34} \\ 0 & a_{42}^{(1)} & a_{43}^{(1)} & a_{44}^{(1)} \\ 0 & a_{12}^{(1)} & a_{13}^{(1)} & a_{14}^{(1)} \\ 0 & a_{22}^{(1)} & a_{23}^{(1)} & a_{24}^{(1)} \end{pmatrix}$$

Pivot Identification   *Row Interchange (2nd and 4th)*

$$\to \begin{pmatrix} a_{31} & a_{32} & a_{33} & a_{34} \\ 0 & a_{42}^{(1)} & a_{43}^{(1)} & a_{44}^{(1)} \\ 0 & 0 & a_{13}^{(2)} & a_{14}^{(2)} \\ 0 & 0 & a_{23}^{(2)} & a_{24}^{(2)} \end{pmatrix}.$$

Gaussian Elimination

Step 3.

$$A^{(2)} \begin{pmatrix} a_{31} & a_{32} & a_{33} & a_{34} \\ 0 & a_{42}^{(1)} & a_{43}^{(1)} & a_{44}^{(1)} \\ 0 & 0 & a_{13}^{(2)} & a_{14}^{(2)} \\ 0 & 0 & \boxed{a_{23}^{(2)}} & a_{24}^{(2)} \end{pmatrix} \to \begin{pmatrix} a_{31} & a_{32} & a_{33} & a_{34} \\ 0 & a_{42}^{(1)} & a_{43}^{(1)} & a_{44}^{(1)} \\ 0 & 0 & a_{23}^{(2)} & a_{24}^{(2)} \\ 0 & 0 & a_{13}^{(2)} & a_{14}^{(2)} \end{pmatrix}$$

Pivot Identification   *Row Interchange (3rd and 4th)*

$$\to \begin{pmatrix} a_{31} & a_{32} & a_{33} & a_{34} \\ 0 & a_{42}^{(1)} & a_{43}^{(1)} & a_{44}^{(1)} \\ 0 & 0 & a_{23}^{(2)} & a_{24}^{(2)} \\ 0 & 0 & 0 & a_{14}^{(3)} \end{pmatrix}.$$

Gaussian Elimination

5.2. LU Factorization Using Gaussian Elimination

General process. *k*th **Step:** Set $A^{(0)} = A$. Then to obtain the matrix $A^{(k)} = (a_{ij}^{(k)})$ from $A^{(k-1)}$ at the previous step, do the following:

1. Identify the largest element in magnitude among all the elements of column k below row $(k-1)$ of the matrix $A^{(k-1)}$. Let it be $a_{r_k,k}^{(k-1)}$.

2. Interchange the rows r_k and k to bring $a_{r_k,k}^{(k-1)}$ to the diagonal position.

3. Apply Gaussian elimination without row interchanges with $a_{r_k,k}^{(k-1)}$ as the pivot to the submatrix consisting of rows k through n and columns k through n.

GEPP in Terms of Matrix Multiplications

Observe that

- row interchange is equivalent to premultiplying the matrix by a suitable permutation matrix;

- Gaussian elimination is equivalent to premultiplying the matrix by an elementary matrix.

So, we can write

$$
\begin{aligned}
A^{(1)} &= M_1 P_1 A; \; A^{(2)} = M_2 P_2 A^{(1)} \\
&\vdots \\
A^{(n-1)} &= M_{n-1} P_{n-1} A^{(n-2)}.
\end{aligned}
$$

For $n = 4$, the complete process is

$$A = \begin{pmatrix} \times & \times & \times & \times \\ \times & \times & \times & \times \\ \times & \times & \times & \times \\ \times & \times & \times & \times \end{pmatrix}.$$

Step 1. $A \xrightarrow{P_1} P_1 A \xrightarrow{M_1} M_1 P_1 A = \begin{pmatrix} \times & \times & \times & \times \\ 0 & \times & \times & \times \\ 0 & \times & \times & \times \\ 0 & \times & \times & \times \end{pmatrix} = A^{(1)}$.

Step 2. $A^{(1)} \xrightarrow{P_2} P_2 A^{(1)} \xrightarrow{M_2} M_2 P_2 A^{(1)} = M_2 P_2 M_1 P_1 A = \begin{pmatrix} \times & \times & \times & \times \\ 0 & \times & \times & \times \\ 0 & 0 & \times & \times \\ 0 & 0 & \times & \times \end{pmatrix} = A^{(2)}$.

Step 3. $A^{(2)} \xrightarrow{P_3} P_3 A^{(2)} \xrightarrow{M_3} M_3 P_3 A^{(2)} = M_3 P_3 M_2 P_2 M_1 P_1 A = \begin{pmatrix} \times & \times & \times & \times \\ 0 & \times & \times & \times \\ 0 & 0 & \times & \times \\ 0 & 0 & 0 & \times \end{pmatrix} = A^{(3)}$.

LU Factorization from GEPP

We will now show that Gaussian elimination with partial pivoting yields a factorization of A in the form

$$PA = LU,$$

where P is a *permutation matrix*, L is a *unit lower triangular* matrix, and U is an *upper triangular matrix*. This will be shown in two steps.

1. First, it will be shown that Gaussian eliminations with partial pivoting directly yields the factorization $MA = U$, were M is a permuted elementary matrix and U is an upper triangular matrix.

$n = 4$: Since $A^{(3)}$ is upper triangular, we set $U = A^{(3)}$. Then from Step 3, we have

$$U = A^{(3)} = M_3 P_3 M_2 P_2 M_1 P_1 A = MA,$$

where

$$M = M_3 P_3 M_2 P_2 M_1 P_1.$$

For an $n \times n$ matrix:

$$M = M_{n-1} P_{n-1} M_{n-2} P_{n-2} \ldots M_2 P_2 M_1 P_1; \quad U = A^{(n-1)}.$$

2. Second, it will be shown how to extract the matrices P and L from $MA = U$ factorization, so that we have $PA = LU$.

$n = 4$:

$$\begin{aligned} U &= M_3 P_3 M_2 P_2 M_1 P_1 A \\ &= M_3 (P_3 M_2 P_3)(P_3 P_2 M_1 P_2 P_3)(P_3 P_2 P_1) A \quad \text{(note that } P_3^2 = P_2^2 = I\text{)} \\ &= M_3' M_2' M_1' P A, \end{aligned}$$

where

$$M_3' = M_3, \quad M_2' = P_3 M_2 P_3, \quad M_1' = P_3 P_2 M_1 P_2 P_3$$

and

$$P = P_3 P_2 P_1.$$

So, setting $L = (M_1')^{-1}(M_2')^{-1}(M_3')^{-1}$, we have $LU = PA$.

For an $n \times n$ matrix: The matrices P and L are given by

$$P = P_{n-1} P_{n-2} \ldots P_2 P_1,$$

$$L = (M_1')^{-1}(M_2')^{-1} \ldots (M_{n-1}')^{-1}.$$

5.2. LU Factorization Using Gaussian Elimination

Constructing the Matrix L

The matrix L is unit lower triangular and easily computable. Observe that

- each M_i' is the same as M_i except that the multipliers are now permuted (this is an effect of multiplication by permutation matrices);
- $(M_i')^{-1}$ is the same as M_i' except that the multipliers are now negated.

Thus, *as in the case of Gaussian elimination without pivoting, we see that the matrix L is a unit lower triangular matrix.*

Example 5.13. Consider Example 5.6 again, *this time with partial pivoting*.

Step 1. Permuted $A \equiv \begin{pmatrix} 4 & 5 & 6 \\ 2 & 2 & 3 \\ 1 & 2 & 4 \end{pmatrix} = \begin{pmatrix} 0 & 1 & 0 \\ 1 & 0 & 0 \\ 0 & 0 & 1 \end{pmatrix} \begin{pmatrix} 2 & 2 & 3 \\ 4 & 5 & 6 \\ 1 & 2 & 4 \end{pmatrix} = P_1 A.$

Gaussian Elimination: Multiply first row of the *permuted A* by $-\frac{1}{2}$ and $-\frac{1}{4}$ and add it to the second and third rows, respectively, to obtain $A^{(1)}$.

$$A^{(1)} = \begin{pmatrix} 4 & 5 & 6 \\ 0 & -\frac{1}{2} & 0 \\ 0 & \boxed{\frac{3}{4}} & \frac{5}{2} \end{pmatrix} = \begin{pmatrix} 1 & 0 & 0 \\ -\frac{1}{2} & 1 & 0 \\ -\frac{1}{4} & 0 & 1 \end{pmatrix} \begin{pmatrix} 4 & 5 & 6 \\ 2 & 2 & 3 \\ 1 & 2 & 4 \end{pmatrix} = M_1 P_1 A.$$

Step 2. Permuted $A^{(1)} \equiv \begin{pmatrix} 4 & 5 & 6 \\ 0 & \frac{3}{4} & \frac{5}{2} \\ 0 & -\frac{1}{2} & 0 \end{pmatrix} = \begin{pmatrix} 1 & 0 & 0 \\ 0 & 0 & 1 \\ 0 & 1 & 0 \end{pmatrix} \begin{pmatrix} 4 & 5 & 6 \\ 0 & -\frac{1}{2} & 0 \\ 0 & \frac{3}{4} & \frac{5}{2} \end{pmatrix} = P_2 A^{(1)}.$

Gaussian elimination: Multiply second row of the *permuted $A^{(1)}$* by $\frac{2}{3}$ and add it to the third row to obtain $A^{(2)}$.

$$A^{(2)} = \begin{pmatrix} 4 & 5 & 6 \\ 0 & \frac{3}{4} & \frac{5}{2} \\ 0 & 0 & \frac{5}{3} \end{pmatrix} = \begin{pmatrix} 1 & 0 & 0 \\ 0 & 1 & 0 \\ 0 & \frac{2}{3} & 1 \end{pmatrix} \begin{pmatrix} 4 & 5 & 6 \\ 0 & \frac{3}{4} & \frac{5}{2} \\ 0 & -\frac{1}{2} & 0 \end{pmatrix} = M_2 P_2 A^{(1)} = M_2 P_2 M_1 P_1 A.$$

- **Factorization $MA = U$.**

  Gaussian elimination: Multiply second row of the *permuted $A^{(1)}$* by $\frac{2}{3}$ and add it to the third row to obtain $A^{(2)}$.

  $$M = M_2 P_2 M_1 P_1 = \begin{pmatrix} 0 & 1 & 0 \\ 0 & -\frac{1}{4} & 1 \\ 1 & -\frac{2}{3} & \frac{2}{3} \end{pmatrix}, \quad U = \begin{pmatrix} 4 & 5 & 6 \\ 0 & \frac{3}{4} & \frac{5}{2} \\ 0 & 0 & \frac{5}{3} \end{pmatrix} = A^{(2)}.$$

- **Factorization $PA = LU$.**

  $$P = P_2 P_1 = \begin{pmatrix} 0 & 1 & 0 \\ 0 & 0 & 1 \\ 1 & 0 & 0 \end{pmatrix}.$$

Chapter 5. Gaussian Elimination and LU Factorization

$$L = (M'_1)^{-1} (M'_2)^{-1} = \begin{pmatrix} 1 & 0 & 0 \\ \frac{1}{4} & 1 & 0 \\ \frac{1}{2} & -\frac{2}{3} & 1 \end{pmatrix}, \quad U = A^{(2)} = \begin{pmatrix} 4 & 5 & 6 \\ 0 & \frac{3}{4} & \frac{5}{2} \\ 0 & 0 & \frac{5}{3} \end{pmatrix}. \quad \blacksquare$$

Storage Scheme for LU Factorization by GEPP

- The multipliers can be stored in the appropriate places of the lower triangular part of A (below the diagonal) as they are computed.

- A can be overwritten with each $A^{(k)}$ as soon as the latter is formed, and thus the final upper triangular matrix $U = A^{(n-1)}$ will be stored in the upper triangular part of A (including the diagonal).

- The permutation indices r_k have to be stored in a separate single subscripted integer array.

In view of our above discussion, we can now formulate the following *practical algorithm* for LU factorization with partial pivoting.

ALGORITHM 5.2. LU Factorization Using GEPP.

Input: An $n \times n$ matrix A.
Outputs: (i) An upper triangular matrix U, (ii) the permutation indices r_k needed to form the permutation matrix P, and (iii) the multipliers m_{ik} needed to form the unit lower triangular matrix L. The result is $PA = LU$.
Storage: The storage arrangements for U and the multipliers are the same as those of Algorithm 5.1. The permutation indices are stored in a separate array.
For $k = 1, 2, \ldots, n-1$ do

1. (*Find the pivot row.*) Find r_k so that $|a_{r_k,k}| = \max\limits_{k \le i \le n} |a_{ik}|$. Save r_k. If $a_{r_k,k} = 0$, then stop. Otherwise, continue.

2. (*Interchange the rows r_k and k.*) $a_{kj} \leftrightarrow a_{r_k,j}$ $(j = k, k+1, \ldots, n)$.

3. (*Form the multipliers.*) $a_{ik} \equiv m_{ik} = -\dfrac{a_{ik}}{a_{kk}}$ $(i = k+1, \ldots, n)$

4. (*Update the entries.*) $a_{ij} \equiv a_{ij} + m_{ik}a_{kj} = a_{ij} + a_{ik}a_{kj}$ $(i = k+1, \ldots, n; \; j = k+1, \ldots, n)$.

End

Flop-count. Algorithm 5.2 requires about $2\frac{n^3}{3}$ flops and $O(n^2)$ comparisons. (Note that the search for the pivot at step k requires $(n - k)$ comparisons.)

5.2. LU Factorization Using Gaussian Elimination

Note: *Algorithm 5.2 does not give the matrices L and P explicitly. However, these can be constructed easily, as explained above, from the multipliers and the permutation indices, respectively.*

Example 5.14. Let
$$A = \begin{pmatrix} 1 & 2 & 4 \\ 4 & 5 & 6 \\ 7 & 8 & 9 \end{pmatrix}.$$

k = 1:

1. *The pivot entry is* 7: $r_1 = 3$.
2. *Interchange rows* 3 *and* 1:
$$A \equiv \begin{pmatrix} 7 & 8 & 9 \\ 4 & 5 & 6 \\ 1 & 2 & 4 \end{pmatrix}.$$
3. *Form the multipliers*:
$$m_{21} = -\frac{4}{7}, \quad m_{31} = -\frac{1}{7}.$$
4. *Update*:
$$A \equiv \begin{pmatrix} 7 & 8 & 9 \\ 0 & \frac{3}{7} & \frac{6}{7} \\ 0 & \boxed{\frac{6}{7}} & \frac{19}{7} \end{pmatrix}.$$

k = 2:

1. *The pivot entry is* $\frac{6}{7}$: $r_2 = 3$.
2. *Interchange rows* 2 *and* 3:
$$A \equiv \begin{pmatrix} 7 & 8 & 9 \\ 0 & \frac{6}{7} & \frac{19}{7} \\ 0 & \frac{3}{7} & \frac{6}{7} \end{pmatrix}.$$
3. *Form the multiplier*:
$$m_{32} = -\frac{1}{2}.$$
4. *Update*:
$$A \equiv U = \begin{pmatrix} 7 & 8 & 9 \\ 0 & \frac{6}{7} & \frac{19}{7} \\ 0 & 0 & -\frac{1}{2} \end{pmatrix}.$$

Form L and P:
$$P = \begin{pmatrix} 0 & 0 & 1 \\ 1 & 0 & 0 \\ 0 & 1 & 0 \end{pmatrix}, \quad L = \begin{pmatrix} 1 & 0 & 0 \\ -m_{31} & 1 & 0 \\ -m_{21} & -m_{32} & 1 \end{pmatrix} = \begin{pmatrix} 1 & 0 & 0 \\ \frac{1}{7} & 1 & 0 \\ \frac{4}{7} & \frac{1}{2} & 1 \end{pmatrix}. \quad \blacksquare$$

MATCOM and MATLAB Notes: MATCOM program **PARPIV** computes M and U such that $MA = U$. MATLAB command **lu** in the form $[L, U, P] = lu(A)$ computes L, U, and P such that $PA = LU$.

5.2.5 Gaussian Elimination with Complete Pivoting (GECP)

In Gaussian elimination with complete pivoting, at the kth step, the search for the pivot is made among all the entries of the submatrix below the first $(k-1)$ rows. Set $A^{(0)} = A$. Thus, to obtain $A^{(k)}$ from $A^{(k-1)}$, $k = 1, \ldots, n$, do the following:

- Identify the largest element in magnitude among all the elements of the submatrix obtained by deleting the first $(k-1)$ rows and $(k-1)$ columns. Let it be $a_{rs}^{(k-1)}$.

- Interchange rows k and r followed by the interchange of columns k and s.

- Apply Gaussian elimination scheme without row interchange with $a_{rs}^{(k-1)}$ as the pivot to the submatrix consisting of rows k through n and columns k through n.

In terms of matrix multiplications, this then means

$$A^{(k)} = M_k P_k A^{(k-1)} Q_k,$$

where M_k is an elementary matrix and P_k is the permutation matrix obtained by interchanging rows k and r of the identity matrix. Similarly for the matrix Q_k. The matrix $A^{(k)}$ has zeros on the kth column below the (k,k) entry. The matrix M_k can of course be computed in two smaller steps as before.

At the end of the $(n-1)$th step, the matrix $A^{(n-1)}$ is an upper triangular matrix.

Obtaining factorization: $PAQ = LU$. Set

$$A^{(n-1)} = U. \tag{5.4}$$

Define

$$Q = Q_1 \cdots Q_{n-1}, \quad P = P_{n-1} P_{n-2} \ldots P_1, \tag{5.5}$$

and

$$L = P(M_{n-1} P_{n-1} \ldots M_1 P_1)^{-1}. \tag{5.6}$$

Then it can be shown (see Golub and Van Loan (1996)) that

$$PAQ = LU,$$

where P and Q are both *permutation matrices* and L is *unit triangular* and U is *upper triangular*.

A Practical Scheme for GECP

Remarks similar to those in the case of partial pivoting hold. Storage space does not have to be wasted by explicitly forming matrices P_k, Q_k $P_k A^{(k-1)} Q_k$, M_k, and $M_k P_k A^{(k-1)} Q_k$. It is enough to save the indices and the multipliers.

Here is a practical scheme for **complete pivoting,** which does not show the explicit formation of the matrices P_k, Q_k, M_k, $M_k A$, and $P_k A Q_k$. Note that partial pivoting is just a special case of complete pivoting.

5.2. LU Factorization Using Gaussian Elimination

ALGORITHM 5.3. LU Factorization Using GECP.

Input: An $n \times n$ matrix A.
Outputs: (i) An upper triangular matrix U, (ii) permutation indices r_k and s_k from which permutation matrices P and Q can be formed, and (iii) the multipliers m_{ik} from which the lower triangular matrix L can be constructed. The result is $PAQ = LU$.
Storage: The storage schemes for U and the multipliers are the same as GEPP. The indices r_k and s_k are saved in separate arrays.

For $k = 1, 2, \ldots, n - 1$ do

1. Find the *pivot indices* r_k and s_k such that $|a_{r_k, s_k}| = \max\{|a_{ij}| : i, j \geq k\}$, and save r_k and s_k.

   If $a_{r_k, s_k} = 0$, then stop. Otherwise, continue.

2. *(Interchange the rows r_k and k.)* $a_{kj} \leftrightarrow a_{r_k, j}$ $(j = k, k+1, \ldots, n)$.

3. *(Interchange the columns s_k and k.)* $a_{ik} \leftrightarrow a_{i, s_k}$ $(i = 1, 2, \ldots, n)$.

4. *(Form the multipliers.)* $a_{ik} \equiv m_{ik} = -\dfrac{a_{ik}}{a_{kk}}$ $(i = k+1, \ldots, n)$.

5. *(Update the entries of A.)* $a_{ij} \equiv a_{ij} + m_{ik} a_{kj} = a_{ij} + a_{ik} a_{kj}$ $(i = k+1, \ldots, n; j = k+1, \ldots, n)$.

End

Note: *Algorithm 5.3 does not explicitly give the matrices L, P, and Q; they have to be formed, respectively, from the multipliers m_{ik} and the permutation indices r_k and s_k.*

Example 5.15. Triangularize

$$A = \begin{pmatrix} 0 & 1 & 1 \\ 1 & 2 & \boxed{3} \\ 1 & 1 & 1 \end{pmatrix}$$

using complete pivoting.

$k = 1$:

1. The pivot entry is 3: $r = 2, s = 3$.
2. and 3. Interchange rows 1 and 2 followed by interchange of columns 1 and 3:

$$A \equiv \begin{pmatrix} \boxed{3} & 2 & 1 \\ 1 & 1 & 0 \\ 1 & 1 & 1 \end{pmatrix}.$$

4. and 5. Perform Gaussian elimination taking the entry 3 as pivot.

$$A \equiv A^{(1)} = \begin{pmatrix} 3 & 2 & 1 \\ 0 & \tfrac{1}{3} & -\tfrac{1}{3} \\ 0 & \tfrac{1}{3} & \boxed{\tfrac{2}{3}} \end{pmatrix}; \quad m_{21} = -\tfrac{1}{3},\; m_{31} = -\tfrac{1}{3}.$$

k = 2:
1. The pivot entry is $\frac{2}{3}$: $r = 3, s = 3$.
2. and 3. Interchange rows 2 and 3 followed by interchange of columns 2 and 3:

$$A \equiv \begin{pmatrix} 3 & 1 & 2 \\ 0 & \boxed{\frac{2}{3}} & \frac{1}{3} \\ 0 & -\frac{1}{3} & \frac{1}{3} \end{pmatrix}.$$

4. and 5. Perform Gaussian elimination taking the entry $\frac{2}{3}$ as pivot.

$$A \equiv A^{(2)} = \begin{pmatrix} 3 & 1 & 2 \\ 0 & \frac{2}{3} & \frac{1}{3} \\ 0 & 0 & \frac{1}{2} \end{pmatrix} = U; \quad m_{32} = \frac{1}{2}.$$

Readers are invited to compute the unit lower triangular matrix L and the permutation matrices P and Q such that $PAQ = LU$, using (5.5) and (5.6). ∎

Flop-count. Algorithm 5.3 requires $2\frac{n^3}{3}$ flops and $O(n^3)$ comparisons.

MATCOM Note: MATCOM program **COMPIV** computes $MAQ = U$.

5.2.6 Summary of Gaussian Elimination and LU Factorizations

Gaussian elimination schemes without pivoting, with partial pivoting, and with complete pivoting, when carried out to completion, yield, respectively,

- $A = LU$ (Gaussian elimination without pivoting);
- $PA = LU$ (GEPP);
- $PAQ = LU$ (GECP).

Here L is *unit lower triangular*, U is *upper triangular*, and P and Q are *permutation matrices*.

5.3 Stability of Gaussian Elimination

We have seen before that the computed matrices L and U obtained by Gaussian elimination without pivoting can be such that the product LU can be completely different from A. In fact, $\frac{\|L\|\|U\|}{\|A\|}$ can be **arbitrarily large**. [Exercise 5.15b] Specifically, the following result can be proved (See Higham (2002, pp. 164–165), Demmel (1997, pp. 47–49)).

Theorem 5.16 (round-off error bound for Gaussian elimination). *The computed matrices L and U obtained by Gaussian elimination without pivoting satisfy*

$$A + E = LU,$$

where

$$\|E\| \leq n\mu \|L\| \|U\|.$$

5.3. Stability of Gaussian Elimination

Since $U = A^{(n-1)}$, the stability of Gaussian elimination is better understood by measuring the *growth of the elements* in the reduced matrices $A^{(k)}$. (Note that although pivoting keeps the multipliers bounded by unity, the elements in the reduced matrices still can grow arbitrarily.)

Definition 5.17. The **growth factor** ρ *is the ratio of the largest element (in magnitude) of $A, A^{(1)}, \ldots, A^{(n-1)}$ to the largest element (in magnitude) of A:*

$$\rho = \frac{\max(\alpha, \alpha_1, \alpha_2, \ldots, \alpha_{n-1})}{\alpha},$$

where $\alpha = \max_{i,j} |a_{ij}|$ *and* $\alpha_k = \max_{i,j} |a_{ij}^{(k)}|$.

Now, if *partial pivoting* is used, then

- $|l_{ij}| \leq 1$ for all $i \geq j$, since these l_{ij} are the multipliers;
- $|u_{ij}| \leq \rho \max_{i,j} |a_{ij}|$.

We then have the following error bound (Exercise 5.17) with partial pivoting (for details, see Chapter 14, available online at *www.siam.org/books/ot116*), noting that the infinity norm does not depend on the sign of the matrix entries.

Theorem 5.18 (round-off error property for GEPP). *The matrices L and U computed by Gaussian elimination with partial pivoting satisfy*

$$LU = A + E,$$

where

$$\|E\|_\infty \leq n^3 \mu \rho \|A\|_\infty.$$

The question, therefore, arises, How large ρ can be? To answer the question, we start with an example.

Example 5.19.
$$A = \begin{pmatrix} 0.0001 & 1 \\ 1 & 1 \end{pmatrix}.$$

1. Gaussian elimination without pivoting gives

$$A^{(1)} = U = \begin{pmatrix} 0.0001 & 1 \\ 0 & -10^4 \end{pmatrix},$$

$$\max |a_{ij}^{(1)}| = 10^4, \ \max |a_{ij}| = 1,$$

$$\rho = \text{the growth factor} = 10^4.$$

2. Gaussian elimination with partial pivoting yields

$$A^{(1)} = U = \begin{pmatrix} 1 & 1 \\ 0 & 1 \end{pmatrix},$$

$$\max |a_{ij}^{(1)}| = 1, \ \max |a_{ij}| = 1,$$

$$\rho = \text{the growth factor} = 1. \quad \blacksquare$$

The question next is, *How large the growth factor ρ in each case can be for an arbitrary matrix?* We answer this question in the following.

Growth Factor for GEPP

For Gaussian elimination with partial pivoting, $\rho \leq 2^{n-1}$ (Exercise 5.15(a)):

ρ *can be as big as* 2^{n-1}.

Unfortunately, one can construct matrices for which this bound is attained. Consider the following example:

$$A = \begin{pmatrix} 1 & 0 & 0 & \cdots & 0 & 1 \\ -1 & 1 & 0 & \cdots & 0 & 1 \\ \vdots & \ddots & \ddots & & & \vdots \\ \vdots & & \ddots & \ddots & & \vdots \\ \vdots & & & \ddots & \ddots & \vdots \\ -1 & \cdots & \cdots & \cdots & -1 & 1 \end{pmatrix},$$

that is,

$$a_{ij} = \begin{cases} 1 & \text{for } j = i, n, \\ -1 & \text{for } j < i, \\ 0 & \text{otherwise.} \end{cases} \tag{5.7}$$

Wilkinson (1965, p. 212) has shown that the growth factor ρ for this matrix with partial pivoting is 2^{n-1}. To see this, take the special case with $n = 4$.

$$A = \begin{pmatrix} 1 & 0 & 0 & 1 \\ -1 & 1 & 0 & 1 \\ -1 & -1 & 1 & 1 \\ -1 & -1 & -1 & 1 \end{pmatrix},$$

$$A^{(1)} = \begin{pmatrix} 1 & 0 & 0 & 1 \\ 0 & 1 & 0 & 2 \\ 0 & -1 & 1 & 2 \\ 0 & -1 & -1 & 2 \end{pmatrix}, \quad A^{(2)} = \begin{pmatrix} 1 & 0 & 0 & 1 \\ 0 & 1 & 0 & 2 \\ 0 & 0 & 1 & 4 \\ 0 & 0 & -1 & 4 \end{pmatrix}, \quad A^{(3)} = \begin{pmatrix} 1 & 0 & 0 & 1 \\ 0 & 1 & 0 & 2 \\ 0 & 0 & 1 & 4 \\ 0 & 0 & 0 & 8 \end{pmatrix}.$$

Thus the growth factor is

$$\rho = \frac{8}{1} = 2^3 = 2^{4-1}.$$

Remarks. Note that this is not the only matrix for which $\rho = 2^{n-1}$. Higham and Higham (1989) have identified a set of matrices for which $\rho = 2^{n-1}$. The matrix

$$B = \begin{pmatrix} 0.7248 & 0.7510 & 0.5241 & 0.7510 \\ 0.7317 & 0.1889 & 0.0227 & -0.7510 \\ 0.7298 & -0.3756 & 0.1150 & 0.7511 \\ -0.6993 & -0.7444 & 0.6647 & -0.7500 \end{pmatrix} \tag{5.8}$$

is such a matrix. Wright (1993) has identified a matrix arising in solutions of a class of two-point boundary value problems which have exponential growth with partial pivoting. Also, Foster (1994) has discovered a class of linear systems arising in solutions of integral equations which have large growth factors with partial pivoting.

Examples of the above type are rare in practice. Indeed, in many practical examples, the elements of the matrices $A^{(k)}$ very often continue to decrease in size. *Thus, though GEPP is not unconditionally stable in theory, in practice it is considered to be a stable algorithm in general.*

Growth Factor for GECP

For Gaussian elimination with complete pivoting,

$$\rho \leq \{n \cdot 2^1 \cdot 3^{\frac{1}{2}} \cdot 4^{\frac{1}{3}} \cdots n^{\frac{1}{n-1}}\}^{1/2}.$$

This is a slowly growing function of n. Furthermore, in *practice* this bound is never attained. *Indeed, there was an unproven conjecture by Wilkinson (1965, p. 213) that the growth factor for complete pivoting was bounded by n for real n × n matrices.* Later Cryer (1968) conjectured that $\rho \leq n$ with equality holding if and only if A is a *Hadamard matrix*. An $n \times n$ matrix is a Hadamard matrix if its elements are ± 1 and $HH^T = nI$. Unfortunately, Wilkinson's conjecture has recently been settled by Gould (1991) negatively for an arbitrary matrix A. Gould exhibited a 13×13 matrix for which GECP gave the growth factor $\rho = 13.0205$. Edelman (1992b) also gave a counterexample to this conjecture by discovering a matrix of order 25 for which $\rho = 32.986341$. In spite of these recent results, *GECP is a stable algorithm.*

The conjecture regarding the growth factor ρ with complete pivoting for Hadamard matrices has been further investigated recently by several mathematicians. What seems to be important in settling this conjecture for Hadamard matrices is to determine the pivot structures and values of the minors of Hadamard matrices. Several results have been obtained in this direction. See the papers of Day and Peterson (1988), Koukouvinos, Mitrouli, and Seberry (2000, 2001), Koukouvinos et al. (2007), and Edelman and Friedman (1998). In a recent interesting paper, Kravvaritis and Mitrouli (2009) have shown that "The growth factor of a Hadamard matrix of order 16 is 16."

Growth factor of Gaussian elimination without pivoting. For Gaussian elimination without pivoting, *ρ can be arbitrarily large*, except for a few special cases, as we shall see later, such as *symmetric positive definite and diagonally dominant matrices.* Thus *Gaussian elimination without pivoting is, in general, a completely unstable algorithm.*

Posteriori stability test. In order to assess the stability of a computed LU factorization, one can either compute the growth factor or the backward error $(A - LU)$ itself. Both will require $O(n^3)$ flops. However, using a norm estimator algorithm (see Chapter 6), one can estimate $\|A - LU\|_1$, in $O(n^2)$ flops (see Higham (2002, pp. 181–182)).

5.4 Summary and Table of Comparisons

For easy reference we now review the most important aspects of this chapter.

5.4.1 Elementary Lower Triangular Matrix

An $n \times n$ matrix M of the form $M = I + m e_k^T$, where $m_k = (0, 0, \ldots, 0, m_{k+1,k}, \ldots, m_{n,k})^T$, is called an elementary lower triangular matrix of type k.

If M is as given above, then $M^{-1} = I - m_k e_k^T$.

5.4.2 LU Factorization

A factorization of A in the form $A = LU$, where L is unit lower triangular and U is upper triangular, is called an *LU factorization* of A. An LU factorization of matrix A does not always exist. If the leading principal minors of A are all different from zero, then the LU factorization of A exists and is unique (Theorem 5.8).

The LU factorization of a matrix A, when it exists, is achieved using elementary lower triangular matrices. The process is called *Gaussian elimination without row interchanges* or *Gaussian elimination without pivoting* (Algorithm 5.1).

The process is efficient, requiring only $2\frac{n^3}{3}$ flops, but *is unstable for arbitrary matrices. Its use is not recommended in practice unless A is symmetric positive definite or column diagonally dominant because, in these cases, the growth factors are 1 and less than or equal to 2, respectively; see* Chapter 6. For decomposition of A into LU in a stable way, row interchanges *(Gaussian elimination with partial pivoting)* (Algorithm 5.2) or both row and column interchanges *(Gaussian elimination with complete pivoting)* (Algorithm 5.3) to identify an appropriate pivot at each step will be needed. Gaussian elimination with partial and complete pivoting yield factorizations $PA = LU$ and $PAQ = LU$, respectively, where P and Q are permutation matrices.

5.4.3 Stability of Gaussian Elimination

Aspects of stability, instability, and practical stability in terms of the growth factors of the Gaussian elimination scheme and the associated round-off results are given in Section 5.3.

5.4.4 Table of Comparisons

We now summarize in Table 5.1 the efficiency and stability properties of these computations. We assume that A is $n \times n$.

Concluding remarks. *Gaussian elimination without pivoting is unstable in general; Gaussian elimination with partial pivoting is stable in practice; Gaussian elimination with complete pivoting is stable.*

5.5 Suggestions for Further Reading

The topics covered in this chapter are standard and can be found in any numerical linear algebra book. The books by Golub and Van Loan (1996), Stewart (1973, 1998b), and Higham (2002) are rich sources of further knowledge in this area. A fair treatment of these topics also appears in some numerical analysis books, such as Atkinson (1989), Heath

Table 5.1. *Table of comparisons of LU factorization methods.*

Problem	Method	Flop-count (Approximate)	Stability
Factorization: $A = LU$	Gaussian elimination without pivoting	$\dfrac{2n^3}{3}$	Unstable in general
Factorization: $PA = LU$	Gaussian elimination with partial pivoting	$\dfrac{2n^3}{3} +$ ($O(n^2)$ comparisons)	Stable in practice
Factorization: $PAQ = U$	Gaussian elimination with complete pivoting	$\dfrac{2n^3}{3} +$ ($O(n^3)$ comparisons)	Stable

(2002), Kincaid and Cheney (2002), Forsythe, Malcolm, and Moler (1977), Forsythe and Moler (1967), Kahaner, Moler, and Nash (1988), Moler (2004), Conte and de Boor (1980), Burden and Faires (2004), and Van Loan (2000). An interesting earlier paper on the stability of Gaussian elimination is Trefethen and Schreiber (1990).

Exercises on Chapter 5

(Use MATLAB, whenever appropriate and necessary.)

5.1 (a) Show that an elementary lower triangular matrix of type k defined by (5.1) has the form
$$M_k = I + me_k^T,$$
where $m = (0, 0, \ldots, 0, m_{k+1,k}, \ldots, m_{n,k})^T$.

(b) Show that the inverse of M_k in (a) is given by
$$M_k^{-1} = I - me_k^T.$$

(c) Show that the elementary matrix M_1 defined by (5.2) is such that $M_1 a$, where $a = (a_{11}, a_{21}, \ldots, a_{n1})^T$, is a multiple of e_1.

5.2 (a) Given
$$a = \begin{pmatrix} 0.00001 \\ 1 \end{pmatrix}.$$

Using three-digit arithmetic, find an elementary matrix M such that Ma is a multiple of e_1.

(b) Using your computations in (a), find the LU factorization of
$$A = \begin{pmatrix} 0.00001 & 1 \\ 1 & 2 \end{pmatrix}.$$

(c) Let $\hat{L}$ and $\hat{U}$ be the computed L and U in part (b). Find

(i) $\dfrac{\|A - \hat{L}\hat{U}\|_F}{\|A\|_F}$,   (ii) $\dfrac{\|\hat{L}\|_F \|\hat{U}\|_F}{\|A\|_F}$.

5.3 Show that the pivots $a_{11}, a_{22}^{(1)}, \ldots, a_{n-1,n-1}^{(n-2)}$ are nonzero if and only if the first $(n-1)$ leading principal minors of A are nonsingular.

Hint: Let A_r denote the rth leading principal minor of A. Then show that
$$\det(A_r) = a_{11} a_{22}^{(1)} \ldots, a_{rr}^{(r-1)}.$$

5.4 Assuming that LU factorization of A exists, prove that

(a) (*LDU factorization.*) A can be written in the form
$$A = LDU_1,$$
where D is diagonal and L and U_1 are unit lower and upper triangular matrices, respectively.

(b) (*LDL^T factorization.*) If A is symmetric, then
$$A = LDL^T.$$

(c) Using (b), prove that if A is symmetric and positive definite, then
$$A = HH^T,$$
where H is a lower triangular matrix with positive diagonal entries. (*This is known as the Cholesky decomposition.*)

5.5 Assuming that LU factorization of A exists, develop an algorithm to compute U by rows and L by columns directly from the equation $A = LU$.

This is known as *Doolittle reduction*.

5.6 Develop an algorithm to compute the factorization $A = LU$, where U is unit upper triangular and L is lower triangular. This is known as *Crout reduction*.
Hint: Derive the algorithm from the equation $A = LU$.

5.7 Compare the Doolittle and Crout reductions with Gaussian elimination without pivoting with respect to flop-count and storage requirements.

5.8 A matrix G of the form
$$G = I - ge_k^T$$
is called a *Gauss–Jordan* matrix. Show that, given a vector x with the property that $e_k^T x \neq 0$, there exists a Gauss–Jordan matrix G such that

$$Gx \text{ is a multiple of } e_k.$$

Develop an algorithm to construct Gauss–Jordan matrices $G_1, G_2, \ldots, G_n$ successively such that $(G_n G_{n-1}, \ldots, G_2 G_1) A$ is a diagonal matrix. This is known as *Gauss–Jordan reduction*.

Derive conditions under which Gauss–Jordan reduction can be carried to completion.

Give a flop-count for the algorithm and compare it with those of Gaussian elimination, Crout reduction, and Doolittle reduction.

5.9 Given
$$A = \begin{pmatrix} 1 & 2 & 3 \\ 2 & 5 & 4 \\ 3 & 4 & 5 \end{pmatrix},$$
find LU factorization of A using Gaussian elimination, Doolittle reduction, and Crout reduction.

5.10 Apply the Gauss–Jordan reduction to A of Exercise 5.9.

5.11 Prove that the matrix L in each of the factorizations $PA = LU$ and $PAQ = LU$, obtained by using Gaussian elimination with partial and complete pivoting, respectively, is unit lower triangular.

5.12 Given
$$A = \begin{pmatrix} 0 & 1 & 0 & 0 \\ 0 & 0 & 1 & 0 \\ 0 & 0 & 0 & 1 \\ 2 & 3 & 4 & 5 \end{pmatrix},$$
find a permutation matrix P, a unit lower triangular matrix L, and an upper triangular matrix U such that $PA = LU$.

5.13 (a) Find permutation matrices P and Q and a unit lower triangular matrix L and an upper triangular matrix U such that $PA = LU$ and $PAQ = LU$ for each of the following matrices.

(i) $A = \begin{pmatrix} 1 & \frac{1}{2} & \frac{1}{3} \\ \frac{1}{2} & \frac{1}{3} & \frac{1}{4} \\ \frac{1}{3} & \frac{1}{4} & \frac{1}{5} \end{pmatrix}$, (ii) $A = \begin{pmatrix} 100 & 99 & 98 \\ 98 & 55 & 11 \\ 0 & 1 & 1 \end{pmatrix}$,

(iii) $A = \begin{pmatrix} 1 & 0 & 1 \\ -1 & 1 & 1 \\ -1 & -1 & 1 \end{pmatrix}$, (iv) $A = \begin{pmatrix} 0.00003 & 1.566 & 1.234 \\ 1.5660 & 2.000 & 1.018 \\ 1.2340 & 1.018 & -3.000 \end{pmatrix}$,

(v) A of the form (5.7) with $n = 5$.

(b) For each of the matrices in (a), find M and U such that $MA = U$.

(c) Compute the growth factor in each case and verify the results on upper bounds of the growth factor in each case given in Section 5.3.

(d) Estimate the backward error for each of the factorizations.

5.14 (a) Consider the 5×5 matrix
$$A = \begin{pmatrix} 1 & 0 & 0 & 0 & 0.1 \\ 0 & 1 & 0 & 0 & 0.1 \\ 0 & 0 & 1 & 0 & 0.1 \\ 0 & 0 & 0 & 1 & 0.1 \\ 0.1 & 0.1 & 0.1 & 0.1 & 0.1 \end{pmatrix}.$$
Find the LU factorizations using both Gaussian elimination and GEPP. How many flops are needed? How many flops will be needed if A is an $n \times n$ row matrix?

(b) Repeat (a) with the permuted matrix
$$A' = \begin{pmatrix} 1 & 0.1 & 0.1 & 0.1 & 0.1 \\ 0.1 & 1 & 0 & 0 & 0 \\ 0.1 & 0 & 1 & 0 & 0 \\ 0.1 & 0 & 0 & 1 & 0 \\ 0.1 & 0 & 0 & 0 & 1 \end{pmatrix}$$
and compare your answers with those obtained in (a).

5.15 (a) Prove that the growth factor $\rho \leq 2^{n-1}$ for GEPP applied to an $n \times n$ matrix.

(b) Construct a small example to show that for GE without pivoting the ratio $\frac{\|L\|\|U\|}{\|A\|}$ can be arbitrarily large.

5.16 (a) Formulate algorithms for LU factorization of an $m \times n$ matrix using Gaussian elimination without and with partial pivoting.

Show that each algorithm requires about $mn^2 - \frac{n^3}{3}$ flops.

(b) Apply your algorithms to

(i) $A = \begin{pmatrix} 0.00001 & 1 & 1 \\ 1 & 1 & 1 \\ 1 & 1 & 1 \end{pmatrix}$, (ii) $A = \text{rand}(5, 2)$.

5.17 Using the following result on inner product computation of the form
$$\text{fl}\left(\sum_{i=1}^{k} x_i y_i\right) = \sum_{i=1}^{k} x_i y_i (1 + \delta_i), \quad |\delta_i| \leq k\mu,$$
show that $A + E = LU$, where $|E| \leq n\mu|L||U|$. Hence prove Theorem 5.18 (consult Demme (1997) or Higham (2002), if necessary).

MATLAB and MATCOM Programs and Problems on Chapter 5

> **Note on MATCOM**
>
> MATCOM is a MATLAB-based interactive software package containing implementation of all major algorithms of Chapters 4 through 12.
> For each problem, there is more than one algorithm so that students can compare different algorithms for the same problem with respect to accuracy, speed, etc. A chapterwise listing of MATCOM programs is given in Appendix C. MATCOM is available from the book's webpage at www.siam.org/books/ot116.

M5.1 Based on Algorithm 5.1, write a MATLAB program, called **lugewp**, to compute L and U such that $A = LU$ and the associated growth factor gf: $[L, U, gf] = $ **lugewp** (A).

Test data:

(i) $A = \begin{pmatrix} 10^{-15} & 1 \\ 1 & 1 \end{pmatrix}$, (ii) $A = \begin{pmatrix} 1 & 1 \\ 0.00001 & 1 \end{pmatrix}$,

(iii) $A = \begin{pmatrix} 10 & 1 & 1 \\ 1 & 10 & 1 \\ 1 & 1 & 20 \end{pmatrix}$, (iv) the matrix A in (5.7) with $n = 10$,

(v) $A = 20 \times 20$ Hilbert matrix, (vi) the matrix A in (5.8).

Print in each case

(i) $\dfrac{\||L||U|\|_F}{\|A\|_F}$, (ii) $\dfrac{\|A - LU\|_F}{\|L\|_F \|U\|_F}$, (iii) $\dfrac{\|A - LU\|_F}{\|A\|_F}$,

and (iv) the growth factor.

Write your observations.

M5.2 Based on Algorithm 5.2, write a MATLAB program, called **lugepp**, to compute (i) P, L and U such that $PA = LU$, using partial pivoting, and (ii) the associated growth factor gf:

$$[L, U, P, gf] = \text{lugepp}(A).$$

Print $\dfrac{\|L\|_F \|U\|_F}{\|A\|_F}$, $\dfrac{\|PA - LU\|_F}{\|A\|_F}$, and the growth factor for each of the matrices A of Problem M5.1. Explain why these results are different.

M5.3 Based on Algorithm 5.3, write a MATLAB program, called **lugecp**, to compute P, Q, L, and U such that $PAQ = LU$, and the associated growth factor gf:

$$[L, U, P, Q, gf] = \text{lugecp}(A).$$

Print $\dfrac{\|L\|_F \|U\|_F}{\|A\|_F}$, $\dfrac{\|PAQ - LU\|_F}{\|A\|_F}$, and the growth factor for each of the matrices of Problem M5.1. Explain why these results are different.

M5.4 Write a MATLAB program, called **GSJOR**, to implement Gauss–Jordan scheme outlined in Exercise 5.8 and apply your program to the matrices of Problem M5.1.

M5.5 (*Experiment on the growth factor for GEPP.*) Plot the growth factors for GEPP of 500 randomly generated matrices of varying dimension. Write down your observations.

M5.6 *Random triangular matrices usually become more and more ill-conditioned as the dimensions increase.* However, the lower triangular matrices L from LU factorization of a matrix A using GEPP are believed to have low condition numbers. Perform an experiment to verify this statement, as follows: Take a random matrix of order 125 and compute its LU factorization using **lugepp** and plot the entries of the inverse of L with magnitude greater than or equal to 1. Then change the signs of the subdiagonal entries of L randomly to create another lower triangular matrix $\tilde{L}$ and plot the entries of the inverse of $\tilde{L}$ with magnitude greater than or equal to 1. Compute $\max_{i,j} |L_{ij}^{-1}|$ and $\max_{i,j} |\tilde{L}_{ij}^{-1}|$. Repeat the above experiment with random matrices with entries uniformly distributed in $[-1, 1]$.

M5.7 Using MATCOM program **parpiv** on each of the matrices of Exercise M5.1, print $\|M\|_F$, $\|U\|_F$, $\|MA - U\|_F$, and $\frac{\|U\|_F}{\|MA\|_F}$, and write your observations.

M5.8 Repeat Problem M5.7 with MATCOM program **compiv** and print $\|M\|_F$, $\|U\|_F$, $\|MAQ - U\|_F$, and $\frac{\|U\|_F}{\|MA\|_F}$.

Chapter 6
Numerical Solutions of Linear Systems

Background Material Needed

- Vector and matrix norms (Sections 2.5.1 and 2.5.2)
- Special matrices (Section 2.4)
- Condition numbers and properties (Sections 4.5–4.7)
- Solutions of triangular systems (Sections 4.1.3 and 4.1.4)
- LU factorizations and stability properties (Sections 5.2 and 5.3)

6.1 Introduction

In this chapter we will discuss methods for numerically solving the linear system

$$Ax = b,$$

where A is an $n \times n$ matrix and x and b are n-vectors. A and b are given and x is unknown. The problem arises in a very wide variety of applications. *As a matter of fact, it might be said that numerical solutions of almost all practical engineering and applied science problems routinely require solution of a linear system problem.* (See Sections 6.3 and 6.12.)

We shall discuss methods for nonsingular linear systems only in this chapter. The case where the matrix A is not square or the system has more than one solution is treated in Chapter 8.

A method called **Cramer's Rule**, taught in an elementary undergraduate linear algebra course, is of high significance from a theoretical point of view.

Cramer's Rule is, however, not at all practical from a computational viewpoint. For example, solving a linear system with 20 equations and 20 unknowns by Cramer's Rule, using the usual definition of determinant, would require more than a million years even on a fast computer (Forsythe, Malcolm, and Moler (1977, p. 30)). For an $n \times n$ system, it will require about $O(n!)$ flops.

Two types of methods are normally used for numerical computations:

(1) **direct methods.**

(2) **iterative methods.**

The direct methods consist of a finite number of steps, and one needs to perform all of the steps in a given method before the solution is obtained. On the other hand, iterative methods are based on computing a sequence of approximations to the solution x and a user can stop whenever a certain desired accuracy is obtained or a certain number of iterations are completed. *The iterative methods are used primarily for large and sparse systems.* We will consider iterative methods in Chapter 12.

The organization of this chapter is as follows:

In Section 6.2 we state the basic theoretical results (without proofs) on the existence and uniqueness of solutions for linear systems.

In Section 6.3 we discuss several engineering applications giving rise to linear systems problems mostly without any special structures.

In Section 6.4 we discuss LU factorization methods for solving arbitrary linear systems.

In Section 6.5 we consider the effects of scaling on solutions of linear systems.

Section 6.6 summarizes the discussions of Sections 6.4 and 6.5.

Computations of the inverse and the determinant are discussed in Section 6.7.

Section 6.8 discusses the effects of the condition number on the accuracy of the solution.

In Section 6.9 we discuss computing and estimating the condition number of a matrix.

Results of componentwise perturbations are given in Section 6.10.

Iterative refinement is discussed in Section 6.11.

Section 6.12 is devoted to the study of numerical solutions of special linear systems: *positive definite, Hessenberg, diagonally dominant, tridiagonal, and block tridiagonal. Some practical applications giving rise to these systems are also discussed here.*

6.2 Basic Results on Existence and Uniqueness

Consider the system of m equations in n unknowns:

$$a_{11}x_1 + a_{12}x_2 + \cdots + a_{1n}x_n = b_1,$$
$$a_{12}x_1 + a_{22}x_2 + \cdots + a_{2n}x_n = b_2,$$
$$\vdots$$
$$a_{m1}x_1 + a_{m2}x_2 + \cdots + a_{mn}x_n = b_m.$$

In matrix form, the system is written as

$$Ax = b,$$

where

$$A = \begin{pmatrix} a_{11} & a_{12} & \cdots & a_{1n} \\ a_{21} & a_{22} & \cdots & a_{2n} \\ \vdots & & & \\ a_{m1} & a_{m2} & \cdots & a_{mn} \end{pmatrix}, \quad x = \begin{pmatrix} x_1 \\ x_2 \\ \vdots \\ x_n \end{pmatrix}, \quad b = \begin{pmatrix} b_1 \\ b_2 \\ \vdots \\ b_m \end{pmatrix}.$$

Given an $m \times n$ matrix A and an m-vector b, if there exists a vector x satisfying $Ax = b$, then we say that the system is **consistent**. Otherwise, it is **inconsistent**. It is natural to ask when a given system $Ax = b$ is consistent and, if it is consistent, how many solutions there are, and when the solution is unique. To this end, we state the following theorem. Proof can be found in any linear algebra textbook.

Theorem 6.1 (existence and uniqueness theorem for a nonhomogeneous system).

1. *The system $Ax = b$ is consistent if and only if $b \in R(A)$; in other words,* $\text{rank}(A) = \text{rank}(A, b)$.

2. *If the system is consistent and the columns of A are linearly independent, then the solution is unique.*

3. *If the system is consistent and the columns of A are linearly dependent, then the system has an infinite number of solutions.*

6.3 Some Applications Giving Rise to Linear Systems Problems

It is probably not an overstatement that linear systems problems arise in almost all practical applications. We will give examples here from electrical, mechanical, chemical, and civil engineering. We start with a simple problem—an electric circuit.

6.3.1 An Electric Circuit Problem

Consider the diagram of an electrical circuit shown in Figure 6.1. We would like to determine the amount of current between the nodes A_1, A_2, A_3, A_4, A_5, and A_6. The famous

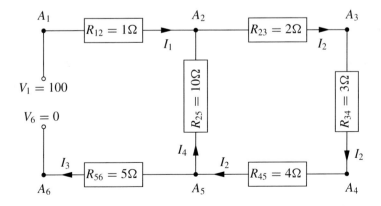

Figure 6.1. *An electric circuit.*

Kirchhoff's current law tells us that *the algebraic sum of all currents entering a node must be zero*. Applying this law at node A_2, A_5, A_3, and A_4, respectively, we have

$$I_1 - I_2 + I_4 = 0, \tag{6.1}$$

$$I_2 - I_3 - I_4 = 0, \tag{6.2}$$

$$I_2 - I_2 = 0, \tag{6.3}$$

$$I_2 - I_2 = 0. \tag{6.4}$$

Now consider the voltage drop around each closed loop of the circuit, $A_1A_2A_3A_4A_5A_6A_1$, $A_1A_2A_5A_6A_1$, $A_2A_3A_4A_5A_2$. Kirchhoff's voltage law tells us that *the net voltage drop around each closed loop is zero*. Thus at the loop $A_1A_2A_3A_4A_5A_6A_1$, substituting the values of resistances and voltages, we have

$$I_1 + 9I_2 + 5I_3 = 100. \tag{6.5}$$

Similarly, at $A_1A_2A_5A_6A_1$ and $A_2A_3A_4A_5A_2$ we have, respectively,

$$I_1 - 10I_4 + 5I_3 = 100, \tag{6.6}$$

$$9I_2 + 10I_4 = 0. \tag{6.7}$$

Note that (6.6) + (6.7) = (6.5). Thus we have four equations in four unknowns:

$$I_1 - I_2 + I_4 = 0, \tag{6.8}$$

$$I_2 - I_3 - I_4 = 0, \tag{6.9}$$

$$I_1 - 10I_4 + 5I_3 = 100, \tag{6.10}$$

$$9I_2 + 10I_4 = 0. \tag{6.11}$$

Equations (6.8)–(6.11) can be written as

$$\begin{pmatrix} 1 & -1 & 0 & 1 \\ 0 & 1 & -1 & -1 \\ 1 & 0 & 5 & -10 \\ 0 & 9 & 0 & 10 \end{pmatrix} \begin{pmatrix} I_1 \\ I_2 \\ I_3 \\ I_4 \end{pmatrix} = \begin{pmatrix} 0 \\ 0 \\ 100 \\ 0 \end{pmatrix},$$

the solution of which yields the current between the nodes.

6.3.2 Analysis of a Processing Plant Consisting of Interconnected Reactors

Many mathematical models are based on conservation laws such as conservation of mass, conservation of momentum, and conservation of energy. In mathematical terms, these conservation laws lead to conservation or balance or continuity equations, which relate the behavior of a system or response of the quantity being modeled to the properties of the system and the external forcing functions or stimuli acting on the system.

As an example, consider a chemical processing plant consisting of six interconnected chemical reactors (Figure 6.2), with different mass flow rates of a component of a mixture into and out of the reactors. We are interested in knowing the concentration of the mixture

6.3. Some Applications Giving Rise to Linear Systems Problems

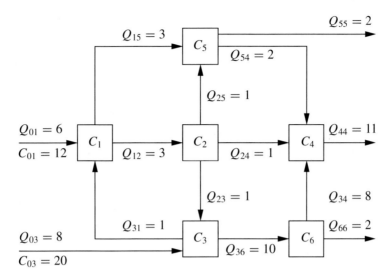

Figure 6.2. *Processing plant with interconnected reactors.*

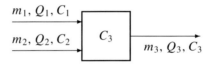

Figure 6.3. *Sketch of a reactor with two incoming and one outgoing flows.*

at different reactors. The example here is similar to that given in Chapra and Canale (2002, pp. 307–308). Application of conservation of mass to all these reactors results in a linear system of equations as shown below, consisting of six equations in six unknowns. The solution of the system will tell us the concentration of the mixture at each of these reactors.

Steady state, completely mixed reactor. Consider first a reactor with two flows coming in and one flow going out, as shown in Figure 6.3. Application of the steady state conservation of mass to the above reactor gives us

$$m_1 + m_2 = m_3. \tag{6.12}$$

Noting that

$$m_i = Q_i \cdot C_i,$$

where

m_i = mass flow rate of the mixture at the inlet and outlet sections i, $i = 1, 2, 3$,
Q_i = volumetric flow rate at the section i, $i = 1, 2, 3$,
C_i = density or concentration at the section i, $i = 1, 2, 3$,
we get from (6.12)

$$Q_1 C_1 + Q_2 C_2 = Q_3 C_3. \tag{6.13}$$

For given inlet flow rates and concentrations, the outlet concentration C_3 can be found from (6.13). Under steady state operation, this outlet concentration also represents the spatially uniform or homogeneous concentration inside the reactor. Such information is necessary for designing the reactor to yield mixtures of a specified concentration. For details, see Chapra and Canale (2002).

Referring now to Figure 6.2, where we consider the plant consisting of six reactors, we have the following equations (derived similarly to that of (6.13)). The derivation of each of these equations is based on the fact that the net mass flow rate into the reactor is equal to the net mass flow out of the reactor.

For reactor 1,

$$6C_1 - C_3 = 72. \tag{6.14}$$

(Note that for this reactor, flow at the inlet is $72 + C_3$ and flow at the outlet is $6C_1$.)

Similarly, for reactor 2, 3, 4, 5 and 6, we have, respectively,

$$3C_1 - 3C_2 = 0, \tag{6.15}$$
$$-C_2 + 11C_3 = 160, \tag{6.16}$$
$$C_2 - 11C_4 + 2C_5 + 8C_6 = 0, \tag{6.17}$$
$$3C_1 + C_2 - 4C_5 = 0, \tag{6.18}$$
$$10C_3 - 10C_6 = 0. \tag{6.19}$$

Equations (6.14)–(6.19) can be rewritten in matrix form as

$$\begin{pmatrix} 6 & 0 & -1 & 0 & 0 & 0 \\ 3 & -3 & 0 & 0 & 0 & 0 \\ 0 & -1 & 11 & 0 & 0 & 0 \\ 0 & 1 & 0 & -11 & 2 & 8 \\ 3 & 1 & 0 & 0 & -4 & 0 \\ 0 & 0 & 10 & 0 & 0 & -10 \end{pmatrix} \begin{pmatrix} C_1 \\ C_2 \\ C_3 \\ C_4 \\ C_5 \\ C_6 \end{pmatrix} = \begin{pmatrix} 72 \\ 0 \\ 160 \\ 0 \\ 0 \\ 0 \end{pmatrix} \tag{6.20}$$

or

$$AC = D.$$

The ith coordinate of the unknown vector C represents the mixture concentration C_i at reactor i of the plant. The solution of the system gives the required concentrations.

6.3.3 Linear Systems Arising from Ordinary Differential Equations: A Case Study on a Spring-Mass Problem

Consider a system of three masses suspended vertically by a series of springs, as shown below, where k_1, k_2, and k_3 are the spring constants, and x_1, x_2, and x_3 are the displacements of each spring from its equilibrium position.

6.3. Some Applications Giving Rise to Linear Systems Problems

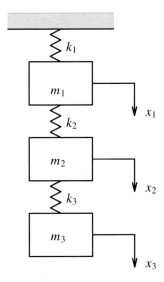

Referring to the above diagram, the equations of motion, by **Newton's second law**, which states that the force acting on a mass m is equal to m times acceleration, are

$$m_1 \frac{d^2 x_1}{dt^2} = k_2(x_2 - x_1) + m_1 g - k_1 x_1,$$

$$m_2 \frac{d^2 x_2}{dt^2} = k_3(x_3 - x_2) + m_2 g - k_2(x_2 - x_1),$$

$$m_3 \frac{d^2 x_3}{dt^2} = m_3 g - k_3(x_3 - x_2).$$

Suppose we are interested in knowing the displacements of these springs when the system eventually returns to the steady state, that is, when the system comes to rest. Then, by setting the second-order derivatives to zero, we obtain the following system of three equations in three unknowns, x_1, x_2, and x_3, in matrix form:

$$\begin{pmatrix} k_1 + k_2 & -k_2 & 0 \\ -k_2 & k_2 + k_3 & -k_3 \\ 0 & -k_3 & k_3 \end{pmatrix} \begin{pmatrix} x_1 \\ x_2 \\ x_3 \end{pmatrix} = \begin{pmatrix} m_1 g \\ m_2 g \\ m_3 g \end{pmatrix}$$

or

$$Kx = w.$$

The matrix K is called the *stiffness matrix*. As in this case, very often in practice this matrix is *symmetric tridiagonal*.

6.3.4 Linear Systems Arising from Partial Differential Equations: A Case Study on Temperature Distribution

Many engineering problems are modeled by partial differential equations. Numerical approaches to these equations typically require discretization by means of difference equations;

that is, partial derivatives in the equations are replaced by approximate differences. This process of discretization in turn gives rise to linear systems. We shall illustrate this with a problem in **heat transfer theory**. See the recent book by Majumdar (2005) for details.

A major objective in a heat transfer problem is to determine the temperature distribution $T(x, y, z, t)$ in a medium resulting from imposed boundary conditions on the surface of the medium. Once this temperature distribution is known, the heat transfer rate at any point in the medium or on its surface may be computed from Fourier's law, which is expressed as

$$q_x = -K \frac{\partial T}{\partial x}, \quad q_y = -K \frac{\partial T}{\partial y}, \quad \text{and } q_z = -K \frac{\partial T}{\partial z},$$

where q_x is the **heat transfer rate** in the x direction, $\frac{\partial T}{\partial x}$ is the **temperature gradient** in the x direction, and the positive constant K is called the **thermal conductivity of the material.** Similarly for the y and z directions.

Consider a homogeneous medium in which temperature gradients exist and the temperature distribution $T(x, y, z, t)$ is expressed in Cartesian coordinates. The **heat diffusion equation** which governs this temperature distribution is obtained by applying conservation of energy over an infinitesimally small differential element, from which we obtain the relation

$$\frac{\partial}{\partial x}\left(K \frac{\partial T}{\partial x}\right) + \frac{\partial}{\partial y}\left(K \frac{\partial T}{\partial y}\right) + \frac{\partial}{\partial z}\left(K \frac{\partial T}{\partial z}\right) + \dot{q} = \rho C_p \frac{\partial T}{\partial t}, \quad (6.21)$$

where ρ is the **density**, C_p is the **specific heat**, and $\dot{q}$ is the **energy** generated per unit volume.

This equation, usually known as the *heat equation*, provides the basic tool for solving heat conduction problems.

It is often possible to work with a simplified form of (6.21). For example, if the thermal conduction is a constant, the heat equation is

$$\frac{\partial^2 T}{\partial x^2} + \frac{\partial^2 T}{\partial y^2} + \frac{\partial^2 T}{\partial z^2} + \frac{\dot{q}}{K} = \frac{1}{\alpha} \frac{\partial T}{\partial t}, \quad (6.22)$$

where $\alpha = K/(\rho C_p)$ is a thermophysical property known as the **thermal diffusivity**.

Under steady state conditions, there can be no changes of energy storage, i.e., the unsteady state term $\frac{\partial T}{\partial t}$ can be dropped, and (6.22) reduces to the three-dimensional **Poisson's equation**:

$$\frac{\partial^2 T}{\partial x^2} + \frac{\partial^2 T}{\partial y^2} + \frac{\partial^2 T}{\partial z^2} + \frac{\dot{q}}{K} = 0. \quad (6.23)$$

If the heat transfer is two-dimensional (e.g., in the x and y directions) and there is no energy generation, then the heat equation reduces to the famous **Laplace's equation**:

$$\frac{\partial^2 T}{\partial x^2} + \frac{\partial^2 T}{\partial y^2} = 0. \quad (6.24)$$

If the heat transfer is unsteady and one-dimensional without energy generation, then the heat equation reduces to

$$\frac{\partial^2 T}{\partial x^2} = \frac{1}{\alpha} \frac{\partial T}{\partial t}. \quad (6.25)$$

6.3. Some Applications Giving Rise to Linear Systems Problems

Finite Difference Scheme

A well-known scheme for numerically solving a partial differential equation is to use **finite differences**. The idea is to discretize the partial differential equation by replacing the partial derivatives with their approximations, i.e., finite differences. We will illustrate the scheme with Laplace's equation in the following.

Let us divide a two-dimensional region into small regions with increments in the x and y directions given as Δx and Δy, as shown in the figure below.

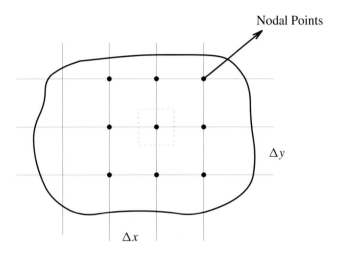

Each nodal point is designated by a numbering scheme i and j, where i indicates the x increment and j indicates the y increment:

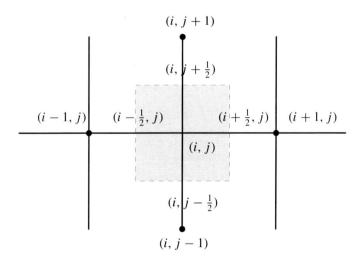

The temperature distribution in the medium is assumed to be represented by the nodal points temperature. The temperature $T_{i,j} = T(x_i, y_j)$ at each *nodal point* (x_i, y_j) (which is symbolically denoted by (i, j) as in the diagram above) is the average temperature of the surrounding hatched region. *As the number of nodal points increases, greater accuracy in representation of the temperature distribution is obtained.*

A finite difference equation suitable for the interior nodes of a steady two-dimensional system can be obtained by considering Laplace's equation at the nodal point i, j as

$$\frac{\partial^2 T}{\partial x^2}\bigg|_{i,j} + \frac{\partial^2 T}{\partial y^2}\bigg|_{i,j} = 0. \tag{6.26}$$

The second derivatives at the nodal point (i, j) can be expressed as

$$\frac{\partial^2 T}{\partial x^2}\bigg|_{i,j} \approx \frac{\frac{\partial T}{\partial x}\big|_{i+\frac{1}{2},j} - \frac{\partial T}{\partial x}\big|_{i-\frac{1}{2},j}}{\Delta x}, \tag{6.27}$$

$$\frac{\partial^2 T}{\partial y^2}\bigg|_{i,j} \approx \frac{\frac{\partial T}{\partial y}\big|_{i,j+\frac{1}{2}} - \frac{\partial T}{\partial y}\big|_{i,j-\frac{1}{2}}}{\Delta x}. \tag{6.28}$$

As shown in the figure, the temperature gradients can be approximated (as derived from the Taylor series) as a linear function of the nodal temperatures as

$$\frac{\partial T}{\partial x}\bigg|_{i+\frac{1}{2},j} \approx \frac{T_{i+1,j} - T_{i,j}}{\Delta x}, \tag{6.29}$$

$$\frac{\partial T}{\partial x}\bigg|_{i-\frac{1}{2},j} \approx \frac{T_{i,j} - T_{i-1,j}}{\Delta x}, \tag{6.30}$$

$$\frac{\partial T}{\partial y}\bigg|_{i,j+\frac{1}{2}} \approx \frac{T_{i,j+1} - T_{i,j}}{\Delta y}, \tag{6.31}$$

$$\frac{\partial T}{\partial y}\bigg|_{i,j-\frac{1}{2}} \approx \frac{T_{i,j} - T_{i,j-1}}{\Delta y}, \tag{6.32}$$

where, $T_{i,j} = T(x_i, y_j)$. Substituting (6.29)–(6.32) into (6.27)–(6.28), we get

$$\frac{\partial^2 T}{\partial x^2}\bigg|_{i,j} \cong \frac{T_{i+1,j} - 2T_{i,j} + T_{i-1,j}}{(\Delta x)^2}, \tag{6.33}$$

$$\frac{\partial^2 T}{\partial y^2}\bigg|_{i,j} \cong \frac{T_{i,j+1} - 2T_{i,j} + T_{i,j-1}}{(\Delta y)^2}. \tag{6.34}$$

Equation (6.26) then gives

$$\frac{T_{i+1,j} - 2T_{i,j} + T_{i-1,j}}{(\Delta x)^2} + \frac{T_{i,j+1} - 2T_{i,j} + T_{i,j-1}}{(\Delta y)^2} = 0.$$

6.3. Some Applications Giving Rise to Linear Systems Problems

Assume $\Delta x = \Delta y$. Then the *finite difference approximation of Laplace's equation for interior regions* can be expressed as

$$T_{i,j+1} + T_{i,j-1} + T_{i+1,j} + T_{i-1,j} - 4T_{i,j} = 0. \tag{6.35}$$

Higher-order approximations for interior nodes and boundary nodes are also obtained in a similar manner.

Example 6.2. An example on heat distribution in a medium. A two-dimensional rectangular plate ($0 \le x \le 1$, $0 \le y \le 1$) is subjected to the uniform temperature boundary conditions (with top surface maintained at 100°C and all other surfaces at 0°C) shown in the figure below, that is, $T(0, y) = 0$, $T(1, y) = 0$, $T(x, 0) = 0$, and $T(x, 1) = 100°$ C. Suppose we are interested only in the values of the temperature at the nine interior nodal points (x_i, y_j), where $x_i = i\Delta x$ and $y_j = j\Delta y$, $i, j = 1, 2, 3$, with $\Delta x = \Delta y = \frac{1}{4}$.

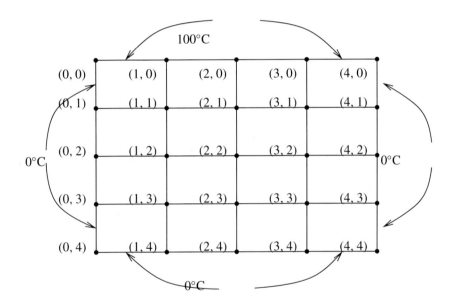

However, we assume symmetry for simplifying the problem. That is, we assume that $T_{33} = T_{13}$, $T_{32} = T_{12}$, and $T_{31} = T_{11}$. We thus have only six unknowns: (T_{11}, T_{12}, T_{13}) and (T_{21}, T_{22}, T_{23}) satisfying the following six equations:

$$\begin{aligned} 4T_{1,1} - 0 - 100 - T_{2,1} - T_{1,2} &= 0, \\ 4T_{2,1} - T_{1,1} - 100 - T_{1,1} - T_{2,2} &= 0, \\ 4T_{1,2} - 0 - T_{1,1} - T_{2,2} - T_{1,3} &= 0, \\ 4T_{2,2} - T_{1,2} - T_{2,1} - T_{1,2} - T_{2,3} &= 0, \\ 4T_{1,3} - 0 - T_{1,2} - T_{2,3} &= 0, \\ 4T_{2,3} - T_{1,3} - T_{2,2} - T_{1,3} - 0 &= 0. \end{aligned} \tag{6.36}$$

After suitable rearrangement, these equations can be written as follows:

$$\begin{pmatrix} 4 & -1 & -1 & 0 & 0 & 0 \\ -2 & 4 & 0 & -1 & 0 & 0 \\ 1 & 0 & 4 & -1 & -1 & 0 \\ 0 & -1 & -2 & 4 & 0 & -1 \\ 0 & 0 & -1 & 0 & 4 & -1 \\ 0 & 0 & 0 & -1 & -2 & 4 \end{pmatrix} \begin{pmatrix} T_{1,1} \\ T_{2,1} \\ T_{1,2} \\ T_{2,2} \\ T_{1,3} \\ T_{2,3} \end{pmatrix} = \begin{pmatrix} 100 \\ 100 \\ 0 \\ 0 \\ 0 \\ 0 \end{pmatrix}. \qquad (6.37)$$

The solution of this system will give us temperatures at the nodal points. ■

6.3.5 Approximation of a Function by a Polynomial: Hilbert System

In Chapter 4 (Section 4.6) we cited an ill-conditioned linear system with the Hilbert matrix. In this section we show how such a system arises. The discussion here has been taken from (Forsythe and Moler (1967, pp. 80–81)).

Suppose a continuous function $f(x)$ defined on the interval $0 \le x \le 1$ is to be approximated by a polynomial $\sum_{i=1}^{n} p_i x^{i-1}$ of degree $n-1$, such that the error

$$E = \int_0^1 \left[\sum_{i=1}^{n} p_i x^{i-1} - f(x) \right]^2 dx$$

is minimized. The coefficients p_i of the polynomial are easily determined by setting

$$\frac{\partial E}{\partial p_i} = 0, \qquad i = 1, \ldots, n.$$

(Note that the error is a differentiable function of the unknowns p_i and that a minimum occurs when all the partial derivatives are zero.) Thus we have

$$\frac{\partial E}{\partial p_i} = 2 \int_0^1 \left[\sum_{j=1}^{n} p_j x^{j-1} - f(x) \right] x^{i-1} dx = 0, \qquad i = 1, \ldots, n,$$

or

$$\sum_{j=1}^{n} \left(\int_0^1 x^{i+j-2} dx \right) p_j = \int_0^1 f(x) x^{i-1} dx, \qquad i = 1, \ldots, n.$$

(To obtain the latter form we have interchanged the summation and integration.)
Letting

$$h_{ij} = \int_0^1 x^{i+j-2} dx = \frac{1}{i+j-1}$$

6.4. LU Factorization Methods

and
$$b_i = \int_0^1 f(x)x^{i-1}\,dx \qquad (i = 1, 2, \ldots, n),$$
we have
$$\sum_{j=1}^n h_{ij} p_j = b_i, \qquad i = 1, \ldots, n.$$
That is, we obtain the linear system $Hp = b$, where $H = (h_{ij})$,
$$b = \begin{pmatrix} b_1 \\ b_2 \\ \vdots \\ b_n \end{pmatrix}.$$

The matrix H is easily identified as the **Hilbert matrix**.

6.4 LU Factorization Methods

Any factorization of the matrix A immediately suggests a method for solving $Ax = b$ or $Ax = B$ *(multiple right-hand sides)*.

In this section, we discuss LU factorization methods. The methods based on QR factorization and singular value decomposition (SVD) will be discussed in Chapter 7.

6.4.1 Solution of the System $Ax = b$ Using LU Factorization

We have seen in Chapter 5 that Gaussian elimination leads to the following factorizations:

- $A = LU$ (Gaussian elimination without pivoting)   (6.38)
- $PA = LU$ (Gaussian elimination with partial pivoting (GEPP))   (6.39)
- $PAQ = LU$ (Gaussian elimination with complete pivoting (GECP)).   (6.40)

These factorizations can then immediately be used to solve $Ax = b$.

Thus, if $A = LU$, then solving $Ax = b$ is equivalent to solving two triangular systems:
$$\begin{cases} Ly = b & \text{(lower triangular)}, \\ Ux = y & \text{(upper triangular)}. \end{cases} \qquad (6.41)$$

If $PA = LU$, then the system $Ax = b$ becomes
$$\begin{cases} Ly = Pb = b' & \text{(lower triangular)}, \\ Ux = y & \text{(upper triangular)}. \end{cases} \qquad (6.42)$$

If $PAQ = LU$, then $Ax = b$ is equivalent to
$$\begin{cases} Lz = Pb = b' & \text{(lower triangular)}, \\ Uy = z & \text{(upper triangular)}, \\ x = Qy. \end{cases} \qquad (6.43)$$

Since GEPP is stable in practice and widely used, we will only state this method for linear systems here.

ALGORITHM 6.1. Solving $Ax = b$ Using GEPP.

Inputs: $A \in \mathbb{R}^{n \times n}, b \in \mathbb{R}^{n \times 1}$.
Output: $x \in \mathbb{R}^{n \times 1}$ such that $Ax = b$.

Step 1. Find the factorization $PA = LU$ by the triangularization algorithm using partial pivoting (Algorithm 5.2).

Step 2. Obtain the solution x to $Ax = b$ as follows:

  2.1. Solve the lower triangular system: $Ly = Pb = b'$.

  2.2. Solve the upper triangular system: $Ux = y$.

Computing b': To compute b', all that is needed is to permute the entries of b according to the permutation indices of the matrix P.

For example, if

$$P = \begin{pmatrix} 0 & 0 & 1 \\ 0 & 1 & 0 \\ 1 & 0 & 0 \end{pmatrix} \text{ and } b = \begin{pmatrix} b_1 \\ b_2 \\ b_3 \end{pmatrix},$$

then

$$b' = Pb = \begin{pmatrix} b_3 \\ b_2 \\ b_1 \end{pmatrix}$$

is obtained just by permuting the first and third components of b.

Flop-count.

- Triangularization process: $\frac{2}{3}n^3$

- Solutions of two triangular systems: $2n^2$ (each system requires n^2 flops).

- Forming the vector b': no flops. (Note that b' is obtained from b just by re-shuffling the entries of b.)

- Total Flops: $\frac{2}{3}n^3 + 2n^2$.

- Furthermore, $O(n^2)$ comparisons will be required to identify the pivots.

6.4. LU Factorization Methods

Example 6.3. Solve $Ax = b$ using partial pivoting:

$$A = \begin{pmatrix} 1 & 2 & 4 \\ 4 & 5 & 6 \\ 7 & 8 & 9 \end{pmatrix}, \quad b = \begin{pmatrix} 7 \\ 15 \\ 24 \end{pmatrix}.$$

Step 1. *Factorization $PA = LU$.* Using the results of Example 5.14, we have

$$L = \begin{pmatrix} 1 & 0 & 0 \\ \frac{1}{7} & 1 & 0 \\ \frac{4}{7} & \frac{1}{2} & 1 \end{pmatrix}, \quad U = \begin{pmatrix} 7 & 8 & 9 \\ 0 & \frac{6}{7} & \frac{19}{7} \\ 0 & 0 & -\frac{1}{2} \end{pmatrix}, \quad P = \begin{pmatrix} 0 & 0 & 1 \\ 1 & 0 & 0 \\ 0 & 1 & 0 \end{pmatrix}.$$

Step 2.

2.1. *Solution of $Ly = b' \Rightarrow y = \begin{pmatrix} 24 \\ 3.5 \\ -0.5 \end{pmatrix}$.*

2.2. *Solution of $Ux = y \Rightarrow x = \begin{pmatrix} 1 \\ 1 \\ 1 \end{pmatrix}$.* ∎

Numerical Stability of the Partial Pivoting Scheme for $Ax = b$

We first state an approximate error bound in terms of the norms of L and U, and then in terms of the growth factor. (See the solution of Exercise 6.40 in Appendix D, as well as Chapter 14, both available online at *www.siam.org/books/OT116*.)

Theorem 6.4 (round-off property). *The computed solution $\hat{x}$ of the linear system $Ax = b$ using LU factorization, obtained by GEPP, satisfies*

$$(A + E)\hat{x} = b,$$

where $|E| \leq 3n\mu |L||U|$.

Round-off Property in Terms of the Growth Factor

In Chapter 5, we discussed stability of Gaussian elimination in terms of the growth factor. Let's therefore interpret the above result in terms of the growth factor for the partial pivoting. For this pivoting scheme, we have $|l_{ij}| \leq 1$ and $|u_{ij}| \leq \rho \max |a_{ij}|$, where ρ is the growth factor. Then from the above theorem, we obtain

$$\boxed{\|E\|_\infty \leq 3\rho n^3 \mu \|A\|_\infty.}$$

Remark. *The quantity $3\rho n^3$ very often grossly overestimates the true error $\|E\|_\infty$. The experiments have shown that $\|E\|_\infty$ is usually $O(\mu)\|A\|_\infty$. So, GEPP for linear systems is backward stable in practice.*

Furthermore, since the quantity ρ is never too large in practice, we conclude that *GEPP for $Ax = b$ is backward stable* for all practical purposes.

6.4.2 Solution of $Ax = b$ Using Factorization: $MA = U$

In Chapter 5, we have seen that as a first step of achieving the factorization $PA = LU$, one obtains the factorization $MA = U$.

This factorization then can directly be used to solve $Ax = b$.

Thus, if $MA = U$, then $Ax = b$ becomes $MAx = Mb$ or $Ux = Mb = b'$, where $M = M_{n-1} P_{n-1} \ldots M_1 P_1$.

Thus, we have the following process for solving $Ax = b$:

Solving $Ax = b$ Using Factorizaton: $MA = U$

Step 1. Obtain the factorization $MA = U$.

Step 2. Implicitly compute $b' = Mb$.

Step 3. Solve the upper triangular system $Ux = b'$.

Remarks. (i) Mathematically and computationally, Algorithm 6.1 and the above process are equivalent.

(ii) One can easily obtain a similar algorithm with complete pivoting.

6.4.3 Solution of $Ax = b$ without Explicit Factorization

It is possible that two steps of solving $Ax = b$ via LU factorization of A can be combined so that only one triangular (upper) system needs to be solved. This can be done by triangularizing the augmented matrix (A, b) rather than the matrix A, as shown below.

ALGORITHM 6.2. Solution of $Ax = b$ Using Partial Pivoting without Explicit Factorization.

A. Triangularization of (A, b).
Inputs: An $n \times n$ matrix A and an n-vector b.
Output: (i) The transformed upper triangular matrix stored in the upper triangular part of A, (ii) the transformed vector stored in b, and (iii) the multipliers stored in the lower-half part of A.

For $k = 1, 2, \ldots, n-1$ do

1. Choose the largest element in magnitude in the column k below the (k, k) entry; call it $a_{r_k,k}$:
$$a_{r_k,k} = \max\{|a_{ik}| : i \geq k\},$$
If $a_{r_k,k} = 0$, Stop.

6.4. LU Factorization Methods

> 2. Interchange the rows k and r_k of A and the kth and r_kth entries of b:
> $$\text{For } j = k, k+1, \ldots, n \text{ do}$$
> $$a_{r_k,j} \leftrightarrow a_{kj}$$
> $$b_{r_k} \leftrightarrow b_k.$$
> End
>
> 3. Form the multipliers:
> $$\text{For } i = k+1, \ldots, n$$
> $$a_{ik} \equiv m_{ik} = -\frac{a_{ik}}{a_{kk}}$$
> End
>
> 4. Update the entries of A in the rows and columns $(k+1)$ through n:
> $$\text{For } i = k+1, \ldots, n \text{ do}$$
> $$\text{For } j = k+1, \ldots, n \text{ do}$$
> $$a_{ij} \equiv a_{ij} + m_{ik} a_{kj}$$
> End
> End
>
> 5. Update the entries of b:
> $$\text{For } i = k+1, \ldots, n \text{ do}$$
> $$b_i \equiv b_i + m_{ik} b_k$$
> End
>
> End

B. Solution of the upper triangular system. Solve the upper triangular system with the upper triangular matrix and the transformed vector obtained from part A using the back substitution algorithm.

MATCOM Note: Algorithm 6.2 has been implemented in the MATCOM program **LINSYSWF**.

Example 6.5.
$$A = \begin{pmatrix} 0 & 1 & 1 \\ 2 & 2 & 3 \\ 4 & 1 & 1 \end{pmatrix}, \quad b = \begin{pmatrix} 2 \\ 6 \\ 3 \end{pmatrix}.$$

A. *Triangularization of (A, b) using GEPP.*

$k = 1$. The pivot entry is $a_{31} = 4$, $r_1 = 3$

Interchange rows 3 and 1 of A and the third and first entries of b:

$$A \equiv \begin{pmatrix} 4 & 1 & 1 \\ 2 & 2 & 3 \\ 0 & 1 & 1 \end{pmatrix}, \quad b \equiv \begin{pmatrix} 3 \\ 6 \\ 2 \end{pmatrix}, \quad m_{21} = -\frac{a_{21}}{a_{11}} = -\frac{1}{2},$$

$$A \equiv A^{(1)} = \begin{pmatrix} 4 & 1 & 1 \\ 0 & \frac{3}{2} & \frac{5}{2} \\ 0 & 1 & 1 \end{pmatrix}, \quad b \equiv b^{(1)} = \begin{pmatrix} 3 \\ \frac{9}{2} \\ 2 \end{pmatrix}.$$

k = 2. The pivot entry is $a_{22} = \frac{3}{2}$; $m_{32} = -\frac{a_{32}}{a_{22}} = -\frac{2}{3}$:

$$A \equiv A^{(2)} = \begin{pmatrix} 4 & 1 & 1 \\ 0 & \frac{3}{2} & \frac{5}{2} \\ 0 & 0 & -\frac{2}{3} \end{pmatrix}, \quad b \equiv b^{(2)} = \begin{pmatrix} 3 \\ \frac{9}{2} \\ -1 \end{pmatrix}.$$

B. *Solution of the upper triangular system.* The back substitution process applied to the triangular system $A^{(2)}x = b^{(2)}$ yields $x_3 = \frac{3}{2}$, $x_2 = \frac{1}{2}$, $x_1 = \frac{1}{4}$, giving $x = \left(\frac{1}{4}, \frac{1}{2}, \frac{3}{2}\right)^T$. ∎

6.4.4 Solving a Linear System with Multiple Right-Hand Sides

Consider the problem

$$AX = B,$$

where $B = (b_1, \ldots, b_m)$ is an $n \times m$ matrix ($m \leq n$) and $X = (x_1, x_2, \ldots, x_m)$. Here b_i and x_i, $i = 1, \ldots, m$, are n-vectors.

Problems of this type arise in many practical applications. For some applications in control, see Arnold (1992), Datta (2003), and Datta and Saad (1991).

The idea will be to factorize A just once and then use this factorization to solve all the subsequent triangular systems. Thus if $PA = LU$, then $AX = B$ is equivalent to two triangular systems each having m equations:

$$LZ = PB = B' \quad \text{and} \quad UX = Z.$$

Set $Z = (z_1, \ldots, z_m)$ and $B' = (b'_1, b'_2, \ldots, b'_m)$.

ALGORITHM 6.3. Solving $AX = B$ (Linear System with Multiple Right-Hand Sides) Using GEPP.

Inputs: $A \in \mathbb{R}^{n \times n}$ and $B \in \mathbb{R}^{n \times m}$.
Output: $X \in \mathbb{R}^{n \times m}$ such that $AX = B$.

Step 1. Factorize A using Gaussian elimination with partial pivoting: $PA = LU$.

Step 2. Solve the m lower triangular systems: $Lz_i = Pb_i = b'_i$, $i = 1, 2, \ldots, m$.

Step 3. Solve the m upper triangular systems: $Ux_i = z_i$, $i = 1, \ldots, m$.

Step 4. Form $X = (x_1, x_2, \ldots, x_m)$.

Flop-count. Algorithm 6.3 requires approximately $2(\frac{n^3}{3} + mn^2)$ flops.

6.5. Scaling

Example 6.6. Solve $AX = B$, where

$$A = \begin{pmatrix} 1 & 2 & 4 \\ 4 & 5 & 6 \\ 7 & 8 & 9 \end{pmatrix}, \qquad B = \begin{pmatrix} 1 & 2 \\ 3 & 4 \\ 5 & 6 \end{pmatrix},$$

Step 1. Using the results of Example 5.14,

$$U = \begin{pmatrix} 7 & 8 & 9 \\ 0 & \frac{6}{7} & \frac{19}{7} \\ 0 & 0 & -\frac{1}{2} \end{pmatrix}, \qquad L = \begin{pmatrix} 1 & 0 & 0 \\ \frac{1}{7} & 1 & 0 \\ \frac{4}{7} & \frac{1}{2} & 1 \end{pmatrix}, \qquad P = \begin{pmatrix} 0 & 0 & 1 \\ 1 & 0 & 0 \\ 0 & 1 & 0 \end{pmatrix}.$$

Step 2. *Solve the two lower triangular systems*

$$\text{(i)} \; Lz_1 = b'_1 \Rightarrow z_1 = \begin{pmatrix} 5 \\ \frac{2}{7} \\ 0 \end{pmatrix}; \quad \text{(ii)} \; Lz_2 = b'_2 \Rightarrow z_2 = \begin{pmatrix} 6 \\ \frac{8}{7} \\ 0 \end{pmatrix}.$$

Step 3. *Solve the two upper triangular systems*

$$Ux_1 = z_1 \Rightarrow x_1 = \begin{pmatrix} \frac{1}{3} \\ \frac{1}{3} \\ 0 \end{pmatrix}; \quad Ux_2 = z_2 \Rightarrow x_2 = \begin{pmatrix} -\frac{2}{3} \\ \frac{4}{3} \\ 0 \end{pmatrix}.$$

Step 4. *Form*

$$X = \begin{pmatrix} \frac{1}{3} & -\frac{2}{3} \\ \frac{1}{3} & \frac{4}{3} \\ 0 & 0 \end{pmatrix}.$$

Note: The vectors $b'_1, \ldots, b'_m$ are obtained just by reshuffling the columns of matrix B according to the permutation indices of P. No matrix multiplication is necessary. ∎

6.5 Scaling

If the entries of matrix A vary widely, then there is a possibility that a very small number needs to be added to a very large number during the process of elimination. This can influence the accuracy greatly, because, "the big one can kill the small one." To circumvent this difficulty, often it is suggested that the rows of A be properly scaled before the elimination process begins. The following simple example illustrates this.

Consider the system

$$\begin{pmatrix} 10 & 10^6 \\ 1 & 1 \end{pmatrix} \begin{pmatrix} x_1 \\ x_2 \end{pmatrix} = \begin{pmatrix} 10^6 \\ 2 \end{pmatrix}.$$

Now apply GEPP. Since 10 is the largest entry in the first column, no interchange is needed. We have, after the first step of elimination,

$$\begin{pmatrix} 10 & 10^6 \\ 0 & -10^5 \end{pmatrix} \begin{pmatrix} x_1 \\ x_2 \end{pmatrix} = \begin{pmatrix} 10^6 \\ 2 - 10^5 \end{pmatrix},$$

which gives $x_2 = 1$, $x_1 = 0$. The exact solution, however, is $\approx \binom{1}{1}$. This happened because "10" was indeed a "false pivot." Note that if the first equation was multiplied by 10^{-6}, then the matrix of the system would be

$$\begin{pmatrix} 10^{-5} & 1 \\ 1 & 1 \end{pmatrix}.$$

Therefore, even choosing the false pivot 10 did not help us. However, if the scaled system is now solved (after modifying the first entry of b appropriately) using partial pivoting, we will then have the accurate solution, as we have seen before.

Scaling of the rows of matrix A is equivalent to finding an invertible diagonal matrix D_1 so that the largest element (in magnitude) in each row of $D_1^{-1}A$ is about the same size. Once such D_1 is found, the solution of the system $Ax = b$ is found by solving the scaled system $\tilde{A}x = \tilde{b}$, where

$$\tilde{A} = D_1^{-1}A, \qquad \tilde{b} = D_1^{-1}b.$$

The process can be easily extended to scale both the rows and columns of A. Mathematically, this is equivalent to finding diagonal matrices D_1 and D_2 such that the largest (in magnitude) element in each row and column of $D_1^{-1}AD_2$ lies between two fixed numbers, say, $\frac{1}{\beta}$ and 1, where β is the base of the number system. Once such D_1 and D_2 are found, the solution of the system $Ax = b$ is obtained by solving the equivalent system $\tilde{A}y = \tilde{b}$, and then computing $x = D_2 y$, where $\tilde{A} = D_1^{-1}AD_2$, $\tilde{b} = D_1^{-1}b$. The above process is known as **equilibration** (Forsythe and Moler (1967)).

Note that the *purpose of scaling is to make the condition number of the scaled matrix $D_1^{-1}AD_2$ considerably smaller than that of A.* In doing so, we might expect a more accurate solution. See more on this in Section 6.8.2.

Thus, *scaling or equilibration is recommended in general, when the entries of the matrix vary widely.* "The round-off error analysis for Gaussian elimination gives the most effective results when a matrix is equilibrated." (Forsythe and Moler (1967))

6.6 Concluding Remarks on the Use of Gaussian Elimination for Linear Systems

- *Gaussian elimination with partial pivoting* is a computationally effective practical scheme for solving modest-size arbitrary linear systems problems. *It is stable in practice and efficient.*

- Gaussian elimination without pivoting should not be used unless matrix A is symmetric positive definite or strictly diagonally dominant (see Section 6.12).

- Scaling is recommended prior to the use of Gaussian elimination if the entries of matrix A vary widely.

6.7 Inverses and Determinant

Associated with the problem of solving the linear system $Ax = b$ are the problems of finding the *determinant and the inverse of the matrix A*. In this section we will see how the determinant and the inverse can be computed using LU factorization.

6.7. Inverses and Determinant

6.7.1 Avoiding Explicit Computation of the Inverses

The inverse of a matrix A very seldom needs to be computed explicitly. Most computational problems involving inverses can be reformulated in terms of solution of linear systems. For example, computing

- $A^{-1}b$ (inverse times a vector) is equivalent to solving the system $Ax = b$;

- $A^{-1}B$ (inverse times a matrix) is equivalent to solving the systems $Ac_i = b_i$, $i = 1, \ldots, m$, if $B = (b_1, b_2, \ldots, b_m)$;

- $b^T A^{-1} c$ (vector times inverse times a vector) is equivalent to solving the system $Ax = c$ followed by computing $b^T x$.

As we will see later in this section, *computing A^{-1} is much more expensive than solving the linear system $Ax = b$.* Thus, all such problems mentioned above can be solved much more efficiently by formulating them in terms of linear systems rather than naively solving them using the matrix inversion.

> *The explicit computation of the inverse should be avoided whenever possible. A linear system should never be solved by explicit computation of the inverse of the system matrix.*

Some Easily Computed Inverses

Before we discuss the computation of A^{-1} for an arbitrary matrix A, we note that the inverses of some special matrices can be easily computed. Here are some examples:

- The inverse of the elementary lower triangular matrix $M = I - me_k^T$ is given by $M^{-1} = I + me_k^T$.

- The inverse of an orthogonal matrix Q is its transpose Q^T.

- The inverse of a nonsingular lower (upper) triangular matrix T is again a lower (upper) triangular matrix and the diagonal entries of the inverse are the reciprocals of the diagonal entries of the matrix T.

6.7.2 The Sherman–Morrison Formula for Matrix Inverse

In many applications once the inverse of a matrix A is computed, it is necessary to find the inverse of another matrix B which differs from A only by a rank-one perturbation. The question naturally arises, *Can the inverse of B be computed without starting all over again?* That is, the question is whether the inverse of B can be found using the inverse of A which has already been computed. The Sherman–Morrison formula shows us how to do this.

Theorem 6.7 (the Sherman–Morrison formula). *If u and v are two n-vectors and A is a nonsingular matrix, then*

$$(A - uv^T)^{-1} = A^{-1} + \alpha(A^{-1}uv^T A^{-1}),$$

where
$$\alpha = \frac{1}{(1 - v^T A^{-1} u)} \quad \text{if } v^T A^{-1} u \neq 1.$$

MATCOM Note: The Sherman–Morrison formula has been implemented in the MATCOM program **SHERMOR**.

Example 6.8. Given
$$A = \begin{pmatrix} 1 & 1 & 1 \\ 2 & 4 & 5 \\ 6 & 7 & 8 \end{pmatrix} \quad \text{and} \quad A^{-1} = \begin{pmatrix} -3 & -1 & 1 \\ 14 & 2 & -3 \\ -10 & -1 & 2 \end{pmatrix}$$
find $(A - uv^T)^{-1}$, where $u = v = (1, 0, 0)^T$.
$$\alpha = \frac{1}{1 - v^T A^{-1} u} = \frac{1}{4}.$$
Thus,
$$(A - uv^T)^{-1} = A^{-1} + \alpha A^{-1} uv^T A^{-1} = \begin{pmatrix} -\frac{3}{4} & -\frac{1}{4} & \frac{1}{4} \\ \frac{7}{2} & -\frac{3}{2} & \frac{1}{2} \\ -\frac{5}{2} & \frac{3}{2} & -\frac{1}{2} \end{pmatrix}. \quad \blacksquare$$

6.7.3 Computing the Inverse of an Arbitrary Nonsingular Matrix

If A is an $n \times n$ nonsingular matrix, then *finding A^{-1} is equivalent to computing X such that $AX = I$, where I is an $n \times n$ matrix.* Thus if $X = (x_1, x_2, \ldots, x_n)$ and $I = (e_1, e_2, \ldots, e_n)$, then $AX = I$ is equivalent to solving n linear systems: $Ax_i = e_i$, $i = 1, \ldots, n$.

If partial pivoting is used to solve these n systems, then we have the following algorithm to compute A^{-1}.

ALGORITHM 6.4. Computing A^{-1} by GEPP.

Input: $A \in \mathbb{R}^{n \times n}$.
Output: A^{-1}.

Step 1. Using Algorithm 6.3, solve n linear systems: $Ax_i = e_i$, $i = 1, \ldots, n$.
Step 2. From $A^{-1} = X = (x_1, x_2, \ldots, x_n)$.

Equivalently, one can compute A^{-1} directly from LU factorization of A. Thus if GEPP is used, then
$$PA = LU, \text{ so } A^{-1} = U^{-1} L^{-1} P.$$

MATCOM Note: MATCOM programs **INLU, INPARPIV,** and **INCOMPIV** compute the inverse using, respectively, LU factorization with no, partial, and complete pivoting.

6.8. Effect of the Condition Number on Accuracy of the Computed Solution

Flop-count. About $\frac{8n^3}{3}$ flops are needed to compute A^{-1} using Algorithm 6.4.

Example 6.9. Consider A of Example 6.6. Using matrices L, U, and P from there, we obtain

$$\begin{bmatrix} Lz_1 = Pe_1 = (0, 1, 0)^T \Rightarrow z_1 = (0, 1, -0.5000)^T, \\ Lz_2 = Pe_2 = (0, 0, 1)^T \Rightarrow z_2 = (0, 0, 1)^T, \\ Lz_3 = Pe_3 = (1, 0, 0)^T \Rightarrow z_3 = (1, -0.1429, -0.50000)^T, \end{bmatrix}$$

$$\begin{bmatrix} Ux_1 = z_1 \Rightarrow x_1 = (1, -2, 1)^T, \\ Ux_2 = z_2 \Rightarrow x_2 = (-4.6670, 6.3333, -2.0000)^T, \\ Ux_3 = z_3 \Rightarrow x_3 = (2.6667, -3.3333, 1.0000)^T. \end{bmatrix}$$

Thus

$$A^{-1} = X = (x_1, x_2, x_3) = \begin{pmatrix} 1 & -4.6670 & 2.6667 \\ -2 & 6.3333 & -3.3333 \\ 1 & -2 & 1.0000 \end{pmatrix}. \blacksquare$$

6.7.4 Computing the Determinant of a Matrix

The determinant of a matrix A is seldom needed in practice. However, if it has to be computed, LU factorization of A again can be used. Thus, if GEPP is used giving $PA = LU$, then $\det(A) = \det(P) \cdot \det(L) \cdot \det(U)$. Now, $\det(P) = (-1)^r$, where r is the number of row interchanges, $\det(L) = 1$, and $\det(U) = a_{11} a_{22}^{(1)} \ldots a_{nn}^{(n-1)}$. Thus,

$$\det(A) = (-1)^r a_{11} a_{22}^{(1)} \cdots a_{nn}^{(n-1)},$$

where r is the number of interchanges.

Example 6.10.

$$A = \begin{pmatrix} 0 & 1 & 1 \\ 1 & 2 & 3 \\ 1 & 1 & 1 \end{pmatrix}.$$

GEPP gives

$$U = \begin{pmatrix} 1 & 2 & 3 \\ 0 & 1 & 1 \\ 0 & 0 & -1 \end{pmatrix}.$$

There was only one interchange; therefore $r = 1$. $\det(A) = (-1) \det(U) = (-1)(-1) = 1$. $\blacksquare$

6.8 Effect of the Condition Number on Accuracy of the Computed Solution

In Chapter 4, we identified the condition number of the system $Ax = b$ by means of perturbation analysis (Theorems 4.19–4.25). Here we discuss the role of the condition number in the accuracy of the solution. Once a solution $\hat{x}$ of the system $Ax = b$ has been computed, it is natural to test how accurate the computed solution $\hat{x}$ is. If the exact solution x is known, then one could, of course, compute the relative error $\frac{\|x-\hat{x}\|}{\|x\|}$ to test the accuracy of $\hat{x}$. However, in most practical situations, the exact solution is not known. In such cases, the most obvious thing to do is to compute the residual $r = b - A\hat{x}$ and see how small the relative residual $\frac{\|r\|}{\|b\|}$ is. Interestingly, we should note that the solution obtained by the Gaussian elimination process in general produces a small residual. (Why?) Unfortunately, a small relative residual does not guarantee the accuracy of the solution. The following example illustrates this fact.

Example 6.11. Let
$$A = \begin{pmatrix} 1.0001 & 1 \\ 1 & 1 \end{pmatrix}, \quad b = \begin{pmatrix} 2.0001 \\ 2 \end{pmatrix}.$$
Let $\hat{x} = \begin{pmatrix} 0 \\ 2 \end{pmatrix}$. Then $r = b - A\hat{x} = \begin{pmatrix} 0.0001 \\ 0 \end{pmatrix}$.

Note that r is small. However, the vector $\hat{x}$ is nowhere close to the exact solution $x = \begin{pmatrix} 1 \\ 1 \end{pmatrix}$. ∎

The above phenomenon can be explained mathematically from the following theorem. The proof can be easily worked out. Theorem 6.12 below gives a **posterior error bound** for the computed solution.

Theorem 6.12 (residual theorem).
$$\frac{\|\hat{x} - x\|}{\|x\|} \leq \text{Cond}(A) \frac{\|r\|}{\|b\|}.$$

Interpretation of Theorem 6.12. Theorem 6.12 tells us that the relative error in x depends not only on the relative residual but also on the condition number of matrix A as well. *The computed solution can be guaranteed to be accurate only when the product of both* Cond(A) *and the relative residual is small.*

Note that in the above example, Cond(A) $= 4.0002 \times 10^4$ (large!). Thus, though the relative residual was small, the computed solution $\hat{x}$ was inaccurate, because Cond(A) is relatively large.

6.8.1 Conditioning and Pivoting

It is natural to wonder if ill-conditioning can be detected during the triangularization process using GEPP. *By a normalized matrix here we mean that* $\|A\|_2 = 1$. Suppose that A and b have been normalized. Then there are certain symptoms for ill-conditioning. These include the following.

Symptoms for Ill-Conditioning

- A small pivot.
- A large computed solution.
- A large residual vector.

Justification. Suppose there is a small pivot. Then this will make A^{-1} large. (Note that if partial pivoting is used, then $A^{-1} = U^{-1}L^{-1}P$.) Similarly, if the computed solution $\hat{x}$ is large, then from $A\hat{x} = \hat{b}$, we have $\|\hat{x}\| = \|A^{-1}\hat{b}\| \leq \|A^{-1}\| \|\hat{b}\|$, showing that $\|A^{-1}\|$ is possibly large. Large $\|A^{-1}\|$, of course, means ill-conditioning, because $\text{Cond}(A) = \|A\| \|A^{-1}\|$ will then be large.

Remark. There are matrices which do not have any of these symptoms but are still ill-conditioned (see Wilkinson (1965, pp. 254–255)).

Example 6.13. Consider the linear system $Ax = b$ with

$$A = \begin{pmatrix} 1 & 0 & 0 \\ 0 & 0.00001 & 0 \\ 0 & 0 & 0.00001 \end{pmatrix}, \quad b = \begin{pmatrix} 0.1 \\ 0.1 \\ 0.1 \end{pmatrix}.$$

Then

$$x = 10^4 \begin{pmatrix} 0.00001 \\ 1 \\ 1 \end{pmatrix}, \text{ which is } \textit{quite large.}$$

$$A^{-1} = 10^5 \begin{pmatrix} 0.00001 & 0 & 0 \\ 0 & 1 & 0 \\ 0 & 0 & 1 \end{pmatrix}, \text{ which is } \textit{large.}$$

Thus, for this example (i) the *computed solution is large,* and (ii) A^{-1} *is large.* A is, *therefore, likely to be ill-conditioned.* It is indeed true: $\text{Cond}(A) = 10^5$. ∎

6.8.2 Conditioning and Scaling

In Section 6.5 we discussed *scaling*, and the message there was that *scaling is in general recommended if the entries of the matrix A vary widely.* Scaling followed by a strategy of pivoting is helpful. We noted there that *scaling has an effect on the condition number of the matrix.* For example, consider

$$A = \begin{pmatrix} 10 & 10^6 \\ 1 & 1 \end{pmatrix}, \quad \text{Cond}(A) = 10^6.$$

However, if the first row of A is scaled to obtain $\tilde{A} = \begin{pmatrix} 0.0001 & 1 \\ 1 & 1 \end{pmatrix}$, then $\text{Cond}(\tilde{A}) = 2$.

The question naturally arises, *Given a matrix A, how can one choose the diagonal matrices D_1 and D_2 such that $\text{Cond}(D_1^{-1} A D_2)$ will be as small as possible?*

This is a complex problem. Some of the historical and well-known papers on this topic include Bauer (1963, 1965), Businger (1968), Skeel (1979, 1981), and van der Sluis (1969). Chapter 7 of Higham (2002) gives a thorough treatment.

6.9 Computing and Estimating the Condition Number

The obvious way to compute the condition number will be to compute it from its definition:

1. Compute A^{-1}.
2. Compute $\|A\|$ and $\|A^{-1}\|$ and multiply them.

We have seen that computing the inverse of A requires about $\frac{8n^3}{3}$ flops. Thus, finding the condition number by explicitly finding A^{-1} is much more expensive than finding the solution of $Ax = b$ itself. On the other hand, to compute $\text{Cond}(A)$, we only need to know $\|A^{-1}\|$, not the inverse itself. *Furthermore, the exact value of* $\text{Cond}(A)$ *itself is seldom needed; an estimate is sufficient.* The question, therefore, arises whether we can get a reasonable estimate of $\|A^{-1}\|$ without computing the inverse of A explicitly. We present an optimization-based algorithm below.

An Optimization Technique for Estimating $\|A^{-1}\|_1$

Hager (1984) has proposed a method for estimating $\|A^{-1}\|$ based on an optimization technique. This technique seems to be quite suitable for randomly generated matrices. Let $A^{-1} = B = (b_{ij})$.

Define a function $f(x)$:

$$f(x) = \|Bx\|_1 = \sum_{i=1}^{n} \left| \sum_{j=1}^{n} b_{ij} x_j \right|.$$

Then

$$\|B\|_1 = \|A^{-1}\|_1 = \max\{f(x) : \|x\|_1 = 1\}.$$

Thus, the problem is to find maximum of the convex function f over the convex set

$$S = \{x \in R^n : \|x\|_1 \leq 1\}.$$

ALGORITHM 6.5. Hager's Norm-1 Condition Number Estimator.

Inputs: An $n \times n$ matrix A and an n-vector b.
Output: An approximation of $\|A^{-1}\|_1$.

Step 0. Set $\rho = \|A^{-1}\|_1 = 0, b = \left(\frac{1}{n}, \frac{1}{n}, \ldots, \frac{1}{n}\right)^T$.

Step 1. Solve $Ax = b$.

Step 2. Test if $\|x\| \leq \rho$. If so, go to Step 6. Otherwise set $\rho = \|x\|_1$ and go to Step 3.

Step 3. Solve $A^T z = y$, where
$$y_i = 1 \quad \text{if } x_i \geq 0, \qquad y_i = -1 \text{ if } x_i < 0.$$

Step 4. Set $j = \arg\max\{|z_i|, i = 1, \ldots, n\}$.

Step 5. If $|z_j| > z^T b$, update $b \equiv e_j$, where e_j is the jth unit vector and return to Step 1. Else go to Step 6.

Step 6. Set $\|A^{-1}\|_1 \approx \rho$. Then compute $\text{Cond}_1(A) = \rho\|A\|_1$.

6.10. Componentwise Perturbations and the Errors

It is well known that the maximum of a convex function is obtained at an extreme point. Hager's method consists in finding this maximum systematically. *Hager (1984) remarks that the algorithm usually stops after two iterations.* An excellent survey of different condition number estimators, including Hager's, and their performance has been given by Higham (2002, Chapter 15). See also earlier papers of Higham (1987, 1990) and O'Leary (1980).

MATCOM, MATLAB, and LAPACK Notes: Algorithm 6.5 has been implemented in the MATCOM program **HAGCOND1**. A block 1-norm estimator due to Higham and Tisseur (2000) is available in MATLAB function **normest1**. MATLAB function **condest** computes a lower bound c for the 1-norm condition number of a square matrix. **rcond** is a LAPACK reciprocal condition estimator. **condest** invokes **normest1**.

Example 6.14. Let

$$A = \begin{pmatrix} 1 & 2 & 3 \\ 3 & 4 & 5 \\ 6 & 7 & 8 \end{pmatrix}, \quad \text{Cond}_1(A) = 2.9575 \times 10^{17}, \text{RCOND} = 3.469447 \times 10^{-18}$$

(A is close to a singular matrix).

Step 0. $b = \left(\frac{1}{3}, \frac{1}{3}, \frac{1}{3}\right)^T$.

Iteration 1. $x = (1.0895, -2.5123, 1.4228)^T$;
$\rho = 5.0245$, $y = (1, -1, 1)^T$, $z = 10^6(2.0271, -3.3785, 1.3514)^T$,
$j = 2$, $|z|_2 = 10^{16}(3.3785) > z^T b = -1.3340$.

Iteration 2. $x = 10^{17}(-1.3564, 2.7128, -1.3564)^T$, $\|x\|_1 = 5.4255 \times 10^{17}$.

Remark. Since $\|x\|_1 > \rho$, we take $\|A^{-1}\|_1 = 5.4255 \times 10^{17}$.
This is an excellent estimate of $\|A^{-1}\|_1$.
Note that $\text{Cond}_1(A) = \rho \times \|A\|_1 = 8.6808 \times 10^{18}$, $\|A^{-1}\|_1 = 1.8014 \times 10^{16}$, and **condest** $(A) = 2.8823 \times 10^{17}$. ■

6.10 Componentwise Perturbations and the Errors

If the componentwise bounds of the perturbations are known, then the following perturbation result obtained by Skeel (1979) holds. In the following, $\|\cdot\|$ stands for infinity norm.

Theorem 6.15. *Let* $Ax = b$ *and* $(A + \Delta A)(x + \Delta x) = b + \delta b$. *Let* $|\Delta A| \leq \epsilon |A|$ *and* $|\delta b| \leq \epsilon |b|$. *Then*

$$\frac{\|\delta x\|}{\|x\|} \leq \epsilon \frac{\||A^{-1}||A||x| + |A^{-1}||b|\|}{(1 - \epsilon \||A^{-1}||A|\|)\|x\|}.$$

Definition 6.16. *The number* $\text{Cond}(A, x) = \frac{\||A^{-1}||A||x|\|}{\|x\|}$ *will be called* **Skeel's condition number** *and* $\text{Cond}_s(A) = \||A^{-1}||A|\|$ *the upper bound of Skeel's condition number.*

An important property of $\text{Cond}(A, x)$: Skeel's condition number is invariant under row-scaling. It can, therefore, be much smaller than the usual condition number $\text{Cond}(A)$. $\text{Cond}(A, x)$ is useful when the column norms of A^{-1} vary widely.

6.11 Iterative Refinement

Suppose a computed solution $\hat{x}$ of the system $Ax = b$ is not acceptable. It is then natural to wonder if $\hat{x}$ can be refined *cheaply* by making use of the triangularization of the matrix A already available.

The following process, known as **iterative refinement**, can be used to refine $\hat{x}$ iteratively up to some desirable accuracy.

Iterative Refinement Algorithm

The process is based on the following simple idea:

Let $\hat{x}$ be a computed solution of the system $Ax = b$. If $\hat{x}$ were an exact solution, then $r = b - A\hat{x}$ would be zero. But in practice we cannot expect that. Let us now try to solve the system again with the computed residual $r (\neq 0)$; that is, let c satisfy

$$Ac = r.$$

Then, $y = \hat{x} + c$ is the exact solution of $Ax = b$, provided that c is the exact solution of $Ac = r$, because

$$Ay = A(\hat{x} + c) = A\hat{x} + Ac = b - r + r = b.$$

It is true that c again will not be an exact solution of $Ac = r$ in practice; however, the above discussion suggests that y might be a better approximation than $\hat{x}$. If so, we can continue the process until a desired accuracy is achieved.

ALGORITHM 6.6. Iterative Refinement.

Inputs: $A \in \mathbb{R}^{n \times n}$, $b \in \mathbb{R}^{n \times 1}$, and tolerance ϵ.
Output: A refined solution.

$$\text{Set } x^{(0)} = \hat{x}.$$

For $k = 0, 1, 2, \ldots$ do

1. Compute the *residual vector* $r^{(k)}$: $r^{(k)} = b - Ax^{(k)}$.

2. Calculate the *correction vector* $c^{(k)}$ by solving the system $Ac^{(k)} = r^{(k)}$, using the same triangularization of A that was used to obtain $x^{(0)}$.

3. *Update the solution:* $x^{(k+1)} = x^{(k)} + c^{(k)}$.

4. *Test for the convergence:* If $\dfrac{\|x^{(k+1)} - x^{(k)}\|_2}{\|x^{(k)}\|_2} < \epsilon$, stop.

End

Remark. If the system is not too ill-conditioned and double precision is used in computing the residuals, then the iterative refinement using Gaussian elimination with pivoting will

6.11. Iterative Refinement

ultimately produce a very accurate solution. The rate of convergence depends upon the condition number (see Higham (2002, Chapter 12)).

MATCOM and LAPACK Notes: Algorithm 6.6 has been implemented in the MATCOM program **ITERREF**.

Example 6.17.

$$A = \begin{pmatrix} 1 & 1 & 0 \\ 0 & 2 & 1 \\ 0 & 0 & 3 \end{pmatrix}, \quad b = \begin{pmatrix} 0.0001 \\ 0.0001 \\ -1.666 \end{pmatrix}.$$

The exact solution is

$$x = \begin{pmatrix} -0.2777 \\ 0.2778 \\ -0.5555 \end{pmatrix} \text{ (correct up to four figures)}.$$

$$x^{(0)} = \begin{pmatrix} 1 \\ 1 \\ 1 \end{pmatrix}.$$

$k = 0$:

$$r^{(0)} = b - Ax^{(0)} = \begin{pmatrix} -1.9999 \\ -2.9999 \\ -4.6666 \end{pmatrix}.$$

The solution of $Ac^{(0)} = r^{(0)}$:

$$c^{(0)} = \begin{pmatrix} -1.2777 \\ -0.7222 \\ -1.5555 \end{pmatrix},$$

$$x^{(1)} = x^{(0)} + c^{(0)} = \begin{pmatrix} -0.2777 \\ 0.2778 \\ -0.5555 \end{pmatrix}.$$

Note that Cond$(A) = 3.8078$. A is *well-conditioned*. ∎

Accuracy Obtained by Iterative Refinement

Suppose that the iteration converges. Then the error at $(k + 1)$th step will be less than the error at the kth step.

Let

$$\frac{\|\hat{x} - x^{(k+1)}\|}{\|\hat{x}\|} \leq c \frac{\|\hat{x} - x^{(k)}\|}{\|\hat{x}\|}.$$

Then if $c \approx 10^{-s}$, *there will be a gain of approximately s figures per iteration*.

Flop-count. The procedure is quite cheap. Since A has already been triangularized to solve the original system $Ax = b$, each iteration requires only $O(n^2)$ flops.

Remarks. Iterative refinement is a very useful technique. GEPP followed by iterative refinement is the most practical approach for solving a linear system accurately. Skeel (1980) has shown that in most cases even one step of iterative refinement is sufficient. See also Higham (2002) and Jankowski and Wozniakowski (1977).

Estimating the condition number from iterative refinement. A very crude estimate of the condition number of matrix A is $10^{t(1-\frac{1}{k})}$ (Rice (1981)), where t is the number of digits and k is the number of iterations performed for the procedure to converge.

Example 6.18 (Stewart (1973, pp. 205)). Consider solving the ill-conditioned system with

$$A = \begin{pmatrix} 7 & 6.990 \\ 4 & 4 \end{pmatrix}, \quad b = \begin{pmatrix} 34.97 \\ 20.00 \end{pmatrix}$$

(the system is ill-conditioned because $\text{Cond}_2(A) = 3.2465 \times 10^3$).

The exact solution is $x = \begin{pmatrix} 2 \\ 3 \end{pmatrix}$.

Let $x^{(0)} = \begin{pmatrix} 1.667 \\ 3.333 \end{pmatrix}$.

$k = 0$:

$$r^{(0)} = b - Ax^{(0)} = \begin{pmatrix} 0.333 \times 10^{-2} \\ 0 \end{pmatrix}.$$

The solution of $Ac^{(0)} = r^{(0)}$:

$$c^{(0)} = \begin{pmatrix} 0.3330 \\ -0.3330 \end{pmatrix},$$

$$x^{(1)} = x^{(0)} + c^{(0)} = \begin{pmatrix} 2 \\ 3 \end{pmatrix}. \quad \blacksquare$$

6.12 Special Systems: Positive Definite, Diagonally Dominant, Hessenberg, and Tridiagonal

In this subsection we will study numerical solutions of the following special systems:

- *Symmetric positive definite system.*

- *Hessenberg system.*

- *Diagonally dominant system.*

- *Tridiagonal and block tridiagonal system.*

Indeed, it is very often said by practicing engineers that there are hardly any systems in practical applications which are not one of the above types. These systems therefore deserve a special treatment. We first give some examples to show how these systems arise in practical applications.

6.12.1 Special Linear Systems Arising from Finite Difference Methods

We have seen in the last section how finite difference schemes for solving partial differential equations lead to linear systems problems. Many times such systems have special properties and structures: some well-known structured matrices arising widely in applications include **tridiagonal**, **diagonally dominant**, **positive definite**, and **block tridiagonal**. We first discuss a situation which gives rise to a tridiagonal system.

A. Tridiagonal Systems

Consider the *one-dimensional* steady conduction of heat such as heat conduction through a wire. In such a case, the temperature remains constant with respect to time. The equation here is $\frac{\partial^2 T}{\partial x^2} = 0$. The difference analogue of this equation is

$$T(x + \Delta x) - 2T(x) + T(x - \Delta x) = 0,$$

where Δx is the increment in x, as shown below.

$$|x_{i+1} - x_i| = \Delta x, \quad i = 0, 1, 2, 3.$$

Using a similar numbering scheme as in Section 6.3.4, the temperature T_i at any point x_i is given by

$$T_{i+1} - 2T_i + T_{i-1} = 0;$$

that is, *the temperature at any point is just the average of the temperatures of the two nearest neighboring points.*

Suppose the domain of the problem is $0 \leq x \leq 1$. Divide now the domain into four segments of equal length. Thus $\Delta x = 0.25$. Suppose that we know the temperature at the end points $x_0 = 0$ and $x_4 = 1$, that is,

$$T_0 = \alpha_1 \quad \text{and} \quad T_4 = \alpha_2.$$

These are then the boundary conditions of the problem.

From the above equations, the temperature at each node $x_0 = 0$, $x_1 = \Delta x$, $x_2 = 2\Delta x$, $x_3 = 3\Delta x$, $x_4 = 1$ is calculated as follows:

$$
\begin{array}{lll}
\text{At } x_0 = 0, & T_0 = \alpha_1 & \text{(given)}, \\
\text{At } x_1 = \Delta x, & T_0 - 2T_1 + T_2 = 0, & \\
\text{At } x_2 = 2\Delta x, & T_1 - 2T_2 + T_3 = 0, & \\
\text{At } x_3 = 3\Delta x, & T_2 - 2T_3 + T_4 = 0, & \\
\text{At } x_4 = 1, & T_4 = \alpha_2 & \text{(given)}.
\end{array}
$$

In matrix form these equations can be written as

$$\begin{pmatrix} 1 & 0 & 0 & 0 & 0 \\ 1 & -2 & 1 & 0 & 0 \\ 0 & 1 & -2 & 1 & 0 \\ 0 & 0 & 1 & -2 & 1 \\ 0 & 0 & 0 & 0 & 1 \end{pmatrix} \begin{pmatrix} T_0 \\ T_1 \\ T_2 \\ T_3 \\ T_4 \end{pmatrix} = \begin{pmatrix} \alpha_1 \\ 0 \\ 0 \\ 0 \\ \alpha_2 \end{pmatrix}. \tag{6.44}$$

This a **tridiagonal system**. The solution of this system will give temperatures at the nodes $x_1, x_2,$ and x_3.

B. Symmetric Tridiagonal and Diagonally Dominant Systems

In order to see how such systems arise, consider now the **unsteady conduction** of heat. This condition implies that the temperature T varies with the time t. The heat equation in this case is

$$\frac{1}{\alpha}\frac{\partial T}{\partial t} = \frac{\partial^2 T}{\partial x^2},$$

where α is *thermal diffusivity* as defined in Section 6.3.4. Let us divide the grid in the (x, t) plane with spacing Δx in the x direction and Δt in the t direction.

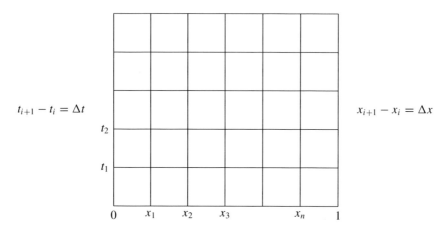

Let the temperature at the nodal point $x_i = i\Delta x$ and $t_j = j\Delta t$, as before, be denoted by T_{ij}. Approximating $\frac{\partial T}{\partial t}$ and $\frac{\partial^2 T}{\partial x^2}$ by the finite differences

$$\frac{\partial T}{\partial t} \approx \frac{1}{\Delta t}(T_{i,j+1} - T_{i,j}),$$

$$\frac{\partial^2 T}{\partial x^2} \approx \frac{1}{(\Delta x)^2}(T_{i+1,j+1} - 2T_{i,j+1} + T_{i-1,j+1}),$$

we obtain the following difference analogue of the heat equation:

$$(1+2C)T_{i,j+1} - C(T_{i+1,j+1} + T_{i-1,j+1}) = T_{i,j}, \quad i = 1, 2, \ldots, n,$$

where $C = \alpha\frac{\Delta t}{(\Delta x)^2}$.

These equations enable us to determine the temperature at time step $j = k + 1$, knowing the temperature at the previous time step $j = k$. Varying i from 1 to n, these equations become

for $i = 1$, $j = k$: $(1+2C)T_{1,k+1} - CT_{2,k+1} = CT_{0,k+1} + T_{1,k}$,

for $i = 2$, $j = k$: $(1+2C)T_{2,k+1} - CT_{3,k+1} - CT_{1,k+1} = T_{2,k}$,

$\vdots$

for $i = n$, $j = k$: $(1+2C)T_{n,k+1} - CT_{n-1,k+1} = T_{n,k} + CT_{n+1,k+1}$.

6.12. Special Systems

Suppose now that the temperatures at the two vertical sides are known, that is,

$$T_{0,t} = T_{w_1} \quad \text{and} \quad T_{n+1,t} = T_{w_2}.$$

Then the above equations can be written in matrix notation as

$$\begin{pmatrix} (1+2C) & -C & 0 & \cdots & \cdots & 0 \\ -C & (1+2C) & -C & 0 & \cdots & 0 \\ \vdots & \ddots & \ddots & \ddots & & \vdots \\ \vdots & & \ddots & \ddots & & -C \\ 0 & \cdots & \cdots & & -C & (1+2C) \end{pmatrix} \begin{pmatrix} T_{1,k+1} \\ T_{2,k+1} \\ \vdots \\ \vdots \\ T_{n,k+1} \end{pmatrix} = \begin{pmatrix} T_{1,k} + CT_{w_1} \\ T_{2,k} \\ \vdots \\ \vdots \\ T_{n,k} + CT_{w_2} \end{pmatrix}.$$

The matrix of the above system is clearly *symmetric, tridiagonal*, and *diagonally dominant* (note that $C > 0$).

For example, when $C = 1$, and we have

$$\begin{pmatrix} 3 & -1 & 0 & \cdots & 0 \\ -1 & 3 & -1 & \cdots & 0 \\ \vdots & \ddots & \ddots & \ddots & \vdots \\ \vdots & & \ddots & \ddots & -1 \\ 0 & \cdots & \cdots & -1 & 3 \end{pmatrix} \begin{pmatrix} T_{1,k+1} \\ T_{2,k+1} \\ \vdots \\ \vdots \\ T_{n,k+1} \end{pmatrix} = \begin{pmatrix} T_{1,k} + T_{w_1} \\ T_{2,k} \\ \vdots \\ \vdots \\ T_{n,k} + T_{w_2} \end{pmatrix}, \quad (6.45)$$

which is of the form

$$Ax = b,$$

where A is symmetric, tridiagonal, and diagonally dominant.

Block Tridiagonal Systems

To see how *block tridiagonal* systems arise in applications, consider the two-dimensional **Poisson's** equation:

$$\frac{\partial^2 T}{\partial x^2} + \frac{\partial^2 T}{\partial y^2} = f(x, y); \quad 0 \le x \le 1, \quad 0 \le y \le 1. \quad (6.46)$$

A discrete analogue to this equation, similar to Laplace's equation derived earlier, is

$$T_{i+1,j} + T_{i-1,j} + T_{i,j+1} + T_{i,j-1} - 4T_{ij} = (\Delta x)^2 f_{ij};$$
$$i = 1, 2, \ldots, n, \quad j = 1, 2, \ldots, n \quad (6.47)$$

This will give rise to a linear system of $(n+2)^2$ variables.

Assume now that the values of T at the four sides of the unit square are known and we are interested in the values of T at the interior grid points, that is, given

$$T_{0,j}; T_{n+1,j} \quad \text{and} \quad T_{i,0}; T_{i,n+1}, \quad j = 0, 1, \ldots, n+1, \quad i = 0, 1, \ldots, n+1, \quad (6.48)$$

we need to find $T_{11}, \ldots, T_{n1}, T_{12}, \ldots, T_{n2}, T_{1n}, \ldots, T_{nn}$. Then we have an $(n^2 \times n^2)$ system with n^2 unknowns which can be written after suitable rearrangement as

$$\begin{pmatrix} A_n & -I_n & & & \\ -I_n & \ddots & \ddots & & \\ & \ddots & \ddots & \ddots & \\ & & \ddots & \ddots & -I_n \\ & & & -I_n & A_n \end{pmatrix} \begin{pmatrix} T_{11} \\ T_{21} \\ \vdots \\ T_{n1} \\ T_{12} \\ \vdots \\ T_{n2} \\ \vdots \\ T_{nn} \end{pmatrix} = \begin{pmatrix} T_{01} + T_{10} - (\Delta x)^2 f_{11} \\ T_{20} - (\Delta x)^2 f_{21} \\ \vdots \\ \vdots \\ T_{n-1,0} - (\Delta x)^2 f_{n-1,1} \\ T_{n+1,1} + T_{n,0} - (\Delta x)^2 f_{n,1} \\ T_{02} - (\Delta x)^2 f_{12} \\ \vdots \end{pmatrix}, \quad (6.49)$$

where

$$A_n = \begin{pmatrix} 4 & -1 & \cdots & 0 \\ -1 & \ddots & \ddots & \vdots \\ \vdots & \ddots & \ddots & -1 \\ 0 & \cdots & -1 & 4 \end{pmatrix}. \quad (6.50)$$

The system matrix above is *block tridiagonal* and each block diagonal matrix A_n is *symmetric, tridiagonal*, and *positive definite*. For details, see Ortega and Poole (1981, pp. 268–272).

6.12.2 Special Linear Systems Arising from Finite Element Methods

We have seen in the last few sections how discretization of differential equations using finite differences gives rise to various types of linear systems problems. The finite element technique is another popular way to discretize differential equations, and this results also in linear systems problems. Just to give a taste to the readers, we illustrate this below by means of a simple differential equation. Interested readers are referred to some well-known books on the subject: Strang and Fix (1973), Becker, Carey, and Oden (1981), Reddy(1993), and Fish and Belytschko (2007).

Variational Formulation of a Two-Point Boundary Value Problem

Let us consider the two-point boundary value problem

$$-u'' + u = f(x), \qquad 0 < x < 1, \qquad (6.51)$$
$$u(0) = u(1) = 0, \qquad (6.52)$$

where $u' = \frac{du}{dx}$ and f is a continuous function on [0,1]. We further assume that f is such that the problem defined by (6.51)–(6.52) has a unique solution.

We introduce the space

$$V = \{v : v \text{ is a continuous function on } [0, 1], v' \text{ is piecewise continuous and bounded on } [0, 1], \text{ and } v(0) = v(1) = 0\}.$$

6.12. Special Systems

Now, if we multiply the equation $-u'' + u = f(x)$ by an arbitrary function $v \in V$ (v is called a test function), integrate by parts the left-hand side, and use the above boundary conditions, we get

$$\int_0^1 (-u''(x) + u(x))v(x)dx = \int_0^1 f(x)v(x)dx,$$

that is,

$$\int_0^1 (u'v' + uv)dx = \int_0^1 f(x)v(x)dx. \tag{6.53}$$

Equation (6.53) can be written as

$$a(u, v) = (f, v) \qquad \text{for every } u \in V,$$

where

$$a(u, v) = \int_0^1 (u'v' + uv)dx$$

and

$$(f, v) = \int_0^1 f(x)v(x)dx.$$

(Notice that the form $a(\cdot, \cdot)$ is symmetric (i.e., $a(u, v) = a(v, u)$) and bilinear.) These two properties will be used later. It can be shown that u is a solution of (6.53) if and only if u is a solution to (6.51)–(6.52).

The Discrete Problem

We now discretize problem (6.53). We start by constructing a finite-dimensional subspace V_n of the space V.

Here, we will consider only the simple case where V_n consists of continuous piecewise linear functions. For this purpose, we let $0 = x_0 < x_1 < x_2 \cdots < x_n < x_{n+1} = 1$ be a partition of the interval $[0,1]$ into subintervals $I_j = [x_{j-1}, x_j]$ of length $h_j = x_j - x_{j-1}$, $j = 1, 2, \ldots, n+1$. With this partition, we associate the set V_n of all functions $v(x)$ that are continuous on the interval $[0,1]$, linear in each subinterval I_j, $j = 1, \ldots, n+1$, and satisfy the boundary conditions $v(0) = v(1) = 0$.

We now introduce the basis functions $\{\phi_1, \phi_2, \ldots, \phi_n\}$ of V_n.

We define $\phi_j(x)$ by

(i) $\phi_j(x_i) = \begin{cases} 1 & \text{if } i = j, \\ 0 & \text{if } i \neq j; \end{cases}$

(ii) $\phi_j(x)$ is a continuous piecewise linear function.

$\phi_j(x)$ can be computed explicitly to yield

$$\phi_j(x) = \begin{cases} \dfrac{x - x_{j-1}}{h_j} & \text{when } x_{j-1} \leq x \leq x_j, \\ \dfrac{x_{j+1} - x}{h_{j+1}} & \text{when } x_j \leq x \leq x_{j+1}. \end{cases}$$

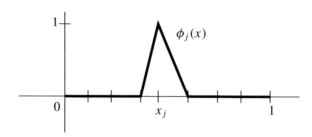

Since $\phi_1, \ldots, \phi_n$ are the basis functions, any function $v \in V_n$ can be written uniquely as

$$v(x) = \sum_{i=1}^{n} v_i \phi_i(x), \qquad \text{where } v_i = v(x_i).$$

We easily see that $V_n \subset V$.

The discrete analogue of problem (6.53) then reads: Find $u_n \in V_n$ such that

$$a(u_n, v) = (f, v) \qquad \forall v \in V_n. \tag{6.54}$$

Now, if we let $u_n = \sum_{i=1}^{n} c_i \phi_i(x)$ and notice that (6.54) is particularly true for every function $\phi_j(x)$, $j = 1, \ldots, n$, we get n equations, namely,

$$a\left(\sum c_i \phi_i, \phi_j\right) = (f, \phi_j) \qquad \forall j = 1, 2, \ldots, n.$$

Now using the linearity of $a(\cdot, \phi_j)$ leads to n linear equations in n unknowns:

$$\sum_{i=1}^{n} c_i a(\phi_i, \phi_j) = (f, \phi_j) \qquad \forall j = 1, 2, \ldots, n,$$

which can be written in the matrix form as

$$Ac = (f_n)_i, \tag{6.55}$$

where $(f_n)_i = (f, \phi_i)$ and $A = (a_{ij})$ is a symmetric matrix given by

$$a_{ij} = a_{ji} = a(\phi_i, \phi_j) \qquad \text{and} \qquad c = (c_1, c_2, \ldots, c_n)^T.$$

The entries of the matrix A can be computed explicitly: We first notice that

$$a_{ij} = a_{ji} = a(\phi_i, \phi_j) = 0 \qquad \text{if } |i - j| \geq 2.$$

(This is due to the local support of the function $\phi_i(x)$.) A direct computation now leads to

$$a_{j,j} = a(\phi_j, \phi_j) = \int_{x_{j-1}}^{x_j} \left[\frac{1}{h_j^2} + \frac{(x - x_{j-1})^2}{h_j^2}\right] dx + \int_{x_j}^{x_{j+1}} \left[\frac{1}{h_{j+1}^2} + \frac{(x_{j+1} - x)^2}{h_{j+1}^2}\right] dx$$

$$= \left[\frac{1}{h_j} + \frac{1}{h_{j+1}}\right] + \frac{1}{3}[h_j + h_{j+1}],$$

$$a_{j,j-1} = \int_{x_{j-1}}^{x_j} \left[-\frac{1}{h_j^2} + \frac{(x_j - x)(x - x_{j-1})}{h_j \, h_j}\right] dx = \frac{-1}{h_j} + \frac{h_j}{6}.$$

Set $a_j = \frac{1}{h_j} + \frac{1}{h_{j+1}} + \frac{1}{3}[h_j + h_{j+1}]$ and $b_j = -\frac{1}{h_j} + \frac{h_j}{6}$.

6.12. Special Systems

Then the system (6.55) can be written as

$$\begin{bmatrix} a_1 & b_1 & & & 0 \\ b_1 & a_2 & & & \\ & \ddots & \ddots & \ddots & \\ & & \ddots & \ddots & b_{n-1} \\ 0 & & & b_{n-1} & a_n \end{bmatrix} \begin{bmatrix} c_1 \\ c_2 \\ \vdots \\ \vdots \\ c_n \end{bmatrix} = \begin{bmatrix} (f_n)_1 \\ (f_n)_2 \\ \vdots \\ \vdots \\ (f_n)_n \end{bmatrix}. \quad (6.56)$$

In the special case of uniform grid $h_j = h = \frac{1}{n+1}$, matrix A then takes the form

$$A = \frac{1}{h}\begin{bmatrix} 2 & -1 & & & 0 \\ -1 & 2 & \ddots & & \\ & \ddots & \ddots & \ddots & \\ & & \ddots & \ddots & -1 \\ 0 & & & -1 & 2 \end{bmatrix} + \frac{h}{6}\begin{bmatrix} 4 & 1 & & & 0 \\ 1 & \ddots & \ddots & & \\ & \ddots & \ddots & 1 & \\ 0 & & 1 & 4 \end{bmatrix}. \quad (6.57)$$

Note that A has a very special structure: it is *tridiagonal* and symmetric *positive definite*.

6.12.3 Symmetric Positive Definite Systems

Definition 6.19 (positive definite matrix). *A symmetric matrix A is positive definite if, for every nonzero vector x, $x^T A x > 0$. Let $x = (x_1, x_2, \ldots, x_n)^T$. Then $x^T A x = \sum_{i,j=1}^{n} a_{ij} x_i x_j$ is called the **quadratic form** associated with A.*

A *positive semidefinite matrix* is similarly defined. A symmetric matrix A is *positive semidefinite* if $x^T A x \geq 0$ for all vectors x.

A commonly used notation for a symmetric positive definite (positive semidefinite matrix) is $A > 0 \, (\geq 0)$.

Some Characterizations and Properties of Positive Definite Matrices

1. A symmetric matrix A is positive definite if and only if all its eigenvalues are positive.

2. A symmetric matrix A is positive definite if and only if all its leading principal minors are positive.

3. If $A = (a_{ij})$ is symmetric positive definite, then $a_{ii} > 0$ for all i.

4. If $A = (a_{ij})$ is symmetric positive definite, then the largest element (in magnitude) of the whole matrix must lie on the diagonal.

5. The sum of two symmetric positive definite matrices is symmetric positive definite.

Remarks. Note that properties 3 and 4 give only necessary conditions for a symmetric matrix to be positive definite. They can serve only as initial tests for positive definiteness. For example, the matrices

$$A = \begin{pmatrix} 4 & 1 & 1 & 1 \\ 1 & 0 & 1 & 2 \\ 1 & 1 & 2 & 3 \\ 1 & 2 & 3 & 4 \end{pmatrix}, \qquad B = \begin{pmatrix} 20 & 12 & 25 \\ 12 & 15 & 2 \\ 25 & 2 & 5 \end{pmatrix}$$

cannot be positive definite, since in matrix A there is a zero entry on the diagonal, and in B the largest entry 25 is not on the diagonal.

The Cholesky Factorization

A more numerically effective way to check the positive definiteness of a symmetric matrix than those given by characterizations 1 and 2 is via Cholesky factorization of A:

Given a symmetric positive definite matrix A, there exists a lower triangular matrix H with positive diagonal entries such that

$$A = HH^T.$$

This factorization is called the **Cholesky factorization**, after the French engineer Andre-Louis Cholesky,[3] who discovered this factorization. H is called the **Cholesky factor**.

The existence of the Cholesky factorization for a symmetric positive definite matrix A can be seen either via LU factorization of A (Exercise 6.30) or by computing matrix H directly from the above relation.

We will not discuss the technique of finding Cholesky factorization via LU decomposition here. However, we note that *Gaussian elimination, even without pivoting, is remarkably stable for positive definite matrices.*

In this case, *it can be shown that the growth factor is exactly equal to 1. Even if there is a small pivot, the elimination scheme does not give rise to the growth in the entries of the subsequent matrices $A^{(k)}$*. For example, consider applying Gaussian elimination without pivoting to the following 2×2 example:

$$A = \begin{pmatrix} 0.00003 & 0.00500 \\ 0.00500 & 1.0000 \end{pmatrix}.$$

There is only one step. The pivot entry is 0.00003. It is *small*. As a result of this the multiplier m_{21} is relatively *large*:

$$m_{21} = -\frac{a_{21}}{a_{11}} = -\frac{0.00500}{0.00003} = -\frac{500}{3}.$$

However, the entries of $A^{(1)}$ did not grow:

$$A^{(1)} = \begin{pmatrix} 0.00003 & 0.00500 \\ 0 & 0.16667 \end{pmatrix}.$$

---

[3] Andre-Louis Cholesky (1875–1918) served as an officer in the French military. His work there involved geodesy and surveying.

6.12. Special Systems

In fact, $\max(a_{ij}^{(1)}) = 0.166667 < \max(a_{ij}) = 1$. Thus the growth factor ρ is 1. This interesting phenomenon of Gaussian elimination without pivoting can be explained through the following result. See Wilkinson (1965).

Theorem 6.20. *Let Gaussian elimination without pivoting be applied to a symmetric positive definite matrix A. Write $A^{(k)}$, the matrix obtained after the kth step, as*

$$A^{(k)} = \left(\begin{array}{c|c} U_{kk} & U_{k,n-k} \\ \hline 0 & W_{n-k,n-k} \end{array} \right).$$

(i) *Then $W_{n-k,n-k}$ is symmetric and positive definite.*

(ii) *If $|a_{ij}| \leq 1$, then $|a_{ij}^{(k)}| \leq 1$.*

The Cholesky Algorithm

We now show how the Cholesky factorization can be computed directly from $A = HH^T$, when $n = 3$. *The general case is analogous.*

$$\underbrace{\begin{pmatrix} a_{11} & a_{12} & a_{13} \\ a_{21} & a_{22} & a_{23} \\ a_{31} & a_{32} & a_{33} \end{pmatrix}}_{A} = \underbrace{\begin{pmatrix} h_{11} & 0 & 0 \\ h_{21} & h_{22} & 0 \\ h_{31} & h_{32} & h_{33} \end{pmatrix}}_{H} \underbrace{\begin{pmatrix} h_{11} & h_{21} & h_{31} \\ 0 & h_{22} & h_{32} \\ 0 & 0 & h_{33} \end{pmatrix}}_{H^T}.$$

1. *Compute the first column of H.* Compare the corresponding entries of the first column of both sides:
$$a_{11} = h_{11}^2 \quad \Rightarrow \quad h_{11} = \sqrt{a_{11}},$$
$$a_{21} = h_{11}h_{21} \quad \Rightarrow \quad h_{21} = \frac{a_{21}}{h_{11}},$$
$$a_{31} = h_{11}h_{31} \quad \Rightarrow \quad h_{31} = \frac{a_{31}}{h_{11}}.$$

2. *Compute the second column of H.* Compare the second and third entries of the second column of both sides:
$$a_{22} = h_{21}^2 + h_{22}^2 \quad \Rightarrow \quad h_{22} = \sqrt{a_{22} - h_{21}^2},$$
$$a_{32} = h_{21}h_{31} + h_{22}h_{32} \quad \Rightarrow \quad h_{32} = \frac{a_{32} - h_{21}h_{31}}{h_{22}}.$$

3. *Compute the third column of H.* Compare the third entry of the third column of both sides:
$$a_{33} = h_{31}^2 + h_{32}^2 + h_{33}^2 \Rightarrow h_{33} = \sqrt{a_{33} - h_{31}^2 - h_{32}^2}.$$

- In general, compute recursively the first through nth columns of H by comparing the entries of the respective column of both sides of $A = HH^T$.

> **ALGORITHM 6.7. The Cholesky Algorithm.**[4]
>
> **Input:** An $n \times n$ symmetric positive definite matrix.
> **Output:** The **Cholesky factor** H, stored over the upper triangular part of A, including the diagonal.
>
> For $k = 1, 2, \ldots, n$ do
> $$a_{kk} \equiv h_{kk} = \sqrt{a_{kk} - \sum_{j=1}^{k-1} h_{kj}^2}$$
> For $i = k+1, \ldots, n$ do
> $$a_{ik} \equiv h_{ik} = \frac{1}{h_{kk}} \left(a_{ik} - \sum_{j=1}^{k-1} h_{ij} h_{kj} \right)$$
> End
> End

This leads to the following algorithm, known as the **Cholesky algorithm**.

Remarks.

1. The matrix H above is computed column by column.

2. In the above pseudocode, $\sum_{j=1}^{0} (\ ) \equiv 0$.

3. The positive definiteness of A will make the quantities under the square-root signs positive.

Round-off property. Let the computed Cholesky factor be denoted by $\hat{H}$. Then it can be shown (Demmel (1989)) that
$$A + E = \hat{H}(\hat{H})^T,$$
where $E = (e_{ij})$, and $|e_{ij}| \leq \frac{(n+1)\mu}{1-(n+1)\mu}(a_{ii}a_{jj})^{1/2}$. Thus, the Cholesky factorization algorithm (Algorithm 6.7) is stable. See also Higham (2002, p. 197).

MATCOM and MATLAB Notes: Algorithm 6.7 has been implemented in the MATCOM program **CHOLES**. Note that the MATLAB program **chol**(A) computes the Cholesky factor R such that $A = R^T R$, where R is upper triangular.

Solution of $Ax = b$ Using the Cholesky Factorization

Having the Cholesky factorization $A = HH^T$ at hand, the positive definite linear system $Ax = b$ can now be solved by solving the lower triangular system $Hy = b$ first, followed by the upper triangular system $H^T x = y$.

---

[4]This algorithm in some fields (such as in statistics) is known as the *square-root algorithm*. A square-root-free algorithm, however, can be developed.

6.12. Special Systems

ALGORITHM 6.8. The Cholesky Algorithm for the Positive Definite System $Ax = b$.

Inputs: A symmetric positive definite matrix $A \in \mathbb{R}^{n \times n}$, $b \in \mathbb{R}^n$.
Output: $x \in \mathbb{R}^n$ such that $Ax = b$.

Step 1. Find the **Cholesky factorization** of $A = HH^T$ (use Algorithm 6.7).
Step 2. Solve the **lower triangular system** for y: $Hy = b$.
Step 3. Solve the **upper triangular system** for x: $H^T x = y$.

Example 6.21. Let
$$A = \begin{pmatrix} 1 & 1 & 1 \\ 1 & 5 & 5 \\ 1 & 5 & 14 \end{pmatrix}, \quad b = \begin{pmatrix} 3 \\ 11 \\ 20 \end{pmatrix}.$$

A. The Cholesky factorization.

1st column ($k = 1$):
$$h_{11} = 1,$$
$$h_{21} = \frac{a_{21}}{h_{11}} = 1, \qquad h_{31} = \frac{a_{31}}{h_{11}} = \frac{1}{1} = 1.$$

(Since the diagonal entries of H have to be positive, we take the $+$ sign for h_{11}).

2nd column ($k = 2$):
$$h_{22} = \sqrt{a_{22} - h_{21}^2} = 2, \qquad h_{32} = \frac{a_{32} - h_{21}h_{31}}{h_{22}} = 2.$$

3rd column ($k = 3$):
$$h_{33} = \sqrt{a_{33} - h_{31}^2 - h_{32}^2} = 3.$$

Thus, $H = \begin{pmatrix} 1 & 0 & 0 \\ 1 & 2 & 0 \\ 1 & 2 & 3 \end{pmatrix}.$

B. Solution of the linear system $Ax = b$.

(1) Solution of $Hy = b \Rightarrow y = (3, 4, 3)^T$.
(2) Solution of $H^T x = y \Rightarrow x = (1, 1, 1)^T$. ∎

Flop-count. (i) The Cholesky algorithm requires $\frac{n^3}{3}$ flops to compute H (*half of the number of flops required to do the same job using LU factorization*) and n square roots. (ii) The solution of each triangular system $Hy = b$ and $H^T x = y$ requires n^2 flops. Thus the solution of the positive definite system $Ax = b$ using the Cholesky algorithm requires $\frac{n^3}{3} + 2n^2$ flops and n square roots.

Round-off property. If $\hat{x}$ is the computed solution of the system $Ax = b$ using the Cholesky algorithm, then it can be shown that $\hat{x}$ satisfies

$$(A + E)\hat{x} = b,$$

where $\|E\|_2 \leq c\mu \|A\|_2$, and c is a small constant depending upon n. Thus the Cholesky algorithm for solving a symmetric positive definite system is stable. See Higham (2002, p. 198).

Relative Error in the Solution by the Cholesky Algorithm

Let $\hat{x}$ be the computed solution of the symmetric positive definite system of $Ax = b$, using the Cholesky algorithm followed by triangular systems solutions as described above. Then it can be shown that

$$\frac{\|x - \hat{x}\|_2}{\|\hat{x}\|_2} \leq \mu \; \text{Cond}(A).$$

Remark. Demmel (1989) has shown that the above bound can be replaced by $O(\mu) \, \text{Cond}(\tilde{A})$, where $\tilde{A} = D^{-1}AD^{-1}$, $D = \text{diag}(\sqrt{a_{11}}, \ldots, \sqrt{a_{nn}})$. The latter may be much better than the previous one, since $\text{Cond}(\tilde{A})$ may be much smaller than $\text{Cond}(A)$.

6.12.4 Hessenberg System

Consider the linear system

$$Ax = b,$$

where A is an upper Hessenberg matrix of order n. Solution of a Hessenberg system arises in several practical applications, including the eigenvector computation of a matrix (see Algorithm 9.8). *Solving a Hessenberg system requires much less computational efforts than solving an arbitrary system.* This is because, at each step of elimination, only one entry needs to be updated, due to the special structure of a Hessenberg matrix. Furthermore, if Gaussian elimination with partial pivoting is used to triangularize A, and if $|a_{ij}| \leq 1$, then $|a_{ij}^{(k)}| \leq k+1$ (Wilkinson (1965, p. 218)). *Thus, the growth factor ρ in this case is less than or equal to n* (Higham (2002, p. 172)).

Growth factor and stability of Gaussian elimination for a Hessenberg system. The growth factor for a Hessenberg matrix using GEPP is bounded by n. *Thus a Hessenberg system can be safely solved using partial pivoting.*

Flop-count. It requires only $3n^2$ flops to solve a Hessenberg system, significantly less than $2\frac{n^3}{3}$ flops required to solve an $n \times n$ system with an arbitrary matrix. This can be seen as follows:

    Triangularization: n^2 flops.
    Solution of the lower triangular system: n^2 flops.
    Solution of the upper triangular system: n^2 flops.
    Total: $3n^2$ flops.

    Thus a Hessenberg system can be solved with only $3n^2$ flops in a stable way using GEPP.

6.12. Special Systems

Example 6.22. Solve the Hessenberg system $Ax = b$ with

$$A = \begin{pmatrix} 1 & 2 & 3 \\ \boxed{2} & 3 & 4 \\ 0 & 5 & 6 \end{pmatrix}, \quad b = (6, 9, 11)^T$$

using *partial pivoting* and compute the *growth factor*.

Step 1. The pivot on the first column is identified as the (2,1) entry. Interchange row 2 with row 1 and overwrite it with A.

$$A \equiv A = \begin{pmatrix} 2 & 3 & 4 \\ 1 & 2 & 3 \\ 0 & 5 & 6 \end{pmatrix}.$$

Now, multiply the first row by $-\frac{1}{2}$ and add it to the second row.

$$A \equiv \begin{pmatrix} 2 & 3 & 4 \\ 0 & \frac{1}{2} & 1 \\ 0 & \boxed{5} & 6 \end{pmatrix}.$$

Multiplier $m_{12} = -\frac{1}{2}$. Permutation row index $r_1 = 2$.

Step 2. The pivot on the second column is identified as the $(3, 2)$ entry of the current matrix A.

Interchange the second and third rows to obtain

$$A \equiv A = \begin{pmatrix} 2 & 3 & 4 \\ 0 & 5 & 6 \\ 0 & \frac{1}{2} & 1 \end{pmatrix}.$$

Multiply the second row by $-\frac{1}{10}$ and add it to the third row to obtain

$$A \equiv A = \begin{pmatrix} 2 & 3 & 4 \\ 0 & 5 & 6 \\ 0 & 0 & \frac{2}{5} \end{pmatrix}.$$

Multiplier $m_{21} = -\frac{1}{10}$. Permutation row index $r_2 = 3$.
So,

$$U = \begin{pmatrix} 2 & 3 & 4 \\ 0 & 5 & 6 \\ 0 & 0 & \frac{2}{5} \end{pmatrix}, \quad L = \begin{pmatrix} 1 & 0 & 0 \\ 0 & 1 & 0 \\ \frac{1}{2} & \frac{1}{10} & 1 \end{pmatrix}, \text{ and } P = \begin{pmatrix} 0 & 1 & 0 \\ 0 & 0 & 1 \\ 1 & 0 & 0 \end{pmatrix}.$$

Computation of the growth factor: $\rho = \frac{\max(6,6,6)}{6} = 1$.
Solution of the system: Solve $Ly = b' = Pb \Rightarrow y = (9, 11, \frac{2}{5})^T$. Solve $Ux = y \Rightarrow x = (1, 1, 1)^T$. ■

6.12.5 Diagonally Dominant Systems

A matrix $A = (a_{ij})$ is **column diagonally dominant** if

$$\begin{aligned}
|a_{11}| &\geq |a_{21}| + |a_{31}| + \cdots + |a_{n1}|, \\
|a_{22}| &\geq |a_{12}| + |a_{32}| + \cdots + |a_{n2}|, \\
&\vdots \\
|a_{nn}| &\geq |a_{1n}| + |a_{2n}| + \cdots + |a_{n-1,n}|.
\end{aligned} \qquad (6.58)$$

If the strict inequalities hold, then A is called a **strictly column diagonally dominant matrix**. A **row diagonally dominant** matrix can be similarly defined.

A column diagonally dominant matrix possesses the attractive property that no row interchanges are necessary at any step during the triangularization procedure using GEPP. The pivot element is already there in the right place.

We show this by taking A as a 3×3 matrix. For a proof in the general case, see the solution to Exercise 6.35(a).

Let

$$A = \begin{pmatrix} a_{11} & a_{12} & a_{13} \\ a_{21} & a_{22} & a_{23} \\ a_{31} & a_{32} & a_{33} \end{pmatrix}$$

be column diagonally dominant. Then a_{11} can be taken as the pivot at the first step, and no interchange of rows is necessary. At the end of Step 1, we have

$$A^{(1)} = \begin{pmatrix} a_{11} & a_{12} & a_{13} \\ 0 & a_{22}^{(1)} & a_{23}^{(1)} \\ 0 & a_{32}^{(1)} & a_{33}^{(1)} \end{pmatrix}.$$

We will now show that

$$|a_{22}^{(1)}| \geq |a_{32}^{(1)}|$$

so that $a_{22}^{(1)}$ will be the pivot at the next step. Observe that

$$a_{32}^{(1)} = a_{32} - a_{12} \times \frac{a_{31}}{a_{11}}. \qquad (6.59)$$

By column diagonal dominance, we have

$$|a_{11}| \geq |a_{21}| + |a_{31}|. \qquad (6.60)$$

Using (6.60) in (6.59), we have

$$\begin{aligned}
|a_{32}^{(1)}| &\leq |a_{32}| + \left|\frac{a_{12}}{a_{11}}\right|(|a_{11}| - |a_{21}|) \\
&= |a_{32}| + |a_{12}| - \left|\frac{a_{12}}{a_{11}}\right||a_{21}| \\
&\leq |a_{22}| - \left|\frac{a_{12}}{a_{11}}\right||a_{21}|
\end{aligned}$$

(since by the column dominance of the second column of A, $|a_{22}| \geq |a_{12}| + |a_{32}|$).

6.12. Special Systems

Thus,
$$a_{32}^{(1)} \leq \left| a_{22} - \frac{a_{12}}{a_{11}} \cdot a_{21} \right| = \left| a_{22}^{(1)} \right|.$$

The general case can be proved using a similar argument.

Furthermore, it can be shown (Higham (2002, pp. 170–172)) that in this case, *the growth factor is less than or equal to 2*.

Growth Factor and Stability of Gaussian Elimination for Diagonally Dominant Systems

- For a column diagonally dominant matrix, GEPP is identical to Gaussian elimination without pivoting.

- The *growth factor* ρ for a column diagonally dominant matrix is bounded by 2, that is, $\rho \leq 2$.

  Thus, for column *diagonally dominant systems, Gaussian elimination without pivoting is perfectly stable.*

- For row diagonally dominant matrices, the multipliers can be large; however, $\rho \leq 2$ and thus Gaussian elimination without pivoting is still stable.

Example 6.23. Let $A = \begin{pmatrix} 5 & -8 \\ 1 & 10 \end{pmatrix}$. Then $A^{(1)} = \begin{pmatrix} 5 & -8 \\ 0 & \frac{58}{5} \end{pmatrix}$.

The growth factor $\rho = \frac{\max(10, \frac{58}{5})}{10} = \frac{\frac{58}{5}}{10} = 1.16$. ∎

6.12.6 Tridiagonal Systems

The LU factorization of a tridiagonal matrix T, when it exists, may yield L and U having very special simple structures: *both bidiagonal*, L having 1's along the main diagonal and the superdiagonal entries of U the same as those of T. Specifically, write

$$T = \begin{pmatrix} a_1 & b_1 & \cdots & 0 \\ c_2 & \ddots & \ddots & \vdots \\ \vdots & \ddots & \ddots & b_{n-1} \\ 0 & \cdots & c_n & a_n \end{pmatrix} = \begin{pmatrix} 1 & & & 0 \\ \ell_2 & \ddots & & \\ & \ddots & \ddots & \\ 0 & & \ell_n & 1 \end{pmatrix} \begin{pmatrix} u_1 & b_1 & & 0 \\ & \ddots & \ddots & \\ & & \ddots & b_{n-1} \\ 0 & & & u_n \end{pmatrix} = LU.$$
(6.61)

By equating the corresponding elements of the matrices on both sides, the entries $\{\ell_i\}$ and $\{u_i\}$ can be computed from

$$a_1 = u_1; \qquad c_i = \ell_i u_{i-1}, \qquad i = 2, \ldots, n; \qquad a_i = u_i + \ell_i b_{i-1}, \qquad i = 2, \ldots, n.$$

> **ALGORITHM 6.9. Computing LU Factorization of a Tridiagonal Matrix.**
>
> **Input:** The tridiagonal matrix T as given above.
> **Outputs:** The unit lower bidiagonal matrix L and the **upper bidiagonal matrix** U as given above.
>
> Set $u_1 = a_1$
> For $i = 2, \ldots, n$ do
> $$\ell_i = \frac{c_i}{u_{i-1}}$$
> $$u_i = a_i - \ell_i b_{i-1}$$
> End

Solving a Tridiagonal System

Once we have the above simple factorization of T, the solution of the tridiagonal system $Tx = b$ can be found by solving the following two special **bidiagonal** systems in the order (i) $Ly = b$, (ii) $Ux = y$.

Flop-count. *A tridiagonal system can be solved by the above procedure in only $O(n)$ flops*, a very cheap procedure indeed.

Stability of the process. Unfortunately, the above factorization procedure breaks down if any u_i is zero. Even if all u_i are theoretically nonzero, the stability of the process in general cannot be guaranteed. However, as we have seen before in many practical situations, such as in discretizing Poisson's equation, the tridiagonal matrices are symmetric positive definite, in which cases the above procedure is quite stable (see Higham (2002, pp. 173–176)).

In fact, for a symmetric positive definite tridiagonal system case, this procedure should be preferred over the Cholesky factorization technique, as it does not involve computations of any square roots. It is true that the Cholesky factorization of a symmetric positive definite tridiagonal matrix can also be computed in $O(n)$ flops; however, an additional n square roots have to be computed (see Golub and Van Loan (1996, p. 156)).

In the general case, to maintain stability, GEPP should be used.

Furthermore, if the entries of T are so scaled that $|a_i|, |b_i|, |c_i| \leq 1$, then it can be shown (Wilkinson (1965, p. 219)) that the entries of $A^{(k)}$ at each step of GEPP will be bounded by 2, showing that the *growth factor in this case is bounded by* 2. For a proof, see Higham (2002, pp. 173).

Growth factor and stability of Gaussian elimination for a tridiagonal system. The growth factor for GEPP of a tridiagonal matrix is bounded by 2: $\rho \leq 2$.

Thus, GEPP for a tridiagonal system is stable.

If T is symmetric, one naturally wants to take advantage of the symmetry; however, *GEPP does not preserve symmetry*. Bunch (1971) and Bunch and Kaufman (1977) proposed symmetry-preserving algorithms. These algorithms can be arranged to have flop-count

6.12. Special Systems

comparable to that of GEPP and require *less storage* than the latter. For details see the papers by Bunch (1971) and Bunch and Kaufman (1977). See also Bunch and Parlett (1971) and Bunch, Kaufman and Parlett (1976).

Example 6.24. Triangularize

$$A = \begin{pmatrix} 0.9 & 0.1 & 0 \\ 0.8 & 0.5 & 0.1 \\ 0 & 0.1 & 0.5 \end{pmatrix}$$

using (i) the formula $A = LU$, and (ii) Gaussian elimination.

(i) From $A = LU$

$$u_1 = 0.9.$$

$i = 2$:

$$\ell_2 = \frac{c_2}{u_1} = \frac{0.8}{0.9} = \frac{8}{9} = 0.8889; \qquad u_2 = a_2 - \ell_2 b_1 = 0.5 - \frac{8}{9} \times 0.1 = 0.4111.$$

$i = 3$:

$$\ell_3 = \frac{c_3}{u_2} = \frac{0.1}{0.41} = 0.2432; \qquad u_3 = a_3 - \ell_3 b_2 = 0.5 - 0.24 \times 0.1 = 0.4757.$$

Thus,

$$L = \begin{pmatrix} 1 & 0 & 0 \\ 0.8889 & 1 & 0 \\ 0 & 0.2432 & 0.1 \end{pmatrix}, \quad U = \begin{pmatrix} 0.9 & 0.1 & 0 \\ 0 & 0.4111 & 0.1 \\ 0 & 0 & 0.4757 \end{pmatrix}.$$

(ii) Using GEPP

Step 1. Multiplier $m_{21} = -\frac{0.8}{0.9} = -0.89$;

$$A^{(1)} = \begin{pmatrix} 0.9 & 0.1 & 0 \\ 0 & 0.4111 & 0.1 \\ 0 & 0.1 & 0.5 \end{pmatrix}.$$

Step 2. Multiplier $m_{32} = -\frac{0.1}{0.41} = -0.243$;

$$A^{(2)} = \begin{pmatrix} 0.9 & 0.1 & 0 \\ 0 & 0.4111 & 0.1 \\ 0 & 0 & 0.4757 \end{pmatrix} = U,$$

$$L = \begin{pmatrix} 1 & 0 & 0 \\ -m_{21} & 1 & 0 \\ 0 & -m_{32} & 1 \end{pmatrix} = \begin{pmatrix} 1 & 0 & 0 \\ 0.8889 & 1 & 0 \\ 0 & 0.2432 & 1 \end{pmatrix}. \quad \blacksquare$$

Block Tridiagonal Systems

In this section we consider solving the block tridiagonal system $Tx = b$, where T is a block tridiagonal matrix and $b = (b_1, b_2, \ldots, b_n)^T$ is a block vector. The number of components of the block vector b_i is the same as the dimension of the ith diagonal block matrix in T.

A. Block LU Factorization

The factorization procedure given in the beginning of this section may be easily extended to the case of the block tridiagonal Matrix. Let

$$T = \begin{pmatrix} A_1 & B_1 & & & \\ C_2 & \ddots & \ddots & & 0 \\ & \ddots & \ddots & \ddots & \\ & 0 & \ddots & \ddots & B_{N-1} \\ & & & C_N & A_N \end{pmatrix}$$

$$= \begin{pmatrix} I & & & & \\ L_2 & \ddots & & 0 & \\ & \ddots & \ddots & & \\ & 0 & \ddots & \ddots & \\ & & & L_N & I \end{pmatrix} \begin{pmatrix} U_1 & B_1 & \cdots & \cdots & 0 \\ \vdots & \ddots & \ddots & & \vdots \\ & & & \ddots & \vdots \\ \vdots & & & \ddots & B_{N-1} \\ 0 & \cdots & \cdots & \cdots & U_N \end{pmatrix} = LU. \quad (6.62)$$

Then the matrices $L_i, i = 2, \ldots, N$, and $U_i, i = 1, \ldots, N$, can be computed as in the case of a scalar tridiagonal matrix, as shown in Algorithm 6.10

ALGORITHM 6.10. Block LU Factorization of a Block Tridiagonal Matrix.

Input: The block tridiagonal matrix T as given above.
Output: The block unit lower bidiagonal matrix L and the block upper bidiagonal triangular matrix U as given above.

Step 1. Set $U_1 = A_1$.

Step 2. For $i = 2, \ldots, N$ do
   2.1. Solve for L_i: $L_i U_{i-1} = C_i$.
   2.2. Compute U_i: $U_i = A_i - L_i B_{i-1}$.
End

Numerical stability of block LU factorization of a block tridiagonal matrix. It is clear from the statement of Algorithm 6.10 that block LU factorization does not always exist. Even if it does, it may not be stable. However, it can be shown to be stable if T *is a symmetric positive definite matrix that is well-conditioned.* The scheme is also stable if T is a block column diagonally dominant matrix with respect to a subordinate matrix norm (Higham (2002, pp. 251–256)). Note that the block tridiagonal matrix in (6.49)–(6.50) appearing in the solution of Poisson's equation (6.46) is such a matrix. $A \in \mathbb{R}^{n \times m}$ with partitioning $A = (A_{ij})$ is *block diagonally dominant by column for a given norm and for all j if*

$$\|A_{jj}^{-1}\|^{-1} - \Sigma_{i \neq j} \|A_{ij}\| \equiv \gamma_{ij} \geq 0. \quad (6.63)$$

6.12. Special Systems

B. Solution of Block Systems

Once we have the above factorization, we can find the solution x of the block tridiagonal system
$$Tx = b$$
by solving $Ly = b$ and $Ux = y$ successively. The solution of $Ly = b$ can be achieved by *block forward elimination*, and the solution of $Ux = y$ can be computed by *block back substitution*.

ALGORITHM 6.11. Block Forward Elimination.

Inputs: A block unit lower triangular matrix $L = (L_i)$ as in (6.62), and a block vector b.
Output: The block solution vector y such that $Ly = b$.

Set $L_1 y_0 = 0$.
For $i = 1, \ldots, n$ do
    $y_i = b_i - L_i y_{i-1}$.
End

ALGORITHM 6.12. Block Back Substitution.

Inputs: A block upper triangular matrix U as in (6.62), and the block vector y, output from Algorithm 6.11.
Output: The block vector x such that $Ux = y$.

Set $B_N x_{N+1} = 0$.
For $i = N, \ldots, 1$ do
    Solve $U_i x_i = y_i - B_i x_{i+1}$.
End

Example 6.25. Consider the system $Tx = b$ with

$$T = \begin{pmatrix} 4 & -1 & 1 & 0 \\ -1 & 4 & 0 & 1 \\ 1 & 0 & 2 & -1 \\ 0 & 1 & -1 & 2 \end{pmatrix}, \quad b = \begin{pmatrix} 4 \\ 4 \\ 2 \\ 2 \end{pmatrix}.$$

Then

$$A_1 = \begin{pmatrix} 4 & -1 \\ -1 & 4 \end{pmatrix}, \quad A_2 = \begin{pmatrix} 2 & -1 \\ -1 & 2 \end{pmatrix}, \quad B_1 = \begin{pmatrix} 1 & 0 \\ 0 & 1 \end{pmatrix},$$

$$b_1 = \begin{pmatrix} 4 \\ 4 \end{pmatrix}, \quad b_2 = \begin{pmatrix} 2 \\ 2 \end{pmatrix}.$$

Block LU factorization:
Set
$$U_1 = A_1 = \begin{pmatrix} 4 & -1 \\ -1 & 4 \end{pmatrix}.$$

$i = 2$:

(1) Solve for L_2: $U_1 L_2 = I_2 = \begin{pmatrix} 1 & 0 \\ 0 & 1 \end{pmatrix} \Rightarrow L_2 = U_1^{-1} = \begin{pmatrix} 0.2667 & 0.0667 \\ 0.667 & 0.2667 \end{pmatrix}.$

(2) Compute U_2 from $U_2 = A_2 - L_2 B_1 \Rightarrow U_2 = \begin{pmatrix} 1.7333 & -1.0667 \\ -1.0667 & 1.7333 \end{pmatrix}.$

Block forward elimination:
$$y_1 = b_1 - L_1 y_0 = b_1 = \begin{pmatrix} 4 \\ 4 \end{pmatrix}, \quad y_2 = b_2 - L_2 y_1 = \begin{pmatrix} 0.6667 \\ 0.6667 \end{pmatrix}.$$

Block back substitution:
Note that
$$U_2 x_2 = y_2 - B_2 x_3 = \begin{pmatrix} 0.6667 \\ 0.6667 \end{pmatrix} \quad \text{(note that } B_2 x_3 = 0\text{)},$$
$$x_2 = \begin{pmatrix} 1 \\ 1 \end{pmatrix},$$
$$U_1 x_1 = y_1 - B_1 x_2 = \begin{pmatrix} 4 \\ 4 \end{pmatrix} - \begin{pmatrix} 1 \\ 1 \end{pmatrix} = \begin{pmatrix} 3 \\ 3 \end{pmatrix},$$
$$x_1 = \begin{pmatrix} 1 \\ 1 \end{pmatrix}. \quad \blacksquare$$

LU Factorization of a Banded Matrix. We have just seen that LU factorization of a block tridiagonal matrix gives L and U both block bidiagonal. For a general banded matrix, the following result holds (see Golub and Van Loan (1996, p. 152)).

Theorem 6.26. *If $A \in \mathbb{R}^{n \times n}$ is a banded matrix with upper bandwidth p and lower bandwidth q, and if $A = LU$, then U has upper bandwidth p and L has lower bandwidth q.*

The growth factor for a banded matrix. If $A \in \mathbb{R}^{n \times n}$ has upper and lower bandwidths p, then the growth factor is $\rho \leq 2^{2p-1} - (p-1)2^{p-2}$.

Proof. See Bohte (1975). $\square$

A comparison of flop-count, growth factor, and stability of the special linear system solvers is given in Table 6.1.

Block Cyclic Reduction

Frequently in practice, the block tridiagonal matrix of a system may possess some special properties that can be exploited to reduce the system to a single lower-order system by using

Table 6.1. *Comparison of different methods for linear system problem with special matrices*

Matrix Type	Method	Flop-count (Approx.)	Growth Factor ρ	Stability
Symmetric Positive Definite	GEWP	$2\dfrac{n^3}{3}$	$\rho = 1$	Stable
	Cholesky	$\dfrac{n^3}{3} + (n$ square roots$)$	None	Stable
Diagonally Strictly Dominant	GEWP	$2\dfrac{n^3}{3}$	$\rho \leq 2$	Stable
Hessenberg	GEPP	$3n^2$	$\rho \leq n$	Stable
Tridiagonal	GEPP	$O(n)$	$\rho \leq 2$	Stable

a technique called **block cyclic reduction**. For details see the book by Golub and Van Loan (1996, pp. 176–180) and the references therein.

6.13 Review and Summary

For an easy reference, we now state the most important results discussed in this chapter.

6.13.1 Numerical Methods for Arbitrary Linear System Problems

Two types of methods—direct and iterative—are used for solving linear systems. Iterative methods are especially helpful for large and sparse systems. We have discussed here only direct methods using Gaussian elimination. *Iterative methods will be discussed in Chapter 12.*

Gaussian elimination (Section 6.4)

- Gaussian elimination without row interchanges, when it exists, gives a factorization of $A : A = LU$.

  The system $Ax = b$ is then solved first by solving the lower triangular system $Ly = b$ followed by solving the upper triangular system $Ux = y$.

  The method requires $\frac{2n^3}{3}$ flops. *It is unstable for arbitrary matrices, and is not recommended for practical use unless matrix A is symmetric positive definite. The growth factor can be arbitrarily large for an arbitrary matrix.*

- Gaussian elimination with partial pivoting (GEPP) gives the factorization $PA = LU$. Once having this factorization, $Ax = b$ can be solved by solving successively the two triangular systems (i) $Ly = Pb = b'$, (ii) $Ux = y$.

The process requires $2\frac{n^3}{3}$ flops and $O(n^2)$ comparisons. In theory, there are some risks involved, but in practice, this is a stable algorithm. *It is the most widely used practical algorithm for solving a dense linear system.*

- Gaussian elimination with complete pivoting (GECP) gives $PAQ = LU$. Once having this factorization, $Ax = b$ can be solved by solving two triangular systems:

$$Lz = Pb = b', Uy = z,$$

and then recovering x from $x = Qy$. The process requires $2\frac{n^3}{3}$ flops and $O(n^3)$ comparisons. Thus it is more expensive than GEPP, but it is more stable (the growth factor ρ in this case is bounded by a slowly growing function of n, whereas the growth factor ρ with GEPP can be as big as 2^{n-1}).

6.13.2 Special Systems

Symmetric positive definite, diagonally dominant, Hessenberg, and tridiagonal systems have been discussed in Section 6.12.

(a) *Symmetric positive definite system.* The Cholesky factorization algorithm (Algorithm 6.7) computes the factorization of a symmetric positive definite matrix A in the form $A = HH^T$, where H is lower triangular with positive diagonal entries. Once having this factorization, the system $Ax = b$ is solved by first solving the lower triangular system $Hy = b$, followed by solving the upper triangular system $H^T x = y$. The method requires $\frac{n^3}{3}$ flops and n square roots evaluations. It is stable.

(b) *Diagonally dominant system.* Gaussian elimination does not require any pivoting. It is stable ($\rho \leq 2$) (Section 6.12.5).

(c) *Hessenberg system.* GEPP requires only $O(n^2)$ flops to solve an $n \times n$ Hessenberg system. It is stable ($\rho \leq n$) (Section 6.12.4).

(d) *Tridiagonal system.* GEPP requires only $O(n)$ flops. It is stable ($\rho \leq 2$) (Section 6.12.6).

(e) *Block tridiagonal system.* Block LU factorization is stable in two important cases: when A is (i) *block column diagonally dominant*, and (ii) *well-conditioned symmetric positive definite matrix* (Section 6.12.6).

6.13.3 Inverse and Determinant

The inverse and the determinant of a matrix A can be readily computed once a factorization of A is available. In practice, only the factorization $PA = LU$, obtained by GEPP, will be used. The inverse can also be computed by solving the system of equations $AX = I$ (Algorithm 6.3).

Note: *The most problems involving inverses can be recast so that the inverse does not have to be computed explicitly.*

- There are matrices (e.g., as *triangular, orthogonal*, etc.) whose inverses are trivially computed.

- The inverse of a matrix B which differs from a matrix A by a rank-one perturbation only can be readily computed, once the inverse of A has been found, by using the *Sherman–Morrison formula:* Let $B = A - uv^T$. Then $B^{-1} = A^{-1} + \alpha(A^{-1}uv^T A^{-1})$, where $\alpha = \frac{1}{(1-v^T A^{-1} u)}$. There is a generalization of this formula, known as the *Woodbury formula* (see Hager (1988, p. 114)).

6.13.4 The Condition Number and Accuracy of Solution

- Computing the condition number from the definition is clearly expensive; it involves finding the norm of the inverse of A, and finding the inverse of A is about four times the expense of solving the linear system itself.

- Condition number estimator: The optimization-based *Hager's norm-1 condition number estimator* (Algorithm 6.5). has been stated.

- There are *symptoms* exhibited during Gaussian elimination with pivoting such as a *small pivot*, a *large computed solution*, a *large residual*, etc., that merely indicate if a system is ill-conditioned, but these are not sure tests (Section 6.8.1).

- When componentwise perturbations are known, *Skeel's condition number* can be useful, especially when the norms of the columns of the inverse matrix vary widely (Theorem 6.15).

6.13.5 Iterative Refinement

Once a solution has been computed, an inexpensive way to refine the solution iteratively, known as the iterative refinement procedure (Algorithm 6.6), has been described in Section 6.11. The iterative refinement technique is a very useful technique.

6.14 Suggestions for Further Reading

The books on numerical methods in engineering literature routinely discuss how various engineering applications give rise to linear systems problems. We have used Chapra and Canale (2002), O'Neil (1991), and James et al. (1989) in our discussions and found them useful. A recent book of interest is by Majumdar (2005). Direct methods (such as Gaussian elimination, QR factorization, etc.) for linear systems and related problems, discussions on perturbation analysis and conditioning of the linear systems problems, iterative refinement, etc., can be found in any standard matrix computations texts: Golub and Van Loan (1996), Demmel (1997), Trefethen and Bau (1997), Watkins (2002), Hager (1988), and Stewart (1998b). *Stewart's classic book (1973) is still a rich source of knowledge*. Most numerical analysis texts contain some discussions on these topics. In particular, the books by Conte and de Boor (1980), Heath (2002), Van Loan (2000), Kincaid and Cheney (2002), Moler (2004), and Ortega (1990) and Stewart's numerical analysis book (Stewart (1998a)) provide a fair amount of numerical linear algebra treatment. See also Rice (1981). For discussion on solutions of linear systems with special matrices such as *diagonally dominant, Hessenberg, positive definite*, etc., see Wilkinson (1965, pp. 218–220) and Higham (2002).

Two authoritative books on error analysis and perturbation analysis are Wilkinson's classic (1965) and the recent book by Higham (2002). *These two books are must-reads for the interested readers on these topics.* A book devoted entirely to perturbation analysis is Stewart and Sun (1990). Some interesting earlier papers on perturbation analysis and conditioning include Stewart (1993a), Demmel (1987b), Edelman (1988, 1992a), Sun (1991), Zha (1993), McCarthy and Strang (1973), O'Leary (1980), Bischof (1990), Higham (1987), and Pierce and Plemmons (1992).

Exercises on Chapter 6

(Use MATLAB, wherever appropriate and necessary.)

EXERCISES ON SECTIONS 6.3 AND 6.12.1–6.12.6

6.1 An engineer requires 5000, 5500, and 6000 yd^3 of sand, cement, and gravel for a building project. He buys his material from three stores. A distribution of each material in these stores is given as follows:

Store	Sand %	Cement %	Gravel %
1	60	20	20
2	40	40	20
3	20	30	50

How many cubic yards of each material must the engineer take from each store to meet his needs?

6.2 If the input to reactor 1 in the "reactor" problem of Section 6.3.2 is decreased 10%, what is the percent change in the concentration of the other reactors?

6.3 Consider the following circuit diagram:

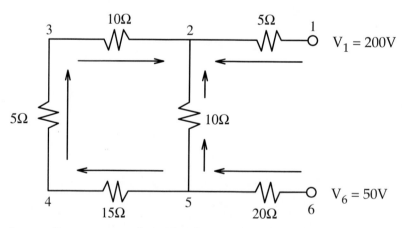

Set up a linear system to determine the current between nodes.

6.4 Using the difference equation (6.47), set up a linear system for heat distribution at the following interior points of a heated plate whose boundary temperatures are held constant:

$$50°\text{C}$$

```
●         ●         ●
(1,1)     (2,1)     (3,1)

●         ●         ●
(1,2)     (2,2)     (3,2)

●         ●         ●
(1,3)     (2,3)     (3,3)
```

$100°\text{C}$ (left side) $75°\text{C}$ (right side)

$$0°\text{C}$$

6.5 Derive the linear system for the finite difference approximation of the elliptic equation

$$\frac{\partial^2 T}{\partial x^2} + \frac{\partial^2 T}{\partial y^2} = f(x, y).$$

The domain is in the unit square, $\Delta x = \Delta y = 0.01$, and the boundary conditions are given by

$$T(x, 0) = 1 - x, \quad T(1, y) = y, \quad T(0, y) = 1, \quad T(x, 1) = 1.$$

6.6 For Exercise 6.5, if

$$f(x, y) = -\pi^2 \sin(\pi x) \sin(\pi y),$$

then the analytic solution to the elliptic equation

$$\frac{\partial^2 T}{\partial x^2} + \frac{\partial^2 T}{\partial y^2} = f(x, y),$$

with the same boundary conditions as in Exercise 6.5, is given by

$$T(x, y) = 1 - x + xy + \left(\frac{1}{2}\right) \sin(\pi x) \sin(\pi y)$$

(Celia and Gray (1992, pp. 105–106)).

(a) Use the finite difference scheme of Section 6.3.4 to approximate the values of T at the interior points with $\Delta x = \Delta y = \frac{1}{n}, n = 4, 8, 16$.

(b) Compare the values obtained in (a) with the exact solution.

6.7 Write down the linear system arising from the finite element method of the solution of the two-point boundary value problem, $-2u'' + 3u = x^2, 0 \le x \le 1, u(0) = u(1) = 0$, using a uniform grid and the same basic functions $\phi_j(x)$ as in Section 6.12.2.

EXERCISES ON SECTIONS 6.4–6.11.

6.8 (a) Solve the linear systems $Ax = b$ with each matrix A of Exercises 5.13 and 5.14 of Chapter 5, taking $b = (1, 1, \ldots, 1)^T$ and using the factorizations (i) $A = LU$, (ii) $MA = U$, and $PA = LU$.

(b) Compute the determinant, the inverse (when it exists), and the growth factor of the above matrices using the factorizations in (a).

6.9 Compute the residual norms for each of the systems of Exercise 6.8 (with all three factorizations) and plot these residuals using separate graphs.

6.10 Solve
$$\begin{pmatrix} 0.00001 & 1 & 1 \\ 3 & 1 & 1 \\ 1 & 2 & 3 \end{pmatrix} \begin{pmatrix} x_1 \\ x_2 \\ x_3 \end{pmatrix} = \begin{pmatrix} 2.0001 \\ 3 \\ 3 \end{pmatrix}$$
using Gaussian elimination without and with partial pivoting and compare the answers.

6.11 Consider m linear systems
$$Ax_i = b_i, \ i = 1, 2, \ldots, m.$$

(a) Develop an algorithm to solve the above systems based on GECP (Gaussian elimination with complete pivoting). Your algorithm should take advantage of the fact that all m systems have the same system matrix A.

(b) Compute the flop-count of this algorithm.

(c) Apply your algorithm to compute A^{-1} and work out a flop-count for this computation.

(d) Apply the algorithm in (c) to compute the inverse of a 5×5 Hilbert matrix.

6.12 Consider the system $Ax = b$, where both A and b are complex. Show how the system can be solved using real arithmetic only. Compare the flop-count in this case with that needed to solve the system with Gaussian elimination using complex arithmetic.

6.13 (*Gaussian elimination with column pivoting.*) Develop an algorithm for LU factorization of a matrix A based on column pivoting instead of row pivoting. Does this factorization always exist? Give reasons for your answer. Show that when it exists, it leads to a factorization, $AP = LU$, where P is a permutation matrix. Apply this factorization to solve each of the systems of Exercise 6.8.

6.14 (a) (*Computing the inverse of a block matrix.*) Let $A = \begin{pmatrix} A_{11} & A_{12} \\ A_{21} & A_{22} \end{pmatrix}$.

Assume that A_{11} and A_{22} are square and that A_{11} and $A_{22} - A_{21}A_{11}^{-1}A_{12}$ are nonsingular.

Let $B = \begin{pmatrix} B_{11} & B_{12} \\ B_{21} & B_{22} \end{pmatrix}$ be the inverse of A. Then show that

$$B_{22} = (A_{22} - A_{21}A_{11}^{-1}A_{12})^{-1}, \quad B_{12} = -A_{11}^{-1}A_{12}B_{22},$$
$$B_{21} = -B_{22}A_{21}A_{11}^{-1}, \quad \text{and } B_{11} = A_{11}^{-1} - B_{12}A_{21}A_{11}^{-1}.$$

(b) How many flops are needed to compute A^{-1} using the results of (a) if A_{11} and A_{22} are, respectively, $m \times m$ and $p \times p$?

(c) Use your result of (a) to compute A^{-1}, where
$$A = \begin{pmatrix} 4 & 0 & -1 & -1 \\ 0 & 4 & -1 & -1 \\ -1 & -1 & 4 & 0 \\ -1 & 1 & 0 & 4 \end{pmatrix}.$$

6.15 Let
$$A = \begin{pmatrix} 1 & 2 & 1 \\ 2 & 4.0001 & 2.0002 \\ 1 & 2.0002 & 2.0004 \end{pmatrix} \quad \text{and} \quad B = \begin{pmatrix} 0 & 2 & 1 \\ 2 & 4.0001 & 2.0002 \\ 1 & 2.0002 & 2.0004 \end{pmatrix}.$$

Write B in the form $B = A - uv^T$, then compute B^{-1} using the Sherman–Morrison formula (Theorem 6.7), knowing
$$A^{-1} = 10^4 \begin{pmatrix} 4.0010 & -2.0006 & 0.0003 \\ -2.0006 & 1.0004 & -0.0002 \\ 0.0003 & -0.0002 & 0.0001 \end{pmatrix}.$$

6.16 Suppose you have solved a linear system with A as the system matrix. Then show how one can solve the augmented system
$$\begin{pmatrix} A & \delta \\ c & \alpha \end{pmatrix} \begin{pmatrix} x \\ x_{n+1} \end{pmatrix} = \begin{pmatrix} b \\ b_{n+1} \end{pmatrix},$$

where A is nonsingular and $n \times n$ and a, b, and c are vectors, using the solution you have already obtained. Apply your result to solve
$$\begin{pmatrix} 1 & 2 & 3 & 1 \\ 4 & 5 & 6 & 1 \\ 1 & 1 & 1 & 1 \\ 0 & 0 & 1 & 2 \end{pmatrix} y = \begin{pmatrix} 6 & 15 & 3 & 1 \end{pmatrix}^T.$$

6.17 Consider the symmetric system $Ax = b$, where
$$A = \begin{pmatrix} 0.4445 & 0.4444 & -0.2222 \\ 0.4444 & 0.4445 & -0.2222 \\ -0.2222 & -0.2222 & 0.1112 \end{pmatrix} \quad b = \begin{pmatrix} 0.6667 \\ 0.6667 \\ -0.3332 \end{pmatrix}.$$

The exact solution of the system is
$$x = \begin{pmatrix} 1 \\ 1 \\ 1 \end{pmatrix}.$$

(a) Make a small perturbation δb in b, keeping A unchanged. Solve the system $Ax' = b + \delta b$. Compare x' with x. Compute Cond(A) and verify the result of Theorem 4.19.

(b) Make a small perturbation ΔA in A such that $\|\Delta A\| < \frac{1}{\|A^{-1}\|}$. Solve the system $(A + \Delta A)x' = b$. Compare x' with x and verify the result of Theorem 4.23. (Hint: $\|A^{-1}\|_2 = 10^4$).

(c) Compute the residual in each case and verify the result of Theorem 6.12 (residual theorem).

6.18 Prove
$$\frac{\|\delta x\|}{\|x + \delta x\|} \le \text{Cond}(A) \frac{\|\Delta A\|}{\|A\|},$$
where $Ax = b$ and $(A + \Delta A)(x + \delta x) = b$.

Verify the above inequality for the system
$$\begin{pmatrix} 1 & \frac{1}{2} & \frac{1}{3} \\ \frac{1}{2} & \frac{1}{3} & \frac{1}{4} \\ \frac{1}{3} & \frac{1}{4} & \frac{1}{5} \end{pmatrix} \begin{pmatrix} x_1 \\ x_2 \\ x_3 \end{pmatrix} = \begin{pmatrix} 1 \\ 1 \\ 1 \end{pmatrix}, \quad \text{with } \Delta A = \begin{pmatrix} 0 & 0 & 0.00003 \\ 0 & 0 & 0 \\ 0 & 0 & 0 \end{pmatrix}.$$

6.19 (a) Construct your own example to show that a small residual does not necessarily guarantee that the solution is accurate.

(b) Give a proof of Theorem 6.12.

6.20 (a) Compute the condition numbers of the following matrices before and after row-scaling and compare the results.

$$\text{(i)} \ A = \begin{pmatrix} 10 & 10^6 \\ 1 & 1 \end{pmatrix}, \quad \text{(ii)} \begin{pmatrix} 2 & -1 & 1 \\ -1 & -10 & -10 \\ 1 & 10^{-10} & 10^{-10} \end{pmatrix}.$$

(b) Solve $Ax = b$ with each of the matrices in (a) before and after scaling (choose b so that exact solution in each case has all entries equal to 1).

6.21 Construct an example to illustrate the phenomenon of the artificial ill-conditioning, that is, ill-conditioning due to improper scaling.

6.22 Let $A = \begin{pmatrix} 1 & 0 \\ 0 & 10^{-5} \end{pmatrix}$.

(a) Calculate A^{-1} and $\text{Cond}_2(A)$.

(b) Find b and x such that $Ax = b$.

(c) Find δb and δx such that $A(x + \delta x) = b + \delta b$ and $\frac{\|\delta b\|_\infty}{\|b\|_\infty}$ is small, but $\frac{\|\delta x\|_\infty}{\|x\|_\infty}$ is large. What is your conclusion?

(d) Now multiply your second equation by 10^5. What is the condition number of this new matrix? What is your conclusion?

6.23 Apply Hager's algorithm (Algorithm 6.5) to the Hilbert matrix of order 5 and then compare the result with those obtained by MATLAB function **condest** and the actual 1-norm condition number.

Exercises on Chapter 6

6.24 Construct an example to verify the statement that Skeel's condition number $\text{Cond}(A, x)$ can be much smaller than $\text{Cond}(A)$.

6.25 Construct an example of an ill-conditioned system that exhibits the symptoms of ill-conditioning stated in Section 6.8.1.

6.26 Apply iterative refinement (Algorithm 6.6) to the system of Exercise 6.10, using

$$x^0 = \begin{pmatrix} 0 \\ 0 \\ 0 \end{pmatrix}.$$

Estimate $\text{Cond}(A)$ from $x^{(1)}$ and $x^{(2)}$ and compare it with the actual condition number.

EXERCISES ON SECTION 6.12.3–6.12.6

6.27 (a) Compute the Cholesky factorization of

$$A = \begin{pmatrix} 1 & 1 & 1 \\ 1 & 1.001 & 1.001 \\ 1 & 1.001 & 2 \end{pmatrix}$$

using (i) Gaussian elimination without pivoting, and (ii) the Cholesky algorithm.

(b) For part (i) verify

$$\max |a_{ij}^{(k)}| \leq \max |a_{ij}^{(k-1)}|, \quad k = 1, 2.$$

What is the growth factor?

(c) Solve the system $Ax = b$, where

$$b = \begin{pmatrix} 3 \\ 3.0020 \\ 4.0010 \end{pmatrix},$$

in each case.

6.28 (a) Show that

$$A = \begin{pmatrix} 4 & -1 & -1 & 0 \\ -1 & 4 & 0 & -1 \\ -1 & 0 & 4 & -1 \\ 0 & -1 & -1 & 4 \end{pmatrix}$$

is a positive definite matrix using both Gaussian elimination and the Cholesky algorithm (Algorithm 6.7).

(b) Compute $\text{Cond}(A)$ from its Cholesky factorization.

6.29 Using the Hilbert matrix of order 10, show that the solution obtained by the Cholesky algorithm may be inaccurate if the positive definite matrix is highly ill-conditioned.

6.30 Prove the existence of the Cholesky factorization of a symmetric positive definite matrix A via LU factorization.

6.31 Let $H = (h_{ij})$ be the Cholesky factorization of a symmetric positive definite matrix $A = (a_{ij})$. Then prove that $h_{ij}^2 \leq \sum_{k=1}^{i} h_{ik}^2 = a_{ii}$.

6.32 Let A be symmetric positive definite. Develop and state an algorithm for computing an upper triangular matrix R such that $A = R^T R$. Apply this algorithm to the matrix A of Exercise 6.27.

6.33 Prove that the growth factor ρ of a symmetric positive definite matrix with Gaussian elimination without pivoting is equal to 1.

6.34 (a) Develop an algorithm to solve a tridiagonal system using GEPP.

(b) Show that the growth factor in this case is bounded by 2. (*Hint:* $\max |a_{ij}^{(1)}| \leq 2 \max |a_{ij}|$.)

(c) Apply your algorithm to find the solution to the tridiagonal system (6.44), with $T_0 = 0°$F and $T_4 = 100°$F.

6.35 (a) Prove that a strictly column diagonally dominant matrix has an LU factorization without pivoting.

(b) Show that the growth factor using Gaussian elimination without pivoting for such a matrix is bounded by 2. (*Hint:* $\max_k \max_{i,j} |a_{ij}^{(k)}| \leq 2 \max_{i,j} |a_{ij}|$.)

(c) Verify the statement of (a) with the matrix of Exercise 6.14.

(d) Construct a 2×2 strictly column diagonally dominant matrix whose growth factor for Gaussian elimination without pivoting is larger than 1 but less than or equal to 2.

(e) Repeat (a)–(d) for a strictly row diagonally dominant matrix.

(f) Construct an example to show that for a strictly row diagonally dominant matrix, the multiplier can be large but the growth factor is still bounded by 2.

6.36 (a) Develop an algorithm for solving a symmetric positive definite tridiagonal linear system using the LDL^T decomposition. Show that this algorithm takes only $O(n)$ flops.

(b) Apply your algorithm to solve the symmetric positive definite system with the matrix A given by (6.50) for $n = 3$, choosing $b = (1, 0, 0)^T$.

6.37 Solve the strictly diagonally dominant system

$$\begin{pmatrix} 10 & 5 & 5 & 5 \\ 5 & 10 & 1 & 2 \\ -1 & 0 & 10 & 2 \\ -1 & -4 & -3 & 10 \end{pmatrix} \begin{pmatrix} x_1 \\ x_2 \\ x_3 \\ x_4 \end{pmatrix} = \begin{pmatrix} 13 \\ 13 \\ 10 \\ 7 \end{pmatrix}$$

using Gaussian elimination without pivoting. Compute the growth factor.

6.38 (a) Develop an algorithm for triangularizing an upper Hessenberg matrix using GEPP by taking advantage of the Hessenberg structure of the matrix. Give flop-count.

(b) Let A be an $n \times n$ upper Hessenberg matrix. Then GEPP gives

$$|a_{ij}^{(k)}| \leq k + 1, \quad \text{if } |a_{ij}| \leq 1.$$

Hence deduce that the growth factor in this case is bounded by n.

(c) Apply your algorithm developed in (a) to solve the following systems:

(i) $\begin{pmatrix} 1 & 2 & 2 & 3 \\ 0.001 & 4 & 5 & 7 \\ 0 & 0.001 & 2 & 3 \\ 0 & 0 & 0.001 & 1 \end{pmatrix} \begin{pmatrix} x_1 \\ x_2 \\ x_3 \\ x_4 \end{pmatrix} = \begin{pmatrix} 6 \\ 7 \\ 8 \\ 9 \end{pmatrix}.$

(ii) $\begin{pmatrix} 1 & 0 & 0 \\ 0.0001 & 2 & 0 \\ 0 & 0.0001 & 3 \end{pmatrix} \begin{pmatrix} x_1 \\ x_2 \\ x_3 \end{pmatrix} = \begin{pmatrix} 1.0001 \\ 2.0001 \\ 3.0000 \end{pmatrix}.$

(iii) $\begin{pmatrix} 0 & 0 & 0 & 1 \\ 1 & 0 & 0 & 2 \\ 0 & 1 & 0 & 3 \\ 0 & 0 & 1 & 4 \end{pmatrix} \begin{pmatrix} x_1 \\ x_2 \\ x_3 \\ x_4 \end{pmatrix} = \begin{pmatrix} 0 \\ 0 \\ 1 \\ 1 \end{pmatrix}.$

(d) Compute the growth factor in each case.

(e) Suppose the data in the above problems are accurate to four digits and you seek an accuracy of three digits in your solution. Identify which of the above problems are ill-conditioned. (Use the result of Section 4.7.2.)

6.39 (*Solution of two-dimensional Poisson's equation.*) Using Algorithms 6.10, 6.11, and 6.12, solve the system (6.49)–(6.50) appearing in the solution of Poisson's equation (6.46), with $n = 3$, and choosing the entries of the right-hand side appropriately. Show that this system is block diagonally dominant and verify inequality (6.63) for this system.

6.40 Using the result of Exercise 5.17, establish the error bound of Theorem 6.4 and also prove the result $\|E\|_\infty \leq 3n^3 \rho \mu \|A\|_\infty$.

MATLAB and MATCOM Programs and Problems on Chapter 6

M6.1 (a) Write MATLAB programs called **linsyswp**, **linsyspp**, and **linsyscp** to solve $Ax = b$ and to compute the growth factor (gf) using Gaussian elimination with no, partial, and complete pivotings, respectively, as follows:

$$(\hat{x}, gf) = \text{lnsyswp}(A, b),$$
$$(\hat{x}, gf) = \text{linsyspp}(A, b),$$
$$(\hat{x}, gf) = \text{linsyscp}(A, b).$$

(b) Using the computed solutions and the growth factors obtained in (a) make the following table for each of the given data set. The function **linsyswf** is available in MATCOM.

Test data for Problem M6.1: Each of the following matrices of order 20: *Hilbert, Pei, Hankel, Vandermonde, a randomly generated matrix*, and a triangular matrix with small diagonal entries. For the Pei matrix, take α close to 1.

Create the vector b in each case such that the solution vector x is a vector with all components equal to 1. Present your results using Table 6.2.

Table 6.2. *Comparison of different methods for the linear systems problems.*

Method	Norm of the Computed Solution $\|\hat{x}\|_\infty$	Relative Error $\dfrac{\|x - \hat{x}\|_2}{\|x\|_2}$	Residual $\|b - A\hat{x}\|_2$	Posterior Error Bound $\text{Cond}_2(A) \times \dfrac{\|b - A\hat{x}\|_2}{\|b\|_2}$	Growth Factor
linsyswp					
linsyspp					
linsyscp					
linsyswf					
$A^{-1}b$					

M6.2 (*Backward error in GEPP.*) Plot the error bounds (i) $3n^3 \rho \mu \|A\|_\infty$, (ii) $\dfrac{\|A\hat{x} - b\|_\infty \|A^{-1}\|_\infty}{\|\hat{x}\|_\infty}$ for solving $Ax = b$ with partial pivoting for five random matrices of dimensions varying from 10 to 1000, by generating b also randomly. Write down your observations.

M6.3 Repeat Problem M6.2 using GECP.

M6.4 Perform an experiment, for both GEPP and GECP, using five random matrices of orders varying from 10 to 100, to verify that the *error bound $3\rho n^3 \mu \|A\|_\infty$ is pessimistic compared to the true error*. Present your results with graphs.

M6.5 Using the MATCOM program **choles** or the MATLAB program **chol**, write a MATLAB program, **linsyschol**, to implement Algorithm 6.8 in the following format:

$$[x] = \text{linsyschol}(A, b).$$

Data: Create a 200×200 lower triangular matrix L with positive diagonal entries taking some of the diagonal entries small enough to be very close to zero, multiply it by L^T, and take $A = LL^T$ as your test matrix A. Create the vector b such that x has all its entries equal to 1.

M6.6 (a) Write a MATLAB program **lutrdg** to implement Algorithm 6.9, then use it to write a program, **linsystrdg**, to solve a tridiagonal system.

(b) (*Solution of one-dimensional heat equation.*) Apply **linsystrdg** to solve the tridiagonal system of the form (6.44) of order 200, choosing $T_0 = 0$ and $T_{199} = 100$.

M6.7 (*The purpose of this exercise is to verify that solving a symmetric positive definite system requires no pivoting to ensure stability in Gaussian elimination.*) Run the program **lynsyswp** to the symmetric positive definite matrix of Problem M6.5 and compute the solution and the growth factor.

M6.8 Using the MATCOM program **hagcond1**, estimate the condition number of each of the following matrices A of order 20: *Hilbert, Pei (with α close to* 1*), randomly generated, Vandermonde, and Hankel,* and then compare your results with the actual condition number obtained by the MATLAB program cond $(A, 1)$. Present your results in the form of a table.

M6.9 (a) Run the program **linsyswp** with the diagonally dominant, symmetric tridiagonal matrix A in (6.50) of order 200 by choosing the right-hand side vector b so that the solution vector x is known a priori. Compare the exact solution x with the computed solution $\hat{x}$.

   (b) (*Implementation of two-dimensional heat equation.*) Using the program **linsyswp**, solve the system (6.45) with $n = 200$, and choosing $k = 0$, $T_{i,0} = 0$, $i = 1, \ldots, n$, and $T_{w1} = 0$ and $T_{w2} = 100$. *(Note that system (6.45) is also symmetric positive definite and tridiagonal.)*

M6.10 Write a MATLAB program, called **lynstrdgpp**, to solve a tridiagonal system with partial pivoting. Apply the algorithm to the data of Problem M6.6.

M6.11 (a) Run the iterative refinement program **iterref** from MATCOM on each of the 50×50 systems: *Hilbert, Pei, Vandermonde, randomly generated, and Hankel,* using the solution obtained from the program **linsyspp** as the initial approximation $x^{(0)}$. For the Pei matrix, take α close to 1.

   (b) Estimate the condition number of each of the above matrices obtained from the iterative refinement procedure and compare them with the actual condition numbers.

M6.12 (a) Write MATLAB programs **bklutrdg, bkforelm, bkbacksub** to implement Algorithm 6.10–6.12, respectively.

   (b) (*Solution of Poisson's equation.*) Using the programs in Problem M6.12(a), write a MATLAB program, **bltrdgls**, to solve the block tridiagonal linear system $Tx = b$. Run your program with the linear system (6.49)–(6.50) by choosing the right-hand side appropriately, with $n = 15$.

Chapter 7

QR Factorization, Singular Value Decomposition, and Projections

Background Material Needed

- Concepts of rank, basis, range, and null space (Sections 2.2.1, 2.3.1, and 2.3.2)
- Special matrices (Section 2.4)
- Vector and matrix norms (Section 2.5)
- Condition number (Sections 4.6 and 4.7)

7.1 Introduction

In Chapter 5, we described *LU factorization* of a matrix, and in Chapter 6 we showed how this factorization is used to solve $Ax = b$. In this chapter, we describe two other important matrix factorizations: *QR* and *singular value decomposition (SVD)*. *These two factorizations play important roles in least-squares solutions* (Chapter 8) *and in many other important matrix and applied computations such as image restoration and image construction, biomedical engineering, etc.*

Recall that a square matrix O is said to be an *orthogonal matrix* if $OO^T = O^T O = I$. Given an $m \times n$ matrix A there exists an $m \times m$ orthogonal matrix Q and an $m \times n$ *upper triangular matrix* R such that $A = QR$. Such a factorization of A is called the *QR factorization*.

We shall prove the existence of QR factorization by actually constructing the matrices Q and R in several different ways so that $A = QR$.

If $m \geq n$, and if the matrix Q is partitioned as $Q = (Q_1, Q_2)$, where Q_1 is the matrix of the first n columns of Q, and if R_1 is defined by $R = \binom{R_1}{0}$, where R_1 is $n \times n$ upper triangular, then $A = Q_1 R_1$.

Thus, if $m \geq n$, A can be factorized into $A = Q_1 R_1$, where Q_1 is $m \times n$ *orthonormal* and R_1 is $n \times n$ *upper triangular*, as shown in Figure 7.1.

This QR factorization is called the "economy size," the "thin," or the **reduced** QR factorization of A.

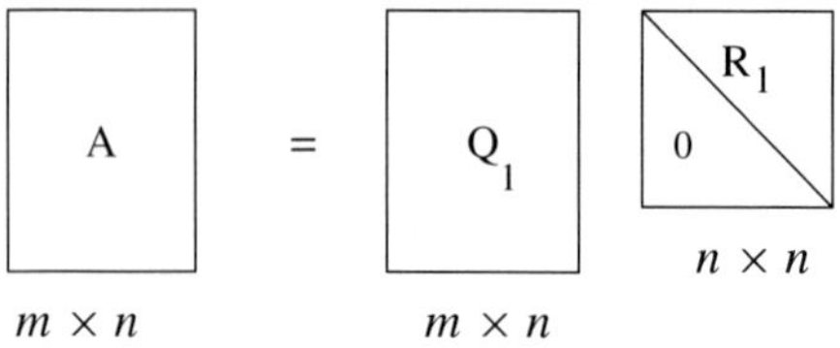

Figure 7.1. *Reduced QR factorization.*

To distinguish between these two types of QR factorization, the factorization $A = QR$ is sometimes called *full QR factorization*. By actual constructions, we will show the following:

Every matrix $A \in \mathbb{R}^{m \times n}$ ($m \geq n$) has a full QR factorizations (and hence also a reduced QR factorization). Moreover, if A has full-rank, then it has a unique reduced QR factorization $A = Q_1 R_1$ with positive diagonal entries of R_1 (Theorem 7.14). If $m < n$, then the factorization can be written as $A = Q(R_1, R_2)$, where R_1 is upper triangular and R_2 is rectangular.

Computing QR Factorization

We will describe the following here:

- Householder's method (Algorithm 7.2).

- Givens' method (Algorithm 7.5).

- The classical (CGS) and modified Gram–Schmidt (MGS) methods (Algorithms 7.7 and 7.8).

The *Householder* and *Givens methods* compute the *full QR factorization* (and therefore the reduced QR factorization), while the *CGS* and *MGS* methods compute the *reduced QR factorization*. These methods are described in Sections 7.2, 7.4, and 7.5, respectively.

The singular value decomposition (SVD) of a matrix $A \in \mathbb{R}^{m \times n}$ is a factorization of A in the form

$$A = U \Sigma V^T,$$

where $U \in \mathbb{R}^{m \times m}$ and $V \in \mathbb{R}^{n \times n}$ are orthogonal and $\Sigma \in \mathbb{R}^{m \times n}$ is diagonal. The SVD has become a computationally viable tool for solving a wide variety of problems arising in many practical applications, including *signal and image processing, biomedical engineering, control engineering,* and others. In Section 7.8, we introduce SVD and discuss its basic properties and applications. The *SVD will be revisited in* Chapter 10. As a preparation for discussing least-squares solutions techniques, which will be described in Chapter 8, we introduce the concept of **orthogonal projection** and its computation using QR factorization and SVD in Sections 7.7 and 7.8.10, respectively. *Complex QR factorization* and *complex SVD* are briefly introduced in Sections 7.3 and 7.8.3, respectively.

7.2 Householder's Matrices and QR Factorization

7.2.1 Definition and Basic Properties

Definition 7.1 (Householder matrix and Householder vector). *A matrix of the form*

$$H = I - \frac{2uu^T}{u^T u}, \tag{7.1}$$

*where u is a nonzero vector in $\mathbb{R}^n$, is called a **Householder matrix** after the celebrated numerical analyst Alston Householder.[5] The vector u determining the Householder matrix H is called the **Householder vector**.*

A Householder matrix is also known as an *elementary reflector* or a *Householder transformation*. We now give a geometric interpretation of a Householder transformation. For the sake of convenience, in the geometric interpretation (see Figure 7.2), we assume that *vector u is such that $u^T u = 1$.*

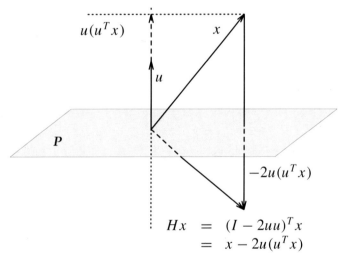

Figure 7.2. *Geometric interpretation of Householder transformation.*

With this geometric interpretation the following results become clear:

- $\|Hx\|_2 = \|x\|_2$ for every $x \in \mathbb{R}^n$. *A reflection does not change the length of the vector.*

---

[5]Alston Householder (1904–1993), an American mathematician, was born in Rockford, Illinois. He was the former Director of the Mathematics and Computer Science Division of the Oak Ridge National Laboratory at Oak Ridge, Tennessee, and a former Professor of Mathematics at the University of Tennessee, Knoxville. A research conference on linear and numerical linear algebra dedicated to Dr. Householder, called the Householder Symposium, is held every three years around the world. See his obituary in *SIAM News*, October 1993.

- H is an orthogonal matrix. $\|Hx\|_2 = \|x\|_2$ for every vector x implies that H is orthogonal.

- $H^2 = I$. Hx reflects x to the other side of P, but $H^2 x = H(Hx)$ reflects it back to x.

- $Hy = y$ for every y for which $v^T u = 0$. Vectors in P cannot be reflected away.

Below we summarize some of the above interesting properties of a Householder matrix and give analytical proofs of some of them.

Theorem 7.2 (properties of a Householder matrix). *Let $H = I - \frac{2uu^T}{u^T u}$ be a Householder matrix with $u \in \mathbb{R}^n$. Then*

(i) *H is symmetric;*

(ii) *H is orthogonal;*

(iii) *$H^2 = I$;*

(iv) *$Hu = -u$;*

(v) *$Hv = v$ if $v^T u = 0$;*

(vi) *if u is chosen to be a vector parallel to $x - y$, where $y \neq x$ but $\|y\|_2 = \|x\|_2$, then $Hx = y$.*

Proof. The proofs are done by direct verification. Let $\beta = \frac{2}{u^T u}$.
Proof of (i). $H^T = (I - \beta uu^T)^T = I - \beta uu^T = H$. Thus, H is symmetric.
Proof of (ii).
$$H^T H = (I - \beta uu^T)(I - \beta uu^T)$$
$$= I - 2\beta uu^T + \beta^2 (u^T u) uu^T \quad \text{(note that } u^T u \text{ is a scalar)}$$
$$= I - 2\beta uu^T + \beta^2 \cdot \frac{2}{\beta} uu^T \quad \text{(note that } u^T u = \frac{2}{\beta}\text{)}$$
$$= I - 2\beta uu^T + 2\beta uu^T = I.$$

Thus, H is *orthogonal*.
Proof of (iii). Since $H = H^T$, we have $H^2 = H \cdot H = H^T H = I$, by parts (i) and (ii).
The proofs of (iv) and (v) are left as Exercise 7.1.
Proof of (vi). Choose $u = x - y$. Now write
$$x = \frac{1}{2}(x + y) + \frac{1}{2}(x - y).$$

Then
$$Hx = \frac{1}{2} H(x + y) + \frac{1}{2} H(x - y) = \frac{1}{2} H(x + y) + \frac{1}{2}(y - x) \qquad (7.2)$$

(by property (iv)).

7.2. Householder's Matrices and QR Factorization

Again, since

$$(x+y)^T(x-y) = x^T x - x^T y + y^T x - y^T y$$
$$= \|x\|_2^2 - \|y\|_2^2 = 0 \quad (\text{since } \|x\|_2 = \|y\|_2),$$

by property (v), we have

$$H(x+y) = x+y.$$

Thus, from (7.2), we have

$$Hx = \frac{1}{2}(x+y) + \frac{1}{2}(y-x) = y. \quad \Box$$

Forming matrix-vector and matrix-matrix products with a Householder matrix. A remarkable computational advantage involving Householder matrices is that *neither a matrix-vector product with a Householder matrix H nor the matrix product HA (or AH) needs to be explicitly formed.* This can be seen from the following computations.

Let $A \in \mathbb{R}^{m \times n}$ and $x \in \mathbb{R}^n$, and let H be an $n \times n$ Householder matrix. Let $\beta = \frac{2}{u^T u}$. Then

1. $Hx = \left(I - 2\frac{uu^T}{u^T u}\right)x = x - \beta u(u^T x)$.

2. $HA^T = (I - \beta u u^t)A^T = A^T - \beta u u^T A^T = A^T - \beta u v^T$, where $v = Au$.

3. $AH = A(I - \beta u u^T) = A - \beta v u^T$, where $v = Au$.

Flop-count for matrix-matrix and matrix-vector products with Householder matrices. From the above statements, it is straightforward to verify the flop-counts for the following computations with Householder matrices (Exercise 7.3):

- Matrix-vector product Hx: $3n$ flops.
- Matrix product HA or AH: $4mn$ flops (using statement 2 or 3 above).
- Explicit computation of the product of Householder matrices: Let $H_i, i = 1, \ldots, r$, be r Householder matrices each of order n. Then computing $Q = H_1 H_2, \ldots, H_r$ requires $4(n^2 r - nr^2 + \frac{r^3}{3})$ flops.
- Matrix product $Q^T C$: Let $C \in \mathbb{R}^{n \times p}$ and let Q be as above. Then the product $Q^T C$ requires

$$\begin{cases} 2pr(2n-r) & (Q \text{ given in factored form as above}), \\ 2pn^2 & (Q \text{ explicitly represented}). \end{cases}$$

We emphasize here that in practice a matrix-matrix or matrix-vector product with a Householder matrix should be computed as shown above, without explicitly forming the Householder matrix H.

Numerical stability. The following round-off properties show that computations with Householder matrices are very stable (Wilkinson (1965, pp. 152–162)). Let $\hat{H}$ denote the computed Householder matrix. Then the following is obtained:

- $\|\hat{H} - H\| = O(\mu)$.
- $\mathrm{fl}(\hat{H}A) = H(A + E); \quad \|E\|_2 = O(\mu\|A\|_2)$.
- $\mathrm{fl}(A\hat{H}) = (A + E)H; \quad \|E\|_2 = O(\mu\|A\|_2)$.

Creating Zeros in a vector with a Householder Matrix

A very useful property of Householder matrices is that, given a nonzero vector x, a Householder matrix H can always be found such that certain specified entries of x can be made zeros. The following result shows how to find the Householder vector u such that Hx is a multiple of e_1.

Theorem 7.3. *Given a nonzero vector $x \ne e_1$, the Householder matrix H defined by the vector $u = x \pm \|x\|_2 e_1$ is such that $Hx = \mp \|x\|_2 e_1$.*

Proof. The proof follows immediately from part (vi) of Theorem 7.2 by choosing $y = \pm \|x\|_2 e_1$. Note that with this choice, $y \ne x$ and $\|y\|_2 = \|x\|_2$. Thus, by part (vi) of Theorem 7.2, we have
$$Hx = y = \pm\|x\|_2 e_1. \quad \square$$

An illustration:

$$x = \begin{pmatrix} \times \\ \times \\ \times \\ \vdots \\ \times \end{pmatrix} \xrightarrow{H} Hx = \begin{pmatrix} \times \\ 0 \\ 0 \\ \vdots \\ 0 \end{pmatrix} = \pm\|x\|_2 e_1.$$

Choosing the sign: While forming $u = x \pm \|x\|_2 e_1$, it is advisable to choose sign (x_1) in place of $\pm$ to avoid catastrophic cancellation in computing the first component of u. Thus the vector u should be formed as follows:

$$u = x + \mathrm{sign}(x_1)\|x\|_2 e_1.$$

If x_1 is zero, just choose sign $(x_1) = +$.

Note: The vector u above differs from x only in the first component—the other components are the same.

Scaling the vector x: Any possibility of overflow or underflow in the computation of $\|x\|_2$ can be avoided by scaling the vector x. For example, the vector u could be determined from the vector $\frac{x}{\max_i\{|x_i|\}}$ rather than from the vector x itself.

Example 7.4. Let
$$x = \begin{pmatrix} 0 \\ 4 \\ 1 \end{pmatrix}.$$

7.2. Householder's Matrices and QR Factorization

ALGORITHM 7.1. Creating Zeros in a Vector with a Householder Matrix.

Input. A nonzero n-vector x.
Output. A vector u such that
$$\left(I - \frac{2uu^T}{u^T u}\right) x = (\times, 0 \ldots 0)^T.$$

Step 1. Find the scaling factor $m = \max(|x_1|, \ldots, |x_n|)$.
Step 2. Scale the vector x as follows:

  For $i = 1, \ldots, n$ do
  $$x_i \equiv \frac{x_i}{m}$$
  End

Step 3. Compute the Householder vector $u = (u_1, \ldots, u_n)^T$:

  $u_1 = x_1 + \|x\|_2 \operatorname{sign}(x_1)$
   For $i = 2, \ldots, n$ do
    $u_i = x_i$
   End

Step 1. $m = 4$.

Step 2. $x_1 \equiv 0$, $x_2 \equiv 1$, $x_3 \equiv 0.25$.

Step 3. $u_1 = x_1 + \|x\|_2 = 0 + 1.0308 = 1.0308$; $u_2 = x_2 = 1$, $u_3 = x_3 = 0.25$. $u = (u_1, u_2, u_3)^T = (1.0308, 1, 0.25)^T$.

Verify: $\left(I - \dfrac{2uu^T}{u^T u}\right) x = (-4.1231, 0, 0)^T$. ∎

Flop-count and round-off property. Creating zeros in a vector by a Householder matrix is a cheap and numerically stable procedure.

It takes only $3n$ flops to create zeros in the positions 2 through n in a vector by using Algorithm 7.1, and it can be shown (Wilkinson (1965, pp. 152–162)) that if $\widehat{H}$ is the computed Householder matrix, then

$$\|H - \widehat{H}\| \leq 10\mu.$$

Moreover,

$$\mathrm{fl}(\widehat{H}x) = H(x + e); \qquad |e| \leq cn^2 \mu \|x\|_2,$$

c is a constant of order unity, and μ is the machine precision.

MATCOM Note: The MATCOM program **HOUSZERO** implements Algorithm 7.1.

7.2.2 Householder's Method for QR Factorization

We will now show how the idea of introducing zeros in a vector using a Householder matrix can be extended to compute a **full QR factorization** of an $m \times n$ matrix ($m \geq n$). The process will yield a factorization

$$A = QR,$$

where Q is $m \times m$ and *orthogonal* and R is $m \times n$ *upper triangular*, as shown in Figure 7.3. (Recall that an $m \times n$ upper triangular matrix is the matrix whose entries below the main diagonal are zero.)

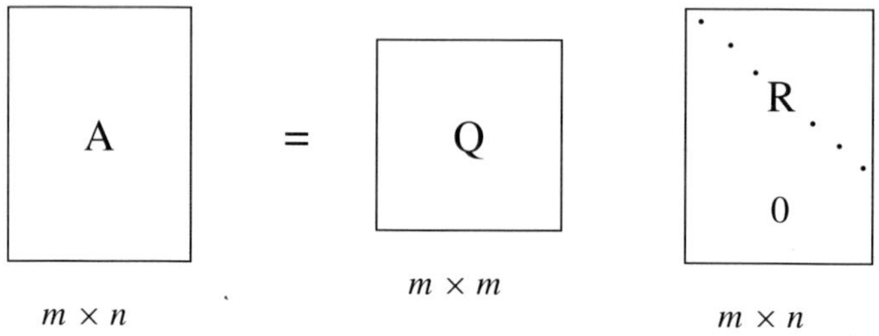

Figure 7.3. *Householder QR factorization.*

The process was introduced by Householder in 1958. In contrast with the Gaussian elimination scheme, the *Householder process can always be carried out to completion.*

> The idea is to reduce the matrix A to an upper triangular matrix R by successively premultiplying A with a series of Householder matrices (which are orthogonal).

For an $m \times n$ matrix, the process will require $s = \min(m-1, n)$ steps.

Let $A \in \mathbb{R}^{m \times n}$ and $m \geq n$. Then $s = n$. Generate successively Householder matrices $H_1, H_2, \ldots, H_n$ such that

$$H_n H_{n-1} \ldots H_2 H_1 A = R,$$

where R is upper triangular.

There are n steps.

Step 1. $A \xrightarrow{H_1} H_1 A = \begin{pmatrix} \times & \times & \ldots & \times \\ 0 & \times & & \times \\ \vdots & \vdots & \ddots & \times \\ 0 & \times & \ldots & \times \end{pmatrix} = A^{(1)}.$

Step 2. $A^{(1)} \xrightarrow{H_2} H_2 A^{(1)} = H_2 H_1 A = \begin{pmatrix} \times & \times & \ldots & \ldots & \times \\ 0 & \times & \ldots & \ldots & \times \\ \vdots & 0 & \times & \ldots & \times \\ \vdots & \vdots & & \ddots & \vdots \\ 0 & 0 & \ldots & \times & \times \end{pmatrix} = A^{(2)}.$

...

7.2. Householder's Matrices and QR Factorization

Step n. $A^{(n-1)} \xrightarrow{H_n} H_n A^{(n-1)} = H_n H_{n-1} \ldots H_2 H_1 A = \begin{pmatrix} \times & \times & \ldots & \times \\ 0 & \ddots & & \times \\ \vdots & & \ddots & \vdots \\ 0 & 0 & \ldots & \times \end{pmatrix} = A^{(n)} = R.$

Forming Q: Set $Q = H_1 H_2 \ldots H_n$. Then from above
$$Q^T A = R.$$

Since each Householder matrix H_i, $i = 1, \ldots, n$, is an orthogonal matrix, so is the matrix Q *(note that the product of orthogonal matrices is an orthogonal matrix)*.

So, premultiplying $Q^T A = R$ by Q, we have $A = QR$.

The construction of Householder matrices is illustrated with $m = 4$ and $n = 3$ in the following.

Construction of Householder Matrices ($m = 4$, $n = 3$)

Step 1. Construction of H_1: Construct a 4×4 Householder matrix H_1 such that
$$H_1 \begin{pmatrix} a_{11} \\ a_{21} \\ a_{31} \\ a_{41} \end{pmatrix} = \begin{pmatrix} \times \\ 0 \\ 0 \\ 0 \end{pmatrix}.$$

Form *implicitly*
$$H_1 A = \begin{pmatrix} \times & \times & \times \\ 0 & * & \times \\ 0 & * & \times \\ 0 & * & \times \end{pmatrix} = A^{(1)}.$$

Step 2. Construction of H_2: Construct a 3×3 Householder matrix $\hat{H}_2$ such that
$$\hat{H}_2 \begin{pmatrix} * \\ * \\ * \end{pmatrix} = \begin{pmatrix} \times \\ 0 \\ 0 \end{pmatrix}.$$

Form *implicitly*
$$H_2 A^{(1)} = \left(\begin{array}{c|c} 1 & 0 \\ \hline 0 & \\ 0 & \hat{H}_2 \\ 0 & \end{array} \right) A^{(1)} = \begin{pmatrix} \times & \times & \times \\ 0 & \times & \times \\ 0 & 0 & \circledast \\ 0 & 0 & \circledast \end{pmatrix} = A^{(2)}.$$

Step 3. Construction of H_3: Construct a 2×2 Householder matrix $\hat{H}_3$ such that
$$\hat{H}_3 \begin{pmatrix} \circledast \\ \circledast \end{pmatrix} = \begin{pmatrix} \times \\ 0 \end{pmatrix}.$$

Form *implicitly*
$$H_3 A^{(2)} = \left(\begin{array}{cc|c} 1 & 0 & 0 \\ 0 & 1 & 0 \\ \hline 0 & & \hat{H}_3 \end{array} \right) A^{(2)} = \begin{pmatrix} \times & \times & \times \\ 0 & \times & \times \\ 0 & 0 & \times \\ 0 & 0 & 0 \end{pmatrix} = A^{(3)} = R.$$

The process is fairly general. Set $A^{(0)} = A$.

Step k.

- Construct a Householder matrix $\hat{H}_k$ of order $m - k + 1$ to annihilate the entries $(k+1, k)$ through (m, k) of the matrix $A^{(k-1)}$ at the previous step.

- Form $H_k A^{(k-1)} = A^{(k)}$ *implicitly*, where $H_k = \mathrm{diag}(I_{k-1}, \hat{H}_k)$. (Note that when $k = 1$, $H_k = \hat{H}_k$.)

Example 7.5.
$$A = \begin{pmatrix} 1 & 1 \\ 0.0001 & 0 \\ 0 & 0.0001 \end{pmatrix}.$$

Step 1. Compute H_1: $H_1 = I - 2u_1 u_1^T / u_1^T u_1$:

$$u_1 = \begin{pmatrix} 1 \\ 0.0001 \\ 0 \end{pmatrix} + \sqrt{1 + (0.0001)^2} \begin{pmatrix} 1 \\ 0 \\ 0 \end{pmatrix} = \begin{pmatrix} 2 \\ 0.0001 \\ 0 \end{pmatrix}.$$

Compute *implicitly*

$$A^{(1)} = H_1 A = \left(I - \frac{2 u_1 u_1^T}{u_1^T u_1}\right) A = \begin{pmatrix} -1 & -1 \\ 0 & -0.0001 \\ 0 & 0.0001 \end{pmatrix}.$$

Step 2. Compute H_2: $H_2 = I - 2u_2 u_2^T / u_2^T u_2$:

$$\hat{u}_2 = \begin{pmatrix} -0.0001 \\ 0.0001 \end{pmatrix} - \sqrt{(-0.0001)^2 + (0.0001)^2} \begin{pmatrix} 1 \\ 0 \end{pmatrix} = 10^{-4} \begin{pmatrix} -2.4141 \\ 1.0000 \end{pmatrix}.$$

$$\hat{H}_2 = \begin{pmatrix} 1 & 0 \\ 0 & 1 \end{pmatrix} - 2\frac{\hat{u}_2 \hat{u}_2^T}{\hat{u}_2^T \hat{u}_2} = \begin{pmatrix} -0.7071 & 0.7071 \\ 0.7071 & 0.7071 \end{pmatrix}, \quad H_2 = \begin{pmatrix} 1 & 0 & 0 \\ 0 & -0.7071 & 0.7071 \\ 0 & 0.7071 & 0.7071 \end{pmatrix},$$

Compute *implicitly*

$$A^{(2)} = H_2 A^{(1)} = H_2 H_1 A = \begin{pmatrix} -1 & -1 \\ 0 & 0.0001 \\ 0 & 0 \end{pmatrix} = R.$$

Form Q and R:

$$Q = H_1 H_2 = \begin{pmatrix} -1 & 0.0001 & -0.0001 \\ -0.0001 & -0.7071 & 0.7071 \\ 0 & 07071 & 0.7071 \end{pmatrix} = (Q_1, Q_2).$$

Note: *In practice, the matrix Q should be formed by implicit matrix multiplication, as shown before.*

$$R = \begin{pmatrix} -1 & -1 \\ 0 & 0.0001 \\ 0 & 0 \end{pmatrix} = \begin{pmatrix} R_1 \\ 0 \end{pmatrix}, \text{ where } R_1 = \begin{pmatrix} -1 & -1 \\ 0 & 0.0001 \end{pmatrix}.$$

7.2. Householder's Matrices and QR Factorization

The full QR Factorization of A: $A = QR$:

$$\begin{pmatrix} 1 & 1 \\ 0.001 & 1 \\ 0 & 0.0001 \end{pmatrix} = \begin{pmatrix} -1 & 0.0001 & -0.0001 \\ -0.0001 & -0.7071 & 0.7071 \\ 0 & 0.7071 & 0.7071 \end{pmatrix} \begin{pmatrix} -1 & -1 \\ 0 & 0.0001 \\ 0 & 0 \end{pmatrix}.$$

$$\quad\quad A \quad\quad\quad\quad\quad\quad\quad Q \quad\quad\quad\quad\quad\quad\quad R$$

The reduced QR Factorization of A:

$$A = Q_1 R_1 = \begin{pmatrix} -1 & 0.0001 \\ -0.0001 & -0.7071 \\ 0 & 0.7071 \end{pmatrix} \begin{pmatrix} -1 & -1 \\ 0 & 0.0001 \end{pmatrix}. \quad\blacksquare$$

$$\quad\quad\quad\quad\quad\quad\quad Q_1 \quad\quad\quad\quad\quad\quad R_1$$

ALGORITHM 7.2. Householder QR Factorization.

Input: An $m \times n$ matrix $A (m \geq n)$.
Outputs: (i) The Householder vectors $u_1, \ldots, u_n$ needed to form Q. (ii) An upper triangular matrix R. The result is $A = QR$.
Storage: (i) R is stored over A in the upper triangular part. (ii) The components $u_{k+1,k}$ through u_{mk} of each u_k are stored in the respective positions of A, and the first component u_{kk} is stored in a separate one-dimensional array.

For $k = 1, 2, \ldots, n$ do

1. Find the vector $u_k = (u_{kk}, \ldots, u_{mk})^T$ defining the Householder matrix $\hat{H}_k$ of order $m - k + 1$ such that

$$\hat{H}_k \begin{pmatrix} a_{k,k} \\ \vdots \\ a_{mk} \end{pmatrix} = \begin{pmatrix} r_{kk} \\ 0 \\ \vdots \\ 0 \end{pmatrix}.$$

2. Store r_{kk} over a_{kk}. $a_{kk} \equiv r_{kk}$.

3. Store the vector u_k as follows:

$$a_{ik} \equiv u_{ik}, \ i = k+1, \ldots, m,$$
$$v_k \equiv u_{kk}.$$

4. Compute $\beta = \dfrac{2}{u_k^T u_k}$.

5. Update the entries of the submatrix of A containing rows k through m and columns k through n, denoted by $A(k:m, k:n)$,

and store these entries over the corresponding entries of A:

$$A(k:m, k:n) \equiv \left(I_{m-k+1} - 2\frac{u_k u_k^T}{u_k^T u_k}\right) A(k:m, k:n)$$
$$= A(k:m, k:n) - \beta u_k u_k^T A(k:m, k:n).$$

End

Note: The algorithm does not produce the matrix Q explicitly. If needed, it has to be formed out of the saved Householder vectors $u_1, \ldots, u_n$.

Example 7.6. Let

$$A = \begin{pmatrix} 0 & 1 & 1 \\ 1 & 2 & 3 \\ 1 & 1 & 1 \end{pmatrix}.$$

$k = 1$: Construct the Householder vector u_1 such that $\left(I_3 - \frac{2u_1 u_1^T}{u_1^T u_1}\right) \begin{pmatrix} 0 \\ 1 \\ 1 \end{pmatrix} = \begin{pmatrix} * \\ 0 \\ 0 \end{pmatrix}$:

$$u_1 = \begin{pmatrix} 0 \\ 1 \\ 1 \end{pmatrix} + \sqrt{2} \begin{pmatrix} 1 \\ 0 \\ 0 \end{pmatrix} = \begin{pmatrix} \sqrt{2} \\ 1 \\ 1 \end{pmatrix}.$$

Update:

$$A \equiv A^{(1)} = H_1 A = \begin{pmatrix} -1.414 & -2.1213 & -2.8284 \\ 0 & \boxed{-0.2071 \quad 0.2929} \\ 0 & \boxed{-1.2071 \quad -1.7071} \end{pmatrix}.$$

$k = 2$: Construct the Householder vector u_2 such that $\left(I_2 - \frac{2u_2 u_2^T}{u_2^T u_2}\right) \begin{pmatrix} -0.2071 \\ -1.2071 \end{pmatrix} = \begin{pmatrix} * \\ 0 \end{pmatrix}$:

$$u_2 = \begin{pmatrix} -0.2071 \\ -1.2071 \end{pmatrix} - 1.2247 \begin{pmatrix} 1 \\ 0 \end{pmatrix} = \begin{pmatrix} -1.4318 \\ -1.2071 \end{pmatrix}.$$

Update: Update the submatrix in the box, $A(2:3, 2:3) \equiv (I_2 - 2\beta u_2 u_2^T) A(2:3, 2:3)$:

$$A(2:3, 2:3) \equiv \begin{pmatrix} 1.2247 & 1.6330 \\ 0 & -0.5774 \end{pmatrix}.$$

Form R: $A \equiv A^{(2)} = H_2 H_1 A = R = \begin{pmatrix} -1.4142 & -2.1213 & -2.8284 \\ 0 & 1.2247 & 1.6330 \\ 0 & 0 & -0.5774 \end{pmatrix}.$

Note: In practical computations using the above storage arrangements, the entries a_{21}, a_{31}, and a_{32} of the above matrix (which are zeros now) will be filled in with $u_{21} = 1$, $u_{31} = 1$, and $u_{32} = -1.2071$, while the entries $u_{11} = \sqrt{2}$ and $u_{22} = -1.4318$ will be stored in a vector v defined by $v = (\sqrt{2}, -1.4318)^T$.

7.2. Householder's Matrices and QR Factorization

Form $Q = H_1 H_2$ by performing the matrix multiplication *implicitly*:

$$Q = \begin{pmatrix} 0 & 0.8165 & 0.5774 \\ -0.7071 & 0.4082 & -0.5774 \\ -0.7071 & -0.4082 & 0.5774 \end{pmatrix}. \quad \blacksquare$$

Flop-count. When $m = n$, Algorithm 7.2 requires approximately $\frac{4}{3}n^3$ flops just to compute the triangular matrix R. This can be seen as follows.

For each k:

- About $4(n - k)$ flops to construct $\hat{H}_k$.
- About $4(n - k)^2$ flops for updating.

$$\text{Total number of flops} = 4 \sum_{k=1}^{n-1} [(n - k)^2 + (n - k)]$$

$$= 4[(n - 1)^2 + (n - 2)^2 + \cdots + 1^2]$$
$$+ 4[(n - 1) + (n - 2) + \cdots + 1]$$
$$= 4 \frac{n(n - 1)(2n - 1)}{6} + 4 \cdot \frac{n(n - 1)}{2}$$
$$\approx \frac{4n^3}{3} \quad \text{(neglecting } O(n^2) \text{ terms)}.$$

Note: *The above count does not include the explicit construction of Q. The matrix Q is available only in factored form. It should be noted that in a majority of practical applications, it is sufficient to have Q in this factored form, and in many applications, Q is not needed at all. If Q is needed explicitly, another $\frac{4}{3}n^3$ flops will be required.*

The approximate flop-count in the case $m \neq n$:

1. $2n^2 \left(m - \frac{n}{3}\right)$ flops if $m \geq n$ (Exercise 7.6(a)).
2. $2m^2 \left(n - \frac{m}{3}\right)$ flops if $m < n$.
3. $4(m^2 n - mn^2 + n^3/3)$ flops to compute Q explicitly (Exercise 7.6(a)).

Round-off property and stability. *In the presence of round-off errors the algorithm computes QR decomposition of a slightly perturbed matrix.* Specifically, it can be shown (Wilkinson (1965, p. 236)) that if $\widehat{R}$ denotes the computed R, then there exists an orthogonal $\widehat{Q}$ such that

$$A + E = \widehat{Q}\widehat{R}.$$

The error matrix E satisfies

$$\|E\|_F \leq \phi(n)\mu\|A\|_F,$$

where $\phi(n)$ is a *slowly growing* function of n and μ is the machine precision.

MATCOM Note: The MATCOM program **HOUSQRN** computes QR factorization of an $m \times n$ matrix A. The MATCOM program **HOUSQR** computes the QR factorization of an $n \times n$ matrix.

7.3 Complex QR Factorization

> If $x \in \mathbb{C}^n$ and $x_1 = re^{i\theta}$, then it is easy to see that the Householder matrix $H = I - \beta vv^*$, where $v = x \pm e^{i\theta} \parallel x \parallel_2 e_1$ and $\beta = \frac{2}{v^*v}$, is such that
> $$Hx = \mp v e^{i\theta} \parallel x \parallel_2 e_1. \qquad (7.3)$$

Using the above formula, the Householder QR factorization method for a real matrix A, described in the last section, can be easily adapted to a complex matrix. In this case, Q is **unitary** and R is **complex upper triangular**. The details are left to the readers. See Golub and Van Loan (1996, p. 233).

The process of **complex QR factorization** of an $m \times n$ matrix, $m \geq n$, using Householder's method requires $8n^2(m - \frac{n}{3})$ real flops.

MATLAB Note: Given a complex $m \times n$ matrix A, a MATLAB program in the form $[Q, R] = \mathbf{qr}(A)$ computes an $m \times n$ complex upper triangular matrix R and an $m \times m$ unitary matrix Q so that $A = QR$. If A is real, Q and R are also real and Q is orthogonal. $[Q, R] = \mathbf{qr}(A, 0)$ produces the **reduced QR** factorization.

7.4 Givens Matrices and QR Factorization

7.4.1 Definition and Basic Properties

Definition 7.7. *A matrix of the form*

$$J(i,j,c,s) = \begin{pmatrix} 1 & 0 & 0 & \cdots & & & & 0 \\ 0 & 1 & 0 & \cdots & & & & 0 \\ \vdots & \vdots & \vdots & & & & \cdots & \vdots \\ 0 & 0 & 0 & \cdots & c & s & \cdots & 0 \\ \vdots & \vdots & \vdots & & & & \cdots & \vdots \\ 0 & 0 & 0 & \cdots & -s & c & \cdots & 0 \\ \vdots & \vdots & \vdots & & & & \cdots & \vdots \\ 0 & 0 & 0 & \cdots & & 0 & \cdots & 1 \end{pmatrix} \begin{matrix} \\ \\ \\ \leftarrow i\text{th} \\ \\ \leftarrow j\text{th} \\ \\ \\ \end{matrix} \text{ rows,}$$

with ith and jth columns indicated,

where $c^2 + s^2 = 1$, is called a **Givens matrix**, *after the numerical analyst Wallace Givens.*[6]

---

[6]Wallace Givens was an American mathematician. His pioneering work done in 1950 on computing the eigenvalues of a symmetric matrix by reducing it to a symmetric tridiagonal form in a numerically stable way forms the basis of many numerically backward stable algorithms developed later. Givens held appointments at many prestigious institutes and research institutions. He died on March 5, 1993, at the age of 82 (see obituary in *SIAM News*, July 1993).

7.4. Givens Matrices and QR Factorization

Since one can choose $c = \cos\theta$ and $s = \sin\theta$ for some θ, a Givens matrix as above can be conveniently denoted by $J(i, j, \theta)$. Geometrically, the matrix $J(i, j, \theta)$ rotates a pair of coordinate axes (ith unit vector as its x-axis and the jth unit vector as its y-axis) through the given angle θ in the (i, j) plane. That is why the Givens matrix $J(i, j, \theta)$ is commonly known as a **Givens rotation** or **plane rotation in the (i, j) plane**. This is illustrated in the following figure.

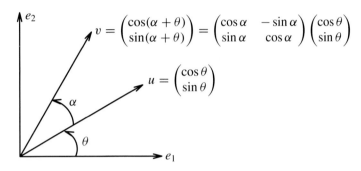

Thus, when an n-vector

$$x = \begin{pmatrix} x_1 \\ x_2 \\ \vdots \\ x_n \end{pmatrix}$$

is premultiplied by the Givens rotation $J(i, j, \theta)$, only the ith and jth components of x are affected; the other components remain unchanged.

Note that since $c^2 + s^2 = 1$, $J(i, j, \theta) \cdot J(i, j, \theta)^T = I$, the rotation $J(i, j, \theta)$ is *orthogonal*.

If $x = \begin{pmatrix} x_1 \\ x_2 \end{pmatrix}$ is a 2-vector, then it is a matter of simple verification that, with

$$c = \frac{x_1}{\sqrt{x_1^2 + x_2^2}}, \qquad s = \frac{x_2}{\sqrt{x_1^2 + x_2^2}},$$

the Givens rotation $J(1, 2, \theta) = \begin{pmatrix} c & s \\ -s & c \end{pmatrix}$ is such that $J(1, 2, \theta)x = \begin{pmatrix} * \\ 0 \end{pmatrix}$.

The above formula for computing c and s might cause some *underflow* or *overflow*. However, the following simple rearrangement of the formula might prevent that possibility.

Computing the Givens Parameters

If $|x_2| \geq |x_1|$, compute $t = \dfrac{x_1}{x_2}$, $s = \dfrac{1}{\sqrt{1 + t^2}}$; $c = st$.

Otherwise, $t = \dfrac{x_2}{x_1}$, $c = \dfrac{1}{\sqrt{1 + t^2}}$; $s = ct$.

(*Note that computations of s and t do not involve θ.*)

Example 7.8.

$$x = \begin{pmatrix} 1 \\ 1 \\ \frac{1}{2} \end{pmatrix}.$$

Since $|x_1| > |x_2|$, we take $t = \frac{1}{2}$, $c = \frac{1}{\sqrt{1+\frac{1}{4}}} = \frac{2}{\sqrt{5}}$; $s = \frac{1}{\sqrt{5}}$.

Verify: $\begin{pmatrix} c & s \\ -s & c \end{pmatrix} x = \begin{pmatrix} \frac{2}{\sqrt{5}} & \frac{1}{\sqrt{5}} \\ -\frac{1}{\sqrt{5}} & \frac{2}{\sqrt{5}} \end{pmatrix} \begin{pmatrix} 1 \\ \frac{1}{2} \end{pmatrix} = \begin{pmatrix} \frac{\sqrt{5}}{2} \\ 0 \end{pmatrix}$. ∎

Implicit Construction of Matrix Product with a Givens Matrix

Because of the special structure of the Givens matrix $J(i, j, \theta)$, which differs from the identity matrix only in four places (i, i), (i, j), (j, i), and (j, j), *matrix multiplication by a Givens matrix does not have to be performed explicitly. It can be done implicitly, as shown in the following algorithm.*

ALGORITHM 7.3. Implicit Construction of JA.

Input: (i) An $m \times n$ matrix $A (m \geq n)$. (ii) The numbers c and s of the Givens matrix $J(i, j, c, s)$
Output: The implicit product JA stored over A.

For $k = 1, \ldots, n$ do

$$a \equiv a_{ik},$$
$$b \equiv a_{jk},$$
$$a_{ik} \equiv ac + bs,$$
$$a_{jk} \equiv -as + bc.$$

End

MATCOM Note: Algorithm 7.3 has been implemented in the MATCOM program **PGIVMUL**.

Zeroing Specified Entries in a Vector

Givens rotations are especially useful in creating zeros in a specified position in a vector. Thus, if $x = (x_1, x_2, \ldots, x_k, \ldots, x_n)^T$, and if we desire to zero x_k only, we can construct the rotation $J(i, k, \theta)$ $(i < k)$ such that $J(i, k, \theta)x$ will have zero in the kth position.

$$x = \begin{pmatrix} x_1 \\ x_2 \\ \vdots \\ x_k \\ \vdots \\ x_n \end{pmatrix} \xrightarrow{J} Jx = \begin{pmatrix} \times \\ \times \\ \vdots \\ 0 \\ \times \\ \vdots \\ \times \end{pmatrix}$$

To construct $J(i, k, \theta)$, first construct a 2×2 Givens rotation $\begin{pmatrix} c & s \\ -s & c \end{pmatrix}$ such that

$$\begin{pmatrix} c & s \\ -s & c \end{pmatrix} \begin{pmatrix} x_i \\ x_k \end{pmatrix} = \begin{pmatrix} * \\ 0 \end{pmatrix}$$

7.4. Givens Matrices and QR Factorization

and then form the matrix $J(i, k, \theta)$ by inserting c into the positions (i, i) and (k, k), and s and $-s$ into the positions (i, k) and (k, i), respectively, and filling the rest of the matrix with entries of the identity matrix.

Example 7.9. Suppose we want to create a zero in the third position of x, that is, $k = 3$.

$$x = \begin{pmatrix} 1 \\ -1 \\ 3 \end{pmatrix}.$$

Choose $i = 2$.

1. Form a 2×2 Givens rotations c and s such that

$$\begin{pmatrix} c & s \\ -s & c \end{pmatrix} \begin{pmatrix} -1 \\ 3 \end{pmatrix} = \begin{pmatrix} * \\ 0 \end{pmatrix}; \quad c = \frac{-1}{\sqrt{10}}, \quad s = \frac{3}{\sqrt{10}}.$$

2. Then

$$J(2, 3, \theta)x = \begin{pmatrix} 1 & 0 & 0 \\ 0 & \frac{-1}{\sqrt{10}} & \frac{3}{\sqrt{10}} \\ 0 & \frac{-3}{\sqrt{10}} & \frac{-1}{\sqrt{10}} \end{pmatrix} \begin{pmatrix} 1 \\ -1 \\ 3 \end{pmatrix} = \begin{pmatrix} 1 \\ \sqrt{10} \\ 0 \end{pmatrix}. \quad \blacksquare$$

Creating Zeros in a Vector Except Possibly in the First Place

Given an n-vector x, if we desire to zero all the entries of x except possibly the first one, we can construct $J(1, 2, \theta)$, $x^{(1)} = J(1, 2, \theta)x$, $x^{(2)} = J(1, 3, \theta)x^{(1)}$, $x^{(3)} = J(1, 4, \theta)x^{(2)}$, etc., so that with

$$P = J(1, n, \theta) \cdots J(1, 3, \theta)J(1, 2, \theta),$$

we will have Px a multiple of e_1. Since each rotation is orthogonal, so is P.

$n = 3$: $\quad x = \begin{pmatrix} \times \\ \times \\ \times \end{pmatrix} \xrightarrow{J(1,2,\theta)} J(1, 2, \theta)x = \begin{pmatrix} \times \\ 0 \\ \times \end{pmatrix} = x^{(1)},$

$$x^{(1)} = \begin{pmatrix} \times \\ 0 \\ \times \end{pmatrix} \xrightarrow{J(1,3,\theta)} J(1, 3, \theta)x^{(1)} = \begin{pmatrix} \times \\ 0 \\ 0 \end{pmatrix} = x^{(2)}$$

Example 7.10. Let

$$x = \begin{pmatrix} 1 \\ -1 \\ 2 \end{pmatrix}.$$

Then

$$x^{(1)} = J(1, 2, \theta)x = \begin{pmatrix} \frac{1}{\sqrt{2}} & \frac{-1}{\sqrt{2}} & 0 \\ \frac{1}{\sqrt{2}} & \frac{1}{\sqrt{2}} & 0 \\ 0 & 0 & 1 \end{pmatrix} \begin{pmatrix} 1 \\ -1 \\ 2 \end{pmatrix} = \begin{pmatrix} \sqrt{2} \\ 0 \\ 2 \end{pmatrix},$$

$$x^{(2)} = J(1, 3, \theta)x^{(1)} = \begin{pmatrix} \sqrt{\frac{2}{6}} & 0 & \frac{2}{\sqrt{6}} \\ 0 & 1 & 0 \\ \frac{-2}{\sqrt{6}} & 0 & \sqrt{\frac{2}{6}} \end{pmatrix} \begin{pmatrix} \sqrt{2} \\ 0 \\ 2 \end{pmatrix} = \begin{pmatrix} \sqrt{6} \\ 0 \\ 0 \end{pmatrix}. \quad \blacksquare$$

Flop-count and round-off property. Creating zeros in a vector using Givens rotations is about twice as expensive as using Householder matrices. To be precise, the process requires only $1\frac{1}{2}$ times as many flops as Householder's, but it requires $O(\frac{n^2}{2})$ square roots, whereas the Householder method requires $O(n)$ square roots. *The process is as stable as the Householder method.*

Creating Zeros in Specified Positions of a Matrix

The idea of creating zeros in specified positions of a vector can be trivially extended to create zeros in specified positions of a matrix as well. Thus, if we wish to create a zero in the $(j,i)[(j > i)]$ position of a matrix A, one way to do this is to construct the rotation $J(i, j, \theta)$ affecting the ith and jth rows only, such that $J(i, j, \theta)A$ will have a zero in the (j, i) position. The procedure then is as follows.

ALGORITHM 7.4. Creating Zeros in a Specified Position of a Matrix Using Givens Rotations.

Input: An $n \times n$ matrix.
Output: The matrix $J(i, j, \theta)A$ with zero in the (j, i) position. The matrix $J(i, j, \theta)A$ is stored over A.

1. Find the Givens parameters $c = \cos\theta$ and $s = \sin\theta$ such that
$$\begin{pmatrix} c & s \\ -s & c \end{pmatrix} \begin{pmatrix} a_{ii} \\ a_{ji} \end{pmatrix} = \begin{pmatrix} * \\ 0 \end{pmatrix}.$$
2. Form $J(i, j, \theta)A$ *implicitly* and store this over A. (Use Algorithm 7.3.)

Remark. Note that there are other ways to do this as well. For example, we can form $J(k, j, \theta)$ affecting the kth and jth rows, such that $J(k, j, \theta)A$ will have a zero in the (j, i) position.

Example 7.11. Let
$$A = \begin{pmatrix} 1 & 2 & 3 \\ 2 & 3 & 4 \\ 4 & 5 & 6 \end{pmatrix}.$$

Create a zero in the (3,1) position using $J(2, 3, \theta)$.

1. Find c and s such that $\begin{pmatrix} c & s \\ -s & c \end{pmatrix} \begin{pmatrix} 2 \\ 4 \end{pmatrix} = \begin{pmatrix} * \\ 0 \end{pmatrix}$; $c = \frac{2}{\sqrt{20}}$, $s = \frac{4}{\sqrt{20}}$.

2. Form
$$A^{(1)} = J(2,3,\theta)A = \begin{pmatrix} 1 & 0 & 0 \\ 0 & \frac{2}{\sqrt{20}} & \frac{4}{\sqrt{20}} \\ 0 & \frac{-4}{\sqrt{20}} & \frac{2}{\sqrt{20}} \end{pmatrix} \begin{pmatrix} 1 & 2 & 3 \\ 2 & 3 & 4 \\ 4 & 5 & 6 \end{pmatrix} = \begin{pmatrix} 1 & 2 & 3 \\ \sqrt{20} & \frac{26}{\sqrt{20}} & \frac{32}{\sqrt{20}} \\ 0 & \frac{-2}{\sqrt{20}} & \frac{-4}{\sqrt{20}} \end{pmatrix}. \blacksquare$$

7.4. Givens Matrices and QR Factorization

MATCOM Note: Algorithm 7.4 has been implemented in the MATCOM program **GIVSZERO**.

Since Givens matrices can be conveniently used to create zeros in a vector, it is natural to think that these matrices can also be used to find the QR factorization of a matrix. *The idea is just like Householder's.* The only difference is that (usually) more than one Givens matrix will be needed to create zeros in desired positions in a column of A. *One way to do this is as follows.*

7.4.2 Givens Method for QR Factorization

Let $A \in \mathbb{R}^{m \times n}$. There are $s = \min(m - 1, n)$ steps.

Step 1. Form an orthogonal matrix $Q_1 = J(1, m, \theta) J(1, m - 1, \theta) \ldots J(1, 2, \theta)$ such that $A^{(1)} = Q_1 A$ has zeros below the $(1, 1)$ entry in the first column.

Step 2. Form an orthogonal matrix $Q_2 = J(2, m, \theta) J(2, m - 1, \theta) \ldots J(2, 3, \theta)$ such that $A^{(2)} = Q_2 A^{(1)}$ has zeros below the $(2, 2)$ entry in the second column.

Step k. Form an orthogonal matrix $Q_k = J(k, m, \theta) \ldots J(k, k + 1, \theta)$ such that $A^{(k)} = Q_k A^{(k-1)}$ has zeros below the (k, k) entry in the kth column.

The *final matrix* $A^{(s)}$ is *upper triangular*.

Obtaining Q and R

Set $R \equiv A^{(s)}$ (upper triangular).

Set $Q = Q_1^T Q_2^T \ldots Q_s^T$ (orthogonal),

where $Q_i = J(i, m, \theta) J(i, m - 1, \theta) \ldots J(i, i + 1, \theta)$ (orthogonal).

Then $Q^T A = Q_s Q_{s-1} \ldots Q_2 Q_1 A = R$

An illustration: Let $m = 4, n = 2$.

$$A \xrightarrow{J(1,2,\theta)} \begin{pmatrix} \times & \times \\ 0 & \times \\ \times & \times \\ \times & \times \end{pmatrix} \xrightarrow{J(1,3,\theta)} \begin{pmatrix} \times & \times \\ 0 & \times \\ 0 & \times \\ \times & \times \end{pmatrix} \xrightarrow{J(1,4,\theta)} \begin{pmatrix} \times & \times \\ 0 & \times \\ 0 & * \\ 0 & * \end{pmatrix} = A^{(1)},$$

$$J(1,2,\theta)A \qquad J(1,3,\theta)J(1,2,\theta)A \qquad J(1,4,\theta)J(1,3,\theta)J(1,2,\theta)A$$

$$A^{(1)} \xrightarrow{J(2,3,\theta)} \begin{pmatrix} \times & \times \\ 0 & \times \\ 0 & 0 \\ 0 & \times \end{pmatrix} \xrightarrow{J(2,4,\theta)} \begin{pmatrix} \times & \times \\ 0 & \times \\ 0 & 0 \\ 0 & 0 \end{pmatrix} = A^{(2)} = R,$$

$$J(2, 3, \theta)A^{(1)} \qquad J(2, 4, \theta)J(2, 3, \theta)A^{(1)}$$

$R = A^{(2)}; \quad Q = Q_1^T Q_2^T = J^T(1, 2, \theta) J^T(1, 3, \theta) J^T(1, 4, \theta) J^T(2, 3, \theta) J^T(2, 4, \theta).$

> **ALGORITHM 7.5. Givens QR Factorization.**
>
> **Input:** An $m \times n$ matrix A.
> **Output:** An $n \times n$ upper triangular matrix R such that $A = QR$. The matrix R is stored over A. Q is formed out of Givens parameters.
>
> For $k = 1, 2, \ldots, \min\{m-1, n\}$ do
>   For $l = k+1, \ldots, m$ do
>
> 1. Find a 2×2 Givens rotation acting on a_{kk} and a_{kl} such that $a_{kl} = 0$:
> $$\begin{pmatrix} c & s \\ -s & c \end{pmatrix} \begin{pmatrix} a_{kk} \\ a_{lk} \end{pmatrix} = \begin{pmatrix} * \\ 0 \end{pmatrix}.$$
> 2. Save the indices k and ℓ and the numbers c and s.
> 3. Form the $m \times m$ Givens rotation $J(k, l, \theta)$ and update $A \equiv J(k, l, \theta) A$ (*by implicitly constructing the product using Algorithm 7.3*).
>
>   End
> End

Remarks. (i) The algorithm does not explicitly produce the matrix Q. If needed, it has to be formed from Givens rotations out of the Givens parameters c and s and the indices k and l.

(ii) The Givens and Householder QR factorizations are intimately related. See Exercise 7.5 and Example 7.12.

Flop-count. The algorithm requires $3n^2 \left(m - \frac{n}{3}\right)$ flops. This count, of course, does not include computation of Q. Thus, *this algorithm is about $1\frac{1}{2}$ times as expensive as the Householder algorithm for QR factorization (Algorithm 7.2)*.

Round-off property. The algorithm is *stable*. It can be shown (Wilkinson (1965, p. 240)) that the computed $\widehat{Q}$ and $\widehat{R}$ satisfy

$$\widehat{R} = \widehat{Q}^T (A + E),$$

where

$$\|E\|_F \leq c \|A\|_F, \quad c \text{ is a constant of order unity}.$$

Example 7.12. Find the QR factorization of A from Example 7.6 using Givens rotations and determine its relationship with that obtained by Householder's method.

$k = 1$. Create Givens rotations $J(1, 2, \theta)$ and $J(1, 3, \theta)$ such that $J(1, 2, \theta) J(1, 3, \theta) A$ has zeros in the $(1, 2)$ and $(1, 3)$ positions:

1. Find c and s such that
   $$\begin{pmatrix} c & s \\ -s & c \end{pmatrix} \begin{pmatrix} 0 \\ 1 \end{pmatrix} = \begin{pmatrix} * \\ 0 \end{pmatrix}; \quad c = 0, \quad s = 1,$$

7.4. Givens Matrices and QR Factorization

$$A \equiv J(1,2,\theta)A = \begin{pmatrix} 0 & 1 & 0 \\ -1 & 0 & 0 \\ 0 & 0 & 1 \end{pmatrix} \begin{pmatrix} 0 & 1 & 1 \\ 1 & 2 & 3 \\ 1 & 1 & 1 \end{pmatrix} = \begin{pmatrix} 1 & 2 & 3 \\ 0 & -1 & -1 \\ 1 & 1 & 1 \end{pmatrix}.$$

2. Find c and s such that

$$\begin{pmatrix} c & s \\ -s & c \end{pmatrix} \begin{pmatrix} 1 \\ 1 \end{pmatrix} = \begin{pmatrix} * \\ 0 \end{pmatrix}; \quad c = \frac{1}{\sqrt{2}}, \quad s = \frac{1}{\sqrt{2}},$$

$$A \equiv J(1,3,\theta)A = \begin{pmatrix} \frac{1}{\sqrt{2}} & 0 & \frac{1}{\sqrt{2}} \\ 0 & 1 & 0 \\ -\frac{1}{\sqrt{2}} & 0 & \frac{1}{\sqrt{2}} \end{pmatrix} \begin{pmatrix} 1 & 2 & 3 \\ 0 & -1 & -1 \\ 1 & 1 & 1 \end{pmatrix} = \begin{pmatrix} \sqrt{2} & \frac{3}{\sqrt{2}} & 2\sqrt{2} \\ 0 & -1 & -1 \\ 0 & -\frac{1}{\sqrt{2}} & -\sqrt{2} \end{pmatrix}.$$

$k = 2$. Create a Givens rotation $J(2,3,\theta)$ such that $J(2,3,\theta)A$ has zeros in (3, 2) position:

$$\begin{pmatrix} c & s \\ -s & c \end{pmatrix} \begin{pmatrix} -1 \\ -\frac{1}{\sqrt{2}} \end{pmatrix} = \begin{pmatrix} * \\ 0 \end{pmatrix}; \quad c = -\frac{\sqrt{2}}{\sqrt{3}}, \quad s = -\frac{1}{\sqrt{3}},$$

$$J(2,3,\theta)A = \begin{pmatrix} 1 & 0 & 0 \\ 0 & -\frac{\sqrt{2}}{\sqrt{3}} & -\frac{1}{\sqrt{3}} \\ 0 & \frac{1}{\sqrt{3}} & -\frac{\sqrt{2}}{\sqrt{3}} \end{pmatrix} \begin{pmatrix} \sqrt{2} & \frac{3}{\sqrt{2}} & 2\sqrt{2} \\ 0 & -1 & -1 \\ 0 & -\frac{1}{\sqrt{2}} & -\sqrt{2} \end{pmatrix}$$

$$= \begin{pmatrix} 1.4142 & 2.1213 & 2.8284 \\ 0 & 1.2247 & 1.6330 \\ 0 & 0 & 0.5774 \end{pmatrix} = R. \quad \blacksquare$$

Relationship with Householder QR factorization. Note that the R matrix in the above example obtained by Givens' method is *essentially* the same as that of the Householder method (see Example 7.6) in the following sense: $R_{\text{Givens}} = DR_{\text{Householder}}$, where $D = \text{diag}(\pm 1, \pm 1, \ldots)$.

MATCOM Note: Algorithm 7.5 has been implemented in the MATCOM program **GIVQR**.

7.4.3 QR Factorization of a Hessenberg Matrix Using Givens Matrices

In several applications, one needs to find the QR factorization of a Hessenberg matrix. For example, the *QR iteration algorithm* (to be described later) for *eigenvalue computation requires QR factorization of a Hessenberg matrix at every iteration* (Section 9.8).

Since an upper Hessenberg matrix has at most $(n-1)$ nonzero subdiagonal entries, we can triangularize A by using only $(n-1)$ Givens rotations. This is illustrated with $n = 4$.

$$\begin{pmatrix} \times & \times & \times & \times \\ \times & \times & \times & \times \\ 0 & \times & \times & \times \\ 0 & 0 & \times & \times \end{pmatrix} \xrightarrow{J(1,2,\theta)} \begin{pmatrix} \times & \times & \times & \times \\ 0 & \times & \times & \times \\ 0 & \times & \times & \times \\ 0 & 0 & \times & \times \end{pmatrix} \xrightarrow{J(2,3,\theta)} \begin{pmatrix} \times & \times & \times & \times \\ 0 & \times & \times & \times \\ 0 & 0 & \times & \times \\ 0 & 0 & \times & \times \end{pmatrix} = A^{(1)},$$

$$A \qquad\qquad\qquad J(1,2,\theta)A \qquad\qquad J(1,2,\theta)J(2,3,\theta)A = A^{(1)}$$

$$A^{(1)} \xrightarrow{J(3,4,\theta)} = \begin{pmatrix} \times & \times & \times & \times \\ 0 & \times & \times & \times \\ 0 & 0 & \times & \times \\ 0 & 0 & 0 & \times \end{pmatrix} = R.$$

> **ALGORITHM 7.6. Givens–Hessenberg QR Factorization.**
>
> **Input:** An $n \times n$ *upper Hessenberg* matrix.
> **Outputs:** An upper triangular matrix R stored over A, and the Givens parameters needed to form Q. The result is $A = QR$.
>
> **Step 1.** For $k = 1, 2, \ldots n - 1$ do
>   **1.1** Find $c = \cos\theta$ and $s = \sin\theta$ such that
> $$\begin{pmatrix} c & s \\ -s & c \end{pmatrix} \begin{pmatrix} a_{kk} \\ a_{k+1,k} \end{pmatrix} = \begin{pmatrix} * \\ 0 \end{pmatrix}.$$
>   **1.2** Save the index k and the numbers c and s.
>   **1.3** Update $A : A \equiv J(k, k+1, \theta)A$.
>   End
>
> **Step 2.** Set $R \equiv A$.

Flop-count. Algorithm 7.6 requires about $3n^2$ flops, compared to $2n^3$ flops required for an arbitrary matrix.

7.5 Classical and Modified Gram–Schmidt Algorithms for QR Factorizations

The *classical Gram–Schmidt[7] process* (CGS) taught in basic linear algebra courses finds, starting from a set of n linearly independent vectors $\{a_k\}$, a set of n *orthonormal vectors* $\{q_k\}$ such that

$$\text{span}\{q_1, q_2, \ldots, q_i\} = \text{span}\{a_1, a_2, \ldots, a_i\}, \quad i = 1, 2, \ldots, n.$$

The vectors $q_1, \ldots, q_n$ are determined as follows.

Step 1.
$$q_1 = \frac{a_1}{\|a_1\|_2} = \frac{a_1}{r_{11}}. \tag{7.4}$$

Step 2. Find a vector q_2 in such a way that q_2 is of unit length and orthogonal to q_1; that is, $\|q_2\| = 1$ and $q_1^T q_2 = 0$. It is easily seen that this will happen if we first define the auxiliary vector
$$\hat{q}_2 = a_2 - r_{12} q_1; \quad r_{12} = q_1^T a_2$$

---

[7] Jorgen Pedersen Gram (1850–1916) was born in Denmark. Despite his career with an insurance company, he pursued mathematical research in several areas of pure and applied mathematics, including probability theory, numerical analysis, and number theory, and managed to influence the Danish Mathematical Society in a positive way. He is best known for his work on the orthogonalization process.

Erhard Schmidt (1876–1959) was a German mathematician. He obtained his doctorate from the University of Göttingen under the supervision of Hilbert and then joined the University of Berlin as a mathematics professor. Later he was appointed as the dean and then vice-chancellor of that university. He also founded the Institute of Applied Mathematics at the University of Berlin. His main research interests were integral equations and Hilbert space. In 1907, in one of his outstanding papers on integral equations, he established what is now called the *Gram–Schmidt orthogonalization process*.

7.5. Classical and Modified Gram–Schmidt Algorithms

and then normalize this vector to obtain

$$q_2 = \frac{\hat{q}_2}{\|\hat{q}_2\|_2} = \frac{\hat{q}_2}{r_{22}}. \tag{7.5}$$

Step 3. Find a vector q_3 such that it is of unit length and orthogonal to both q_1 and q_2. Again, it is easy to see that this will again happen if we define

$$\hat{q}_3 = a_3 - r_{13}q_1 - r_{23}q_2,$$

where $r_{13} = q_1 a_3^T$ and $r_{23} = q_2 a_3^T$, and then take

$$q_3 = \frac{\hat{q}_3}{\|\hat{q}_3\|_2} = \frac{\hat{q}_3}{r_{33}}. \tag{7.6}$$

The process is fairly general and can be continued until all the vectors up to q_n are computed.

Step k. Find $\hat{q}_k = a_k - \sum_{i=1}^{k-1} r_{ik} q_i$, and then normalize to obtain

$$q_k = \hat{q}_k / r_{kk}. \tag{7.7}$$

QR Factorization from the Classical Gram–Schmidt Process

The CGS process just described gives a reduced QR factorization of a matrix A whose columns are $a_1, a_2, \ldots, a_n$; that is, $A = (a_1, a_2, \ldots, a_n) \in \mathbb{R}^{m \times n}$.

To see this, note that (7.4)–(7.7) can be rewritten as

$$a_1 = q_1 r_{11},$$
$$a_2 = q_1 r_{12} + q_2 r_{22},$$
$$\vdots$$
$$a_k = q_1 r_{1k} + q_2 r_{2k} + \cdots + q_k r_{kk}, \quad k = 3, 4, \ldots, n.$$

In matrix form, we can then write

$$(a_1, a_2, \ldots, a_n) = (q_1, q_2, \ldots, q_n) \begin{pmatrix} r_{11} & r_{12} & \cdots & r_{1n} \\ 0 & r_{22} & \cdots & r_{2n} \\ \vdots & \vdots & \ddots & \vdots \\ 0 & 0 & \cdots & r_{nn} \end{pmatrix}$$

or $A = Q_1 R_1$, where $Q_1 = (q_1, q_2, \ldots, q_n)$, and R_1 is the matrix in the parentheses on the right.

Note that R_1 is $n \times n$ *upper triangular* and Q_1 is $m \times n$ *orthonormal*.

An illustration: $m = 4$, $n = 2$. Reduced QR factorization by CGS follows.

$$A = \begin{pmatrix} \times & \times \\ \times & \times \\ \times & \times \\ \times & \times \end{pmatrix} = \begin{pmatrix} \times & \times \\ \times & \times \\ \times & \times \\ \times & \times \end{pmatrix}_{4 \times 2} \begin{pmatrix} \times & \times \\ 0 & \times \end{pmatrix}_{2 \times 2} = Q_1 R_1.$$

ALGORITHM 7.7. Classical Gram–Schmidt (CGS) Method for QR Factorization.

Input: $A = (a_1, a_2, \ldots, a_n) \in R^{m \times n}$, $\text{rank}(A) = n$.
Output: Reduced QR factorization: $A = QR$; $Q \in \mathbb{R}^{m \times n}$ and $R \in \mathbb{R}^{n \times n}$.

For $k = 1, 2, \ldots, n$ do
    For $i = 1, 2, \ldots, k-1$ do
        $r_{ik} \equiv q_i^T a_k$
    End
    $q_k \equiv a_k - \sum_{i=1}^{k-1} r_{ik} q_i$
    $r_{kk} = \|q_k\|_2$
    $q_k \equiv \dfrac{q_k}{r_{kk}}$.
End

Numerical stability. The algorithm, as outlined above, is known to have serious numerical difficulties. *During the computations of the q_k's, cancellation can take place and, as a result, the computed q_k's can be far from orthogonal.* (See later in this section for details.)

The algorithm, however, can be modified to have better numerical properties. The following algorithm, known as the **modified Gram–Schmidt (MGS) algorithm**, computes the QR factorization of A in which, at the kth step, the kth column of Q, and the kth row of R are computed *(note that the Gram–Schmidt algorithm computes the kth columns of Q and R at the kth step).*

ALGORITHM 7.8. Modified Gram–Schmidt (MGS) for QR Factorization.

Input: $A = (a_1, a_2, \ldots, a_n) \in \mathbb{R}^{m \times n}$, $\text{rank}(A) = n$.
Output: Reduced QR factorization of A : $A = QR$, $Q \in \mathbb{R}^{m \times n}$, $R \in \mathbb{R}^{n \times n}$; Q is *orthonormal* and R is *upper triangular*.

Set $q_k = a_k$, $k = 1, 2, \ldots, n$.
For $k = 1, 2, \ldots, n$ do
    $r_{kk} = \|q_k\|_2$
    $q_k \equiv \dfrac{q_k}{r_{kk}}$
    For $j = k+1, \ldots, n$ do
        $r_{kj} \equiv q_k^T q_j$
        $q_j \equiv q_j - r_{kj} q_k$.
    End
End

The above is the **row-oriented modified Gram–Schmidt method**. The column-oriented version can similarly be developed (Exercise 7.9). *The two versions are numerically equivalent.*

7.5. Classical and Modified Gram–Schmidt Algorithms

Flop-count. The flop-count for the MGS algorithm is $2mn^2$, compared to $2(mn^2 - \frac{n^3}{3})$ needed for the Householder method. (Note that MGS works with the full-length column vector at each step, whereas the Householder method deals with successively shorter columns.)

Numerical stability. *Although the MGS process is more efficient than Householder's, it is not as numerically satisfactory as the Householder or Givens method for computing the QR factorization of A.* It can be shown (Björck (1996)) that if the computed Q is denoted by $\tilde{Q}$, then the following comparisons hold:

- *Orthogonality with MGS:* $\tilde{Q}^T \tilde{Q} = I + E$; $\|E\| \approx \mu \operatorname{Cond}(A)$.

- *Orthogonality with Householder:* $\tilde{Q}^T \tilde{Q} = I + E$; $\|E\| \approx \mu$.

  (For more details, see discussions in the next section.)

- *Reorthogonalization:* Orthogonality of the vectors in the matrix Q can be improved by reorthogonalization, and in general one reorthogonalization is sufficient. Indeed, William Kahan, Professor of Mathematics and Computer Science at the University of California at Berkeley, has remarked on this matter that "twice is enough." For more on this, see Björck (1996, pp. 67–69). Unfortunately, however, the reorthogonalization makes the process more expensive—almost double.

MATCOM Notes: Algorithms 7.7 and 7.8 have been implemented in the MATCOM programs **CLGRSCH** and **MDGRSCH**, respectively.

Example 7.13. Find QR factorizations of A of Example 7.5 using both CGS and MGS. Although in this case the CGS and MGS algorithms produce the same results, we use this example to illustrate here how the computational arrangements differ with the same matrix. All computations are performed with four-digit arithmetic.

- **CGS Method**

  $k = 1$:
  $$q_1 = a_1 = \begin{pmatrix} 1 \\ 0.0001 \\ 0 \end{pmatrix}, \quad r_{11} = \|q_1\|_2 = 1,$$

  $$q_1 = \frac{q_1}{r_{11}} = \begin{pmatrix} 1 \\ 0.0001 \\ 0 \end{pmatrix}.$$

  $k = 2$:
  $$r_{12} = 1, \quad q_2 = a_2 - r_{12} q_1 = \begin{pmatrix} 0 \\ -0.7071 \\ 0.7071 \end{pmatrix}, \quad q_1^T q_2 = -7.0711 \times 10^{-5}.$$

  Form Q_1 and R_1:
  $$Q_1 = (q_1, q_2) = \begin{pmatrix} 1 & 0 \\ 0.0001 & -0.7071 \\ 0 & 0.7071 \end{pmatrix}; \quad R_1 = \begin{pmatrix} r_{11} & r_{12} \\ 0 & r_{22} \end{pmatrix} = \begin{pmatrix} 1 & 1 \\ 0 & 1.414 \times 10^{-4} \end{pmatrix}.$$

- **MGS**

$$q_1 = a_1, \qquad q_2 = a_2.$$

k = 1:

$$r_{11} = \|q_1\|_2 = 1, \qquad q_1 = \begin{pmatrix} 1 \\ 0.0001 \\ 0 \end{pmatrix},$$

$$r_{12} = q_1^T q_2, \qquad q_2 = q_2 - r_{12} q_1 = \begin{pmatrix} 0 \\ -0.0001 \\ 0.0001 \end{pmatrix}.$$

k = 2:

$$r_{22} = \|q_2\| = 1.4142 \times 10^{-4}, \qquad q_2 = \frac{q_2}{r_{22}} = \begin{pmatrix} 0 \\ -0.7071 \\ 0.7071 \end{pmatrix}.$$

Form Q_1 and R_1:

$$Q_1 = \begin{pmatrix} 1 & 0 \\ 0.0001 & -0.7071 \\ 0 & 0.7071 \end{pmatrix}; \qquad R_1 = \begin{pmatrix} 1 & 1 \\ 0 & 1.4142 \times 10^{-4} \end{pmatrix}. \qquad \blacksquare$$

Modified Gram–Schmidt versus Classical Gram–Schmidt Algorithms

Mathematically, the CGS and MGS algorithms are equivalent. However, as remarked earlier, *their numerical properties are different.* For example, consider the computation of q_2 by the CGS method, given q_1 with $\|q_1\|_2 = 1$. We have

$$q_2 \equiv a_2 - r_{12} q_1, \quad \text{where } r_{12} = q_1^T a_2.$$

Then it can be shown (Björck (1996)) that

$$\|\text{fl}(q_2) - q_2\| < (1.06)(2m+3)\mu \|a_2\|_2.$$

Since $q_1^T q_2 = 0$, it follows that

$$|q_1^T \text{fl}(q_2)| < (1.06)(2m+3)\mu \|a_2\|_2.$$

This shows that in CGS two computed vectors, q_1 and q_2, can be far from orthogonal. On the other hand, it can be shown (Björck (1996)) that in MGS the loss of orthogonality depends upon the condition number of the matrix A. Specifically, it has been shown that the computed Q, denoted by $\hat{Q}$, satisfies

$$\left\| I - \hat{Q}^T \hat{Q} \right\|_2 \leq \frac{c_1 \mu \operatorname{Cond}_2(A)}{1 - c_2 \mu \operatorname{Cond}_2(A)},$$

assuming that $c_2 \mu \operatorname{Cond}_2(A) < 1$, where c_1 and c_2 are small constants.

Since in MGS the loss of orthogonality depends upon the condition number, one could use column pivoting to maintain orthogonality as much as possible. Thus, as far as finding the QR factorization of A is concerned, neither algorithm can be recommended over the Householder or the Givens method. *With CGS the orthogonality of Q can be completely lost; with MGS the orthogonality of Q may not be acceptable when A is ill-conditioned.*

7.5. Classical and Modified Gram–Schmidt Algorithms

Note that in Example 7.13, the computed $\hat{Q}$ (in four-digit arithmetic) is such that

$$\hat{Q}^T \hat{Q} = \begin{pmatrix} 1 & -0.0001 \\ -0.0001 & 1 \end{pmatrix}.$$

On the other hand, for the same problem using the Householder method, $\hat{Q}^T \hat{Q} = I$ (in four-digit arithmetic).

The following table shows the departure of orthogonality of the Q matrix for the QR factorization of a 5×5 Hilbert matrix (extended precision) using three different methods (Householder, CGS, and MGS).

Table 7.1. *Comparison of QR factorization of a 5×5 Hilbert matrix with CGS and MGS.*

Method	$\|I - \hat{Q}^T \hat{Q}\|_2$
CGS	$O(10^{-7})$
MGS	$O(10^{-12})$
Householder	$O(10^{-16})$

Remark. *Table 7.1 clearly shows the superiority of the Householder method over both the CGS and MGS methods; of the latter two methods, MGS is clearly preferred over CGS.* We now summarize in Table 7.2 the *flop-count* (**for computing R only**) and *stability properties* of the four methods for QR factorization of a matrix A.

Table 7.2. *Comparison of efficiency and stability of QR factorization methods.*

Method	Flop-count	Stability
Householder	$2n^2 \left(m - \dfrac{n}{3} \right)$	Stable
Givens	$3n^2 \left(m - \dfrac{n}{3} \right)$	Stable
CGS	$2mn^2$	Unstable (*possible severe loss of orthogonality*).
MGS	$2mn^2$	Better stability property than CGS, but not as stable as Householder's or Givens' method.

Full QR versus Reduced QR Factorizations and Uniqueness

The CGS and MGS methods give a reduced QR factorization. The questions thus arise: (i) *How does one obtain a full QR factorization from a reduced one?* (ii) *When is a reduced QR factorization unique?*

To obtain a full QR factorization from a reduced one, $A = Q_1 R_1$, *just append an additional $(m - n)$ orthonormal columns to Q_1 so that it becomes an $m \times m$ orthogonal matrix Q and also append a row of zeros to R_1 so that it becomes an $m \times n$ triangular matrix R.*

To answer the question of uniqueness, we turn to the CGS process.

Note that theoretically this process does not break down unless q_k is identically zero; but this will not happen if A is assumed to have full rank (*why?*). The choices of $r_{kk} = \|q_k\|_2 > 0$ were made deliberately to normalize the vectors q_k so that they have unit lengths. But once this choice is made, then all the computations are uniquely determined.

Thus, we can state the following result. Another proof of this result can be obtained via Cholesky decomposition; see Exercise 7.7.

Theorem 7.14 (uniqueness in reduced QR factorization). *Let $A \in \mathbb{R}^{m \times n} (m \geq n)$ have full rank. Then it has a unique reduced QR factorization: $A = Q_1 R_1$ with the diagonal entries of R_1 positive.*

7.6 Solution of $Ax = b$ Using QR Factorization

The QR factorization
$$A = QR$$
immediately leads to the following algorithm for solving $Ax = b$:

Solving $Ax = b$ Using QR.

Step 1. Find the QR factorization of $A : Q^T A = R$.
Step 2. From $b' = Q^T b$.
Step 3. Solve $Rx = b'$.

Note: If Householder's method for QR factorization is used, then the vector b' can be computed *implicitly* from the factored form of $Q^T = H_1 H_2 \ldots H_{n-1}$, as $b' = H_1 H_2 \ldots H_{n-1} b$.

Example 7.15. Consider matrix A from Example 7.6 and $b = (2, 6, 3)^T$.

Step 1. The Householder matrices are given by

$$H_1 = \begin{pmatrix} 0 & -0.7071 & -0.7071 \\ -0.7071 & 0.5000 & -0.5000 \\ -0.7071 & -0.5000 & 0.5000 \end{pmatrix}, \quad H_2 = \begin{pmatrix} 1 & 0 & 0 \\ 0 & -0.1691 & -0.9856 \\ 0 & -0.9856 & 0.1691 \end{pmatrix}.$$

Step 2. Compute b':

$$y_1 = b = \begin{pmatrix} 2 \\ 6 \\ 3 \end{pmatrix}; \quad y_2 = H_1 y_1 = \begin{pmatrix} -6.3640 \\ 0.0858 \\ -2.9142 \end{pmatrix}; \quad b' = y_3 = H_2 y_2 = \begin{pmatrix} -6.3640 \\ 2.8577 \\ -0.5773 \end{pmatrix}.$$

(Note that b' above has been computed without explicitly forming the matrix Q.)

Step 3. Solve: $Rx = b' \Rightarrow x = \begin{pmatrix} 1 \\ 1 \\ 1 \end{pmatrix}$ (using R from Example 7.6). ∎

Flop-count. If the Householder method is used to factor A into QR, then the solution of $Ax = b$ requires $\frac{4}{3}n^3 + O(n^2)$ flops. Thus, solving $Ax = b$ using Householder QR factorization requires roughly *twice* the number of operations as with Gaussian elimination with pivoting. *This explains why the QR approach for solving $Ax = b$ is not used in practice over GEPP.*

Round-off property. It can be shown (Lawson and Hanson (1995)) that the computed solution $\hat{x}$ is the exact solution of $(A+E)\hat{x} = b + \delta b$, where $\|E\|_F \leq (3n^2 + 41n)\mu \|A\|_F + O(\mu^2)$, and $\|\delta b\| \leq (3n^2 + 40n)\mu \|b\| + O(\mu^2)$.

Thus the QR method for solving $Ax = b$ is *stable* and *does not involve any growth factor.*

7.7 Projections Using QR Factorization

Definition 7.16. *The $n \times n$ matrix P_S having the following properties is called the **orthogonal projection** onto a subspace S of $\mathbb{R}^n$.*

(i) $R(P_S) = S$ (the range of P_S is S).

(ii) $P_S^T = P_S$ (P_S is *symmetric*).

(iii) $P_S^2 = P_S$ (P_S is *idempotent*).

A relationship between P_S and $P_{S^\perp}$. *If P_S is the orthogonal projection onto S, then $I - P_S$, denoted by $P_{S^\perp}$, where I is the identity matrix of the same order as P_S, is the orthogonal projection onto $S^\perp$ (Exercise 7.35).*

7.7.1 Orthogonal Projections and Orthonormal Bases

Let $S \subseteq \mathbb{R}^n$ be a subspace. Let $\{v_1, \ldots, v_k\}$ be an orthonormal basis for the subspace S. Form $V = (v_1, v_2, \ldots, v_k)$. Then
$$P_S = VV^T$$
is the orthogonal projection onto S.
Note that V is not unique, but P_S is.

The orthogonal projections onto $R(A)$ and $N(A^T)$. When the subspace S is $R(A)$ or $N(A^T)$ associated with the matrix A, we will denote the unique orthogonal projections onto $R(A)$ and $N(A^T)$ by P_A and P_N, respectively. If A is $m \times n (m \geq n)$ and has full rank, their explicit expressions are given by (Exercise 7.36)

(i) $P_A = A(A^TA)^{-1}A^T$;   (ii) $P_N = I - A(A^TA)^{-1}A^T$.

Remark. It is not *advisable to compute projections* using the above expressions, because the matrix A^TA can be computationally singular.

7.7.2 Projection of a Vector onto the Range and the Null Space of a Matrix

Any vector b can be written as
$$b = b_S + b_{S^\perp},$$
where $b_S \in S$ and $b_{S^\perp} \in S^\perp$. If s is the rank $R(A)$ of a matrix A, then $b_S \in R(A)$ and $b_{S^\perp} \in N(A^T)$. We will therefore denote b_S by b_R and $b_{S^\perp}$ by b_N, meaning that b_R is in the range of A and b_N is in the null space of A^T. The vectors b_R and b_N, in fact, can be expressed in terms of the orthogonal projections onto $R(A)$ and $N(A^T)$, respectively. It can be shown (Exercise 7.36) that
$$b_R = P_A b \quad \text{and} \quad b_N = P_N b.$$
The vectors b_R and b_N are called the *orthogonal projection of b onto $R(A)$* and the *orthogonal projection of b onto $N(A^T)$*, respectively.

From above, we easily see that $b_R^T b_N = 0$.

7.7.3 Orthonormal Bases and Orthogonal Projections onto the Range and Null Space using QR Factorization

Computing orthogonal projections using the explicit formulas above require the matrix inversion $(A^T A)^{-1}$, and therefore can be numerically unstable. *A stable way of computing the projections is either via* (i) *QR factorization or* (ii) *SVD*.

For discussions on finding orthonormal bases and orthogonal projections using the SVD, see Section 7.8.10.

Theorem 7.17. *Let $A = QR$ be the QR factorization of a full-rank $m \times n$ matrix $A(m \geq n)$. Let $Q = (Q_1, Q_2)$, where Q_1 is the matrix of first n columns. Then*

(i) *the columns of Q_1 and Q_2 form orthonormal bases of $R(A)$ and $N(A^T)$, respectively;*

(ii) *the orthogonal projections, P_A and P_N, onto $R(A)$ and $N(A^T)$ are, respectively, $Q_1 Q_1^T$ and $Q_2 Q_2^T$.*

Proof. Assertions (i) follow from the fact already established:
$$\text{span}\{q_1, \ldots, q_i\} = \text{span}\{a_1, \ldots, a_i\}, \quad i = 1, \ldots, n.$$
Assertions (ii) follow from (i) and the definition of the projection. □

Example 7.18. Consider Example 7.5 again.
$$A = \begin{pmatrix} 1 & 1 \\ 0.0001 & 0 \\ 0 & 0.0001 \end{pmatrix}.$$
Using the results of Example 7.5 we have the following:

- *An orthonormal basis of* $\mathbb{R}(A)$: $\left\{ \begin{array}{cc} -1 & 0.0001 \\ 0.0001 & -0.7071 \\ 0 & 0.7071 \end{array} \right\}$.

7.8. Singular Value Decomposition and Its Properties

- *An orthonormal basis of* $N(A^T)$: $\left\{\begin{array}{c} -0.0001 \\ 0.7071 \\ 0.7071 \end{array}\right\}$.

- *The orthogonal projection onto* $R(A)$:

$$P_A = Q_1 Q_1^T = \begin{pmatrix} 1.000 & 0.0000 & 0.0000 \\ 0.0000 & 0.5000 & -0.5000 \\ 0.0000 & -0.5000 & 0.5000 \end{pmatrix}.$$

- *The orthogonal projection onto* $N(A^T)$:

$$P_N = Q_2 Q_2^T = \begin{pmatrix} 0.0000 & -0.0000 & -0.0000 \\ -0.0000 & 0.5000 & 0.5000 \\ -0.0000 & 0.5000 & 0.5000 \end{pmatrix}.$$

- *Orthogonal Projections of b*: Let $b = (1, 1, 1)^T$ and A as above. Then

$$b_R = P_A b = \begin{pmatrix} 1.0001 \\ 0.0001 \\ 0.0001 \end{pmatrix} \quad \text{and} \quad b_N = P_N b = \begin{pmatrix} -0.0001 \\ 0.9999 \\ 0.9999 \end{pmatrix}. \quad \blacksquare$$

QR Factorization with Column Pivoting.

If A is rank-deficient, then QR factorization cannot be used to find a basis for $R(A)$.

To see this, consider the following 2×2 example:

$$A = \begin{pmatrix} 0 & 0 \\ 0 & 1 \end{pmatrix} = \begin{pmatrix} c & -s \\ s & c \end{pmatrix} \begin{pmatrix} 0 & s \\ 0 & c \end{pmatrix} = QR.$$

Rank $(A) = 1 < 2$. So, the columns of Q do not form an orthonormal basis of $R(A)$ nor of its complement.

In this case, one needs to use a modification of the QR factorization process, called *QR factorization with column pivoting*.

The process finds a permutation matrix P and the matrices Q and R such that $AP = QR$. The details are given in Chapter 14, available online at *www.siam.org/books/ot116*. See also Golub and Van Loan (1996, pp. 248–250).

MATLAB command $[Q, R, P] = \mathbf{qr}(A)$ can be used to compute the QR factorization with column pivoting.

Also $[Q, R, E] = \mathbf{qr}(A, 0)$ produces an economy-sized QR factorization in which E is a permutation vector so that $Q * R = A(:, E)$.

7.8 Singular Value Decomposition and Its Properties

We have so far seen two matrix factorizations: LU and QR. In this section, we introduce another very important factorization, called *singular value decomposition (SVD)*. A proof of the SVD (Theorem 7.19) will be deferred until Chapter 10.

7.8.1 Singular Values and Singular Vectors

Theorem 7.19 (SVD theorem). *Let $A \in \mathbb{R}^{m \times n}$. Then there exist orthogonal matrices $U \in \mathbb{R}^{m \times m}$ and $V \in \mathbb{R}^{n \times n}$ such that*
$$A = U \Sigma V^T, \tag{7.8}$$
where $\Sigma = \mathrm{diag}(\sigma_1, \ldots, \sigma_p) \in \mathbb{R}^{m \times n}$, $p = \min(m, n)$, *and* $\sigma_1 \geq \sigma_2 \geq \cdots \geq \sigma_p \geq 0$.

Definition 7.20. *The decomposition $A = U \Sigma V^T$ is called the **singular value decomposition** of A.*

- $U \in \mathbb{R}^{m \times m}$ (orthogonal).
- $V \in \mathbb{R}^{n \times n}$ (orthogonal).
- $\Sigma \in \mathbb{R}^{m \times n}$ (diagonal).

Note: Notice that when $m \geq n$, Σ has the form
$$\Sigma = \left(\begin{array}{c} \begin{matrix} \sigma_1 & & \\ & \ddots & \\ & & \sigma_n \end{matrix} \\ \hline 0_{(m-n) \times n} \end{array} \right).$$

An illustration: $m = 4$, $n = 2$. The following is an SVD of a 4×2 matrix:

$$\underbrace{\begin{pmatrix} \times & \times \\ \times & \times \\ \times & \times \\ \times & \times \end{pmatrix}}_{A} = \underbrace{\begin{pmatrix} \times & \times & \times & \times \\ \times & \times & \times & \times \\ \times & \times & \times & \times \\ \times & \times & \times & \times \end{pmatrix}}_{U} \underbrace{\begin{pmatrix} \times & 0 \\ 0 & \times \\ 0 & 0 \\ 0 & 0 \end{pmatrix}}_{\Sigma} \underbrace{\begin{pmatrix} \times & \times \\ \times & \times \end{pmatrix}}_{V^T}.$$

Definition 7.21. *The diagonal entries $\sigma_1, \sigma_2, \ldots, \sigma_p$ are called the **singular values** of A.*

Definition 7.22. *The columns of U are called the **left singular vectors**, and those of V are called the **right singular vectors**.*

A Convention

For the remainder of this chapter we will assume, without any loss of generality, that $m \geq n$, because if $m < n$, we consider the SVD of A^T, and if the SVD of A^T is $U \Sigma V^T$, then the SVD of A is $V \Sigma^T U^T$. Also the following convention will be used:

- $\sigma_{\max} = \sigma_1 =$ the largest singular value.
- $\sigma_{\min} = \sigma_n =$ the smallest singular value.
- $\sigma_i =$ the ith singular value.
- $\sigma_1 \geq \sigma_2 \geq \cdots \geq \sigma_n \geq 0$.

7.8. Singular Value Decomposition and Its Properties

Notes:

1. When $m \geq n$, we have n singular values. *It can be shown that these are the square roots of the n eigenvalues of the symmetric matrix $A^T A$.*

2. *The singular values of A are uniquely determined while the matrices U and V are not unique in general. (Why?)*

3. *The SVD immediately reveals several matrix properties, including* rank, norms, condition number, *and important information on the structure of a matrix, such as* orthonormal bases of $R(A)$ and $N(A)$ and orthogonal projections *onto $R(A)$ and $N(A)$. (See Sections 7.8.7 and 7.8.10.)*

7.8.2 Computation of the SVD (MATLAB Command)

The computation of the SVD is more expensive than computing the QR factorization either by Householder's or Givens' method. A widely used method, called the Golub–Kahan–Reinsch algorithm, comes in two phases. In Phase I, *the matrix A is reduced to a bidiagonal matrix B by orthogonal equivalence*, and then in Phase II, *the matrix B is further reduced to a diagonal matrix of singular values.* We shall describe this SVD method in detail in Chapter 10. The *method is numerically stable.* For the time being, to use SVD as a computational tool, one can use the MATLAB program **svd**,

$$[U, S, V] = \mathbf{svd}(A),$$

which gives the complete SVD. If only the singular values of A are required, use **svd** (A).

Example 7.23. Let

$$A = \begin{pmatrix} 1 & 2 \\ 2 & 3 \\ 3 & 4 \end{pmatrix}.$$

Then $[U, S, V] = \mathbf{svd}(A)$ gives

$$\Sigma = \begin{pmatrix} 6.5468 & 0 \\ 0 & 0.3742 \\ 0 & 0 \end{pmatrix}_{3 \times 2}, \quad U = \begin{pmatrix} 0.3381 & 0.8480 & 0.4082 \\ 0.5506 & 0.1735 & -0.8165 \\ 0.7632 & -0.5009 & 0.4082 \end{pmatrix}_{3 \times 3},$$

$$V = \begin{pmatrix} 0.5696 & -0.8219 \\ 0.8299 & 0.5696 \end{pmatrix}_{2 \times 2}.$$

There are two singular values: 6.5458, 0.3742. ∎

7.8.3 The SVD of a Complex Matrix

Let $A \in \mathbb{C}^{m \times n}$. Then there exist **unitary matrices** $U \in \mathbb{C}^{m \times m}$ and $V \in \mathbb{C}^{n \times n}$ such that

$$A = U \Sigma V^*,$$

where $\Sigma = \text{diag}(\sigma_1, \ldots, \sigma_p) \in \mathbb{R}^{m \times n}$, $p = \min\{m, n\}$, with $\sigma_1 \geq \sigma_2 \geq \cdots \geq \sigma_p \geq 0$.

MATLAB Note: The function **svd** can be used to produce the SVD of a complex matrix as well. In this case U and V are complex unitary matrices.

7.8.4 Geometric Interpretation of the Singular Values and Singular Vectors

Let S be the unit sphere in $\mathbb{R}^n$. Then the image of S under A is a *hyperellipsoid* E defined by $E = \{Ax : \|x\|_2 = 1\}$.

- The singular values are the lengths of the semi-axes of E.
- The left singular vectors are the unit vectors in the direction of the semi-axes of E.
- The right singular vectors are the unit vectors in S that are the preimages of the semi-axes of E.

Thus, the unitary map V^* preserves the sphere, the diagonal matrix Σ stretches the sphere into a hyperellipsoid, and the unitary map U rotates or reflects the hyperellipsoid, keeping its shape. For more details see Trefethen and Bau (1997).

See Figure 7.4 for an illustration *of a unit ball under the SVD of A.*
Let $A = U\Sigma V^T$, $\Sigma = \text{diag}(3, 0.5)$.

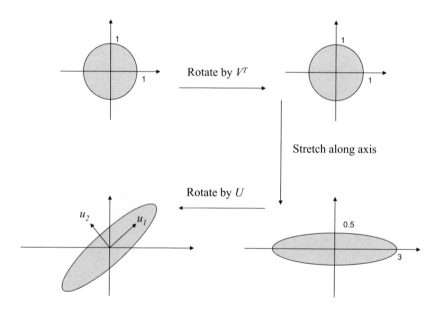

Figure 7.4. *Image of a unit ball under the SVD of a matrix.*

7.8.5 Reduced SVD

If $A = U\Sigma V^T$ is the SVD of $A \in \mathbb{R}^{m \times n}$ ($m \geq n$), then, as in the case of QR factorization, we can write

$$A = U_1 \Sigma_1 V^T,$$

7.8. Singular Value Decomposition and Its Properties

where
$$U_1 = (u_1, u_2, \ldots, u_n) \in \mathbb{R}^{m \times n}, \quad V = (v_1, v_2, \ldots, v_n) \in \mathbb{R}^{n \times n},$$
and
$$\Sigma_1 = \mathrm{diag}\,(\sigma_1, \sigma_2, \ldots, \sigma_n) \in \mathbb{R}^{n \times n}.$$

The above form of SVD is called the **reduced** or the **thin** SVD of A.

Reduced SVD

Any matrix $A \in \mathbb{R}^{m \times n} (m \geq n)$ can be written in the form
$$A = U_1 \Sigma_1 V^T,$$
where U_1 is an $m \times n$ *orthonormal matrix*, Σ_1 is an $n \times n$ *diagonal matrix*, and V is an $n \times n$ *orthogonal matrix*.
In many applications, this reduced form of SVD is sufficient.

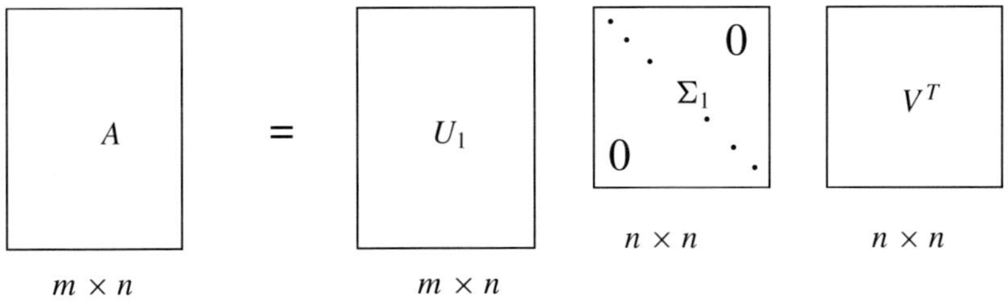

Figure 7.5. *Reduced SVD* $(m > n)$.

MATLAB Note: $[U, S, V] = \mathbf{svd}(A, 0)$ can be used to produce the reduced SVD of A.

7.8.6 Sensitivity of the Singular Values

One reason for the wide applicability of singular values in practical applications is that the singular values are well-conditioned. We state a result here in this context. The proof will be deferred until Chapter 10.

Theorem 7.24 (perturbation theorem for singular values). *Let A and $B = A + E$ be two $m \times n$ matrices ($m \geq n$). Let σ_i, $i = 1, \ldots, n$, and $\tilde{\sigma}_i$, $i = 1, \ldots, n$, be, respectively, the singular values of A and $A + E$ in decreasing order. Then $|\tilde{\sigma}_i - \sigma_i| \leq \|E\|_2$ for each i.*

Example 7.25.
$$A = \begin{pmatrix} 1 & 2 & 3 \\ 3 & 4 & 5 \\ 6 & 7 & 8 \end{pmatrix}, \quad E = \begin{pmatrix} 0 & 0 & 0 \\ 0 & 0 & 0 \\ 0 & 0 & 0.0002 \end{pmatrix}.$$

The singular values of A: $\sigma_1 = 14.5576$, $\sigma_2 = 1.0372$, $\sigma_3 = 0.0000$.

The singular values of $A + E$: $\tilde{\sigma}_1 = 14.5577$, $\tilde{\sigma}_2 = 1.0372$, and $\tilde{\sigma}_3 = 0.0000$, *Absolute error:* $|\tilde{\sigma}_1 - \sigma_1| = 1.1373 \times 10^{-4}$, $|\tilde{\sigma}_2 - \sigma_2| = 0$, and $|\tilde{\sigma}_3 - \sigma_3| = 0$. Since $\|E\|_2 = 0.002$, the statement of Theorem 7.24 is easily verified. ∎

Next, we present a result (without proof) on the perturbation of singular values that uses the Frobenius norm instead of the 2-norm.

Theorem 7.26. *Let A, E, σ_i, and $\tilde{\sigma}_i$, $i = 1, \ldots, n$, be the same as in Theorem 7.24. Then*

$$\sqrt{\sum_{i=1}^{n}(\tilde{\sigma}_i - \sigma_i)^2} \leq \|E\|_F.$$

7.8.7 Norms, Condition Number, and Rank via the SVD

Theorem 7.27. *Let $\sigma_1 \geq \sigma_2 \geq \cdots \geq \sigma_n$ be the n singular values of an $m \times n$ matrix $A (m \geq n)$. Then*

1. $\|A\|_2 = \sigma_1 = \sigma_{\max}$.

2. $\|A\|_F = (\sigma_1^2 + \sigma_2^2 + \cdots + \sigma_n^2)^{\frac{1}{2}}$.

3. $\|A^{-1}\|_2 = \frac{1}{\sigma_n}$, *when A is $n \times n$ and nonsingular.*

4. $\text{Cond}_2(A) = \|A\|_2 \|A^{-1}\|_2 = \frac{\sigma_1}{\sigma_n} = \frac{\sigma_{\max}}{\sigma_{\min}}$ *if A is nonsingular.*

5. rank $(A) =$ *number of nonzero singular values.*

Proof.

1. $\|A\|_2 = \|U\Sigma V^T\|_2 = \|\Sigma\|_2 = \max_i(\sigma_i) = \sigma_1$.

2. $\|A\|_F = \|U\Sigma V^T\|_F = \|\Sigma\|_F = (\sigma_1^2 + \sigma_2^2 + \cdots + \sigma_n^2)^{\frac{1}{2}}$.

   (Note that the 2-norm and Frobenius norm are invariant under orthogonal matrix products).

3. To prove 3 we note that the largest singular value of A^{-1} is $\frac{1}{\sigma_n}$. (Note that when A is invertible, $\sigma_n \neq 0$.) Then the result follows from 1.

4. Item 4 follows from the definition of $\text{Cond}_2(A)$ and the results of 1 and 3.

5. Since the rank of a matrix is invariant under orthogonal matrix multiplication, rank(A) = rank $(U\Sigma V^T)$ = rank (Σ). The matrix Σ being a diagonal matrix, its rank is equal to the number of nonzero diagonal entries. □

Remarks on the SVD condition number.

1. If A is a rectangular matrix having full rank, then $\text{Cond}_2(A) = \frac{\sigma_{\max}}{\sigma_{\min}}$.

2. When A is rank-deficient, $\sigma_{\min} = 0$, and we say that $\text{Cond}_2(A)$ is *infinite*.

7.8. Singular Value Decomposition and Its Properties

3. When A is nearly rank-deficient, $\text{Cond}_2(A)$ is large.

4. $\text{Cond}_2(A)$ measures how far the *hyperellipsoid* $\{Ax : \|x\|_2 = 1\}$ is elongated.

5. Determining rank in presence of round-off errors and noisy data is a nontrivial task. What is more important in practice is to talk about *numerical rank* rather than just the rank of a matrix (see Section 7.8.9). See also discussions on *rank-deficiency* in Theorem 7.29 in this context.

Example 7.28.
$$A = \begin{pmatrix} 1 & 2 \\ 2 & 3 \\ 3 & 4 \end{pmatrix}.$$
Singular values of A are $\sigma_1 = 6.5468$, $\sigma_2 = 0.3742$. $\|A\|_2 = \sigma_1 = 6.5468$; $\|A\|_F = \sqrt{\sigma_1^2 + \sigma_2^2} = 6.5574$; $\text{Cond}_2(A) = \frac{\sigma_1}{\sigma_2} = 17.4975$. ∎

7.8.8 The Distance to Singularity, Rank-Deficiency, and Numerical Rank via the SVD.

We have just seen that the number of nonzero singular values of a matrix is its rank. The rank of a matrix can also be determined using (i) *Gaussian elimination*, and (ii) *QR factorization with column pivoting*. Both are less expensive than computing the singular values, but *are not as numerically reliable as determining it by the SVD*, especially if it is desired to determine the closeness of a full-rank matrix to a nearby rank-deficient one. As an example, consider the celebrated **Kahan matrix** *(which is upper triangular)*:

$$R = \text{diag}(1, s, \ldots, s^{n-1}) \begin{pmatrix} 1 & -c & -c & \cdots & -c \\ & 1 & -c & \cdots & -c \\ & & 1 & & \vdots \\ & & & \ddots & \vdots \\ & & & & 1 \end{pmatrix}, \quad (7.9)$$

with $c^2 + s^2 = 1$; $c, s > 0$.

For $n = 100$, $c = 0.2$, $r_{nn} = s^{n-1} = 0.133$, which is not small; on the other hand, R has a singular value of order 10^{-8}, indicating *that it is nearly singular*.

Since the number of *nonzero* singular values determines the rank of a matrix, we can say that a matrix A is arbitrarily near a matrix of full rank: *just change each zero singular value by a small number ϵ. It is, therefore, more meaningful to know if a matrix is near a matrix of a certain rank, rather than knowing what the rank is. The SVD exactly answers this question.*

Suppose that A has rank r, that is, $\sigma_1 \geq \sigma_2 \geq \cdots \geq \sigma_r > 0$ and $\sigma_{r+1} = \cdots = \sigma_n = 0$. Then the question is how far is A from a matrix of rank $k < r$. The following theorem (Theorem 7.29) can be used to answer the question. This theorem is generally known as the Eckart–Young theorem (see Eckart and Young (1939)).

Theorem 7.29 (SVD and nearness to rank-deficiency). *Let $A = U \Sigma V^T$ be the SVD of A. Let $k \leq r = \text{rank}(A)$. Define $A_k = U \Sigma_k V^T$, where $\Sigma_k = \text{diag}(\sigma_1, \ldots, \sigma_k, 0 \ldots 0)_{m \times n}$, where $\sigma_1 \geq \sigma_2 \geq \cdots \geq \sigma_k > 0$.*

Then

(a) A_k has rank k.

(b) *The distance of A_k from A (in the 2-norm) is* $\|A - A_k\|_2 = \sigma_{k+1}$.

(c) *Out of all the matrices of rank k, A_k is the closest to A; that is,*

$$\min_{\text{rank}(B)=k} \|A - B\|_2 = \|A - A_k\|_2.$$

Proof. *Proof of* (a). rank (A_k) = rank$(U\Sigma_k V^T)$ = rank (Σ_k) = k (note that $\sigma_1 \geq \sigma_2 \geq \cdots \geq \sigma_k > 0$).

Proof of (b). Because $A - A_k = U(\Sigma - \Sigma_k)V^T$, we have $\|A - A_k\|_2 = \|U(\Sigma - \Sigma_k)V^T\|_2 = \|(\Sigma - \Sigma_k)\|_2 = \sigma_{k+1}$. Thus, the distance between A and A_k is σ_{k+1}.

Proof of (c). To prove (c), we show that if $B \in \mathbb{R}^{m \times n}$ is any other matrix of rank k, then $\|A - B\|_2 = \sigma_{k+1}$; that is, A_k is closest to A among all other matrices of rank k.

Since B has rank k, the null space of B, $N(B)$, has dimension $n - k$.

Consider now the space $S = \text{span } \{v_1, \ldots, v_{k+1}\}$, where v_1 through v_{k+1} are the right singular vectors of A. Since $N(B)$ and S are both subspaces of $\mathbb{R}^n$ and the sum of their dimensions is greater than n, their intersection must be nonempty. Let z be a unit vector lying in this intersection. Then since $z \in \text{span } \{v_1, \ldots, v_{k+1}\}$, there exist scalars (not all zero) such that $z = c_1v_1 + c_2v_2 + \cdots + c_{k+1}v_{k+1}$.

Furthermore, because $v_1, \ldots, v_{k+1}$ are orthonormal, we must have $|c_1|^2 + |c_2|^2 + \cdots + |c_{k+1}|^2 = 1$. Because z also belongs to $N(B)$, we have $Bz = 0$. So

$$(A - B)z = Az = \sum_{i=1}^{k+1} c_i A v_i = \sum_{i=1}^{k+1} \sigma_i c_i u_i \quad \text{(note that } Av_i = \sigma_i u_i\text{)}.$$

Because $u_1, \ldots, u_{k+1}$ are also orthonormal,

$$\|(A - B)z\|_2^2 = \sum_{i=1}^{k+1} |\sigma_i c_i|^2 \geq \sigma_{k+1}^2 \sum_{i=1}^{k+1} |c_i^2| = \sigma_{k+1}^2.$$

Thus, $\|A - B\|_2 \geq \frac{\|(A-B)z\|_2}{\|z\|_2} \geq \sigma_{k+1}$ (because $\|z\|_2 = 1$). □

Corollary 7.30. *Let A be an $m \times n$ matrix of full rank and let $r = \min(m, n)$. Let $\sigma_1 \geq \sigma_2 \geq \cdots \geq \sigma_r > 0$ be the singular values of A. If C is another $m \times n$ matrix such that $\|C - A\|_2 < \sigma_r$, then C has also full rank.*

Corollary 7.31 (distance to singularity). *The relative distance of a nonsingular matrix A to the nearest singular matrix C is $\frac{1}{\text{Cond}_2(A)}$; that is,*

$$\frac{\|C - A\|_2}{\|A\|_2} = \frac{1}{\text{Cond}_2(A)}.$$

Example 7.32. Consider Example 7.23 again.

$$\sigma_1 = 6.5468, \quad \sigma_2 = 0.3742, \quad k = 1.$$

7.8. Singular Value Decomposition and Its Properties

Then
$$A_1 = U\Sigma_1 V^T = \begin{pmatrix} 1.2608 & 1.8193 \\ 2.0534 & 2.9630 \\ 2.8460 & 4.1067 \end{pmatrix}.$$

Out of all the matrices of rank 1, A_1 is the closest to A. (Verify this by constructing any other arbitrary rank-one matrix of order 3×2 and then computing its 2-norm distance from A.) ∎

Distance of a Matrix from the Nearest Matrix of Lower Rank

The above result states that the smallest nonzero singular value gives the distance from A to the nearest matrix of lower rank. In particular, *for a nonsingular $n \times n$ matrix A, the smallest singular value σ_n gives the measures of the distance of A to the nearest singular matrix*.

Thus, in order to know if a matrix A of rank r is close enough to a matrix of lower rank, look into the smallest nonzero singular value σ_r. If this is very small, then the matrix is very close to a matrix of rank $r - 1$, because there exists a perturbation of size as small as $|\sigma_r|$ which will produce a matrix of rank $r - 1$. In fact, one such perturbation is $u_r \sigma_r v_r^T$.

Example 7.33.

$$A = \begin{pmatrix} 1 & 0 & 0 \\ 0 & 2 & 0 \\ 0 & 0 & 0.0000004 \end{pmatrix}; \quad \text{rank}(A) = 3, \quad \sigma_3 = 0.0000004.$$

$$U = \begin{pmatrix} 0 & 1 & 0 \\ 1 & 0 & 0 \\ 0 & 0 & 1 \end{pmatrix}, \quad V = \begin{pmatrix} 0 & 1 & 0 \\ 1 & 0 & 0 \\ 0 & 0 & 1 \end{pmatrix}$$

$$A' = A - u_3 \sigma_3 v_3^T = \begin{pmatrix} 1 & 0 & 0 \\ 0 & 2 & 0 \\ 0 & 0 & 0 \end{pmatrix}; \quad \text{rank}(A') = 2.$$

The required perturbation $u_3 \sigma_3 v_3^T$ to make A singular is very small:

$$10^{-6} \begin{pmatrix} 0 & 0 & 0 \\ 0 & 0 & 0 \\ 0 & 0 & 0.4000 \end{pmatrix}. \quad \blacksquare$$

Note: The following is a Frobenius-norm analogue of Theorem 7.29

Theorem 7.34 (low-rank approximation in Frobenius norm). *Let B be any matrix of the same order as A and let A_k be the same as in Theorem 7.29. Then*

$$\|B - A\|_F^2 \geq \|A_k - A\|_F^2.$$

7.8.9 Numerical Rank

In practical applications that need singular values we have to know when to accept a computed singular value to be "zero." Of course, if it is of an order of "round-off zeros" (machine epsilon), we can declare it to be zero.

However, if the data share a large relative error, it should also be taken into consideration. A practical criterion would be the following:

Accept a computed singular value to be zero if it is less than or equal to $10^{-t}\|A\|_\infty$, where the entries of A are correct to t digits.

Having defined a tolerance $\delta = 10^{-t}\|A\|_\infty$ for a zero singular value, we can have the following convention for the numerical rank of a matrix (see Golub and Van Loan (1996, p. 261)):

A has "numerical rank" r if the computed singular values $\tilde{\sigma}_1, \tilde{\sigma}_2, \ldots, \tilde{\sigma}_n$ satisfy

$$\tilde{\sigma}_1 \geq \tilde{\sigma}_2 \geq \cdots \geq \tilde{\sigma}_r > \delta \geq \tilde{\sigma}_{r+1} \geq \cdots \geq \tilde{\sigma}_n. \tag{7.10}$$

Thus, roughly, to determine **numerical rank** of a matrix A, count the "large" singular values only. If this number is r, then A has numerical rank r.

Remark. Note that finding the numerical rank of a matrix will be tricky if there is no suitable gap between a set of singular values.

7.8.10 Orthonormal Bases and Projections from the SVD

Theorem 7.35. *Let $A = U\Sigma V^T$ be the SVD of $A \in \mathbb{R}^{m \times n}$ ($m \geq n$) and let r be the rank of A. Partition*

$$U = (U_1, U_2) \quad \text{and} \quad V = (V_1, V_2),$$

where U_1 and V_1 consist of the first r columns of U and V, respectively. Then

(a) *the columns of U_1 form an orthonormal basis of $R(A)$;*

(b) *the columns of V_2 form an orthonormal basis of $N(A)$;*

(c) *orthogonal projection onto $R(A) = U_1 U_1^T$;*

(d) *orthogonal projection onto $N(A^T) = U_2 U_2^T$;*

(e) *orthogonal projection onto $N(A) = V_2 V_2^T$;*

(f) *orthogonal projection onto $R(A^T) = V_1 V_1^T$.*

Proof. The proofs of (a) and (b) follow, respectively, from the fact that $R(\Sigma) = \{e_1, \ldots, e_r\} \in \mathbb{R}^m$ and $N(\Sigma) = \{e_{r+1}, \ldots, e_n\} \in \mathbb{R}^n$.

The proof of (c) follows from (a) and the definition of the projection. The proofs of (d)–(f) are similar. $\square$

7.9. Some Practical Applications of the SVD

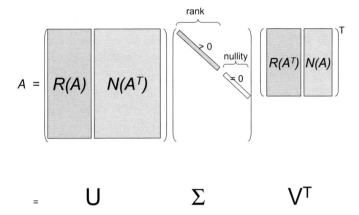

Figure 7.6. *Representation of range and null space of a matrix by SVD.*

Figure 7.6 shows a representation of the range and null space of a matrix by SVD.

Example 7.36. Let A be the same as in Example 7.18. From the SVD of A, we have

$$U = \begin{pmatrix} -1 & 0 & . & -0.0001 \\ -0.0000 & -0.7071 & . & 0.7071 \\ -0.0000 & 0.7071 & . & 0.7071 \\ U_1 & & & U_2 \end{pmatrix}, \quad V = \begin{pmatrix} -0.7071 & . & -0.7071 \\ -0.7071 & . & 0.7071 \\ V_1 & & V_2 \end{pmatrix}.$$

- An orthonormal basis of $R(A)$ = the columns of U_1

$$= \left\{ \begin{array}{cc} -1 & 0 \\ -0.0000 & -0.7071 \\ -0.0000 & 0.7071 \end{array} \right\}.$$

- An orthonormal basis of $N(A)$ = the column of $V_2 = \left\{ \begin{array}{c} -0.7071 \\ 0.7071 \end{array} \right\}.$

- P_A = orthogonal projection onto $R(A) = U_1 U_1^T$

$$= \begin{pmatrix} 1.0000 & 0.0000 & 0.0000 \\ 0.0000 & 0.5000 & -0.5000 \\ 0.0000 & -0.50000 & 0.50000 \end{pmatrix}.$$

- P_N = orthogonal projection onto $N(A^T) = U_2 U_2^T$

$$= \begin{pmatrix} 0.0000 & -0.0000 & -0.0000 \\ -0.0000 & 0.5000 & 0.50000 \\ -0.0000 & 0.5000 & 0.50000 \end{pmatrix}. \quad \blacksquare$$

MATCOM Note: MATCOM program **ORTHOPROJ** computes the projections using SVD.

7.9 Some Practical Applications of the SVD

Part (c) of Theorem 7.29 has some important consequences in practical applications. Matrix A_k, the best rank-k approximation of A, is given by $A = U \Sigma_k V^T$, which can be written as

$$A_k = \sigma_1 u_1 v_1^T + \cdots + \sigma_k u_k v_k^T.$$

The storage of matrix A requires mn locations, whereas matrix A_k can be stored using only $(m+n)k$ locations, thus resulting in a considerable savings when k is small. This fact can be conveniently exploited in image processing and other applications. As illustrated below, even such low-rank approximations of A are useful in practical applications.

Image compression. An image can be represented by an $m \times n$ matrix A whose (i, j)th entry corresponds to the brightness of the pixel (i, j). The idea of image compression is to compress the image represented by a very large matrix to the one which corresponds to a lower-order approximation of A but whose quality is still acceptable to a user.

As an example, we present in Figures 7.7 and 7.8 the different low-rank approximations of the portrait of a child. The matrix A here is of dimension 250×312.

Image restoration. The idea of image restoration is to restore the original image from a blurry image contaminated by "noises." It can be shown that the "noises" correspond to the small singular values. Thus elimination of these small singular values will result in rank-k approximations A_k corresponding to the noise-free images. The necessity of image restoration arises in clinical diagnosis and other practical applications. In Figure 7.9 (supplied by James Nagy) we illustrate the idea with an example of a planet. Matrix A here is of dimension 256×256. The singular values less than 0.0055 are discarded.

A Biomedical Application (Extracting Fetal ECG from Maternal ECG)

Here we show how the same type of idea as above can be used in a biomedical application of extracting fetal ECG from the maternal ECG. It can be assumed (see Vandewalle and De Moor (1988)) that this relationship is linear, and, indeed, each measurement signal $m_i(t)$ can be written as a linear combination of r source signals $s_i(t)$ and additive noise signal $n_i(t)$. This leads to the following equations:

$$\begin{aligned} m_1(t) &= t_{11}s_1(t) + t_{12}s_2(t) + \cdots + t_{1r}s_r(t) + n_1(t), \\ m_2(t) &= t_{21}s_1(t) + t_{22}s_2(t) + \cdots + t_{2r}s_r(t) + n_2(t), \\ &\vdots \\ m_p(t) &= t_{p1}s_1(t) + t_{p2}s_2(t) + \cdots + t_{pr}s_r(t) + n_r(t) \end{aligned} \quad (7.11)$$

or

$$\boldsymbol{m}(t) = T\boldsymbol{s}(t) + \boldsymbol{n}(t), \quad (7.12)$$

where

$$T = (t_{ij}) \quad \text{and} \quad \boldsymbol{n}(t) = (n_1(t), n_2(t), \ldots, n_r(t))^T. \quad (7.13)$$

Matrix T is called the **transfer matrix** and depends upon the geometry of the body, the positions of the electrodes and sources, and the conductivities of the body tissues.

The problem now is to get an estimate of the source signals $\boldsymbol{s}(t)$ knowing only $\boldsymbol{m}(t)$, and from that estimate separate out the estimate of fetal source signals.

Let each measurement consist of q samples. Then the measurements can be stored in a matrix M of order $p \times q$.

We now show that the SVD of M can be used to get estimates of the source signals. Let

$$M = U\Sigma V^T \quad (7.14)$$

be the SVD of M.

7.9. Some Practical Applications of the SVD

(a) Original image.

(b) Compressed image with $k = 2$.

(c) Compressed image with $k = 5$.

Figure 7.7. *Original and compressed images.*

Then the $p \times q$ matrix $\hat{S}$ defined by

$$\hat{S} = U^T M \tag{7.15}$$

will contain p estimates of the source signals. Next, we need to extract the estimates of the fetal source signals from $\hat{S}$; let this be called $\hat{S}_F$.

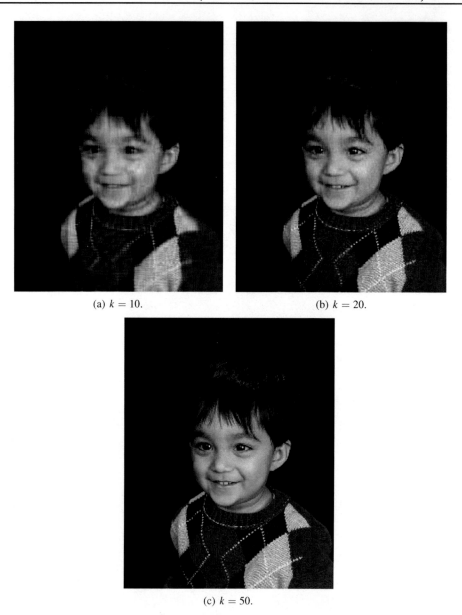

Figure 7.8. *Compressed image with* (a) $k = 10$, (b) $k = 20$, *and* (c) $k = 50$.

Partition the matrix of singular values Σ of M as follows:

$$\Sigma = \begin{pmatrix} \Sigma_m & 0 & 0 \\ 0 & \Sigma_F & 0 \\ 0 & 0 & \Sigma_0 \end{pmatrix}, \qquad (7.16)$$

7.9. Some Practical Applications of the SVD

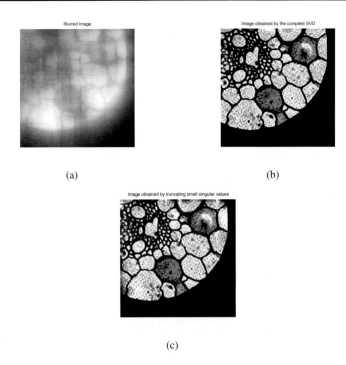

Figure 7.9. (a) *Blurred image,* (b) *image with complete SVD, and* (c) *image with truncated SVD.*

where Σ_M contains r_m large singular values, associated with the maternal heart; Σ_F contains r_f singular values, those smaller ones associated with the fetal heart; and Σ_0 contains the remaining singular values associated with noise, etc.

Let $U = (U_M, U_F, U_0)$ be a conformable partitioning of U. Then, obviously, we have

$$\hat{S} = U^T M = \begin{pmatrix} U_M^T \\ U_F^T \\ U_0^T \end{pmatrix} M \\ = \begin{pmatrix} U_M^T M \\ U_F^T M \\ U_0^T M \end{pmatrix} = \begin{pmatrix} \hat{S}_M \\ \hat{S}_F \\ \hat{S}_0 \end{pmatrix}. \quad (7.17)$$

Thus $\hat{S}_F = U_F^T M$.

Once $\hat{S}_F$ is determined, we can also construct a matrix F containing only the contributions of fetus in each measured signal, as follows:

$$F = U_F \hat{S}_F = \sum_{i=r_m+1}^{r_m+r_f} \sigma_i u_i v_i^T,$$

where u_i and v_i are the ith column of U and V, and σ_i is the ith singular value of M. The signals in $\hat{S}_F$ are called the principal fetal signals.

The above method has been automated and an online adaptive algorithm to compute the U matrix has been designed. For the details of the method and test results, see the paper by Callaerts, DeMoor, Vandewalle, and Sansen (1990).

Note that if the SVD of M is given by

$$M = U\Sigma V^T = (U_1, U_2) \begin{pmatrix} \Sigma_1 & 0 \\ 0 & \Sigma_2 \end{pmatrix} \begin{pmatrix} V_1^T \\ V_2^T \end{pmatrix},$$

then S can be estimated by $\hat{S} = U_1^T M$.

7.10 Geometric Mean and Generalized Triangular Decompositions

In this section, we briefly state two other decompositions related to SVD and QR factorization. Motivated by their applications to control theory, signal processing, and numerical solutions to important optimization problems, these decompositions were recently discovered by William Hager and his colleagues.

Geometric mean decomposition (GMD). (See Jiang, Hager, and Li (2005).) Given $A \in \mathbb{C}^{m \times n}$ of rank k, there exist two orthonormal matrices, P and Q, and a real upper triangular matrix of order k such that

$$A = QRP^*,$$

where the diagonal entries of R are all equal to the geometric mean of the positive singular values of A:

$$r_{ii} = \overline{\sigma} = \left(\prod_{\sigma_j > 0} \sigma_j\right)^{\frac{1}{k}}, \ 1 \leq i \leq k.$$

Here the $\{\sigma_j\}$ are the singular values of A, and $\overline{\sigma}$ is the geometric mean of the positive singular values.

The above decomposition is called **geometric mean decomposition**, or GMD, the term coined by William Hager and his colleagues. These authors have also developed two computational algorithms for GMD: one is SVD-based, and the other is a direct algorithm combining the Lanczos method with the Householder QR factorization.

Generalized triangular decomposition (GTD). (See Jiang, Hager, and Li (2008).) The GTD is an extension of the GMD. It can be shown that the diagonal entries of R in GMD satisfy Weyl's multiplicative majorization conditions:

$$\prod_{i=1}^{r} |r_i| \leq \prod_{i=1}^{r} \sigma_i, \ 1 \leq r \leq k; \quad \prod_{i=1}^{k} |r_1| = \prod_{i=1}^{k} \sigma_i,$$

where r_i is the largest (in magnitude) diagonal entry of R.

There exist an SVD-based algorithm and MATLAB codes for GTD. For details, the readers are referred to the above papers. For other relevant papers on this topic, the website of William Hager (www.math.ufl.edu/~hager) can be consulted.

7.11 Review and Summary

Three major topics have been discussed in this chapter:

- QR factorization
- singular value decomposition (SVD)
- orthogonal projections

7.11.1 QR Factorization

An $m \times n$ matrix A can always be factored into $A = QR$, where Q is $m \times m$ orthogonal and R is $n \times n$ upper triangular.

If $m \geq n$, and $Q = (Q_1, Q_2)$, where Q_1 is the matrix of the first n columns of Q and $R = \binom{R_1}{0}$, where R_1 is $n \times n$ upper triangular, then $A = Q_1 R_1$. This factorization is known as a reduced QR factorization. The following methods for QR factorization have been described:

- Householder's method (Algorithm 7.2)
- Givens method (Algorithm 7.5)
- Classical (CGS) and modified Gram–Schmidt (MGS) processes (Algorithms 7.7 and 7.8)

The Householder and Givens methods yield a full QR factorization of $A : A = QR$, $Q \in \mathbb{R}^{m \times m}$, $R \in \mathbb{R}^{n \times n}$, from which a reduced QR factorization can be easily obtained.

The CGS and MGS produce a reduced QR factorization.

The Householder and Givens methods have excellent numerical properties: Both are stable. However, Givens QR method is slightly more expensive than the Householder QR method.

In the *CGS process, orthogonality of the vectors of the Q matrix might completely be lost*. The MGS process has better numerical property than the CGS process; however, it is not as stable as the Householder or Givens method. (See Table 7.2.)

7.11.2 The SVD, GMD, and GTD

Let $A \in \mathbb{R}^{m \times n}$. Then the decomposition

$$A = U \Sigma V^T,$$

where $U \in \mathbb{R}^{m \times m}$ and $V \in \mathbb{R}^{n \times n}$ are orthogonal and $\Sigma = \text{diag}(\sigma_1, \sigma_2, \ldots, \sigma_p)$, $p = \min(m, n)$ and $\sigma_1 \geq \sigma_2 \geq \cdots \geq \sigma_p \geq 0$, is called the *SVD* of A. The numbers $\sigma_1, \ldots, \sigma_p$ are called the *singular values*. The columns of U are the *left singular vectors* and the columns of V are the *right singular vectors*. The singular values and singular vectors are extremely useful in determining many important properties of a matrix: rank, norms, condition number, etc. (Theorem 7.27). The SVD, in particular, is *the most reliable technique for determining the rank-deficiency and nearness to singularity* (Theorems 7.29 and 7.34).

Many practical computations arising in applications such as *signal* and *image processing* and *statistics*, and others can be made highly storage efficient and simplified greatly using the SVD (see Section 7.9).

Two other decompositions, related to SVD and QR, called *GMD* and *GTD*, have been briefly introduced in Section 7.10. They have applications in control theory, signal processing, and numerical optimization.

7.11.3 Projections

Orthogonal projections onto $\mathbb{R}(A)$ and $N(A^T)$ are frequently needed computational tasks arising in *least-squares solutions* and other practical applications. These projections can be computed using both QR factorization (Theorem 7.17) and the SVD (Theorem 7.35). Again, *the SVD techniques are most numerically reliable, especially if A is nearly rank-deficient.*

- (Projections via QR factorization.) Let $A \in \mathbb{R}^{m \times} (m \geq n)$ and $A = QR$, where $Q = (Q_1, Q_2)$. Then $P_A = Q_1 Q_1^T$, $P_N = Q_2 Q_2^T$.

- (Projections via SVD.) Let $A = U \Sigma V^T$, where $U = (U_1, U_2)$ and $V = (V_1, V_2)$. Then $P_A = U_1 U_1^T$, $P_N = U_2 U_2^T$.

7.12 Suggestions for Further Reading

The topics treated in this chapter are fairly standard and have been discussed in all major textbooks on these subjects: Demmel (1997), Trefethen and Bau (1997), Watkins (2002), and Hager (1988). For advanced treatment of these topics, see Golub and Van Loan (1996), Higham (2002), Stewart (1998b), Lawson and Hanson (1995), and Björck (1996). Stewart (1973) is still a rich source of knowledge on the basic topics of numerical linear algebra. For details of QR factorization with column pivoting, see Golub and Van Loan (1996). For an associated topic on *rank-revealing QR factorizations* (to be discussed in Chapter 13), see Golub and Van Loan (1996), Chan (1987), and Hong and Pan (1992). For the use of SVD in bioelectric imaging of the brain, see Major and Sidman (1991). For applications of QR factorization to statistics, see Hammarling (1985) and Thisted (1988). For applications of QR decomposition and SVD to search engines, see the book by Berry and Browne (2005). For matrix methods in data mining and pattern recognition, see Eldén (2007). For results on perturbation analysis of QR factorization, see Stewart (1977b) and Zha (1993). For details of deblurring images using SVD and other matrix techniques, see the books by Hansen, Nagy, and O'Leary (2006) and O'Leary (2009). Some earlier books on image processing include Andrews and Hunt (1988) and Jain (1989). See also Bojanczyk (1995). For the computation of the SVD of a complex matrix, see Businger and Golub (1969).

Exercises on Chapter 7

EXERCISES ON SECTIONS 7.2–7.6

7.1 Let $H = I - \frac{2uu^T}{u^T u}$ be a Householder matrix. Then prove that (i) $Hu = -u$, and (ii) $Hv = v$ if $v^T u = 0$.

7.2 Let x be an n-vector. Develop an algorithm to compute a Householder matrix $H = I - 2\frac{uu^T}{u^Tu}$ such that Hx has zeros in the positions $(r+1)$ through n; $r < n$.

How many flops will be required to implement this algorithm?

Given $x = (1, 2, 3)^T$, apply your algorithm to construct H such that Hx has a zero in the third position.

7.3 (a) Develop algorithms for *implicitly computing* (i) HA, (ii) AH, (iii) $H_1H_2\ldots H_rC$, and *explicitly computing* (iv) $Q = H_1H_2\ldots H_r$, where the matrices H and $H_i, i = 1, \ldots, r$, are Householder matrices of order n, and A and C are arbitrary rectangular matrices of appropriate sizes.

(b) Compute flop-counts for each of the above computations and verify the counts given in Section 7.2.

(c) Develop an algorithm for implicitly computing AJ, where J is a Givens matrix. What is the flop-count?

7.4 (a) Given the Householder vector $u = (1, 1, 1)^T$ and

$$A = \begin{pmatrix} 1 & 2 \\ 3 & 4 \\ 5 & 6 \end{pmatrix},$$

compute HA and A^TH using the algorithms developed in Exercise 7.3(a).

(b) Given the Householder vectors $u_1 = (1, 1, 1)^T$, $u_2 = (1, 2, 3)$, $u_3 = (1, 0, 0)$, and

$$C = \begin{pmatrix} 1 & 1 \\ 1 & 2 \\ 1 & 3 \end{pmatrix},$$

compute both implicitly and explicitly $H_1H_2H_3C$ and compare the computational efforts. (Here u_i, $i = 1, 2, 3$, are Householder vectors associated with the Householder matrices H_1, H_2, and H_3, respectively.)

7.5 (a) Let $A = \hat{Q}\hat{R}$ be a QR factorization of a nonsingular matrix A. Define a diagonal matrix $D = \text{diag}(d_{11}, \ldots, d_{nn})$ such that $d_{ii} = 1$ if $\hat{r}_{ii} > 0$ and $d_{ii} = -1$ if $\hat{r}_{ii} < 0$. Define now $Q = \hat{Q}D^{-1}$ and $R = D\hat{R}$. Then show that $A = QR$. What is the significance of this result?

(b) Find QR factorizations of

$$A = \begin{pmatrix} 10 & 1 & 1 & 1 \\ 2 & 10 & 1 & 1 \\ 1 & 1 & 10 & 1 \\ 1 & 1 & 1 & 10 \end{pmatrix}$$

using (i) the Householder method, and (ii) the Givens method. Establish a relationship between these two factorization using the results of (a).

7.6 (a) Show that it requires $2n^2(m - \frac{n}{3})$ flops to compute R in the QR factorization of an $m \times n$ matrix A $(m \geq n)$ using Householder's method, and that if Q is needed, then the count is $4(m^2n - mn^2 + \frac{n^3}{3})$ flops.

(b) Show that the flop-counts for both CGS and MGS are $2mn^2 (m \geq n)$.

7.7 Using the Cholesky decomposition of $A^T A$, where A is $m \times n (m \geq n)$ and has full rank, prove that A has a unique reduced QR factorization with positive diagonal entries of R. What are the computational drawbacks of this approach for finding a reduced QR factorization numerically? Illustrate the difficulties with an example.

7.8 Given $v = (1, 2, 3)^T$, find a Givens matrix $J(1, 3, \theta)$ such that the third component of $J(1, 3, \theta)v$ is zero. Repeat the process by creating the Givens matrix $J(2, 3, \theta)$.

7.9 Develop a column-oriented version of MGS for QR factorization. Show that the row version (Algorithm 7.8) and the column-oriented version are numerically equivalent.

7.10 Show that the flop-count to compute R in the QR factorization of an $m \times n$ matrix A ($m \geq n$) using Givens rotations is about $3n^2(m - \frac{n}{3})$.

7.11 Let A be $m \times n$. Let $s = \min(m, n)$. Show that the orthogonal matrix
$$Q^T = Q_s Q_{s-1} \ldots Q_2 Q_1,$$
where each Q_i is the product of $(m - i)$ Givens rotations, can be computed with $4n(m^2 - \frac{n^2}{3})$ flops.

7.12 Let
$$A = \begin{pmatrix} 0 & 0 & 0 \\ 0.0001 & 0 & 0 \\ 0 & 0.0001 & 0 \\ 0 & 0 & 0.0001 \end{pmatrix}.$$

(a) Find the reduced QR factorizations of A using (i) the Householder method, (ii) the Givens method, and (iii) CGS and MGS methods. Compare the results.

(b) From the reduced QR factorization obtained from MGS and CGS methods, find a full QR factorization.

7.13 Based on the statement of Section 7.3, develop an algorithm for complex QR factorization. Use your algorithm to find QR factorizations of

$$\text{(i) } A = \begin{pmatrix} 1+i & 2 \\ 3 & 1-i \end{pmatrix}, \quad \text{(ii) } A = \begin{pmatrix} 1 & 1 & 1 \\ 1+i & 1 & 1 \\ 1 & 1-i & 0 \end{pmatrix}.$$

Print $\|A - QR\|$, $\|Q^T Q\|$, $\|Q^* Q\|$ in each case.

7.14 Find QR factorization of the Hessenberg matrix H obtained by the MATLAB command **hess**, $H = $ **hess** (rand(10)), using both Householder's and Givens' methods, and compare the results. Write down your observation on the uniqueness of this decomposition.

7.15 Suppose you have computed the QR factorization of an $n \times n$ matrix A. Develop now an efficient algorithm for computing the QR factorization of $B = A + uv^T$, where u and v are two n-vectors, by making use of the QR factorization of A at hand. How many flops are needed? How does this flop-count compare with those for finding the QR factorization of B without knowing a priori the QR factorization of A?

EXERCISES ON SECTION 7.8

7.16 (a) Let A be $m \times n$, and let U and V be orthogonal. Then, using the definition of singular values, prove that the singular values of A and $V^T A V$ are the same. What about the singular vectors?

(b) How are the singular vectors of A related with those of $U^T A V$?

(c) How are the singular values of a symmetric matrix related to its eigenvalues?

7.17 Let σ be a singular value of A with multiplicity ℓ; that is, $\sigma_i = \sigma_{i+1} = \cdots = \sigma_{i+\ell-1}$. Let $U \Sigma V^T$ be the SVD of A. Then construct $\tilde{U}$ and $\tilde{V}$ such that $\tilde{U} \Sigma (\tilde{V})^T$ is also an SVD.

7.18 (a) Using the MATLAB command **svd**, find the SVD of the following matrices:

$$A = \begin{pmatrix} 1 & 2 \\ 3 & 4 \\ 5 & 6 \end{pmatrix}, \qquad A = \begin{pmatrix} 1 & 2 & 3 \end{pmatrix},$$

$$A = \begin{pmatrix} 1 \\ 1 \\ 1 \end{pmatrix}, \qquad A = \operatorname{diag}(1, 0, 2, 0, -5),$$

$$A = \begin{pmatrix} 1 & 1 \\ \epsilon & 0 \\ 0 & \epsilon \end{pmatrix}, \qquad \epsilon = 10^{-5}.$$

(b) Using the results of (a), find (i) rank, (ii) $\|\cdot\|_2$ and $\|\cdot\|_F$, (iii) orthonormal bases for $\mathbb{R}(A)$ and $N(A^T)$, (iv) P_A and P_N, (v) and b_R and b_N by choosing a vector b appropriately for each A.

7.19 For an $m \times n$ matrix A, prove the followings results using the SVD of A:

(i) rank $(A^T A)$ = rank $(A A^T)$ = rank (A) = rank (A^T).

(ii) $A^T A$ and $A A^T$ have the same nonzero eigenvalues.

(iii) If the eigenvectors u_1 and u_2 of $A^T A$ are orthogonal, then $A u_1$ and $A u_2$ are orthogonal.

7.20 Let A be an invertible matrix. Then show that $\|A\|_2 = 1$ if and only if A is a multiple of an orthogonal matrix.

7.21 Let U have orthonormal columns. Then using SVD, prove that

(i) $\|AU\|_2 = \|A\|_2$,

(ii) $\|AU\|_F = \|A\|_F$,

(iii) $\|Ax\|_2 / \|x\|_2$ is maximized if $x = v_1$ and minimized if $x = v_n$, where v_1 and v_n are the singular vectors associated with the largest and smallest singular values of A, respectively.

7.22 Let $U \Sigma V^T$ be the SVD of A. Then prove that $\|U^T A V\|_F^2 = \sum_{i=1}^{p} \sigma_i^2$, where σ_i are the singular values of A.

7.23 Let A be an $m \times n$ matrix.

(a) Using the SVD of A, prove that
  (i) $\|A^T A\|_2 = \|A\|_2^2$;
  (ii) $\text{Cond}_2(A^T A) = (\text{Cond}_2(A))^2$; if A has full rank.
  (iii) $\text{Cond}_2(A) = \text{Cond}_2(U^T A V)$, where U and V are orthogonal.

(b) Let $\text{rank}(A_{m \times n}) = n$, and let $B_{m \times r}$ be a matrix obtained by deleting $(n-r)$ columns from A. Then prove that $\text{Cond}_2(B) \leq \text{Cond}_2(A)$.

7.24 Prove that if A is an $m \times n$ matrix with rank r, and if B is another $m \times n$ matrix satisfying $\|A - B\|_2 < \sigma_r$, then B has at least rank r.

7.25 Consider the matrix A in Example 7.5.

(a) Finding the singular values of A, show that this matrix has rank 2, but is close to a matrix of rank 1.

(b) Find a matrix A_1 of rank 1 such that out of all the matrices of rank 1, A_1 is the closest to A.

(c) Find $\|A - A_1\|_2$ and verify Theorem 7.29.

(d) Find $\|A - A_1\|_F$ and verify Theorem 7.34 by taking B as a random matrix of order 3×2.

7.26 Let A and B be $n \times n$ real matrices. Let Q be an orthogonal matrix such that $\|A - BQ\|_F \leq \|A - BX\|_F$ for any orthogonal matrix X. Then prove that $Q = VU^T$, where $A^T B = U \Sigma V^T$.

7.27 Given
$$A = \begin{pmatrix} 1 & 2 & 3 \\ 2 & 3 & 4 \\ 5 & 6 & 7 \end{pmatrix},$$
use the result of Exercise 7.26 to show that the orthogonal matrix
$$Q = \begin{pmatrix} -0.2310 & -0.3905 & 0.8912 \\ -0.4824 & 0.8414 & 0.2436 \\ 0.8449 & 0.3736 & 0.3827 \end{pmatrix}$$
is such that $\|A - Q\|_F \leq \|A - X\|_F$, where $X = \{$ The set of all 3×3 orthogonal matrices$\}$.

7.28 (a) Let
$$A = \begin{pmatrix} 1 & 2 \\ 1 & 3 \\ 1 & 4 \end{pmatrix}.$$
Express A in terms of its singular values and singular vectors.

(b) Compute $(A^T A)^{-1}$ using the SVD of A.

7.29 (a) Generate randomly a matrix A of order 16×4 by using the MATLAB command **rand** (16,4). Then verify using the MATLAB command **rank** that rank $(A) = 4$. Now run the following MATLAB command: $[U, S, V] = $ **svd** (A). Set $S(4, 4) = 0$; compute $B = U*S*V'$. What is rank (B)?

(b) Construct a matrix C of order 16×4 of rank 3. Verify that (i) $\|C - A\|_F^2 \geq \|B - A\|_F^2$, (ii) $\|C - A\|_2^2 \geq \|B - A\|_2^2$ using the MATLAB command **norm**.

(c) What is the distance of B from A?

(d) Find a matrix D of rank 2 that is closest to A.

7.30 Let
$$A = \begin{pmatrix} 1 & 1 & 1 & 1 \\ 0 & 0.0001 & 1 & 1 \\ 0 & 0 & 0.0001 & 1 \\ 0 & 0 & 0 & 1 \end{pmatrix}.$$

Find the distance of A from the nearest singular matrix. Find a perturbation which will make A singular. Compare the size of this perturbation with $|\sigma_4|$.

7.31 Let $A = U \Sigma V^T$ be the SVD of a randomly generated 15×10 matrix $A = $ rand (15,10), obtained by using the MATLAB command $[U, S, V] = $ **svd**(A). Set $S(8, 8) = S(9, 9) = S(10, 10) = 0$. Compute $B = U * S * V'$.

Find the best approximation of the matrix B in the form $B \approx \sum_{i=1}^{r} x_i y_i^T$ such that $\|B - \sum_{i=1}^{r} x_i y_i^T\|_2 = $ minimum, where x_i and y_i are vectors, and r is the rank of B.

7.32 For matrices A and B in Exercise 7.31, find an orthogonal matrix Q such that $\|A - BQ\|_2$ is minimized. (*Hint:* $Q = VU^T$, where $A^T B = U \Sigma V^T$. Use the MATLAB command **svd** to solve this problem.)

7.33 (a) Develop an algorithm for solving $Ax = b$ using the SVD of A. Compare the efficiency of this algorithm with those of Gaussian elimination and QR factorization processes for solving $Ax = b$.

(b) Prove that a small singular value of A signals the sensitivity of the solution of $Ax = b$. Construct an example to demonstrate this.

7.34 Prove that the singular values of a symmetric positive definite matrix A are the same as its eigenvalues. What are the relationships between the eigenvectors and right and left singular vectors?

EXERCISES ON SECTIONS 7.7 AND 7.8.10 (PROJECTION PROBLEMS)

7.35 Prove that if P_s is the orthogonal projection onto S, then $I - P_s$ is the orthogonal projection onto $S^\perp$.

7.36 Prove that (i) $P_A = A(A^T A)^{-1} A^T$, (ii) $P_N = I - A(A^T A)^{-1} A^T$, assuming that A has full rank, and (iii) $b_R = P_A b$, $b_N = P_N b$. What are the computational drawbacks of computing P_A using these expressions? Illustrate the difficulties with a numerical example.

7.37 Write a detailed proof of Theorem 7.17.

7.38 For matrix A in Exercise 7.4(a), and with $b = (1, 2, 3)^T$, using both QR and SVD, compute (i) P_A and P_N, (ii) b_R and b_N.

7.39 Given
$$A = \begin{pmatrix} 1 & 3 \\ 2 & 4 \\ 3 & 5 \end{pmatrix}$$
and choosing B randomly of appropriate order, compute B_R and B_N.

MATLAB and MATCOM Programs on Chapter 7

Data set for Problems M7.1, M7.2, and M7.3:

(i) $A = 20 \times 15$ random matrix,

(ii) $u = $ ones $(15, 1)$,

(iii) $u_1 = u_2 = u_3 = (1, 2, 3, 4, \ldots, 15)^T$.

M7.1 Write a MATLAB program **hmat** in the following format to *compute implicitly Householder matrix multiplications AH and HA^T*:

$[C] = $ **hmat** (A, u)

- A – an $m \times n$ matrix (input).
- u – the Householder vector defining the Householder matrix H (input).
- C – the output matrix

M7.2 Write a MATLAB program **hhmat** in the following format to implicitly implement the product $H_1 H_2 H_3 A^T$:

$[C] = $ **hhmat** (A, u_1, u_2, u_3)

- A – an $m \times n$ matrix (input).
- u_1, u_2, u_3 – the Householder vectors defining H_1, H_2, and H_3, respectively (input).
- C – the output matrix.

M7.3 Use the program **hhmat** to compute $H_1 H_2 H_3$.

M7.4 (*The purpose of this exercise is to compare the accuracy and efficiency of different methods for QR factorization of a matrix.*)

(a) Compute the QR factorization for each matrix A of the following data set as Follows:

(i) $[Q, R] = $ **qr**(A) from MATLAB or $[Q, R] = $ **housqr** (A) or **housqrn** (MATCOM implementations of Householder's method).

(ii) $[Q, R] = $ **givqr** (A) (Givens QR implementation from MATCOM).

(iii) $[Q, R] = $ **clgrsch**(A) (classical Grass–Schmidt implementation from MATCOM).

(iv) [Q, R] = **mdgrsch**(A) from MATCOM (modified Gram–Schmidt implementation from MATCOM).

(b) Using the results of (a), complete Table 7.3 for each matrix A. $\hat{Q}$ and $\hat{R}$ stand for the computed Q and R.

Data set:

(i) $A = $ rand (25),

(ii) A is a Hilbert matrix of order 25,

(iii) $A = \begin{pmatrix} 1 & 1 & 1 \\ 10^{-4} & 0 & 0 \\ 0 & 10^{-4} & 0 \\ 0 & 0 & 10^{-4} \end{pmatrix}$,

(iv) a Vandermonde matrix of order 25.

Table 7.3. *Comparison of different QR factorization methods.*

Method	$\|(\hat{Q})^T \hat{Q} - I\|_F$	$\dfrac{\|A - \hat{Q}\hat{R}\|_F}{\|A\|_F}$
housqr		
givqr		
clgrsch		
mdgrsch		

M7.5 Using **givqr** from MATCOM, which uses Givens' method, and **qr** from MATLAB, which uses Householder's method, find the QR factorization of each of these matrices below and verify the statement of Exercise 7.5(a) which shows how Q and R matrices of each method are related.

Test data:

(i)
$$A = \begin{pmatrix} 1 & 1 & 1 & 1 \\ 0 & 0.99 & 1 & 1 \\ 0 & 0 & 0.99 & 1 \\ 0 & 0 & 0 & 0.99 \end{pmatrix},$$

(ii) $A = $ The *Wilkinson bidiagonal matrix* of order 20.

M7.6 (a) Write MATLAB programs **orthqr, orthqrp**, and **orthsvd** to compute the orthonormal basis for the range of a matrix A using (i) QR factorization, (ii) QR factorization with partial pivoting, and (iii) the SVD, respectively.

(b) Repeat Exercise M.7.6(a) to compute the orthonormal basis for the null space of A, that is, write the programs **nullqr, nullqrp**, and **nullsvd**.

(c) Compare the results of each of the three programs in both cases for the test data given below.

Test data:

(i)
$$A = \begin{pmatrix} 0 & 0 & 1 & 1 \\ 0 & 0 & 0 & 0 \\ 1 & 1 & 1 & 0 \\ 1 & 1 & 1 & 0 \end{pmatrix},$$

(ii) $A = $ a randomly generated matrix of order 20.

M7.7 Compute the *rank* of each of the following matrices using (i) the MATLAB command **rank** (which uses the singular values of A) and (ii) the MATLAB commands $[Q, R, E] = qr(A)$, which computes the QR factorization of A using column pivoting, and (iii) the MATLAB command $[Q, R] = qr(A)$.

Test data:

(i) *The Kahan matrix (7.9), with* $n = 100$, *and* $c = 0.2$.

(ii) A 15×10 matrix A created as follows: $A = xy^T$, where

$$x = \text{round}(10 * \text{rand}(15, 1)), \qquad y = \text{round}(10 * \text{rand}(10, 1)).$$

M7.8 Write a MATLAB program, **covsvd**, to compute $(A^T A)^{-1}$ using the SVD and test it with the 20×20 Hilbert matrix. Compare your results with those obtained by **varcovar** from MATCOM, which is based on QR factorization.

M7.9 Let A be a 20×20 Hilbert matrix and let b be a vector generated such that all entries of the vector x of $Ax = b$ are equal to 1.

Solve $Ax = b$ using (i) QR factorization and (ii) SVD. Compare the accuracy and flop-count with those obtained by **linsyswf** from MATCOM or using MATLAB command $A \backslash b$.

Chapter 8

Least-Squares Solutions to Linear Systems

Background Material Needed

- Cholesky factorization algorithm (Algorithm 6.7)
- Householder's QR factorization algorithm (Algorithm 7.2)
- Orthonormal basis and projections (Section 7.7)
- Iterative refinement algorithm for linear systems (Algorithm 6.6)
- Perturbation analysis for linear systems and condition number (Sections 4.6 and 4.7)

8.1 Introduction

In Chapter 6 we discussed methods for solving the linear system

$$Ax = b,$$

where A was assumed to be square and nonsingular. However, in several practical situations, such as in statistical applications, geometric modeling, and signal processing, one needs to solve a system where matrix A is either a nonsquare matrix or is a square matrix but singular. In such cases, solutions may not exist at all; in cases where there are solutions, there may be infinitely many. For example, when A is $m \times n$ and $m > n$, we have an *overdetermined* system (that is, the number of equations is greater than the number of unknowns), and an overdetermined system typically has no solution. In contrast, an *underdetermined* system ($m < n$) typically has an infinite number of solutions.

In these cases, the best one can hope for is to find a vector x which will make Ax as close as possible to the vector b. In other words, we seek a vector x such that $r(x) = \|Ax - b\|$ is minimized. When the Euclidean norm $\| \cdot \|_2$ is used, this solution is referred to as a *least-squares solution* to the system $Ax = b$. The term "least-squares solution" is justified *because it is a solution that minimizes the Euclidean norm of the residual vector* and, by definition, the square of the Euclidean norm of a vector is just the sum of squares of the components of the vector. The problem of finding least-squares solutions to the linear system $Ax = b$ is known as the **linear least-squares problem**. The linear least-squares problem is formally defined as follows.

> **Statement of the Linear Least-Squares Problem**
>
> Given a real $m \times n$ matrix A of rank $k \leq \min(m,n)$ and a real vector b, find a real n-vector x such that the function $r(x) = \|Ax - b\|_2$ is minimized.

If the least-squares problem has more than one solution, the one having the minimum Euclidean norm is called the **minimum-length solution** or the **minimum-norm solution**.

This chapter is devoted to the study of such problems. The organization of the chapter is as follows.

In Section 8.2 we give a *geometric interpretation* of the least-squares problem.

In Section 8.3 we prove a theorem on the *existence and uniqueness of the solution* of an overdetermined least-squares problem. Some applications leading to the least-squares problem are discussed in Section 8.4.

In Section 8.5, we define the *pseudoinverse* of a full-rank matrix A and give the expressions for the *condition number* of a rectangular matrix in terms of the pseudoinverse and for the *unique least-squares solution* in terms of the pseudoinverse.

In Section 8.6 we analyze *the sensitivity of the least-squares problems* due to perturbations in data. We prove only a simple result (Theorem 8.10) there and state other results without proofs (Theorems 8.13 and 8.16).

Section 8.7 deals with *computational methods* for both full-rank and rank-deficient overdetermined problems.

Underdetermined least-squares problems are considered in Section 8.8.

In Section 8.9 an *iterative improvement procedure* for refining an approximate solution to a least-squares problem is presented.

8.2 Geometric Interpretation of the Least-Squares Problem

Let A be an $m \times n$ matrix with $m > n$. Then A is a linear mapping of $\mathbb{R}^n \to \mathbb{R}^m$. $R(A)$ is a subspace of $\mathbb{R}^m$. Every vector $u \in R(A)$ can be written as $u = Ax$ for some $x \in \mathbb{R}^n$. Let $b \in \mathbb{R}^m$. Because $\|\cdot\|_2$ is a Euclidean norm, $\|b - Ax\|_2$ is the distance between the end points of b and Ax. It is clear that this distance is minimal if and only if $b - Ax$ is perpendicular to $R(A)$ (see Figure 8.1). In that case, $\|b - Ax\|_2$ is the distance from the end point of b to the "plane" $R(A)$.

From this interpretation, it is easy to understand that a *solution of the least-squares problem to the linear system $Ax = b$ always exists*. This is because one can project b onto the "plane" $R(A)$ to obtain a vector $u \in R(A)$, and there is $x \in \mathbb{R}^n$ such that $u = Ax$. This x is a solution.

Because $b - Ax$ is perpendicular to $R(A)$ and every vector in $R(A)$ is a linear combination of column vectors of A, the vector $b - Ax$ is orthogonal to every column of A. That is,

$$A^T(b - Ax) = 0$$

or

$$A^T Ax = A^T b.$$

8.3. Existence and Uniqueness

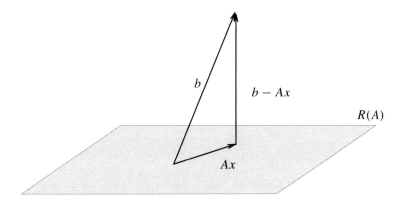

Figure 8.1. *Geometric interpretation of the least-squares solution.*

Definition 8.1. *Let $A \in \mathbb{R}^{m \times n}$. The system of n equations*
$$A^T A x = A^T b$$
*is called the **normal equations**.*

8.3 Existence and Uniqueness

From the geometric configuration above, we have just seen that *a least-squares solution always exists and satisfies the normal equations.* We shall now give an *analytical proof* of the existence and uniqueness result and derive some equivalent expressions for least-squares solutions. *We assume that the system $Ax = b$ is overdetermined; that is, A is of order $m \times n$, where $m > n$.* Least-squares solutions to an underdetermined system will be discussed in Section 8.8. An overdetermined system $Ax = b$ can be represented graphically as shown in Figure 8.2.

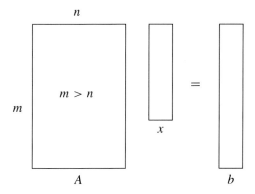

Figure 8.2. *An overdetermined system.*

8.3.1 Existence and Uniqueness Theorem

Theorem 8.2. (i) *Given $A \in \mathbb{R}^{m \times n}$ ($m \geq n$) and $b \in \mathbb{R}^m$, a vector $x \in \mathbb{R}^n$ is a least-squares solution to $Ax = b$ if and only if x satisfies the normal equations*

$$A^T A x = A^T b. \tag{8.1}$$

(ii) *The least-squares solution, when it exists, is unique if and only if A has full rank.*

Proof. *Proof of* (i) We denote the residual $r = b - Ax$ by $r(x)$ to emphasize that given A and b, r is a function of x. Let y be an n-vector. Then $r(y) = b - Ay = r(x) + Ax - Ay = r(x) + A(x - y)$. So,

$$\|r(y)\|_2^2 = \|r(x)\|_2^2 + 2(x - y)^T A^T r(x) + \|A(x - y)\|_2^2.$$

First assume that x satisfies

$$A^T A x = A^T b,$$

that is, $A^T r(x) = 0$. Then from the above, we have

$$\|r(y)\|_2^2 = \|r(x)\|_2^2 + \|A(x - y)\|_2^2 \geq \|r(x)\|^2,$$

implying that x is a least-squares solution.

Next assume that x does not satisfy the normal equations; that is, $A^T r(x) \neq 0$. Set $A^T r(x) = z \neq 0$. Define now a vector y such that

$$y = x + \mu z.$$

Then

$$r(y) = r(x) + A(x - y) = r(x) - \mu A z,$$
$$\|r(y)\|_2^2 = \|r(x)\|_2^2 + \mu^2 \|Az\|_2^2 - 2\mu A^T r(x) z^T$$
$$= \|r(x)\|_2^2 + \mu^2 \|Az\|_2^2 - 2\mu \|z\|_2^2 < \|r(x)\|_2^2$$

for any $\mu > 0$ if $Az = 0$, and for $0 < \mu < \frac{2\|z\|_2^2}{\|Az\|_2^2}$ if $Az \neq 0$. This implies that x is not a least-squares solution.

Proof of (ii) To prove uniqueness, all we need to show then is that matrix $A^T A$ is nonsingular when A has full rank, and vice versa (note that the matrix $A^T A$ is square and, therefore, the system (8.1) has a unique solution if and only if $A^T A$ is nonsingular).

We prove this by contradiction. First, suppose that A has full rank, but $A^T A$ is singular. Then $A^T A x = 0$ for some nonzero vector x, meaning that $x^T A^T A x = 0$. That is, $Ax = 0$, implying that A is rank-deficient, which is a contradiction.

Conversely, suppose that A is rank-deficient but $A^T A$ is nonsingular. Since A is rank-deficient, there exists a nonzero x such that $Ax = 0$, showing that $A^T A x = 0$. This implies that $A^T A$ is singular, which is again a contradiction. $\square$

8.3.2 Normal Equations, Projections, and Least-Squares Solutions

Theorem 8.3 (least-squares and orthogonal projection). *Let $A \in \mathbb{R}^{m \times n}$, $m \geq n$, and $b \in \mathbb{R}^m$. Let A have full rank, and let x be a least-squares solution to $Ax = b$. Then x satisfies*
$$Ax = P_A b,$$
where P_A is the orthogonal projection of A onto $R(A)$.

Proof. Since x is a least-squares solution, it must satisfy the following: $A^T A x = A^T b$ (part (i) of Theorem 8.2). Also, $A^T A$ is nonsingular, because A has full rank (proof of part (ii) of Theorem 8.2). That is, $x = (A^T A)^{-1} A^T b$. So,
$$Ax = A(A^T A)^{-1} A^T b = P_A b.$$
(Note that $P_A = A(A^T A)^{-1} A^T$.) □

Remark. The converse of Theorem 8.3 is also true, and is left as an exercise (Exercise 8.3).

Theorem 8.4 (least-squares residual equation). *Let $r = b - Ax$. Then $A^T r = 0$ if and only if x is a least-squares solution.*

Proof. The proof follows immediately from the normal equations. □

Summarizing the above results, we have the following.

Equivalent Expressions for Least-Squares Solutions

The vector x is a least-squares solution to $Ax = b$, where $A \in \mathbb{R}^{m \times n}$, $m \geq n$, if and only if any of the following equivalent conditions hold:

(i) $A^T A x = A^T b$ (normal equations). The solution x is unique if and only if A has full rank.

(ii) $Ax = P_A b$, where P_A is the orthogonal projection onto $R(A)$.

(iii) $A^T r = 0$, where $r = b - Ax$.

Example 8.5. Let
$$A = \begin{pmatrix} 1 & 2 \\ 2 & 3 \\ 4 & 5 \end{pmatrix}, \quad b = \begin{pmatrix} 3 \\ 5 \\ 9 \end{pmatrix}.$$

1. *Using the normal equations:* $A^T A x = A^T b$.
$$A^T A = \begin{pmatrix} 21 & 28 \\ 28 & 38 \end{pmatrix}, \quad A^T b = \begin{pmatrix} 49 \\ 66 \end{pmatrix}, \quad x = \begin{pmatrix} 1 \\ 1 \end{pmatrix}.$$

2. *Using orthogonal projection:* $Ax = P_A b$.

$$P_A = \begin{pmatrix} 0.7143 & 0.4286 & -0.1429 \\ 0.4286 & 0.3571 & 0.2143 \\ -0.1429 & 0.2143 & 0.9286 \end{pmatrix}, \quad b' = P_A b = \begin{pmatrix} 3 \\ 5 \\ 9 \end{pmatrix}, \quad x = \begin{pmatrix} 1 \\ 1 \end{pmatrix}. \quad \blacksquare$$

8.4 Some Applications of the Least-Squares Problem

In this section, we describe two well-known real-life problems that give rise to the least-squares problem.

8.4.1 Polynomial-Fitting to Experimental Data

A well-known example of how normal equations arise in practical applications is the fitting of a polynomial to a set of experimental data.

Engineers and scientists gather data from experiments. A meaningful representation of the collected data is needed to make meaningful decisions for the future.

Let $(x_1, y_1), (x_2, y_2), \ldots, (x_n, y_n)$ be a set of paired observations. Suppose that the mth ($m \leq n$) degree polynomial

$$y(x) = a_0 + a_1 x + a_2 x^2 + \cdots + a_m x^m \tag{8.2}$$

is the "best fit" for this set of data. One strategy for this "best fit" is to minimize the sum of the squares of the residuals

$$E = \sum_{i=1}^{n}(y_i - a_0 - a_1 x_i - a_2 x_i^2 - \cdots - a_m x_i^m)^2. \tag{8.3}$$

We must then have

$$\frac{\partial E}{\partial a_i} = 0, \quad i = 1, \ldots, m.$$

Now,

$$\left. \begin{aligned} \frac{\partial E}{\partial a_0} &= -2 \sum_{i=1}^{n}(y_i - a_0 - a_1 x_i - a_2 x_i^2 - \cdots - a_m x_i^m), \\ \frac{\partial E}{\partial a_1} &= -2 \sum_{i=1}^{n} x_i (y_i - a_0 - a_1 x_i - a_2 x_i^2 - \cdots - a_m x_i^m), \\ &\vdots \\ \frac{\partial E}{\partial a_m} &= -\sum_{i=1}^{n} x_i^m (y_i - a_0 - a_1 x_i - \cdots - a_m x_i^m). \end{aligned} \right\} \tag{8.4}$$

8.4. Some Applications of the Least-Squares Problem

Setting these equations to zero, we have

$$\left.\begin{aligned} a_0 n + a_1 \sum x_i + a_2 \sum x_i^2 + \cdots + a_m \sum x_i^m &= \sum y_i, \\ a_0 \sum x_i + a_1 \sum x_i^2 + \cdots + a_m \sum x_i^{m+1} &= \sum x_i y_i, \\ &\vdots \\ a_0 \sum x_i^m + a_1 \sum x_i^{m+1} + \cdots + a_m \sum x_i^{2m} &= \sum x_i^m y_i \end{aligned}\right\} \quad (8.5)$$

(where $\sum$ denotes the summation from $i = 1$ to n).

Setting $\sum x_i^k = S_k$, $k = 0, 1, \ldots, 2m$, and denoting the entries of the right-hand side, respectively, by $b_0, b_1, \ldots, b_m$, the system of equations (8.5) can be written as

$$\begin{pmatrix} S_0 & S_1 & \cdots & S_m \\ S_1 & S_2 & \cdots & S_{m+1} \\ \vdots & & & \vdots \\ S_m & S_{m+1} & \cdots & S_{2m} \end{pmatrix} \begin{pmatrix} a_0 \\ a_1 \\ \vdots \\ a_m \end{pmatrix} = \begin{pmatrix} b_0 \\ b_1 \\ \vdots \\ b_m \end{pmatrix}. \quad (8.6)$$

(Note that $S_0 = n$.)

This is a system of $(m + 1)$ equations in $(m + 1)$ unknowns $a_0, a_1, \ldots, a_m$. This is really a system of *normal equations*. To see this, define

$$V = \begin{pmatrix} 1 & x_1 & \cdots & x_1^m \\ 1 & x_2 & & x_2^m \\ \vdots & \vdots & \ddots & \vdots \\ 1 & x_n & & x_n^m \end{pmatrix}, \quad y = \begin{pmatrix} y_1 \\ y_2 \\ \vdots \\ y_n \end{pmatrix}. \quad (8.7)$$

Then system (8.6) becomes equal to

$$V^T V a = V^T y = b, \quad (8.8)$$

where $a = (a_0, a_1, \ldots, a_m)^T$ and $b = (b_0, b_1, \ldots, b_m)^T$.

If the x_i's are all distinct, then matrix V has full rank.

The matrix V is known as the **Vandermonde matrix**. From our discussion in the previous section, we see that a is the least-squares solution to the system $Va = b$. If the x_i's are all distinct, then x is unique.

Example 8.6. Suppose that an electrical engineer has gathered the following experimental data consisting of the measurement of the current in an electric wire for various voltages:

x = voltage	0	2	5	7	9	13	24
y = current	0	6	7.9	8.5	12	21.5	35

We would like to derive the normal equations for the above data corresponding to the best fit of the data to (a) a straight line (b) a quadratic, and would like to see a comparison of the predicted results with the actual result when $x = 5$.

Case 1. *Straight-line fit*: $m = 1$.

$$V = \begin{pmatrix} 1 & 0 \\ 1 & 2 \\ 1 & 5 \\ 1 & 7 \\ 1 & 9 \\ 1 & 13 \\ 1 & 24 \end{pmatrix}, \quad y = \begin{pmatrix} 0 \\ 6 \\ 7.9 \\ 8.5 \\ 12.0 \\ 21.5 \\ 35.0 \end{pmatrix}.$$

The normal equations $V^T V a = V^T y = b$ are

$$\begin{pmatrix} 7 & 60 \\ 60 & 904 \end{pmatrix} a = 10^3 \begin{pmatrix} 0.0906 \\ 1.3385 \end{pmatrix}.$$

The solution of these equations is

$$a_0 = 0.6831, \quad a_1 = 1.4353.$$

The value of $a_0 + a_1 x$ at $x = 5$ is $0.6831 + 1.4345 \times 5 = 7.8596$.

Case 2. *Quadratic fit*: $m = 2$.

$$V = \begin{pmatrix} 1 & 0 & 0 \\ 1 & 2 & 4 \\ 1 & 5 & 25 \\ 1 & 7 & 49 \\ 1 & 9 & 81 \\ 1 & 13 & 169 \\ 1 & 24 & 576 \end{pmatrix}.$$

The normal equations are

$$V^T V a = V^T y = b$$

or

$$\begin{pmatrix} 7 & 60 & 904 \\ 60 & 904 & 17226 \\ 904 & 17226 & 369940 \end{pmatrix} a = 10^4 \begin{pmatrix} 0.0091 \\ 0.1338 \\ 2.5404 \end{pmatrix}.$$

The solution of these normal equations is

$$a = \begin{pmatrix} a_0 \\ a_1 \\ a_2 \end{pmatrix} = \begin{pmatrix} 0.8977 \\ 1.3695 \\ 0.0027 \end{pmatrix}.$$

The value of $a_0 + a_1 x + a_2 x^2$ at $x = 5$ is 7.8127.

Note: *The use of a higher-degree polynomial may not necessarily give the best result. The matrix of the normal equations in this case may be very ill-conditioned: Indeed, it is well known that the Vandermonde matrices become progressively ill-conditioned as the order of matrices increases. Note that in Case 2, $\text{Cond}(V^T V) = 2.3260 \times 10^5$, whereas in Case 1, $\text{Cond}(V^T V) = 302.2199$.* ■

8.4.2 Predicting Future Sales

Suppose that the number of units b_i of a product sold by a company in the district i of a town depends upon the population a_{i1} (in thousands) of the district and the per capita income a_{i2} (in thousands of dollars). The table below (taken from Neter, Wasserman, and Kutner (1983)), compiled by the company, shows the sales in five districts, as well as the corresponding population and per capita income.

District i	Sales b_i	Population a_{i1}	Per Capita Income a_{i2}
1	162	274	2450
2	120	180	3254
3	223	375	3802
4	131	205	2838
5	67	86	2347

Suppose the company wants to use the above table to predict future sales and believes (from past experience) that the following relationship between b_i, a_{i1}, and a_{i2} is appropriate:

$$b_i = x_1 + a_{i1}x_2 + a_{i2}x_3.$$

If the data in the table have satisfied the above relation, we have

$$162 = x_1 + 274x_2 + 2450x_3,$$
$$120 = x_1 + 180x_2 + 3254x_3,$$
$$223 = x_1 + 375x_2 + 3802x_3,$$
$$131 = x_1 + 205x_2 + 2838x_3,$$
$$67 = x_1 + 86x_2 + 2347x_3,$$

or

$$Ax = b,$$

where

$$A = \begin{pmatrix} 1 & 274 & 2450 \\ 1 & 180 & 3254 \\ 1 & 375 & 3802 \\ 1 & 205 & 2838 \\ 1 & 86 & 2347 \end{pmatrix}, \quad b = \begin{pmatrix} 162 \\ 120 \\ 223 \\ 131 \\ 67 \end{pmatrix}, \quad x = \begin{pmatrix} x_1 \\ x_2 \\ x_3 \end{pmatrix}.$$

The above is an overdetermined system of five equations in three unknowns.

The least-squares solution of the problem will give us predictions of sales (see Example 8.20).

8.5 Pseudoinverse and the Least-Squares Problem

Assume that A is $m \times n$ ($m \geq n$) and has *full rank*. So, $A^T A$ is invertible.
Denote $(A^T A)^{-1} A^T = A^{\dagger}$.

Definition 8.7. *The matrix*

$$A^{\dagger} = (A^T A)^{-1} A^T$$

is called the pseudoinverse *or the* Moore-Penrose generalized inverse *of A. It therefore follows from Theorem* 8.2 *that*

$$\text{the unique least-squares solution } x = A^\dagger b.$$

Clearly, the above definition of the *pseudoinverse generalizes the ordinary definition of the inverse of a square matrix A*. Note that when A is square and invertible,

$$A^\dagger = (A^T A)^{-1} A^T = A^{-1}(A^T)^{-1}A^T = A^{-1}.$$

An excellent reference on the subject is the classic book by Rao and Mitra (1971). Some other books of interests on generalized inverses include Guorong, Yimin, and Sanzheng (2004) and Campbell and Meyer (1979).

Having defined the generalized inverse of a rectangular matrix, we now define the *condition number* of such a matrix as $\text{Cond}(A) = \|A\| \|A^\dagger\|$.

Definition 8.8. *If an* $m \times n$ *matrix A has full rank, then* $\text{Cond}(A) = \|A\| \|A^\dagger\|$.

Note: If not explicitly stated, all the norms used in the rest of this chapter are 2-norms, and $\text{Cond}(A)$ is the condition number with respect to the 2-norm. That is, $\text{Cond}(A) = \|A\|_2 \|A^\dagger\|_2$.

Example 8.9.

$$A = \begin{pmatrix} 1 & 2 \\ 2 & 3 \\ 4 & 5 \end{pmatrix}, \quad b = \begin{pmatrix} 3 \\ 5 \\ 9 \end{pmatrix}.$$

Thus, A has full rank; rank $(A) = 2$.

$$A^\dagger = (A^T A)^{-1} A^T = \begin{pmatrix} -1.2857 & -0.5714 & 0.8571 \\ 1 & 0.5000 & -0.5000 \end{pmatrix},$$

$$\text{Cond}_2(A) = \|A\|_2 \|A\|_2^\dagger = 7.6656 \times 2.0487 = 15.7047.$$

The unique least-squares solution $x = A^\dagger b = \begin{pmatrix} 1 \\ 1 \end{pmatrix}$. ∎

8.6 Sensitivity of the Least-Squares Problem

In this section we study the sensitivity of a least-squares solution to perturbations in data; that is, *we investigate how a least-squares solution changes with respect to small changes in the data*. This study is important in understanding the different behaviors of different methods for solving the least-squares problem that will be discussed in the next section. We consider two cases: *perturbation in vector b and perturbation in matrix A*. The results in this section are normwise perturbation results. For componentwise perturbation results, see Björck (1996).

8.6. Sensitivity of the Least-Squares Problem

Case 1: Perturbation in vector b

Here we assume that vector b has been perturbed to $\hat{b} = b + \delta b$, but A has remained unchanged.

Theorem 8.10 (least-squares right perturbation theorem). *Let x and $\hat{x}$, respectively, be the unique least-squares solutions to the original and the perturbed problems. Then if $\|b_R\| \neq 0$,*

$$\text{Relative change} = \frac{\|\hat{x} - x\|}{\|x\|} \leq \text{Cond}(A) \frac{\|\delta b_R\|}{\|b_R\|}.$$

Here

$$\text{Cond}(A) = \|A\| \, \|A^\dagger\|,$$

and b_R and δb_R are, respectively, the projections of vectors b and δb onto $R(A)$.

Proof. Since x and $\hat{x}$ are the unique least-squares solutions to the original and the perturbed problems, we have

$$x = A^\dagger b, \quad \hat{x} = A^\dagger (b + \delta b).$$

Thus,

$$\hat{x} - x = A^\dagger b + A^\dagger \delta b - A^\dagger b = A^\dagger \delta b. \tag{8.9}$$

Let δb_N denote the projection of δb onto the orthogonal complement of $R(A)$. That is,

$$\delta b = \delta b_R + \delta b_N.$$

Since δb_N lies in the orthogonal complement of $R(A) = N(A^T)$, we have $A^T(\delta b_N) = 0$. So

$$\hat{x} - x = A^\dagger \delta b = A^\dagger(\delta b_R + \delta b_N)$$
$$= A^\dagger(\delta b_R) + A^\dagger(\delta b_N) = A^\dagger \delta b_R + (A^T A)^{-1} A^T \delta b_N = A^\dagger \delta b_R.$$

Again, since x is the unique least-squares solution, we have

$$Ax = b_R,$$

from which (again taking norms on both sides) we get

$$\|x\| \geq \frac{\|b_R\|}{\|A\|}. \tag{8.10}$$

Combining (8.9) and (8.10), we have the theorem. □

Interpretation of Theorem 8.10. Theorem 8.10 tells us that *if only vector b is perturbed, then, as in the case of linear systems,* $\text{Cond}(A) = \|A\| \, \|A^\dagger\|$ *serves as the condition number in the sensitivity analysis of the unique least-squares solution.* If this number is large, then even with a small relative error in the projection of b onto $R(A)$, we can have a drastic change in the least-squares solution. On the other hand, if this number is small and the relative error in the projection of b onto $R(A)$ is also small, then the least-squares solution will not change much. *Note that it is the smallness of the relative error in the projection of b onto $R(A)$, namely, $\frac{\|\delta b_R\|}{\|b_R\|}$, that plays the role here, not merely the smallness of $\|\delta b_R\|$.*

Example 8.11. An insensitive least-squares problem.

$$A = \begin{pmatrix} 1 & 2 \\ 0 & 1 \\ 1 & 0 \end{pmatrix}, \quad b = \begin{pmatrix} 1 \\ 1 \\ 1 \end{pmatrix}, \quad \delta b = 10^{-4} \begin{pmatrix} 1 \\ 1 \\ 1 \end{pmatrix},$$

$$b_R = P_A b = \begin{pmatrix} 0.8333 & 0.3333 & 0.1667 \\ 0.3333 & 0.3333 & -0.3333 \\ 0.1667 & -0.3333 & 0.8333 \end{pmatrix} \begin{pmatrix} 1 \\ 1 \\ 1 \end{pmatrix} = \begin{pmatrix} 1.3333 \\ 0.3333 \\ 0.6667 \end{pmatrix},$$

$$\delta b_R = P_A \delta b = 10^{-3} \begin{pmatrix} 0.13333 \\ 0.03333 \\ 0.06667 \end{pmatrix}.$$

Thus, according to Theorem 8.10, an upper bound for the relative error is

$$\frac{\|\delta b_R\|}{\|b_R\|} \cdot \text{Cond}(A) = 10^{-4} \times 2.4495.$$

So, *we expect that the least-squares solution will not be perturbed much.* The following computations show that this is indeed the case:

$$x = A^\dagger b = \begin{pmatrix} 0.6667 \\ 0.3333 \end{pmatrix}, \quad \hat{x} = A^\dagger (b + \delta b) = \begin{pmatrix} 0.6667 \\ 0.3334 \end{pmatrix}.$$

So, the relative error

$$\frac{\|\hat{x} - x\|}{\|x\|} = 10^{-4}. \quad \blacksquare$$

Example 8.12. A sensitive least-squares problem.

$$A = \begin{pmatrix} 1 & 1 \\ 10^{-4} & 0 \\ 0 & 10^{-4} \end{pmatrix}, \quad b = \begin{pmatrix} 2 \\ 10^{-4} \\ 10^{-4} \end{pmatrix}, \quad x = \begin{pmatrix} 1 \\ 1 \end{pmatrix}.$$

Suppose that

$$\delta b = 10^{-3} \begin{pmatrix} 1 \\ 0.1 \\ 0 \end{pmatrix}. \quad \text{Then } \hat{x} = \begin{pmatrix} 1.5005 \\ 0.5005 \end{pmatrix}.$$

The product $\text{Cond}_2(A) \cdot \frac{\|\delta b_R\|}{\|b_R\|} = 7.088$. *Since an upper bound for the relative error is 7.0888, there might be a substantial change in the solution.* Indeed, this is the case. The relative error in the solution is 0.5000; on the other hand the relative error in b is 5.0249×10^{-4}. $\blacksquare$

Case 2. Perturbation in matrix A

The analysis here is much more complicated than in the previous case. We will state the result here (without proof) and the major consequences of the result. Let the perturbation E of the matrix be small enough so that

$$\text{rank}(A) = \text{rank}(A + E).$$

8.6. Sensitivity of the Least-Squares Problem

Let x and $\hat{x}$ denote the unique least-squares solutions, respectively, to the original and the perturbed problem. Let E_A and E_N denote the projections of E onto $R(A)$ and onto the orthogonal complement of $R(A)$, respectively. Then if $b_R \neq 0$, we have the following theorem (see Stewart (1973, p. 223)).

Theorem 8.13 (least-squares left perturbation theorem). *Let x and $\hat{x}$ be the unique least-squares solutions to $Ax = b$ and $(A + E)\hat{x} = b$, and let $\text{rank}(A + E)$ be the same as $\text{rank}(A)$. Then*

$$\frac{\|\hat{x} - x\|}{\|x\|} < 2\,\text{Cond}(A)\frac{\|E_A\|}{\|A\|} + 4(\text{Cond}(A))^2 \frac{\|E_N\|}{\|A\|}\frac{\|b_N\|}{\|b_R\|} + O\left(\frac{\|E_N\|}{\|A\|}\right)^2.$$

Interpretation of Theorem 8.13. Theorem 8.13 tells us that *in the case where only matrix A is perturbed, the sensitivity of the unique least-squares solution, in general, depends upon squares of the condition number of A. However, if $\|E_N\|$ or $\|b_N\|$ is zero or small, then the sensitivity will depend only on Cond(A).* Note that the residual $r = b - Ax$ is zero if $b_N = 0$.

Two Examples with Different Sensitivities

We now present two examples with the same matrix A, but with different b, to illustrate the different sensitivities of the least-squares problem in different cases. *In the first example, $(\text{Cond}(A))^2$ serves as the condition number of the problem; in the second example, $\text{Cond}(A)$ serves as the condition number.*

Example 8.14. Sensitivity depending upon the square of the condition number. Let A be the same as in Example 7.18 and let

$$b = \begin{pmatrix} 1 \\ 1 \\ 1 \end{pmatrix}.$$

Then, using the results of Example 7.18, we have

$$b_R = P_A b = \begin{pmatrix} 1.0001 \\ 0.0001 \\ 0.0001 \end{pmatrix}, \quad b_N = \begin{pmatrix} -0.0001 \\ 0.9999 \\ 0.9999 \end{pmatrix}, \quad \frac{\|b_N\|}{\|b_R\|} = 1.4140.$$

Let

$$E = 10^{-4} \begin{pmatrix} 0 & -0.0001 \\ 0 & 0.9999 \\ 0 & 0.9999 \end{pmatrix}.$$

Then

$$A + E = \begin{pmatrix} 1 & 1 \\ 0.0001 & 0.0001 \\ 0 & 0.0002 \end{pmatrix}, \quad E_N = 10^{-4}\begin{pmatrix} 0 & -10^{-4} \\ 0 & 0.9999 \\ 0 & 0.9999 \end{pmatrix},$$

$$\frac{\|E_N\|}{\|A\|} = \frac{\|E\|}{\|A\|} = 9.999 \times 1.4140 \times 10^{-5}.$$

The product of $\frac{\|E_N\|}{\|A\|}$ and $\frac{\|b_N\|}{\|b_R\|}$ is rather small; however, $(\text{Cond}(A))^2 = 2 \times 10^8$ is large. Thus, there might be a drastic departure of the solution of the perturbed problem from the solution of the original problem. This is indeed true, as the following computations show:

$$\hat{x} = 10^3 \begin{pmatrix} -4.999 \\ 5 \end{pmatrix}, \quad x = \begin{pmatrix} 0.5 \\ 0.5 \end{pmatrix}.$$

Relative error: $\frac{\|\hat{x}-x\|}{\|x\|} = 9.999 \times 10^3$ (large!).
   Note that

$$\frac{\|E_N\|}{\|A\|} \cdot \frac{\|b_N\|}{\|b_R\|} \cdot (\text{Cond}_2(A))^2 = 9.999 \times 1.4140 \times 10^{-5} \times 2 \times 10^8 = 2.8277 \times 10^4. \quad \blacksquare$$

Example 8.15. Sensitivity depending upon the condition number. Let A and E be the same as in the previous example, but

$$b = \begin{pmatrix} 2 \\ 0.0001 \\ 0.0001 \end{pmatrix}.$$

In this case,

$$b_R = P_A b = \begin{pmatrix} 1 & 10^{-4} & 10^{-4} \\ 10^{-4} & 0.5000 & -0.5000 \\ 10^{-4} & -0.5000 & 0.5000 \end{pmatrix} b = b.$$

Thus, $b_N = 0$.
   So, according to Theorem 8.13, the *square of* Cond(A) *does not have any effect; the least-squares solution is affected only by* Cond(A). We verify this as follows:

$$x = A^\dagger b = \begin{pmatrix} 1 \\ 1 \end{pmatrix}, \quad \hat{x} = (A+E)^\dagger b = \begin{pmatrix} 1.4999 \\ 0.5000 \end{pmatrix}.$$

The *relative error* $\frac{\|\hat{x}-x\|}{\|x\|} = 0.5000$. Note that $\text{Cond}_2(A) = 1.4142 \times 10^4$, and $\frac{\|E_A\|}{\|A\|} = 10^{-4}$; thus, the *predicted upper bound of the relative error* is about 1.4142. $\blacksquare$

Residual sensitivity. We have just seen that the sensitivities of the least-squares solutions due to perturbations in the matrix A are different for different right-hand side vectors b; however, the following theorem shows that the residual sensitivity always depends upon the condition number of matrix A. We state the result in a somewhat simplified and crude form. See Golub and Van Loan (1996, pp. 242–244) for a precise statement and a proof.

Theorem 8.16 (least-squares residual sensitivity theorem). Let r and $\hat{r}$ denote the residuals, respectively, for the original and the perturbed least-squares problems; that is,

$$r = b - Ax, \quad \hat{r} = b - (A+E)(\hat{x}).$$

8.6. Sensitivity of the Least-Squares Problem

Then

$$\frac{\|\hat{r} - r\|}{\|b\|} \le \frac{\|E_N\|}{\|A\|} + 2\operatorname{Cond}(A)\frac{\|E_N\|}{\|A\|} + O\left(\frac{\|E_N\|}{\|A\|}\right)^2.$$

Interpretation of Theorem 8.16. *The above result tells us that the sensitivity of the residual depends at most on the condition number of A. One the other hand, as we have seen before, for the nonzero residual problem, it is the square of the condition number that measures the sensitivity.*

Example 8.17. Let A, b, and E be the same as in Example 8.14. Then

$$x = \begin{pmatrix} 0.5 \\ 0.5 \end{pmatrix}, \quad \hat{x} = 10^3 \begin{pmatrix} -4.9999 \\ 5 \end{pmatrix},$$

$$r = b - Ax = \begin{pmatrix} -0.0001 \\ 0.9999 \\ 0.9999 \end{pmatrix}, \quad \hat{r} = b - (A+E)\hat{x} = \begin{pmatrix} -0.0001 \\ 0.9999 \\ 0 \end{pmatrix}.$$

Thus, the relative residual $\frac{\|\hat{r}-r\|}{\|b\|} = 0.5773$. Note that $\operatorname{Cond}(A) \cdot \frac{\|E_N\|}{\|A\|} = 1.4142$. ∎

Sensitivity of the pseudoinverse. The following result, due to Wedin (1973), shows that it is $\operatorname{Cond}(A)$ again that plays a role in the sensitivity analysis of the pseudoinverses of a matrix.

Theorem 8.18 (pseudoinverse sensitivity theorem). *Let A be $m \times n$, where $m \ge n$. Let $A^\dagger$ and $\tilde{A}^\dagger$ be, respectively, the pseudoinverse of A and of $\tilde{A} = A + E$. Then, provided that $\operatorname{rank}(A) = \operatorname{rank}(\tilde{A})$, we have*

$$\frac{\left\|\tilde{A}^\dagger - A^\dagger\right\|}{\left\|\tilde{A}^\dagger\right\|} \le \sqrt{2}\operatorname{Cond}(A)\frac{\|E\|}{\|A\|}.$$

Example 8.19.

$$A = \begin{pmatrix} 1 & 2 \\ 2 & 3 \\ 4 & 5 \end{pmatrix}, \quad E = 10^{-4}A = \begin{pmatrix} 0.0010 & 0.0020 \\ 0.0020 & 0.0030 \\ 0.0040 & 0.0050 \end{pmatrix},$$

$$A^\dagger = \begin{pmatrix} -1.2857 & -0.5714 & 0.8571 \\ 1 & 0.5000 & -0.5000 \end{pmatrix},$$

$$A + E = \tilde{A} = \begin{pmatrix} 1.001 & 2.002 \\ 2.002 & 3.003 \\ 4.004 & 5.005 \end{pmatrix}, \quad \tilde{A}^\dagger = \begin{pmatrix} -1.2844 & -0.5709 & 0.8563 \\ 0.9990 & 0.499995 & -0.4995 \end{pmatrix}.$$

So,

$$\frac{\left\|\tilde{A}^\dagger - A^\dagger\right\|}{\left\|\tilde{A}^\dagger\right\|} = 10^{-4} \approx \frac{\|E\|}{\|A\|}.$$

Note that
$$\text{Cond}(A) = 15.7047. \quad \blacksquare$$

8.7 Computational Methods for Overdetermined Problems: Normal Equations, QR, and SVD Methods

The following least-squares solution methods are described in this section:

- The normal equations method (Algorithm 8.1).

- The QR factorization methods using Householder and modified Gram–Schmidt (MGS) processes. (Algorithms 8.2 and 8.3).

- The SVD method (Algorithm 8.4).

8.7.1 The Normal Equations Method

We have already seen in Section 8.3.2 that when A is $m \times n$ ($m \geq n$) and has full rank, the unique least-square solution x satisfies the normal equation: $A^T A x = A^T b$. Indeed, this approach of solving a least-squares problem had been a popular method for many years among statisticians. We now show how to implement this approach numerically. Since A has full rank, $A^T A$ is symmetric and positive definite, and it admits the *Cholesky decomposition*: $A^T A = H H^T$. Therefore, the normal equations approach for solving the least-squares problem can be stated as follows.

ALGORITHM 8.1. Least-Squares Solution Using Normal Equations.

Inputs: (i) An $m \times n$ ($m > n$) matrix A of *full rank*, and (ii) an m-vector b.
Output: A unique least-squares solution x.

Step 1. Form $c = A^T b$.

Step 2. Find the Cholesky factorization of $A^T A = H H^T$.

Step 3. Solve the triangular systems in the sequence $Hy = c, \quad H^T x = y$.

Flop-count. The above method for solving the full-rank least-squares problem requires about $mn^2 + \frac{n^3}{3}$ flops. This can be seen as follows: mn^2 flops for computing $A^T A$ and $A^T b$, $\frac{n^3}{3}$ flops for computing the Cholesky factorization of $A^T A$, and $2n^2$ flops to solve two triangular systems.

Example 8.20. Normal equation solution of the problem on predicting future sales (Section 8.4.2).

Step 1. Form $c = A^T b = \begin{pmatrix} 703 \\ 182230 \\ 2164253 \end{pmatrix}$.

8.7. Computational Methods for Overdetermined Problems

Step 2. Form $A^T A = \begin{pmatrix} 5 & 1120 & 14691 \\ 1120 & 297522 & 3466402 \\ 14691 & 3466402 & 44608873 \end{pmatrix}$;

$$H = 10^3 \begin{pmatrix} 0.0022 & 0 & 0 \\ 0.5009 & 0.2160 & 0 \\ 6.5700 & 0.8132 & 0.8846 \end{pmatrix}.$$

Step 3. Solve the two triangular systems:

$$y = \begin{pmatrix} 314.3912 \\ 114.6376 \\ 6.1934 \end{pmatrix}, \quad x = \begin{pmatrix} 7.0325 \\ 0.5044 \\ 0.0070 \end{pmatrix} \quad \blacksquare$$

The following table compares the prediction of sales in each district, obtained by the least-squares solution, with the actual value. The prediction for district i is computed as $a_i^T x$, where a_i^T is the ith row of A, $i = 1, 2, 3, 4$.

District	Prediction of Sales	Actual Sale
1	162.4043	162
2	120.6153	120
3	222.8193	223
4	130.3140	131
5	66.8471	67

Suppose that the company would like to predict, using the above results, the sales in a district with the population 220,000 and per capita income of $2,500. Then the best prediction using the given model is

$$\begin{pmatrix} 1 & 220 & 2500 \end{pmatrix} \begin{pmatrix} 7.0325 \\ 0.5044 \\ 0.0070 \end{pmatrix} = 135.5005.$$

Numerical Difficulties with the Normal Equations Method

The normal equations method, though easy to understand and implement, may give rise to numerical difficulties in certain cases.

First, *we might lose some significant digits during the explicit formation of $A^T A$, and the computed matrix $A^T A$ may be far from positive definite; computationally, it may even be singular.* Indeed, it has been shown by Stewart (1973, pp. 225–226) that unless Cond(A) is less than $10^{\frac{t}{2}}$, where it is assumed that $A^T A$ has been computed exactly and then rounded to t significant digits, the matrix $A^T A$ may fail to be positive definite or even may not be nonsingular. The following simple example illustrates this fact. (Note that $t \leq 16$ for a 32-bit machine.)

Example 8.21. Consider the matrix A from Example 8.12. Let $t = 8$. The columns of A are linearly independent. In exact arithmetic, we have

$$A^T A = \begin{pmatrix} 1 + 10^{-8} & 1 \\ 1 & 1 + 10^{-8} \end{pmatrix}.$$

Since $t = 8$, we will get

$$A^T A = \begin{pmatrix} 1 & 1 \\ 1 & 1 \end{pmatrix},$$

which is *singular*. Note that $\text{Cond}(A) = 1.4142 \times 10^4 > 10^{\frac{t}{2}} = 10^4$. ∎

Second, *the normal equations approach may, in certain cases, introduce more errors than those which are inherent in the problem.* This is seen as follows.

From the perturbation analysis done in Chapter 6, we easily see that if $\hat{x}$ is the solution obtained by the normal equations method, then (Exercise 8.17)

$$\frac{\|\hat{x} - x\|}{\|x\|} \approx \mu \, \text{Cond}(A^T A) = \mu \, (\text{Cond}(A))^2.$$

Thus, the accuracy of the least-squares solution using normal equations will depend upon the square of the condition number of matrix A. However, we have just seen in the last section that the sensitivity of the least-squares problem in certain cases, such as when the residual is zero, depends only on the condition number of A (see Theorems 8.10 and 8.13). *Thus, in these cases, the normal equations method will introduce more errors in the solution than what is warranted by the data.* Having said this, we note that there exist some modifications of normal equations method which are numerically stable. See Foster (1991).

MATCOM Note: Algorithm 8.1 has been implemented in the MATCOM program **LSFRNME**.

8.7.2 QR Factorization Method

In this section, we will show how the factorization $A = QR$ can be used to solve the least-squares problem. Let $A \in \mathbb{R}^{m \times n}$ ($m \geq n$) have full rank.

> **Idea:** Reduce the least-squares problem for a full matrix A to an upper triangular linear system problem using QR factorization of A.

From Theorem 8.3, we know that the unique least-squares solution x satisfies $Ax = P_A b$.

Let $Q_1 R_1 = A$ be the *reduced QR factorization* of A. Then $P_A = Q_1 Q_1^T$.

So, $Ax = P_A b = Q_1 Q_1^T b$. Multiplying both sides by Q_1^T, we have $Q_1^T Ax = Q_1^T b$, or $R_1 x = Q_1^T b = c$. Thus, *finding the least-squares solution x to $Ax = b$ reduces to solving the upper triangular system $R_1 x = c$.*

8.7. Computational Methods for Overdetermined Problems

Least-Squares Solution Using Reduced QR Factorization

1. Find the *reduced* QR factorization of A: $A = Q_1 R_1$.
2. Compute $c = Q_1^T b$.
3. Solve the upper triangular system $R_1 x = c$.

An Alternative Derivation of the Least-Squares Solution and Expression for the Residual

Let $A = QR$ be the *full* QR factorization of A. That is,

$$Q^T A = \begin{pmatrix} R_1 \\ 0 \end{pmatrix} \begin{matrix} n \\ m-n \end{matrix} \quad \text{and} \quad Q^T b = \begin{pmatrix} c \\ d \end{pmatrix} \begin{matrix} n \\ m-n \end{matrix}.$$

Then $\|Ax - b\|_2^2 = \|Q^T Ax - Q^T b\|_2^2 = \|R_1 x - c\|^2 + \|d\|^2$. *(Note that the 2-norm is preserved by orthogonal matrix multiplication.)* Thus, x is the least-squares solution if x satisfies $R_1 x = c$. Also, note that if $Q = (Q_1, Q_2)$, then matrix R_1 and vector c are given by

$$R_1 = Q_1^T A \quad \text{and} \quad c = Q_1^T b.$$

To obtain an expression for the residual, we note that when x is the least-squares solution, $\|Ax - b\|_2 = \|d\|_2$. Again, $d = Q_2^T b$. Thus we have the following.

The Least-Squares Residual Norm Using QR Factorization

Let $A = QR = (Q_1, Q_2)R$. Then the *least-squares residual* is given by
$$\|r\|_2 = \|Ax - b\|_2 = \|Q_2^T b\|_2.$$

Example 8.22.

$$A = \begin{pmatrix} 1 & 1 \\ 10^{-4} & 0 \\ 0 & 10^{-4} \end{pmatrix}, \quad b = \begin{pmatrix} 2 \\ 10^{-4} \\ 10^{-4} \end{pmatrix}.$$

Step 1. $A = QR$ (see Example 7.5).

$$Q = (Q_1, Q_2) = \begin{pmatrix} -1 & 0.0001 & \cdot & -0.0001 \\ -0.0001 & -0.7071 & \cdot & 0.7071 \\ 0 & 0.7071 & \cdot & 0.7071 \end{pmatrix},$$

$$R = \begin{pmatrix} -1 & -1 \\ 0 & 0.0001 \\ \cdots & \cdots \\ 0 & 0 \end{pmatrix} = \begin{pmatrix} R_1 \\ \cdots \\ 0 \end{pmatrix}.$$

Step 2.
$$c = Q_1^T b = \begin{pmatrix} -2 \\ 0.0001 \end{pmatrix}.$$

Step 3. Solution of the system: $R_1 x = c$ is $x = (1, 1)^T$. The unique least-squares solution is $\begin{pmatrix} 1 \\ 1 \end{pmatrix}$.

Step 4. The residual norm $= \|r\|_2 = \|d\|_2 = \|Q_2^T b\|_2 = 0$. ∎

Use of Householder Matrices

If Householder's method is used to factorize A into QR, then the vector $\begin{pmatrix} c \\ d \end{pmatrix} = Q^T b$ can be computed implicitly as

$$\text{For} \quad k = 1, 2, \ldots, n \text{ do}$$
$$b = H_k b,$$

where H_k, $k = 1, \ldots, n$, are the Householder matrices such that $Q^T = H_n H_{n-1} \cdots H_2 H_1$. Thus, *the matrix Q does not have to be formed explicitly*. The idea was due to Golub (1965).[8]

ALGORITHM 8.2. The Householder–Golub Method for the Full-Rank Least-Squares Problem.

Inputs: (i) An $m \times n$ ($m \geq n$) matrix A of *full rank* (rank $(A) = n$). (ii) An m-vector b.
Outputs: (i) The *unique least-squares solution* x to $Ax = b$. (ii) The *residual norm*.

Step 1. Apply the Householder QR factorization method (Algorithm 7.2) to A. Obtain R_1 and the Householder matrices $H_1, H_2, \ldots, H_n$.
Step 2. Form $H_n H_{n-1} \cdots H_2 H_1 b = \begin{pmatrix} c \\ d \end{pmatrix}$, where c is an n-vector, by *computing the product implicitly*.
Step 3. Solve $R_1 x = c$.
Step 4. Form the residual norm: $\|d\|_2$.

Example 8.23. Consider Example 8.22 again.

Step 1. $R_1 = \begin{pmatrix} -1 & -1 \\ 0 & 0.0001 \end{pmatrix}.$

---

[8]Gene H. Golub (1932–2007) was Fletcher-Jones Professor of Computer Science at Stanford University. Golub made everlasting contributions in many areas of numerical linear algebra related to SVD, least-squares, and related topics. He was a coauthor of the celebrated numerical linear algebra book *Matrix Computations*. Golub was a member of both the National Academy of Sciences and the National Academy of Engineering. Several conferences around the world were held to remember Gene Golub on February 29, 2008, the date that would have been his 76th birthday. For more details on the life of Gene Golub, see the obituary of Gene Golub by Trefethen (2007), an interview with Gene Golub by Higham (2008), and a *New York Times* article published December 10, 2007.

8.7. Computational Methods for Overdetermined Problems

Step 2. $H_2 H_1 b = \begin{pmatrix} -2 \\ 0.0001 \\ 0 \end{pmatrix}$. Then $c = \begin{pmatrix} -2 \\ 0.001 \end{pmatrix}$.

Step 3. Solve $R_1 x = c \Rightarrow x = \begin{pmatrix} 1 \\ 1 \end{pmatrix}$.

Step 4. Residual norm $= 0$. ∎

Flop-count. Since the cost of Algorithm 8.2 is dominated by the cost of the QR factorization of A, the overall flop-count for the full-rank least-squares solution using Householder's QR method is (Exercise 8.15) $2n^2(m - \frac{n}{3})$. Thus, the *method is about* twice *as expensive as the normal equations method*. Note that the normal equations method requires about $(n^2 m + \frac{n^3}{3})$ flop.

Round-off error and stability. The method is *stable*. It has been shown in Lawson and Hanson (1995, p. 90) that the computed solution $\hat{x}$ is such that it minimizes
$$\left\| (A + E)\hat{x} - (b + \delta b) \right\|_2,$$
where E and δb are small. Specifically,
$$\|E\|_F \leq c\mu n \|A\|_F + O(\mu^2), \quad \|\delta b\|_2 \leq c\mu \|b\|_2 + O(\mu^2),$$
where $c \approx (6m - 3n + 41)$ and μ is the machine precision. That is, the *computed solution is the exact least-squares solution of a nearby problem*.

MATCOM Note: Algorithm 8.2 has been implemented in the MATCOM program **LSFRQRH**.

Use of Givens Rotations

We can, of course, use the Givens rotations to decompose A into QR and then use this decomposition to solve the least-squares problem. However, as we have seen before, the use of Givens rotations will be more expensive than the use of Householder matrices. Recall that computations of Givens rotations require evaluations of square roots; however, there are "square-root-free" Givens rotations, introduced by Gentleman (1973), which can be used to solve the least-squares problem. The square-root-free Givens rotations are also known as *fast Givens rotations*. For details, see Golub and Van Loan (1996, pp. 218–220).

Use of the MGS Method in the Least-Squares Solution

We have seen in Chapter 7 that, in general, MGS is not fully satisfactory for QR factorization of A; however, it has turned out to be numerically effective for the least-squares solution if the vector $c = Q^T b$ is computed by finding the QR factorization of the augmented matrix (A, b) rather than of A itself. *The least-squares solution with the matrix Q obtained directly from the QR factorization of A may not be accurate,* due to the possible departure from orthogonality of the computed Q. Thus if
$$(A, b) = (Q_1, q_{n+1}) \begin{pmatrix} R_1 & z \\ 0 & \rho \end{pmatrix},$$

then
$$Ax - b = (A, b)\begin{pmatrix} x \\ -1 \end{pmatrix} = (Q_1, q_{n+1})\begin{pmatrix} R_1 & z \\ 0 & \rho \end{pmatrix}\begin{pmatrix} x \\ -1 \end{pmatrix} = Q_1(R_1 x - z) - \rho q_{n+1}.$$

If q_{n+1} is orthogonal to Q_1, then $\|Ax - b\|_2$ will be a minimum when $R_1 x = z$. Thus, the least-squares solution can be obtained by solving $R_1 x = z$. The residual r will be given by $r = \rho q_{n+1}$. Details can be found in Björck (1996). The above discussion leads to the following least-squares algorithm.

ALGORITHM 8.3. Least-Squares Solution by MGS.

Inputs: $A \in \mathbb{R}^{m \times n} (m \geq n)$, rank $(A) = n$; $b \in \mathbb{R}^m$.
Output: A *unique least-squares* solution x.

1. Apply MGS (Algorithm 7.8) to A to obtain $Q_1 = (q_1, \ldots, q_n)$ and R_1.

2. For $k = 1, \ldots, n$ do
$$\delta_k = q_k^T b$$
$$b \equiv b - \delta_k q_k$$
   End

3. Solve $R_1 x = (\delta_1, \ldots, \delta_n)^T$.

Example 8.24. Consider solving the least-squares problem using the MGS with the data of Example 8.22. The exact solution is $x = \binom{1}{1}$. The QR factorization of A using MGS is given by

$$Q_1 = \begin{pmatrix} 1 & 0 \\ 0.0001 & -0.7071 \\ 0 & 0.7071 \end{pmatrix}, \quad R_1 = \begin{pmatrix} 1 & 1 \\ 0 & 0.0001 \end{pmatrix}.$$

If we now form $c = Q_1^T b$ and solve $R_1 x = c$, we obtain $x = \binom{2}{0}$. On the other hand, if we obtain x using Algorithm 8.3, we get $(\delta_1, \delta_2) = (2, 0.0001)$, and the solution of $R_1 x = (\delta_1, \delta_2)^T$ is $x \approx \binom{1}{1}$. ∎

Round-off property and flop-count. It has been shown by Björck and Paige (1992) that the MGS process for the least-squares problem is numerically equivalent to the Householder method applied to $\begin{pmatrix} 0 & 0 \\ A & b \end{pmatrix}$; that is,

$$H_n H_{n-1} \cdots H_2 H_1 \begin{pmatrix} 0 & 0 \\ A & b \end{pmatrix} = \begin{pmatrix} R & c_1 \\ 0 & c_2 \end{pmatrix}.$$

From this equivalence, it follows that *the MGS method is backward stable for the least-squares problem.* The method is slightly more expensive than the Householder method. It requires about $2mn^2$ flops, compared to the $2mn^2 - 2\frac{n^3}{3}$ flops needed by the Householder method.

MATCOM Note: Algorithm 8.3 has been implemented in the MATCOM program **LSFRMGS**.

8.7.3 The SVD Method

In Chapter 7, we have seen that the SVD of A can be used to compute the orthogonal projection onto $R(A)$. So it is natural to think of using the SVD to compute the least-squares solution.

> **Idea:** Reduce the least-squares problem of a full matrix A to a diagonal problem using the SVD of A.

Consider the *reduced SVD* of A: $A = U_1 \Sigma_1 V^T$. Then the orthogonal projection of A onto $R(A)$ is $P_A = U_1 U_1^T$. Since the lease-squares solution x satisfies $Ax = P_A b$ (Theorem 8.3), we have
$$Ax = P_A b = U_1 U_1^T b.$$
Multiplying both sides by U_1^T, we obtain
$$U_1^T A x = U_1^T b \quad \text{(note that } U_1^T U_1 = I_{n\times n}\text{)}.$$
That is,
$$U_1^T A V V^T x = U_1^T b \quad \text{(note that } V V^T = I_{n\times n}\text{)}$$
or $\Sigma_1 y = b'$, where $y = V^T x$ and $b' = U_1^T b$.

Thus, using the reduced SVD of A, solving the least-squares problem is reduced to the solution of the diagonal system $\Sigma_1 y = b'$.

This observation leads us to the following.

> **ALGORITHM 8.4. Least-Squares Solution via Reduced SVD.**
>
> **Inputs:** $A \in \mathbb{R}^{m \times n} (m \geq n)$, $b \in \mathbb{R}^m$; A is of *full rank*.
> **Output:** The unique least-squares solution x.
>
> **Step 1.** Find the reduced SVD of A: $A = U_1 \Sigma_1 V^T$.
> **Step 2.** Compute $b' = U_1^T b$.
> **Step 3.** Solve the diagonal system $\Sigma_1 y = b'$.
> **Step 4.** Obtain the least-squares solution $x = V y$.

Example 8.25. A, b are the same as in Example 8.22.

Step 1. The reduced SVD of $A = U_1 \Sigma_1 V^T$:
$$U_1 = \begin{pmatrix} -1 & 0 \\ 0 & -0.7071 \\ 0 & 0.7071 \end{pmatrix}, \quad \Sigma_1 \begin{pmatrix} 1.4142 & 0 \\ 0 & 0.0001 \end{pmatrix}, \quad V = \begin{pmatrix} -0.7071 & -0.7071 \\ -0.7071 & 0.7071 \end{pmatrix}.$$

Step 2. $b' = U_1^T b = \begin{pmatrix} -2 \\ 0 \end{pmatrix}$.

Step 3. $y = \begin{pmatrix} -1.4142 \\ 0 \end{pmatrix}$.

Step 4. $x = V y = \begin{pmatrix} 1 \\ 1 \end{pmatrix}$. ∎

Rank-Deficient Least-Squares Solutions using SVD.

If $A \in \mathbb{R}^{m \times n}$, $m \geq n$, and rank$(A) = k < n$, then there are infinitely many least-squares solutions. *The QR factorization with column pivoting (see Chapter 14, available online at www.siam.org/books/ot116) can be used to solve the problem* (Exercise 8.16). *However, the best numerically reliable way to solve a rank-deficient least-squares problem is via SVD.* Of importance in practical applications is to compute the one with *minimum-norm*. We now describe how to do this. To show that in the rank-deficient case there are infinitely many solutions, we will use the full SVD.

Let $A = U\Sigma V^T$ be the full SVD of A. Then we have

$$\begin{aligned}\|Ax - b\|_2 &= \|(U\Sigma V^T x - b)\|_2 \\ &= \|U(\Sigma V^T x - U^T b)\|_2 \\ &= \|\Sigma y - \hat{b}\|_2,\end{aligned}$$

where $V^T x = y$ and $U^T b = \hat{b}$.

Again,

$$\|\Sigma y - \hat{b}\|_2 = \sum_{i=1}^{k} |\sigma_i y_i - \hat{b}_i|^2 + \sum_{i=k+1}^{m} |\hat{b}_i|^2,$$

where k is the rank of A. Thus the vector

$$y = \begin{pmatrix} y_1 \\ y_2 \\ \vdots \\ y_n \end{pmatrix}$$

that minimizes $\|\Sigma y - \hat{b}_2\|_2$ is given by

$$y_i = \frac{\hat{b}_i}{\sigma_i}, \qquad i = 1, \ldots, k,$$
$$y_i = \text{arbitrary}, \qquad i = k+1, \ldots, n.$$

(Note that when $k < n$, y_{k+1} through y_m do not appear in the above expression and therefore do not have any effect on the residual.) Of course, once y is computed, the solution to the original problem can be recovered from $x = Vy$.

Remark. Since finding rank is a tricky matter in practical computations, we will use *numerical rank* $\hat{r}$, as defined in Section 7.8.9.

Since corresponding to each (computationally) "zero" singular value σ_i, y_i can be set arbitrarily, *in the rank-deficient case, we will have infinitely many solutions to the least-squares problem.* There are instances where this rank-deficiency is actually desirable because it provides a rich family of solutions which might be used for optimizing some other aspects of the original problem.

Thus, an algorithm for finding infinitely many least-squares solutions in the rank-deficient case, using SVD, can be stated as follows.

8.7. Computational Methods for Overdetermined Problems

ALGORITHM 8.5. Rank-Deficient Least-Squares Solutions Using SVD.

Inputs: $A \in \mathbb{R}^{m \times n}$, $b \in \mathbb{R}^m$; A is a *numerically rank-deficient* matrix with tolerance δ.
Output: A *family* of least-squares solutions $\{x\}$.

Step 1. Find the SVD of A:
$$A = U\Sigma V^T.$$

Step 2. Form $\hat{b} = U^T b = \begin{pmatrix} \hat{b}_1 \\ \hat{b}_2 \\ \vdots \\ \hat{b}_m \end{pmatrix}$.

Step 3. Determine the numerical rank $\hat{r}$ of A using the tolerance δ *(see Section 7.8.9)*.

Step 4. Compute $y = \begin{pmatrix} y_1 \\ \vdots \\ y_n \end{pmatrix}$ choosing

$$y_i = \begin{cases} \dfrac{\hat{b}_i}{\sigma_i}, & i = 1, 2, \ldots, \hat{r}, \\ \text{arbitrary}, & i = \hat{r}+1, \ldots, n. \end{cases}$$

Step 5. Compute the family of least-squares solutions $\{x\}$ as
$$x = Vy.$$

Remark. *Algorithm 8.5 can be used to compute the least-squares solutions in both the full-rank and the rank-deficient cases. Note that in the full-rank case, the family has just one number.*

Example 8.26.
$$A = \begin{pmatrix} 1 & 1 \\ 0 & 10^{-7} \\ 0 & 10^{-7} \end{pmatrix}, \quad b = \begin{pmatrix} 2 \\ 0 \\ 0 \end{pmatrix}.$$

Step 1. $A = U\Sigma V^T$, where

$$U = \begin{pmatrix} 1 & 0 & 0 \\ 0 & -0.7071 & -0.7071 \\ 0 & -0.7071 & 0.7071 \end{pmatrix}, \quad \Sigma = \begin{pmatrix} 1.4142 & 0 \\ 0 & 0 \\ 0 & 0 \end{pmatrix}, \quad V = \begin{pmatrix} 0.7071 & 0.7071 \\ 0.7071 & -0.7071 \end{pmatrix}.$$

Step 2. $\hat{b} = U^T b = \begin{pmatrix} 2 \\ 0 \\ 0 \end{pmatrix}$.

Step 3. $\hat{r} = 1$.

Step 4. $y = \begin{pmatrix} y_1 \\ y_2 \end{pmatrix} = \begin{pmatrix} 1.4142 \\ \text{arbitrary} \end{pmatrix}$.

The family of least-squares solutions is given by $x = Vy$. By choosing different values of y_2, we will obtain different solutions. For example, when $y_2 = 0$, we have $x = \begin{pmatrix} 1 \\ 1 \end{pmatrix}$. ∎

Flop-count. Using the Golub–Kahan–Reinsch method to compute the SVD of A, to be described later in Chapter 10, it takes about $4mn^2 + 8n^3$ flops to solve the least-squares problem, when A is $m \times n$ and $m \geq n$. (In deriving this flop-count, it is noted that the complete vector $\hat{b}$ does not need to be computed; only the columns of U that correspond to the nonzero singular values are needed in computation.)

MATCOM Note: Algorithm 8.5 has been implemented in the MATCOM program **LSSVD**.

An Expression for the Minimum-Norm Least-Squares Solution

It is clear from Step 4 of Algorithm 8.5 above that in the rank-deficient case, the minimum 2-norm least-squares solution is the one that is obtained by setting $y_i = 0$ whenever $\sigma_i = 0$ (numerically). Thus, from above, we have the following expression for the minimum 2-norm solution.

Minimum-Norm Least-Squares Solution of a Rank-Deficient Least-Squares Problem Using SVD

$$x = \sum_{i=1}^{\hat{r}} \frac{u_i^T b}{\sigma_i} v_i, \qquad (8.11)$$

where $\hat{r} = $ numerical rank(A).

Example 8.27.

$$A = \begin{pmatrix} 1 & 2 & 3 \\ 2 & 3 & 4 \\ 1 & 2 & 3 \end{pmatrix}, \qquad b = \begin{pmatrix} 6 \\ 9 \\ 6 \end{pmatrix}.$$

1. The singular values are $\sigma_1 = 7.5358$, $\sigma_2 = 0.4597$, and $\sigma_3 = 0$. A is rank-deficient, $\hat{r} = 2$.

2. The singular vectors corresponding to the nonzero singular values are

    $u_1 = (0.4956, 0.7133, 0.4956)^T$, $u_2 = (0.5044, -0.7008, 0.5044)^T$;
    $v_1 = (0.3208, 0.5470, 0.7732)^T$, $v_2 = (-0.8546, -0.1847, 0.4853)^T$.

The minimum 2-norm least-squares solution is $x = \frac{u_1^T b}{\sigma_1} v_1 + \frac{u_2^T b}{\sigma_2} v_2 = (1, 1, 1)^T$. ∎

8.8. Underdetermined Linear Systems

MATCOM Note: The MATCOM program **MINNMSVD** computes the minimum-norm solution using the SVD.

8.7.4 Solving the Linear System Using the SVD

Note that the idea of using the SVD in the solution of the least-squares problem can be easily applicable for determining whether a linear system $Ax = b$ has a solution and, if so, how to compute it.

Thus if
$$A = U\Sigma V^T,$$
then $Ax = b$ is equivalent to $\Sigma y = b'$, where $y = V^T x$ and $b' = U^T b$.

Thus, to solve $Ax = b$ using SVD, do the following steps:

Step 1. Compute the SVD of A: $A = U\Sigma V^T$.

Step 2. Compute the vector $b' = U^T b$.

Step 3. Solve the *diagonal* system $\Sigma y = b'$.

Step 4. Obtain the solution $x = Vy$.

However, this approach is much more expensive than the Gaussian elimination and QR methods. *That is why, in practice, the SVD is not generally used to solve a linear system.*

8.8 Underdetermined Linear Systems

Let A be $m \times n$ and $m < n$. Then the system
$$Ax = b$$
has more equations than unknowns. Such a system is called an *underdetermined system*. An underdetermined system can be illustrated graphically, as shown in Figure 8.3.

Underdetermined systems, though arising in a variety of practical applications, are unfortunately not widely discussed in the literature. An excellent source is the survey paper by Cline and Plemmons (1976). *An underdetermined system has either no solution or an*

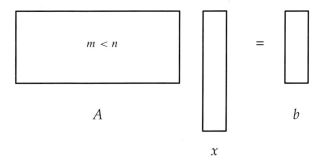

Figure 8.3. *Underdetermined system.*

infinite number of solutions. In case A has full rank, solutions do exist and the general solution can be written as follows:

$$x = A^T(AA^T)^{-1}b + (I - A^T(AA^T)^{-1}A)y, \text{ where } y \text{ is arbitrary.}$$

By setting $y = 0$, we obtain the *minimum-norm solution*:

$$x = A^T(AA^T)^{-1}b,$$

which is the *minimum-norm normal equations solution* to the full-rank underdetermined problem.

The normal equations approach for the minimum-norm solution to the underdetermined full-rank least-squares problem as obtained above will have the same disadvantages as in the case of full-rank overdetermined problem.

MATCOM Note: The MATCOM program **MNUDNME** implements the above solution method.

8.8.1 The QR Approach for the Minimum-Norm Solution

Decomposing A^T, instead of A, into QR,

$$Q^T A^T = \begin{pmatrix} R \\ 0 \end{pmatrix},$$

the system $Ax = b$ becomes

$$(R^T, 0^T) \begin{pmatrix} y_R \\ y_N \end{pmatrix} = b, \text{ where } y = Q^T x = \begin{pmatrix} y_R \\ y_N \end{pmatrix}.$$

The unique minimum-norm least-squares solution is obtained by setting

$$y_N = 0.$$

The above discussion leads to the following algorithm.

ALGORITHM 8.6. Minimum-Norm Solution to the Full-Rank Underdetermined Problem Using QR Factorization.

Inputs: $A \in \mathbb{R}^{m \times n} (m < n)$ with *full-rank m*; $b \in \mathbb{R}^n$.
Output: The *minimal 2-norm solution* to $Ax = b$.

Step 1. Find the QR factorization of A^T:

$$Q^T A^T = \begin{pmatrix} R_1 \\ 0 \end{pmatrix}, \; R_1 \in \mathbb{R}^{m \times m}.$$

Step 2. Partition $Q = (Q_1, Q_2)$, $Q_1 \in \mathbb{R}^{n \times m}$.

Step 3. Solve for y_R: $R_1^T y_R = b$.

Step 4. Form the *minimum-norm solution* $x = Q_1 y_R$.

8.8. Underdetermined Linear Systems

A note on implementation: If we use the Householder method to compute the QR factorization of A, the product $Q_1 y_R$ should be computed from the factored form of Q as the product of Householder matrices.

Flop-count. Using Householder orthogonalization, $2m^2 n - 2\frac{m^3}{3}$ flops will be required to implement Algorithm 8.6 (Exercise 8.15).

Round-off property. It has been shown (Lawson and Hanson (1995 p. 93)) that the *computed vector $\hat{x}$ is close to the exact minimum-length least-squares solution of a perturbed problem.* That is, there exist a matrix E and a vector $\hat{x}$ such that $\hat{x}$ is the minimum-length solution of

$$(A + E)\hat{x} \approx b,$$

where

$$\|E\|_F \leq (6n - 3m + 41)m\mu \|A\|_F + O(\mu^2).$$

MATCOM Note: Algorithm 8.6 has been implemented in the MATCOM program **MNUDQRH**.

Example 8.28. Consider the underdetermined system with A and b as follows:

$$A = \begin{pmatrix} 1 & 2 & 3 \\ 2 & 3 & 4 \end{pmatrix}, \quad b = \begin{pmatrix} 6 \\ 9 \end{pmatrix}.$$

Step 1. Find QR factorization of A^T using $[Q, R] = qr(A^T)$.

Step 2.

$$Q_1 = \begin{pmatrix} -0.2673 & 0.8729 \\ -0.5345 & 0.2182 \\ -0.8018 & -0.4364 \end{pmatrix}, \quad R_1 = \begin{pmatrix} -3.7417 & -5.3452 \\ 0 & 0.6547 \end{pmatrix}.$$

Step 3. $y_R = \begin{pmatrix} -1.6036 \\ 0.6547 \end{pmatrix}.$

Step 4. The minimum-norm solution is $x = Q_1 y_R = \begin{pmatrix} 1 \\ 1 \\ 1 \end{pmatrix}.$ ∎

8.8.2 The SVD Approach for the Minimum-Norm Solution

Let $A = \sum_{i=1}^{\hat{r}} \sigma_i u_i v_i^T$ be the SVD of A, where $\hat{r}$ is the numerical rank of A. Then, as in case of the rank-deficient overdetermined system, the minimum-norm solution x to the underdetermined system is given by

$$x = \sum_{i=1}^{\hat{r}} \frac{u_i^T b}{\sigma_i} v_i.$$

Remark. *The SVD approach is recommended if A is nearly rank-deficient.*

Example 8.29. Consider Example 8.28 again. From the SVD of A, we get

1. $\sigma_1 = 6.5468$, $\sigma_2 = 0.3742$, $\sigma_3 = 0$;
2. $u_1 = (-0.5696, -0.8219)^T$, $u_2 = (-0.8219, 0.5696)^T$;
3. $v_1 = (-0.3381, -0.5506, -0.7632)^T$, $v_2 = (0.8480, 0.1735, -0.5009)^T$;
4. the *minimum-norm solution* $x = \dfrac{u_1^T b}{\hat{\sigma}_1} v_1 + \dfrac{u_2^T b}{\hat{\sigma}_2} v_2 = \begin{pmatrix} 1 \\ 1 \\ 1 \end{pmatrix}$. ∎

A comparison of different least-squares methods is given in Table 8.1.

Table 8.1. *Comparison of different least-squares methods.*

Problem	Method	Flop-count	Numerical Properties
Overdetermined Full-Rank	Normal Equations	$mn^2 + \dfrac{n^3}{3}$	(1) Difficulties with formation of $A^T A$ (2) Produces more errors in the solution than what is warranted by data, in certain cases
Overdetermined Full-Rank	Householder-QR	$2mn^2 - 2\dfrac{n^3}{3}$	**Stable:** The computed solution is the **exact** solution of a nearby problem
Overdetermined Full-Rank	MGS-QR	$2mn^2$	Almost as stable as Householder-QR
Overdetermined Rank-Deficient	Householder-QR with Column Pivoting	$2mr - r^2(m+n) + \dfrac{2r^3}{3}$. where $r = \text{rank}(A)$	**Mildly stable:** The computed minimum-norm solution is **close** to the minimum-norm solution of a perturbed problem
Underdetermined Full-Rank	Normal Equations	$m^2 n + \dfrac{m^3}{3}$	Same difficulties as in the case of the overdetermined problem
Undetermined Full-Rank	Householder-QR	$2m^2 n - 2\dfrac{m^3}{3}$	Same as the rank-deficient overdetermined problem
Overdetermined Full-Rank	SVD	$4mn^2 + 8n^3$	**Stable**

8.9 Least-Squares Iterative Refinement

It is natural to wonder if a computed least-squares solution x can be improved cheaply in an iterative manner, as was done in the case of a linear system. A natural analogue of the iterative

8.9. Least-Squares Iterative Refinement

refinement procedure for the linear system problem described in a section of Chapter 6 can be easily worked out. This is left as an exercise for the readers (Exercise 8.22).

An analysis by Golub and Wilkinson (1966) reveals that the method is satisfactory only when the residual vector $r = b - Ax$ is sufficiently small. A successful procedure used widely in practice now follows. The method is based upon an interesting observation made by Golub that the least-squares solution x and the corresponding residual vector r satisfy the linear system

$$\begin{pmatrix} I_m & A \\ A^T & 0 \end{pmatrix} \begin{pmatrix} r \\ x \end{pmatrix} = \begin{pmatrix} b \\ 0 \end{pmatrix}, \quad A \in \mathbb{R}^{m \times n}, \ b \in \mathbb{R}^m.$$

ALGORITHM 8.7. Iterative Refinement for Least-Squares Solutions.

Inputs: An $m \times n$ matrix A of full rank, and an n-vector b.
Output: A refined least-squares solution and residual.

Step 1. Set $r^{(0)} = 0$, $x^{(0)} = 0$.
Step 2. For $k = 1, 2, \ldots$ do
  2.1. Compute $\begin{pmatrix} r_1^{(k)} \\ r_2^{(k)} \end{pmatrix} = \begin{pmatrix} b \\ 0 \end{pmatrix} - \begin{pmatrix} I & A \\ A^T & 0 \end{pmatrix} \begin{pmatrix} r^{(k)} \\ x^{(k)} \end{pmatrix}.$
  2.2. Solve the system $\begin{pmatrix} I & A \\ A^T & 0 \end{pmatrix} \begin{pmatrix} c_1^{(k)} \\ c_2^{(k)} \end{pmatrix} = \begin{pmatrix} r_1^{(k)} \\ r_2^{(k)} \end{pmatrix}.$
  2.3. Update the solution and the residual:
$$\begin{pmatrix} r^{(k+1)} \\ x^{(k+1)} \end{pmatrix} = \begin{pmatrix} r^{(k)} \\ x^{(k)} \end{pmatrix} + \begin{pmatrix} c_1^{(k)} \\ c_2^{(k)} \end{pmatrix}.$$
End

Implementation of Step 2. Since the matrix $\begin{pmatrix} I & A \\ A^T & 0 \end{pmatrix}$ is of order $m + n$, the above scheme would be quite expensive when m is large. However, using QR decomposition $Q^T A = \begin{pmatrix} R_1 \\ 0 \end{pmatrix}$, the system

$$\begin{pmatrix} I & A \\ A^T & 0 \end{pmatrix} \begin{pmatrix} c_1 \\ c_2 \end{pmatrix} = \begin{pmatrix} r_1 \\ r_2 \end{pmatrix}$$

can be transformed into

$$Q^T c_1 + \begin{pmatrix} R_1 \\ 0 \end{pmatrix} c_2 = Q^T r_1, \quad (R_1^T, 0) Q^T c_1 = r_2.$$

This shows that the above augmented system can be solved by solving two triangular systems and two matrix-vector multiplications as follows:

1. Form $Q^T r_1 = \begin{pmatrix} r_1' \\ r_2' \end{pmatrix}$.

2. Solve for c_2': $R_1^T c_2' = r_2'$.

3. Solve for c_2: $R_1 c_2 = r_1' - c_2'$.

4. Form $c_1 = Q \begin{pmatrix} c_2' \\ r_2' \end{pmatrix}$.

Flop-count. With the above formulation each iteration will require only $8mn - 2n^2$ flops, assuming that the Householder method has been used and that Q has not been formed explicitly. Note that for the matrix-vector multiplications in steps 1 and 4, Q *does not need to be formed explicitly;* these products can be obtained if Q is known only in implicit form, for example, as the product of Householder matrices.

Round-off error. It can be shown (Björck (1996)) that, using extended precision in computing step 2.1 of Algorithm 8.7, the initial rate of improvement of the solution is linear with rate
$$\frac{\|x^{(s)} - x\|_2}{\|x^{(s-1)} - x\|_2} < c\mu \; \text{Cond}(A), \quad s = 2, 3, \ldots,$$
and c is an error constant, depending upon m and n.

An interpretation of the result and remarks. The above result tells us that *the iterative refinement procedure is quite satisfactory. It is even more satisfactory for least-squares problems with large residuals.* Note that for these problems, $(\text{Cond}(A))^2$ serves as the condition number. However, the above result shows that the error at an iterative refinement step depends upon the condition number of A. The procedure "may give solutions to full single precision accuracy even when the initial solution may have no correct significant figures" (Björck (1996, p. 123)). For a well-conditioned matrix, the convergence may occur even in one iteration. Björck and Golub (1967) have shown that with an 8×8 ill-conditioned Hilbert matrix, three digits of accuracy per step both for the solution and the residual can be obtained.

Example 8.30.
$$A = \begin{pmatrix} 1 & 2 \\ 2 & 3 \\ 3 & 4 \end{pmatrix}, \qquad b = \begin{pmatrix} 3 \\ 5 \\ 9 \end{pmatrix},$$
$$r^{(0)} = 0, \qquad x^{(0)} = 0, \qquad k = 0.$$

Step 1.
$$\begin{pmatrix} r_1^{(0)} \\ r_2^{(0)} \end{pmatrix} = \begin{pmatrix} b \\ 0 \end{pmatrix} = \begin{pmatrix} 3 \\ 5 \\ 9 \\ 0 \\ 0 \end{pmatrix}.$$

Step 2. Solve the system
$$\begin{pmatrix} I & A \\ A^T & 0 \end{pmatrix} \begin{pmatrix} c_1^{(0)} \\ c_2^{(0)} \end{pmatrix} = \begin{pmatrix} r_1^{(0)} \\ r_2^{(0)} \end{pmatrix},$$
$$\begin{pmatrix} c_1^{(0)} \\ c_2^{(0)} \end{pmatrix} = \begin{pmatrix} 0.3333 \\ -0.6667 \\ 0.3333 \\ 3.3333 \\ -0.3333 \end{pmatrix}.$$

8.10. Review and Summary

Step 3. Update the solution and the residual:

$$\begin{pmatrix} r^{(1)} \\ x^{(1)} \end{pmatrix} = \begin{pmatrix} r^{(0)} \\ x^{(0)} \end{pmatrix} + \begin{pmatrix} c_1^{(0)} \\ c_2^{(0)} \end{pmatrix} = \begin{pmatrix} 0.3333 \\ -0.6667 \\ 0.3333 \\ 3.3333 \\ -0.3333 \end{pmatrix}.$$

The computations of $c_1^{(0)}$ and $c_2^{(0)}$ are shown below.

$$r_1^{(0)} = b, \quad r_2^{(0)} = \begin{pmatrix} 0 \\ 0 \end{pmatrix}.$$

$$\begin{pmatrix} r'_1 \\ r'_2 \end{pmatrix} = Q^T b = \begin{pmatrix} -10.6904 \\ -0.2182 \\ 0.8165 \end{pmatrix}. \text{ Thus, } r'_1 = \begin{pmatrix} -10.6904 \\ 0.2182 \end{pmatrix}, \quad r'_2 = (0.8165).$$

$$c'_2 = \begin{pmatrix} 0 \\ 0 \end{pmatrix}, \quad c_2^{(0)} = \begin{pmatrix} 3.3333 \\ -0.3333 \end{pmatrix}, \text{ and } c_1^{(0)} = \begin{pmatrix} 0.3333 \\ -0.6667 \\ 0.3333 \end{pmatrix}.$$

Note that $x^{(1)} = \begin{pmatrix} 3.3333 \\ -0.3333 \end{pmatrix}$ is the same least-squares solution as obtained by the QR and normal equation methods. ∎

MATCOM Note: Algorithm 8.7 has been implemented in the MATCOM program **LSITRN2**. The linear system analogue least-squares iterative refinement process has been implemented in the MATCOM program **LSITRN1**.

8.10 Review and Summary

8.10.1 Existence and Uniqueness

The least-squares solution x to the problem $Ax = b$ always exists. *In the overdetermined case, it is unique if A has full rank* (Theorem 8.2).

8.10.2 Overdetermined Problems

For *full-rank* problems, we have discussed the following methods:

- The normal equations (Algorithm 8.1)
- The QR method (Algorithms 8.2 and 8.3)
- The SVD method (Algorithm 8.4)

The normal equations method is easy to implement but has some numerical difficulties.
The QR approach can be implemented using *Householder, Givens, and the modified Gram–Schmidt methods*. The Householder QR method is the most efficient among all the QR methods, and if *the matrix A is well-conditioned, the method is recommended as a general-purpose least-squares solver.* For *nearly rank-deficient matrices, the QR method with column pivoting should be used.*

For rank-deficient problems, there are basically two choices:

- The QR method with pivoting (Exercise 8.16)
- The SVD method (Algorithm 8.5).

The SVD method, although more expensive than all other methods, is most reliable to deal with rank-deficiency or near rank-deficiency.

8.10.3 The Underdetermined Problem

The underdetermined lease-squares problem is discussed in Section 8.8. We have described two methods for the *minimum-norm solution to an underdetermined problem:* the QR algorithm (Algorithm 8.6) and the SVD method (Section 8.8.2).

8.10.4 Perturbation Analysis

The results of perturbation analyses vary for different cases of the perturbations in the data.

- If only *b is perturbed*, then $\text{Cond}(A) = \|A\| \, \|A^\dagger\|$ serves as the condition number for the unique least-squares solution (Theorem 8.10).

- If only *A is perturbed*, then the sensitivity of the unique least-squares solution, in general, depends upon the square of the condition number (Theorem 8.13). In certain cases, such as when the residual is zero, the sensitivity depends only on the condition number of A.

8.10.5 Iterative Refinement

As in the case of the linear system problem, it is possible to improve the accuracy of a computed least-squares solution in an iterative fashion. An algorithm which is a natural analogue to the one for the linear system (Section 6.9) is satisfactory only when the residual vector $r = b - Ax$ is sufficiently small. A widely used algorithm due to Björck is presented in Section 8.9. This algorithm requires the solution of an augmented system of order $m + n$ (where A is $m \times n$). It is shown how to solve the system in a rather inexpensive way using QR factorization of A.

The solution obtained by this iterative refinement algorithm is quite satisfactory.

8.10.6 Comparison of Least-Squares Methods

We summarize the speed, stability, and accuracy of the least-squares methods as follows:

- *The normal equations method:* Fastest but in certain cases might have numerical difficulties.

- *The QR approach:* More expensive than the normal equations method, but is stable and can be used as a general-purpose least-squares problem solver.

- *The SVD approach:* Most expensive but most reliable, especially while dealing with rank-deficient, nearly rank-deficient, and underdetermined problems.

8.11 Suggestions for Further Reading

Techniques of least-squares solutions are covered in any numerical linear algebra and some numerical analysis texts. The emphasis in most books is on the overdetermined problems. For a thorough treatment of the subject we refer the readers to Golub and Van Loan (1996) and Stewart (1973, 1998b). The book by Gill, Murray, and Wright (1991) also contains detailed discussions on perturbation analyses of the least-squares problems.

Two authoritative books completely devoted to the subject are Lawson and Hanson (1995) and Björck (1996). *These two books are must-reads for anyone interested in further study on the subject.* The book by Lawson and Hanson, in particular, gives the proofs of the round-off error analyses of the various algorithms described in the present book. See also Higham (2002) in this context.

Any book on regression analysis in statistics contains applications of least-squares problem in statistics. We have, in particular, used the book by Neter, Wasserman, and Kutner (1983). A classical survey paper of Golub (1969) contains an excellent exposition of numerical linear algebra techniques for least-squares problems and singular value decomposition problems arising in statistics and elsewhere. A paper by Stewart (1987) is also interesting to read.

These papers, along with other papers in the area by Golub, Björck, Stewart, etc., representing the most fundamental contributions in this area, are highly recommended. For details, see the list of references in Björck (1996) and Lawson and Hanson (1995) as well as the bibliography of this book. For more on underdetermined problems, see the papers by Cline and Plemmons (1976) and Arioli and Laratta (1985). For more on least-squares by MGS, see Plemmons (1974).

Exercises on Chapter 8

EXERCISES ON SECTIONS 8.2–8.6

8.1 (a) Prove that $A^T A$ is symmetric and positive definite if and only if A has full rank.

  (b) Show that the residual vector $r = b - Ax$ is orthogonal to all vectors in $R(A)$.

8.2 Prove that x is a least-squares solution to $Ax = b$ if and only if
$$Ax = b_R \quad \text{and} \quad b - Ax = b_N,$$
where b_R and b_N are, respectively, the range-space and column-space components of the vector b.

8.3 Let a vector x satisfy $Ax = P_A b$, where A is $m \times n$ ($m \geq n$) and has full rank. Then prove that x is a unique least-squares solution to $Ax = b$.

8.4 Using least-squares, fit a straight line and a quadratic to the data

x	0	1	3	5	7	9	12
y	10	12	18	15	20	25	36

Compare your results.

Compute the condition number of the associated Vandermonde matrix in each case.

8.5 Find the condition number of each of the following matrices using both generalized inverse and singular values and compare your results:

$$A = \begin{pmatrix} 0.0001 & 1 \\ 1 & 1 \end{pmatrix}, \quad A = \begin{pmatrix} 1 & 1 \\ 0.0001 & 0 \\ 0 & 0.0001 \end{pmatrix},$$

$$A = \begin{pmatrix} 7 & 6.990 \\ 4 & 4 \end{pmatrix}, \quad A = \begin{pmatrix} 1 & 2 & 3 \\ 3 & 4 & 5 \\ 0 & 7 & 8 \end{pmatrix}.$$

(Compute the generalized inverse from its definition given in Section 8.5.)

8.6 Let A and $\tilde{A}$ have full rank. Let x and $\tilde{x}$ be, respectively, the unique least-squares solutions to the problems $Ax = b$ and $\tilde{A}\tilde{x} = b$, where $\tilde{A} = A + E$. Then prove that

$$\frac{\|\tilde{x} - x\|}{\|\tilde{x}\|} \leq \text{Cond}(A)\frac{\|E\|}{\|A\|}\left(1 + \frac{\|b\|}{\|\tilde{x}\|}\right) + (\text{Cond}(A))^2\frac{\|E\|}{\|A\|}\left(1 + \frac{\|E\|}{\|A\|}\right).$$

(*Hint*: Apply the perturbation analysis of the linear systems with normal equations.)

8.7 Verify the inequality of Exercise 8.6 with the following data:

$$A = \begin{pmatrix} 1 & 2 \\ 3 & 4 \\ 5 & 6 \end{pmatrix}, \quad b = \begin{pmatrix} 3 \\ 7 \\ 11 \end{pmatrix}, \quad E = 10^{-4}A.$$

8.8 Verify the inequality of Theorem 8.16 in each of the following cases.

(a) $A = \begin{pmatrix} 1 & 2 \\ 3 & 4 \\ 5 & 6 \end{pmatrix}, E = 10^{-4}A.$ (b) $A = \begin{pmatrix} 1 & 1 \\ 10^{-4} & 0 \\ 0 & 10^{-4} \end{pmatrix}, E = 10^{-4}A.$

(c) $A = \begin{pmatrix} 1 & 1 \\ 0 & 1 \\ 0 & 0 \end{pmatrix}, E = 10^{-3}A.$

8.9 Work out a proof of Theorem 8.13 (least-squares left perturbation theorem).

8.10 Let

$$A = \begin{pmatrix} 1 & 1 \\ 2 & 3 \\ 0 & 1 \end{pmatrix}, \quad b = \begin{pmatrix} 0 \\ 5 \\ 1 \end{pmatrix}.$$

(a) Find the unique least-squares solution x using
  (i) $x = A^\dagger b$,
  (ii) the normal equations method,
  (iii) the Householder and the Givens QR factorization methods,
  (iv) the CGS and MGS methods.
(b) Find Cond(A).
(c) Show that for this problem the sensitivity of the least-squares problem, when only A is perturbed, depends upon Cond(A).

(d) Let $\Delta A = A = 10^{-4} A$ and let $\tilde{x} = \tilde{A}^\dagger b$, where $\tilde{A} = A + E$. Find $\frac{\|\tilde{x}-x\|}{\|x\|}$ and verify the inequality of Theorem 8.13 for this problem.

(e) Find r and $\hat{r}$ and verify the inequality of Theorem 8.16.

8.11 (a) If A is of order $m \times n$ and has full rank, then, using the pseudoinverse, prove that $\text{Cond}_2(A^T A) = \text{Cond}_2^2(A)$.

(b) Construct your own example where the sensitivity of the least-squares problem will depend upon the square of the condition number of the matrix. (Show all your work.)

EXERCISES ON SECTION 8.7

8.12 Consider the following well-known ill-conditioned matrix (Björck (1996)):

$$A = \begin{pmatrix} 1 & 1 & 1 \\ \epsilon & 0 & 0 \\ 0 & \epsilon & 0 \\ 0 & 0 & \epsilon \end{pmatrix}, \quad |\epsilon| \ll 1.$$

(a) Choose an ϵ small, so that $\text{rank}(A) = 3$. Then compute $\text{Cond}_2(A)$ to check that A is ill-conditioned.

(b) Find the least-squares solution to

$$Ax = \begin{pmatrix} 3 \\ \epsilon \\ \epsilon \\ \epsilon \end{pmatrix}$$

using
  (i) the normal equations method,
  (ii) the Householder, CGS, and MGS QR factorization methods.

(c) Change b to

$$b' = \begin{pmatrix} 3 \\ 0 \\ 0 \\ 0 \end{pmatrix}.$$

Keep A unchanged. Find an upper bound for the relative change in the least-squares solution.

(d) Change A to $A' = A + \Delta A$, where $\Delta A = 10^{-3} A$. Keep b unchanged. Find an upper bound for the relative change in the least-squares solution.

(e) Find the maximum departure from orthogonality of the computed columns of the Q matrix using the CGS and MGS methods.

(f) Compute the least-squares solution of the problem in (b) using the SVD.

8.13 (*Square-root-free Cholesky.*) Given a symmetric positive definite matrix A, develop an algorithm for finding the Cholesky decomposition of A without any square roots:

$$A = LDL^T,$$

where L is a unit lower triangular matrix and D is a diagonal matrix with positive diagonal entries.

Apply your algorithm to solve the full-rank least-squares problem based on solving normal equations

8.14 (a) Construct an example to show that the MGS will not yield an accurate result if the matrix Q obtained from the algorithm is explicitly used to solve a full-rank overdetermined least-squares problem.

(b) Do the example now with Algorithm 8.3 and compare the results.

8.15 Show that the flop-count for solving the least-squares problem for an $m \times n$ overdetermined system using the Householder QR method requires about $2n^2 m - 2\frac{n^3}{3}$ flops, and the corresponding count for the underdetermined system is $2m^2 n - 2\frac{m^3}{3}$.

8.16 (a) (*Least-squares solution using QR with column pivoting.*) Consider the QR factorization with column pivoting of $A \in \mathbb{R}^{m \times n}$, $m \geq n$, with rank $r < n$:

$$AP = QR, \quad \text{where } R = \begin{pmatrix} R_{11} & R_{12} \\ 0 & 0 \end{pmatrix} \begin{matrix} r \\ m-r \end{matrix}$$
$$\quad\quad\quad\quad\quad\quad\quad\quad\quad\quad r \quad n-r$$

Develop an expression for least-squares solutions to $Ax = b$ based on this factorization. Give a flop-count for this computation.

(b) Show that in the rank-deficient case, a least-squares solution cannot be a minimum-norm solution unless R_{12} is zero.

(c) Using the MATLAB function $[Q, R, P] = QR(A)$ for QR factorization with column pivoting, find the minimum-norm solution to $Ax = b$, where

$$A = \begin{pmatrix} 1 & 0 \\ 0 & 0 \\ 0 & 0 \end{pmatrix}, \quad b = \begin{pmatrix} 1 \\ 0 \\ 0 \end{pmatrix}.$$

(d) Work out the above example using the SVD of A.

8.17 Prove that the relative error obtained by the normal equations method is proportional to the square of the condition number of the matrix.

8.18 Consider the *complete orthogonal decomposition* of A: $A = Q \begin{pmatrix} R & 0 \\ 0 & 0 \end{pmatrix} V^T$.

(a) Show that $A^\dagger = V \begin{pmatrix} R^{-1} & 0 \\ 0 & 0 \end{pmatrix} Q^T$, by verifying the properties of the Pseudoinverse $A^\dagger$ given in Section 10.3.4 (page 368).

(b) Obtain an expression for the minimum-norm solution to $Ax = b$.

(c) Find the minimum-norm solution to the least-squares problem with

$$A = \begin{pmatrix} 1 & 1 & 1 \\ 1 & 1 & 1 \\ 1 & 1 & 1 \end{pmatrix}, \quad b = \begin{pmatrix} 3 \\ 3 \\ 3 \end{pmatrix}$$

using the results of (a).

EXERCISES ON SECTIONS 8.8 AND 8.9

8.19 Develop an algorithm based on QR factorization with MGS to compute the minimum-norm solution to the underdetermined system $Ax = b$, where A is $m \times n$, $m < n$. Give a flop-count for the algorithm. Apply your algorithm to

$$\begin{pmatrix} 1 & 2 & 3 \\ 4 & 5 & 6 \end{pmatrix} \begin{pmatrix} x_1 \\ x_2 \\ x_3 \end{pmatrix} = \begin{pmatrix} 6 \\ 15 \end{pmatrix}.$$

8.20 Develop an algorithm based on QR factorization with column pivoting to find a solution to the underdetermined problem.

8.21 Prove that the minimum-norm solution to an underdetermined system can be obtained by projecting any solution to the system onto $R(A^T)$. That is, if $P_A^\dagger$ is the orthogonal projection onto $R(A^T)$, then the minimum-norm solution x is given by $x = P_A^\dagger y$, where y is any solution. Using the above formula, compute the minimum-norm solution to the system

$$\begin{pmatrix} 1 & 10^{-4} & 0 & 0 \\ 1 & 0 & 10^{-4} & 0 \\ 1 & 0 & 0 & 10^{-4} \end{pmatrix} \begin{pmatrix} x_1 \\ x_2 \\ x_3 \\ x_4 \end{pmatrix} = \begin{pmatrix} 1 \\ 1 \\ 1 \end{pmatrix}.$$

8.22 Consider the *natural* algorithm for iterative refinement to improve a computed least-squares solution with $x^{(1)} = (0, \ldots, 0)^T$.

For $k = 1, 2, \ldots$ do

Step 1. $r^{(k)} = b - Ax^{(k)}$.

Step 2. Solve the least-squares problem: Find $c^{(k)}$ such that $\|Ac^{(k)} - r^{(k)}\|_2$ is minimum.

Step 3. Correct the solution
$$x^{(k+1)} = x^{(k)} + c^{(k)}.$$

End

   (a) Apply three iterations of this algorithm to each of the following problems:

      (i) $A = \begin{pmatrix} 1 \\ 2 \\ 3 \end{pmatrix}$, $b = \begin{pmatrix} 3 \\ 5 \\ 9 \end{pmatrix}$.

      (ii) $A = \begin{pmatrix} 1 & 1 \\ 10^{-4} & 0 \\ 0 & 10^{-4} \end{pmatrix}$, $b = \begin{pmatrix} 2 \\ 10^{-4} \\ 10^{-4} \end{pmatrix}$.

      (iii) $A = \begin{pmatrix} 1 & 1 \\ 10^{-4} & 0 \\ 0 & 10^{-4} \end{pmatrix}$, $b = \begin{pmatrix} 1 \\ 1 \\ 1 \end{pmatrix}$.

   (b) What is the relationship of this algorithm with the iterative algorithm using the augmented system (Algorithm 8.7)?

8.23 Apply Algorithm 8.7 to each of the problems of Exercise 8.22 and compare the results.

8.24 Apply Algorithm 8.7 to the least-squares problem with the 7×7 *Hilbert matrix* and $b = \left(1, \frac{1}{2}, \frac{1}{3}, \frac{1}{4}, \frac{1}{5}, \frac{1}{6}, \frac{1}{7}\right)^T$.

Tell how many digits of accuracy per iteration step were obtained in both the solution and the residual.

8.25 In many applications, only the diagonal entries of the variance-covariance matrix $X = (A^T A)^{-1}$ are needed. Show how these diagonal entries can be computed from $A = QR$ using only $\frac{2}{3}n^3$ flops. Illustrate the computation of X with a numerical example of order 5×2.

8.26 Develop Algorithm 8.7 in detail by incorporating the implementation of Step 2 as shown in the text. Apply this algorithm now to each of the problems of Exercise 8.22.

MATLAB Programs and Problems on Chapter 8

M8.1 Consider the following set of data points:

x	0	1	2	3	4	5	6	7	8	9
y	1	2.9	6.8	12	20.5	30.9	42.9	51.5	73	90.5

Using the MATLAB command **vander** and the operation "\" compute the least-squares fit of the data to polynomials of degrees 1 through 4.

Plot the original data point and the least-squares fits using the MATLAB commands **plot** and **polyval** and compare the results.

M8.2 (*Study of sensitivities of the least-squares problem.*) Let

$$A = \begin{pmatrix} 1 & 1 \\ 10^{-3} & 0 \\ 0 & 10^{-3} \\ 0 & 0 \end{pmatrix}, \quad b = \begin{pmatrix} 1 \\ 1 \\ 1 \\ 1 \end{pmatrix},$$

$$E_A = 10^{-4} \begin{pmatrix} 0 & -0.0001 \\ 0 & 0.0009 \\ 0 & 0.0003 \\ 0 & 0.0001 \end{pmatrix}, \quad \delta b = a0^{-4} \begin{pmatrix} 0.0001 \\ 0.0001 \\ 0.0001 \\ 0.0001 \end{pmatrix}.$$

Using the MATLAB commands **pinv, cond, norm, orth, null**, etc., verify the inequalities of Theorems 8.10, 8.13, 8.16, and 8.18 on different sensitivities of least-squares problems.

Test Data for Problems M8.3, M8.4, and M8.5:

1. A randomly generated matrix of order 100.

Exercises on Chapter 8

2. Hilbert matrix of order 20.

3. $\begin{pmatrix} 1 & 1 & 1 \\ \epsilon & 0 & 0 \\ 0 & \epsilon & 0 \\ 0 & 0 & \epsilon \end{pmatrix}$; ϵ is such that $\text{fl}(1 + \epsilon^2) = 1$.

For each of these matrices, *generate b so that the least-squares solution x in each case has all entries equal to* 1.

M8.3 (*Implementation of the least-squares QR algorithm using Givens rotations.*) Using **givqr** and **bcksub**, from MATCOM, write a MATLAB program to implement the QR algorithm using Givens rotations for the full-rank overdetermined least-squares problem $[\hat{x}]$ = **lsfrqrg** (A,b).

M8.4 (*Implementation of the SVD algorithm for full-rank overdetermined least-squares problems.*) Write a MATLAB program, called **lsfrsvd**, to implement Algorithm 8.4 using reduced SVD as follows:

$$[\hat{x}] = \textbf{lsfrsvd}\ (A, b).$$

M8.5 (*The purpose of this exercise is to compare the accuracy and residuals of different least-squares methods for full-rank overdetermined problems.*)

(a) Compute the least-squares solution $\hat{x}$ for each data set using the following:

  (i) $[\hat{x}]$ = **lsfrmgs** (A,b) (least-squares using MGS).
  (ii) $[\hat{x}]$ = **lsfrqrh** (A,b) (least-squares using Householder QR).
  (iii) $[\hat{x}]$ = **lsfrqrg** (A,b) (least-squares using Givens QR).
  (iv) $[\hat{x}]$ = **lsfrnme** (A,b) (least-squares using normal equations).
  (v) $[\hat{x}]$ = **pinv** (A) * b (least-squares using generalized inverse).
  (vi) $[\hat{x}]$ = **lsfrsvd** (A) (least-squares using SVD).

  Note: lsfrmgs, lsfrqrh, and **lsfrnme** are all available in MATCOM. **pinv** is a MATLAB command for computing the generalized inverse of a matrix.

(b) Using the results of (a), make one table for each data set in the following format shown in Table 8.2. Note also that the vector *x* has all entries equal to 1. Write your observations.

M8.6 Using **housqr** from MATCOM or $[Q, R] = qr(A)$ from MATLAB, and **backsub** from MATCOM, write a MATLAB program, called **lsrdqrh**(A,b), to compute the *minimum-norm least-squares solution $\hat{x}$ to the rank-deficient overdetermined problem* $Ax = b$, and the corresponding residual $\hat{r}$, using Householder QR factorization of A:

$$[\hat{x}, \hat{r}] = \textbf{lsrdqrh}\ (A, b).$$

Test data: A 20×2 matrix with all entries equal to 1, and *b* a vector with all entries equal to 2.

Table 8.2. *Comparison of different methods for the full-rank overdetermined least-squares problem.*

Method	$\|x - \hat{x}\|_2 / \|x\|_2$	$\|A\hat{x} - b\|_2$
lsfrmgs		
lsfrqrh		
lsfrqrg		
lsfrnme		
generalized-inverse		
lsfrsvd		

M8.7 Using the MATLAB function [U,S,V] = **svd** (A) write a MATLAB program, called **lsrdsvd**, to compute the *minimum-norm least-squares solution* $\hat{x}$ to the rank-deficient overdetermined system $Ax = b$:

$$[\hat{x}] = \text{lsrdsvd } (A, b).$$

Use the same test data as in Problem M8.6 and compare the results with those of **lsrdqrh**.

M8.8 Run the programs **mnudnme** (*least-squares solution for the underdetermined full-rank problem using normal equations*) and **mnudqrh** (*least-squares solution for the underdetermined full-rank problem using Householder QR factorization*) from MATCOM on the following sets of data to compute the minimum-norm solution $\hat{x}$ to the full-rank underdetermined problem $Ax = b$, and compare the results.

$$A = \begin{pmatrix} 1 & 2 & 3 & 4 & 5 & 6 \end{pmatrix}, \quad A = \begin{pmatrix} 1 & 1 & 1 & 1 & 1 & 1 & 1 & 1 \\ 0 & 1 & 2 & 3 & 4 & 5 & 6 & 7 \end{pmatrix},$$

$$A = \begin{pmatrix} 10 & 10 & 10 & 10 & 10 & 10 & 10 \\ 0 & 1 & 0 & 0 & 0 & 0 & 0 \\ 0 & 1 & 0 & 0 & 0 & 0 & 0 \end{pmatrix}.$$

Construct b for each A so that the minimum-norm solution x has all its entries equal to 1.

M8.9 Run the programs **lsitrn2** (based on Algorithm 8.7) from MATCOM on the 20×20 Hilbert matrix A and construct b randomly. How do these results compare with those obtained by the algorithm developed in Exercise 8.22?

M8.10 (a) Compute $(A^T A)^{-1}$ for each of the following matrices A:

(i) Compute explicitly $(A^T A)^{-1}$ using MATLAB command **inv**.

(ii) Run the program **varcovar** from MATCOM.

(b) Compare the results of (i) and (ii).

Test data:

- $A = $ The 20×20 Hilbert matrix.
- $A = \begin{pmatrix} 1 & 1 & 1 \\ 10^{-3} & 0 & 0 \\ 0 & 10^{-3} & 0 \\ 0 & 0 & 10^{-3} \end{pmatrix}.$
- $A = $ A 30×30 randomly generated matrix.

Chapter 9

Numerical Matrix Eigenvalue Problems

Background Material Needed

- Norm properties of matrices (Section 2.5)
- The QR factorization of an arbitrary and a Hessenberg matrix using Householder and Givens transformations (Algorithms 7.2 and 7.6)
- Linear system solutions with arbitrary, Hessenberg, and triangular matrices (Sections 6.4 and 6.12)
- The condition number and its properties (Sections 4.6 and 4.7)

9.1 Introduction

This chapter is devoted to the study of the numerical matrix eigenvalue problem. The problem is a very important practical problem and arises in a variety of application areas, including engineering, physics, chemistry, statistics, and economics.

Since the eigenvalues of a matrix A are the zeros of the characteristic polynomial $\det(A - \lambda I)$, one would naively think of computing the eigenvalues of A by finding its characteristic polynomial and then computing its zeros by a standard root-finding method. Unfortunately, *eigenvalue computation via the characteristic polynomial is not a practical approach.*

*A standard practical algorithm for finding the eigenvalues of a matrix is the **QR iteration method** with a single or double shift.* Several applications do not need knowledge of the whole spectrum. A few selected eigenvalues, usually a few largest or smallest ones, suffice. A classical method, based on implicit powering of A, known as the **power method** is useful for this purpose.

The organization of this chapter is as follows.

Section 9.2 is devoted to the discussions of how the eigenvalue problem arises in some practical applications such as *stability analyses* of a system of differential and difference equations, *vibration analysis, transient behavior of an electrical circuit, the buckling problem, and principal component analysis in statistics.*

In Section 9.3 some classical results on eigenvalues locations such as Geršgorin's disk theorem are stated and proved.

Section 9.4 describes the power method, the inverse power method, the Rayleigh Quotient iteration, etc., for finding a selected number of eigenvalues and the corresponding eigenvectors.

In Section 9.5, we describe two powerful methods, the Householder and Givens methods, for transforming an arbitrary matrix to a Hessenberg matrix by orthogonal similarity. Numerical difficulties with diagonal similarity transformation and the *difficulties of computing the eigenvalues of a matrix via the characteristic polynomial and the Jordan canonical form* are highlighted.

Eigenvalue and eigenvector sensitivity are discussed in Sections 9.6 and 9.7. The most important result in this section is the Bauer–Fike theorem (Theorem 9.37).

Section 9.8 *is the most important section of this chapter.* The QR iteration method with and without shifts and their implementations are described in this section.

The Hessenberg-inverse iteration is described in Section 9.9.

9.2 Eigenvalue Problems Arising in Practical Applications

The problem of finding eigenvalues and eigenvectors arises in a wide variety of practical applications in science and engineering. The words "eigenvalue" and "eigenvector" are derived from the German word "*eigenwerte*." As we have seen before, the mathematical models of many engineering problems are systems of differential and difference equations, and the solutions of these equations are often expressed in terms of the eigenvalues and eigenvectors of the matrices of these systems. Furthermore, many important characteristics of physical and engineering systems, such as *stability*, often can be determined only by knowing the nature and location of the eigenvalues. We will give a few representative examples in this section.

9.2.1 Stability Problems for Differential and Difference Equations

A homogeneous linear system of differential equations with constant coefficients of the form

$$\dot{x}(t) = Ax(t), \quad x(0) = x_0, \tag{9.1}$$

where

$$A = (a_{ij})_{n \times n} \quad \text{and} \quad \dot{x}(t) = \frac{d}{dt}\begin{pmatrix} x_1(t) \\ \vdots \\ x_n(t) \end{pmatrix},$$

arises in a wide variety of physical and engineering systems. As can be seen from the proof of Theorem 9.2 below, the solution of this system is intimately related to the eigenvalue problem for matrix A. Many interesting and desirable properties of physical and engineering systems can be studied just by knowing the location or the nature of the eigenvalues of matrix A. Stability is one such property.

Definition 9.1. *An equilibrium solution of (9.1) is the vector x_e such that $Ax_e = 0$. An equilibrium solution x_e is asymptotically stable if there exists a $\delta > 0$ such that $\|x(t) - $*

9.2. Eigenvalue Problems Arising in Practical Applications

$x_e \| \to 0$ as $t \to \infty$, whenever $\|x_0 - x_e\| \le \delta$. System (9.1) *is asymptotically stable if the equilibrium solution* $x_e = 0$ *is asymptotically stable. A system that is not asymptotically stable will be called unstable.*

Asymptotic stability guarantees that if the system is perturbed from the position of equilibrium a little bit, then it will eventually return to that position after making small oscillations.

Mathematical Criteria for Asymptotic Stability

Theorem 9.2 (stability theorem for a homogeneous system of differential equations). *A necessary and sufficient condition for system* (9.1) *to be asymptotically stable is that the eigenvalues of matrix A all have negative real parts. It is unstable if at least one eigenvalue has a positive real part.*

Proof. We will sketch the proof in the case when A is diagonalizable, that is, in the case where there exists a nonsingular matrix X such that $X^{-1}AX = D = \text{diag}(\lambda_1, \ldots, \lambda_n)$. In this case, $e^{At} = Xe^{Dt}X^{-1} = X\,\text{diag}\,(e^{\lambda_1 t}, \ldots, e^{\lambda_n t})X^{-1}$.

Again if $\lambda_j = \alpha_j + i\beta_j$, $j = 1, 2, \ldots, n$, then $e^{\lambda_j t} = e^{\alpha_j t}e^{i\beta_j t}$, and $e^{\lambda_j t} \to 0$, when $t \to \infty$, if and only if $\alpha_j < 0$. □

Remark. The proof of Theorem 9.2 in the general case is obtained by using the *Jordan canonical form* of A (Theorem 9.28).

Stability of a Nonhomogeneous System

Many practical situations give rise to mathematical models of the form

$$\dot{x}(t) = Ax(t) + b, \tag{9.2}$$

where b is a constant vector. The stability of such a system is also governed by the eigenvalues of A. This can be seen as follows.

Let $\bar{x}(t)$ be an equilibrium solution of (9.2). Define $z(t) = x(t) - \bar{x}(t)$. Then

$$\dot{z}(t) = \dot{x}(t) - (\dot{\bar{x}})(t) = Ax(t) + b - A\bar{x}(t) - b = A(x(t) - \bar{x}(t)) = Az(t).$$

Thus, $x(t) \to \bar{x}(t)$ if and only if $z(t) \to 0$. The following theorem therefore follows from Theorem 9.2.

Theorem 9.3 (stability theorem for a nonhomogeneous system of differential equations). (i) *An equilibrium solution of* (9.2) *is asymptotically stable if and only if all the eigenvalues of A have negative real parts.* (ii) *An equilibrium solution is unstable if at least one eigenvalue has a positive real part.*

Stability of a System of Difference Equations

Like the system of differential equations (9.2), there are practical systems which are modeled by systems of difference equations of the form $x_{k+1} = Ax_k + b$.

A well-known mathematical criterion for the asymptotic stability of such a system is given in the following theorem. We leave the proof to the reader.

Theorem 9.4 (stability theorem for a nonhomogeneous system of difference equations). *The system*
$$x_{k+1} = Ax_k + b$$
is asymptotically stable if and only if all the eigenvalues of A are inside the unit circle. It is unstable if at least one eigenvalue has a magnitude greater than 1.

Summarizing, to determine the stability and asymptotic stability of a system modeled by a system of first order ordinary differential or difference equations, all we need to know is if the eigenvalues of A are in the left half plane or inside the unit circle, respectively. The explicit knowledge of the eigenvalues is *not* needed.

Example 9.5. A European arms race. Consider the arms race of 1909–1914 between two European alliances.

Alliance 1: France and Russia. Alliance 2: Germany and Austria–Hungary.

The two alliances went to war against each other. Let's try to explain this historical fact through the notion of stability.

First consider the following crude (but simple) mathematical model of war between two countries:

$$\frac{dx_1}{dt} = k_1 x_2 - \alpha_1 x_1 + g_1, \qquad \frac{dx_2}{dt} = k_2 x_1 - \alpha_2 x_2 + g_2,$$

where

$$x_i(t) = \text{war potential of the country } i, \ i = 1, 2$$
$$g_i(t) = \text{the grievances that country } i \text{ has against the other}, \ i = 1, 2.$$

The quantities g_i, α_i, and k_i, $i = 1, 2$, are all positive constants. $\alpha_i x_i$ denotes the cost of armaments of the country i. This mathematical model is due to L. F. Richardson and is known as the **Richardson model**.

Note that this simple model is realistic in the sense that the rate of change of the war potential of one country depends upon the war potential of the other country, the grievances that one country has against its enemy country, and the cost of the armaments the country can afford. While the first two factors cause the rate to increase, the last factor certainly has a slowing effect (that is why we have a *negative sign* associated with that term).

In matrix form, this model can be written as $\dot{x}(t) = Ax(t) + g$, where

$$A = \begin{pmatrix} -\alpha_1 & k_1 \\ k_2 & -\alpha_2 \end{pmatrix}, \qquad x(t) = \begin{pmatrix} x_1(t) \\ x_2(t) \end{pmatrix}, \qquad g = \begin{pmatrix} g_1 \\ g_2 \end{pmatrix}.$$

The eigenvalues of A are

$$\lambda = \frac{-(\alpha_1 + \alpha_2) \pm \sqrt{(\alpha_1 - \alpha_2)^2 + 4k_1 k_2}}{2}.$$

Thus the equilibrium solution $x(t)$ is asymptotically stable if $\alpha_1 \alpha_2 - k_1 k_2 > 0$, and unstable if $\alpha_1 \alpha_2 - k_1 k_2 < 0$. This is because, when $\alpha_1 \alpha_2 - k_1 k_2 > 0$, both the eigenvalues will have negative real parts; if it is negative, then one eigenvalue will have positive real part.

9.2. Eigenvalue Problems Arising in Practical Applications

For the above European arms race, the estimates of α_1, α_2 and k_1, k_2 were made under some realistic assumptions: $\alpha_1 = \alpha_2 = 0.2$; $k_1 = k_2 = 0.9$. (For details of how these estimates were obtained, see Braun (1978).) The main assumptions are that *both the alliances have roughly the same strength, and α_1 and α_2 are the same as Great Britain, which is usually taken to be the reciprocal of the lifetime of the British Parliament (five years)*.

With these values of α_1, α_2 and k_1, k_2, we have

$$\alpha_1\alpha_2 - k_1k_2 = \alpha_1^2 - k_1^2 = -0.7700.$$

Thus the equilibrium is unstable. In fact, the two eigenvalues are 1.4000 and -2.2000. ■

For a general model of Richardson's theory of arms races and the role of eigenvalues there, see Luenberger (1979, pp. 209–214).

9.2.2 Phenomenon of Resonance

Vibrating structures such as buildings, bridges, and highways, sometimes experience a dangerous oscillation, called *resonance*, causing partial or complete destruction of the structures. Some classical and recent events that *possibly might have been caused by resonances* include[9]

- the fall of the Tacoma Narrows Bridge in the state of Washington in the United States;
- the fall of the Broughton Suspension Bridge in England;
- the wobbling of the Millennium Bridge over the River Thames in London.

A general model of vibrating structures is a system of second-order differential equations:

$$M\ddot{x}(t) + D\dot{x}(t) + Kx(t) = 0, \tag{9.3}$$

where M, K, and D are, respectively, known as the *mass*, *stiffness*, and *damping* matrices. Substituting $x(t) = ue^{\lambda t}$ leads to the "quadratic eigenvalue problem"

$$(\lambda^2 M + \lambda D + K)u(t) = 0 \text{ or } P(\lambda)u(t) = 0, \quad \text{where } P(\lambda) = \lambda^2 M + \lambda D + K.$$

In many practical instances, matrices M, K, and D are symmetric, and furthermore $M = M^T > 0$ and $K = K^T \geq 0$. Assuming M is nonsingular, the above quadratic eigenvalue problem reduces to the standard eigenvalue problem:

$$\begin{pmatrix} 0 & I \\ -M^{-1}K & -M^{-1}D \end{pmatrix} \begin{pmatrix} x \\ \lambda x \end{pmatrix} = \lambda \begin{pmatrix} x \\ \lambda x \end{pmatrix}$$

or $Az = \lambda z$, where $z = \begin{pmatrix} x \\ \lambda x \end{pmatrix}$.

---

[9]The commonly accepted explanation for the collapse of the Tacoma Narrows Bridge has been recently challenged by scientists who believed that there may be something more to it. See the papers by Lazer and McKenna (1990). For a complete story of the collapse of the Tacoma Bridge, see Braun (1978, pp. 167–169).

Thus, if each of the matrices M, K, and D is of order n, then there are $2n$ eigenvalues of A, and these $2n$ eigenvalues are the same as those of $P(\lambda)$ (see Section 11.8). These eigenvalues are related to the *natural frequencies of the structure* (see Section 11.8 for specifics).

The resonance occurs when a frequency of the external force becomes equal or very close to a natural frequency, as explained below (see also section 11.7.2).

As in the case of (9.1), the system of equations (9.3) can also be solved by knowledge of the eigenvalues and eigenvectors of the pencil $P(\lambda)$. Specifically, if the eigenvalues λ_k are all distinct and z_k are the corresponding eigenvectors, then we can write

$$x(t) = \sum_{k=1}^{2n} \alpha_k z_k e^{\lambda_k t},$$

where α_k are scalars.

Suppose now the system is excited by an external time-harmonic force of the form $f(t) = f_o e^{i\omega t}$, with the frequency ω. Then a particular solution in this case is given by

$$x_p(t) = e^{i\omega t} \sum_{j=1}^{2n} \frac{y_j f_o}{i\omega - \lambda_j} z_k,$$

where y_j are the left eigenvectors of $P(\lambda)$. *This shows that as $i\omega$ approaches a particular eigenvalue λ_j, the response of the system becomes unbounded and the system approaches a resonance condition (see more on this in* Chapter 11*).*

In each of the above cases, a periodic force of very large amplitude was generated, and the frequency of this force became equal or close to one of the natural frequencies. In the case of the Broughton Bridge, the large response was set up by soldiers marching in cadence over the bridge. In the case of the Tacoma Bridge, it was wind (see Figure 9.1). *Because of what happened with the Broughton Bridge, soldiers are no longer permitted to march in cadence over a bridge.* In the case of the Millennium Bridge, it was again the pedestrian-induced movements (see Figure 9.2). *This bridge was closed only two days after its opening, because on its opening day in June* 2000 *the bridge started to wobble due to the weight of several thousand people who came to see the bridge.* For more on this event, visit *www.arup.com/MillenniumBridge/Challenge/#*.

9.2.3 Buckling Problem (a Boundary Value Problem)

Consider a thin, uniform beam of length l. An axial load P is applied to the beam at one of the ends (see Figure 9.3).

We are interested in knowing how and when the beam buckles.

Let y denote the vertical displacement of a point of the beam which is at a distance x from the (deflection) left support. Suppose that both ends of the beam are simply supported, i.e., $y(0) = y(l) = 0$.

Using the relationship between the curvature $\frac{d^2y}{dx^2}$ and the internal moment M, we obtain the *bending moment equation* $EI\frac{d^2y}{dx^2} = -Py$, where E is the *modulus of elasticity* and I is the *area moment of inertia* of column cross section.

9.2. Eigenvalue Problems Arising in Practical Applications

Figure 9.1. *Fall of the Tacoma Bridge.*

Figure 9.2. *Millennium Bridge.*

Let the interval $[0, l]$ be partitioned into n subintervals of equal length h, with $x_0, x_1, \ldots, x_n$ as the points of division. That is, $0 = x_0 < x_1 < x_2 < \cdots < x_j < \cdots < x_{n-1} < x_n = l$.

Let
$$\left.\frac{d^2 y}{dx^2}\right|_{x=x_i} \approx \frac{y_{i+1} - 2y_i + y_{i-1}}{h^2}, \quad \text{where } h = \frac{1}{n}. \tag{9.4}$$

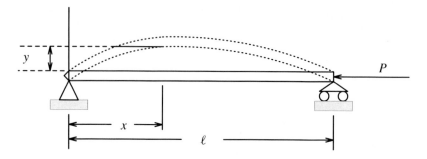

Figure 9.3. *Buckling of a beam.*

Substituting this approximation of $\frac{d^2y}{dx^2}$ into the bending moment equation above and taking into account the given boundary conditions, we obtain the following **symmetric tridiagonal matrix eigenvalue problem**:

$$\begin{pmatrix} 2 & -1 & 0 & \cdots & 0 \\ -1 & 2 & -1 & \cdots & 0 \\ \vdots & \ddots & \ddots & \ddots & \vdots \\ \vdots & \ddots & \ddots & \ddots & -1 \\ 0 & 0 & 0 & -1 & 2 \end{pmatrix} \begin{pmatrix} y_1 \\ y_2 \\ \vdots \\ \vdots \\ y_n \end{pmatrix} = \lambda \begin{pmatrix} y_1 \\ y_2 \\ \vdots \\ \vdots \\ y_n \end{pmatrix}, \qquad (9.5)$$

where $\lambda = \frac{Ph^2}{EI}$.

Each value of λ determines a load $P = \frac{\lambda EI}{h^2}$ which is called a **critical load**. These critical loads are the ones which are of practical interest, because they determine the possible onset of the buckling of the beam.

In particular, the smallest value of P is of primary importance, since the bending associated with larger values of P may not be obtained without failure occurring under the action of the lowest critical value of P.

9.2.4 Simulating Transient Current for an Electric Circuit

(See Chapra and Canale (2002).) Given an electric circuit consisting of four loops (see Figure 9.4), suppose we are interested in the transient behavior of the electric circuit. In particular, *we want to know the oscillation of each loop with respect to the other.*

Kirchhoff's voltage law applied to each loop gives the following.

Loop 1:

$$-L_1 \frac{di_1}{dt} - \frac{1}{C_1} \int_{-\infty}^{t} (i_1 - i_2) dt = 0. \qquad (9.6)$$

9.2. Eigenvalue Problems Arising in Practical Applications

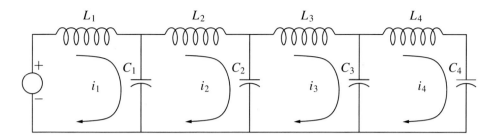

Figure 9.4. *Transient current for electric current.*

Loop 2:
$$-L_2 \frac{di_2}{dt} - \frac{1}{C_2} \int_{-\infty}^{t} (i_2 - i_3) dt + \frac{1}{C_1} \int_{-\infty}^{t} (i_1 - i_2) dt = 0. \tag{9.7}$$

Loop 3:
$$-L_3 \frac{di_3}{dt} - \frac{1}{C_3} \int_{-\infty}^{t} (i_3 - i_4) dt + \frac{1}{C_2} \int_{-\infty}^{t} (i_2 - i_3) dt = 0. \tag{9.8}$$

Loop 4:
$$-L_4 \frac{di_4}{dt} - \frac{1}{C_4} \int_{-\infty}^{t} i_4 dt + \frac{1}{C_3} \int_{-\infty}^{t} (i_3 - i_4) dt = 0. \tag{9.9}$$

The system of ordinary differential equations given above can be differentiated and rearranged to give

$$L_1 \frac{d^2 i_1}{dt^2} + \frac{1}{C_1}(i_1 - i_2) = 0, \tag{9.10}$$

$$L_2 \frac{d^2 i_2}{dt^2} + \frac{1}{C_2}(i_2 - i_3) - \frac{1}{C_1}(i_1 - i_2) = 0, \tag{9.11}$$

$$L_3 \frac{d^2 i_3}{dt^2} + \frac{1}{C_3}(i_3 - i_4) - \frac{1}{C_2}(i_2 - i_3) = 0, \tag{9.12}$$

$$L_4 \frac{d^2 i_4}{dt^2} + \frac{1}{C_4} i_4 - \frac{1}{C_3}(i_3 - i_4) = 0. \tag{9.13}$$

Assume
$$i_j = A_j \sin(\omega t), \quad j = 1, 2, 3, 4. \tag{9.14}$$

From (9.10), we have

$$-L_1 A_1 \omega^2 \sin \omega t + \frac{1}{C_1} A_1 \sin \omega t - \frac{1}{C_1} A_2 \sin \omega t = 0$$

or

$$\left(\frac{1}{C_1} - L_1 \omega^2\right) A_1 - \frac{1}{C_1} A_2 = 0. \tag{9.15}$$

Similarly, from (9.11), (9.14), we obtain, respectively, the following equations:

$$-\frac{1}{C_1}A_1 + \left(\frac{1}{C_1} + \frac{1}{C_2} - L_2\omega^2\right)A_2 - \frac{1}{C_2}A_3 = 0, \tag{9.16}$$

$$-\frac{1}{C_2}A_2 + \left(\frac{1}{C_2} + \frac{1}{C_3} - L_3\omega^2\right)A_3 - \frac{1}{C_3}A_4 = 0, \tag{9.17}$$

$$-\frac{1}{C_3}A_3 + \left(\frac{1}{C_3} + \frac{1}{C_4} - L_4\omega^2\right)A_4 = 0. \tag{9.18}$$

The above is an eigenvalue problem. To see it more clearly, consider the special case

$$C_1 = C_2 = C_3 = C_4 = C \quad \text{and} \quad L_1 = L_2 = L_3 = L_4 = L.$$

Assuming $\lambda = LC\omega^2$, and noting that $i_j = A_j \sin \omega t, j = 1, \ldots, 4$, we obtain the following eigenvalue problem:

$$\begin{pmatrix} 1 & -1 & 0 & 0 \\ -1 & 2 & -1 & 0 \\ 0 & -1 & 2 & -1 \\ 0 & 0 & -1 & 2 \end{pmatrix} \begin{pmatrix} i_1 \\ i_2 \\ i_2 \\ i_3 \end{pmatrix} = \lambda \begin{pmatrix} i_1 \\ i_2 \\ i_3 \\ i_4 \end{pmatrix}. \tag{9.19}$$

The solution of this eigenvalue problem will give us the natural frequencies ($w_i^2 = \lambda_i/LC$). Moreover the knowledge of the eigenvectors can be used to study the circuit's physical behavior such as the natural modes of oscillation.

These eigenvalues and the corresponding normalized eigenvectors (in four-digit arithmetic) are $\lambda_1 = 0.1206$, $\lambda_2 = 1$, $\lambda_3 = 2.3473$, $\lambda_4 = 3.5321$.

$$\begin{pmatrix} 0.6665 \\ 0.5774 \\ 0.4285 \\ 0.2280 \end{pmatrix}, \begin{pmatrix} 0.5774 \\ -0.0000 \\ -0.5774 \\ -0.5774 \end{pmatrix}, \begin{pmatrix} -0.4285 \\ 0.5774 \\ 0.2289 \\ -0.6565 \end{pmatrix}, \begin{pmatrix} -0.2280 \\ 0.5774 \\ -0.6565 \\ 0.4285 \end{pmatrix}.$$

From the directions of the eigenvectors we conclude that for λ_1 all the loops oscillate in the same direction. For λ_3 the second and third loops oscillate in the opposite directions from the first and fourth, and so on. This is shown in Figure 9.5.

9.2.5 An Example of the Eigenvalue Problem Arising in Statistics: Principal Component Analysis

Many practical-life applications involving statistical analysis (e.g., stock market or weather prediction) involve a huge amount of data. The volume and complexities of the data in these cases can make the computations required for analysis practically infeasible. In order to handle and analyze such a voluminous amount of data in practice, it is therefore necessary to reduce the data. The basic idea then will be to choose judiciously "k" components from a data set consisting of n measurements on p ($p > k$) original variables, in such a way that much of the information (if not most) in the original p variables is contained in the k chosen components. Such k components are called the first k **principal components** in statistics.

9.2. Eigenvalue Problems Arising in Practical Applications

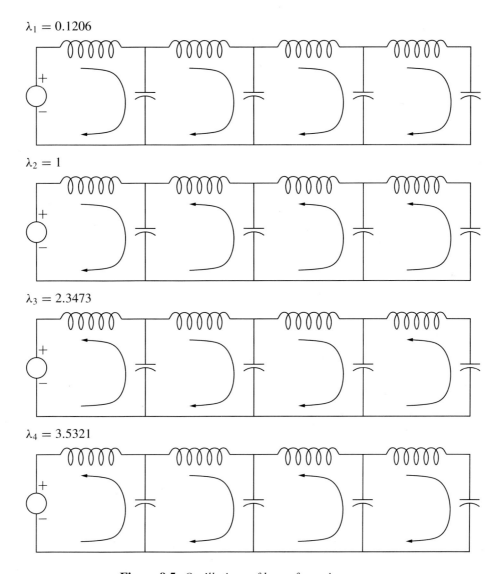

Figure 9.5. *Oscillations of loops from eigenvectors.*

The knowledge of eigenvalues and eigenvectors of the covariance matrix is needed to find these principal components.

Specifically, if Σ is the covariance matrix corresponding to the random vector $X = (X_1, X_2, \ldots, X_p)$, $\lambda_1 \geq \lambda_2 \geq \cdots \geq \lambda_p \geq 0$ are the eigenvalues, and x_1 through x_p are the corresponding eigenvectors of the matrix Σ, then the ith principal component is given by $Y_i = x_i^T X$, $i = 1, 2, \ldots, p$.

Furthermore, the proportion of total population variance due to the ith principal component is given by the ratio $\frac{\lambda_i}{\lambda_1+\lambda_2+\ldots+\lambda_p} = \frac{\lambda_i}{\text{trace}(\Sigma)}$, $i = 1, \ldots, p$.

Note: *The covariance matrix is symmetric positive semidefinite, and therefore its eigenvalues are all nonnegative.*

If the first k ratios constitute the most of the total population variance, then the first k principal components can be used in statistical analysis.

Note that in computing the kth ratio, we need to know only the kth eigenvalue of the covariance matrix; the entire spectrum does not need to be computed.

To end this section, we remark that many real-life practices, such as computing the index of the Dow Jones Industrial Average, can now be better understood and explained through principal component analysis. This is shown in the example below.

A Stock Market Example (Johnson and Wichern (1992))

Suppose that the covariance matrix for the weekly rates of return for stocks of five major companies (Allied Chemical, DuPont, Union Carbide, Exxon, and Texaco) in a given period of time is given by

$$R = \begin{pmatrix} 1.000 & 0.577 & 0.509 & 0.387 & 0.462 \\ 0.577 & 1.000 & 0.599 & 0.389 & 0.322 \\ 0.509 & 0.599 & 1.000 & 0.436 & 0.426 \\ 0.387 & 0.389 & 0.436 & 1.000 & 0.523 \\ 0.462 & 0.322 & 0.426 & 0.523 & 1.000 \end{pmatrix}.$$

The first two eigenvalues of R are $\lambda_1 = 2.857$, $\lambda_2 = 0.809$. The proportion of total population variance due to the first component is approximately $\frac{2.857}{5} = 57\%$. The proportion of total population variance due to the second component is $\frac{0.809}{5} = $ approximately 16%. Thus the first two principal components account for 73% of the total population variance. The eigenvectors corresponding to these principal components are

$$\begin{aligned} x_1^T &= (0.464, 0.457, 0.470, 0.421, 0.421), \\ x_2^T &= (0.240, 0.509, 0.260, -0.526, -0.582). \end{aligned} \quad (9.20)$$

These eigenvectors have interesting interpretations. From the expression of x_1 we see that the first component is a (roughly) equally weighted sum of the five stocks. This component is generally called the *market component*. However, the expression for x_2 tells us that the second component represents a contrast between the chemical stocks and the oil industry stocks. This component will be generally called an *industry component*. Thus, we conclude that about 57% of total variations in these stock returns is due to the market activity and 16% is due to industry activity.

The eigenvalue problem also arises in many other important statistical analysis, for example, in computing the *canonical correlations*. Interested readers are referred to the book by Johnson and Wichern (1992) for further reading.

A final comment: Most eigenvalue problems arising in statistics, such as in *principal component analysis and canonical correlations*, are actually *SVD problems* and should be handled computationally using *singular value decomposition* (see Chapters 7 and 10).

9.3 Localization of Eigenvalues

As we have just seen, in several practical applications explicit knowledge of eigenvalues is not required; all that is required is a knowledge of distribution of the eigenvalues in some

9.3. Localization of Eigenvalues

given regions of the complex plane or estimates of some specific eigenvalues. There are ways such information may be acquired without actually computing the eigenvalues of the matrix. We start with a well-known result of Geršgorin[10] (1931).

9.3.1 The Geršgorin Disk Theorems

Theorem 9.6 (Geršgorin's first theorem). *Let $A = (a_{ij})_{n \times n}$. Define*

$$r_i = \sum_{\substack{j=1 \\ i \neq j}}^{n} |a_{ij}|, \quad i = 1, \ldots, n.$$

Then each eigenvalue λ of A satisfies at least one of the following inequalities:

$$|\lambda - a_{ii}| \leq r_i, \quad i = 1, 2, \ldots, n.$$

In other words, all the eigenvalues of A can be found in the union of disks $\{z : |z - a_{ii}| \leq r_i, \ i = 1, \ldots, n\}$.

Proof. Let λ be an eigenvalue of A and x be an eigenvector associated with λ. Then from $Ax = \lambda x$, we have

$$(\lambda - a_{ii})x_i = \sum_{\substack{j=1 \\ i \neq j}}^{n} a_{ij}x_j, \quad i = 1, \ldots, n,$$

where x_i is the ith component of the vector x. Let x_k be the largest component of x (in absolute value). Then, since $|x_j|/|x_k| \leq 1$ for $j \neq k$, we have from above

$$|\lambda - a_{kk}| \leq \sum_{\substack{j=1 \\ j \neq k}}^{n} |a_{kj}| \frac{|x_j|}{|x_k|} \leq \sum_{\substack{j=1 \\ j \neq k}}^{n} |a_{kj}|.$$

Thus λ is contained in the disk $\{\lambda : |\lambda - a_{kk}| \leq r_k\}$. $\square$

Definition 9.7. *The disks $R_i = \{z : |z - a_{ii}| \leq r_i\}$, $i = 1, \ldots, n$, are called **Geršgorin disks** in the complex plane.*

Example 9.8.

$$A = \begin{pmatrix} 1 & 2 & 3 \\ 3 & 4 & 9 \\ 1 & 1 & 1 \end{pmatrix}.$$

---

[10]Semyon Aranovich Geršgorin (1901–1933) was born in Belarus and educated at St. Petersburg Technological Institute. He was a professor at St. Petersburg Machine-Construction Institute from 1930–1933. His seminal contributions include the results of the convergence of finite difference approximation to the solution of Laplace-type equations and his original results on estimating the eigenvalue of a complex $n \times n$ matrix.

(The eigenvalues of A are 7.3067, $-0.6533 \pm 0.3473i$.)
$$r_1 = 5, \quad r_2 = 12, \quad r_3 = 2.$$
The Geršgorin disks (shown in Figure 9.6) are
$$R_1 : \{z : |z - 1| \leq 5\}, \quad R_2 : \{z : |z - 4| \leq 12\}, \quad R_3 : \{z : |z - 1| \leq 2\}. \quad \blacksquare$$

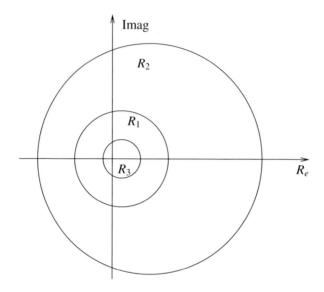

Figure 9.6. *Geršgorin disks of Example* 9.8.

While the above theorem only tells us that the eigenvalues of A lie in the union of n Geršgorin disks, the following theorem gives some more specific information. We state the theorem without proof. Several other generalizations exist. See Horn and Johnson (1985) Brualdi and Mellendorf (1994), and the recent book by Varga (2004).

Theorem 9.9 (Geršgorin's second theorem). *Suppose that r Geršgorin disks are disjoint from the rest. Then exactly r eigenvalues of A lie in the union of the r disks.*

Proof. See Horn and Johnson (1985, pp. 344–345). □

Example 9.10.
$$A = \begin{pmatrix} 1 & 0.1 & 0.2 \\ 0.2 & 4 & 0.3 \\ 0.4 & 0.5 & 8 \end{pmatrix}.$$

The Geršgorin disks are
$$R_1 : \{z : |z - 1| \leq 0.3\}, \quad R_2 : \{z : |z - 4| \leq 0.5\}, \quad R_3 : \{z : |z - 8| \leq 0.9\}.$$

All three disks are disjoint from each other. Therefore, by Theorem 9.9, each disk must contain exactly one eigenvalue of A. This is indeed true. Note that the eigenvalues A are 0.9834, 3.9671, and 8.0495. ∎

9.3.2 Eigenvalue Bounds and Matrix Norms

Simple matrix norms can sometimes be used to obtain useful bounds for the eigenvalues. Here are two examples.

Theorem 9.11. *Let λ be an eigenvalue of A. Then, for any consistent pair of matrix-vector norms,*
$$|\lambda| \leq \|A\|.$$
In particular, $\rho(A)$, the spectral radius of A (largest eigenvalue in magnitude), is bounded by $\|A\|$: $\rho(A) \leq \|A\|$.

Proof. From $Ax = \lambda x$, we have
$$\|\lambda x\| = \|Ax\| \leq \|A\| \, \|x\|$$
or
$$|\lambda| \, \|x\| \leq \|A\| \, \|x\|; \quad \text{that is, } |\lambda| \leq \|A\|. \quad \square$$

Theorem 9.12.
$$\rho(A) \leq \min\left\{ \max_i \sum_{j=1}^n |a_{ij}|, \; \max_j \sum_{i=1}^n |a_{ij}| \right\}.$$

Proof. The proof follows immediately from Theorem 9.11. $\quad \square$

9.4 Computing Selected Eigenvalues and Eigenvectors

We have just seen that in *several applications all one needs to compute is a few largest or smallest eigenvalues and the corresponding eigenvectors.* Examples of such applications include the following.

- **The buckling problem.** It is the smallest eigenvalue that is the most important one here.

- **Vibration analysis of structures**. *A common engineering practice in vibration engineering is to compute just the first few smallest eigenvalues (frequencies) and the corresponding eigenvectors (modes), because it has been seen in practice that the larger eigenvalues and eigenvectors contribute very little to the total response of the system.* The same remarks also hold in the case of control problems modeled by a system of second-order differential equations arising in the finite-element-generated reduced-order model of large flexible space structures (see Inman (2006)).

- **Statistical applications.** In statistical applications, such as those arising in *principal component analysis*, only the first few largest eigenvalues need to be computed. There are other applications where only the dominant and the subdominant eigenvalues and the corresponding eigenvectors play an important role (see Luenberger (1979)).

9.4.1 The Power Method, the Inverse Iteration, and the Rayleigh Quotient Iteration

In this section we will briefly describe two well-known classical methods for finding the dominant eigenvalues and the corresponding eigenvectors of a matrix. *The methods are particularly suitable for* sparse matrices, *because they rely on matrix-vector multiplications only* (and, therefore, the zero entries in a sparse matrix do not get filled in during the process).

Perhaps one of the most famous applications of the power method is its use in computing the **PageRank** of the **Google matrix**. The PageRank is an eigenvector of the Google matrix and measures the relative importance of each element of a hyperlinked set of documents, such as the World Wide Web, within the set. Thus, PageRank is Google's way of deciding a page's importance. The concept of PageRank was developed by Larry Page (hence the name PageRank) and Sergey Brin while they were graduate students at Stanford University.

The *Google matrix* itself is not sparse, but it is a rank-one modification of very sparse matrix. *The matrix size can be as large as few billion.* Though not practically feasible, the power method can, in principle, be used to compute the eigenvector of such a large matrix. But it is still a viable method for computing the PageRank of a modest-sized Google matrix. Many articles computing the PageRank and its relation to the power method (with its several variations) can be found from the Internet.

The Power Method

The **power method** is frequently used to find the **dominant eigenvalue** and the corresponding eigenvector of a matrix. It is so named because it is based on *implicit* construction of the powers of A.

Let the eigenvalues $\lambda_1, \lambda_2, \ldots, \lambda_n$ of A be such that

$$|\lambda_1| > |\lambda_2| \geq |\lambda_3| \geq \cdots \geq |\lambda_n|;$$

that is, λ_1 is the *dominant eigenvalue* of A. Let v_1 be the corresponding eigenvector. Let $\max(g)$ denote the largest element in the vector g.

ALGORITHM 9.1. Power Method.

Input: An $n \times n$ matrix A.
Outputs: Approximate dominant eigenvalue and the corresponding eigenvector.

Step 1. Choose x_0, an initial approximation to the eigenvector.

Step 2. For $k = 1, 2, 3, \ldots$ do
    2.1 Compute $\hat{x}_k = A x_{k-1}$.
    2.2 Normalize $x_k = \hat{x}_k / \max(|\hat{x}_k|)$.
    End

Theorem 9.13. $\{\max(|\hat{x}_k|)\} \to \lambda_1$, *and* $\{x_k\} \to w_1$, *a multiple of* v_1, *as* $k \to \infty$.

9.4. Computing Selected Eigenvalues and Eigenvectors

Proof. From Step 2 of Algorithm 9.1, we have

$$x_k = \frac{A^k x_0}{\max(|A^k x_0|)}.$$

Assume that eigenvectors $v_1, \ldots, v_n$ associated with $\lambda_1, \ldots, \lambda_n$ are linearly independent. We can then write $x_0 = \alpha_1 v_1 + \alpha_2 v_2 + \cdots + \alpha_n v_n$, $\alpha_1 \neq 0$. So,

$$A^k x_0 = A^k(\alpha_1 v_1 + \alpha_2 v_2 + \cdots + \alpha_n v_n) = \alpha_1 \lambda_1^k v_1 + \alpha_2 \lambda_2^k v_2 + \cdots + \alpha_n \lambda_n^k v_n$$

$$= \lambda_1^k \left[\alpha_1 v_1 + \alpha_2 \left(\frac{\lambda_2}{\lambda_1}\right)^k v_2 + \cdots + \alpha_n \left(\frac{\lambda_n}{\lambda_1}\right)^k v_n \right].$$

Since λ_1 is the dominant eigenvalue, $(\frac{\lambda_i}{\lambda_1})^k \to 0$ as $k \to \infty$, $i = 2, 3, \ldots, n$. Thus,

$$x_k = \frac{A^k x_0}{\max(|A^k x_0|)} \to c v_1 \quad \text{and} \quad \{\max(|\hat{x}_k|)\} \to \lambda_1. \quad \square$$

Example 9.14.

$$A = \begin{pmatrix} 1 & 2 & 3 \\ 2 & 3 & 4 \\ 3 & 4 & 5 \end{pmatrix}; \quad x_0 = (1, 1, 1)^T.$$

The eigenvalues of A are 0, -0.6235, and 9.6235. The normalized eigenvector corresponding to the largest eigenvalue 9.6233 is $(0.3851, 0.5595, 0.7339)^T$.

$k = 1$:

$$\hat{x}_1 = A x_0 = \begin{pmatrix} 6 \\ 9 \\ 12 \end{pmatrix}; \quad \max(|\hat{x}_1|) = 12, \quad x_1 = \frac{\hat{x}_1}{\max(|\hat{x}_1|)} = \begin{pmatrix} 0.50 \\ 0.75 \\ 1 \end{pmatrix}.$$

$k = 2$:

$$\hat{x}_2 = A x_1 = \begin{pmatrix} 5.00 \\ 7.25 \\ 9.50 \end{pmatrix}; \quad \max(|\hat{x}_2|) = 9.50, \quad x_2 = \frac{\hat{x}_2}{\max(|\hat{x}_2|)} = \begin{pmatrix} 0.5263 \\ 0.7632 \\ 1.0000 \end{pmatrix}.$$

$k = 3$:

$$\hat{x}_3 = A x_2 = \begin{pmatrix} 5.0526 \\ 7.3421 \\ 9.6316 \end{pmatrix}; \quad \max(|\hat{x}_3|) = 9.6316, \quad x_3 = \frac{\hat{x}_3}{\max(|\hat{x}_3|)} = \begin{pmatrix} 0.5246 \\ 0.7623 \\ 1.000 \end{pmatrix}.$$

Thus the sequence $\{\max(\hat{x}_k)\}$ is converging towards the largest eigenvalue 9.6235, and $\{x_k\}$ is converging towards the direction of the eigenvector associated with this eigenvalue. (Note that the normalized dominant eigenvector

$$\begin{pmatrix} 0.3851 \\ 0.5595 \\ 0.7339 \end{pmatrix}$$

is a scalar multiple of x_3.) ∎

Remarks. We have derived the power method under two constraints: (i) $\alpha_1 \neq 0$, and (ii) λ_1 is the *only* dominant eigenvalue. The first constraint ($\alpha_1 \neq 0$) is not really a serious practical constraint, because after a few iterations, round-off errors will almost always make it happen.

As far as the second constraint is concerned, we note that the method still converges when matrix A has more than one dominant eigenvalue. For example, let $\lambda_1 = \lambda_2 = \cdots = \lambda_r$ and $|\lambda_1| > |\lambda_{r+1}| > \cdots > |\lambda_n|$, and assume that the eigenvectors associated with λ_1 are independent. Then we have

$$A^k x_0 = \lambda_1^k \left(\sum_{i=1}^{r} \alpha_i v_i + \sum_{i=r+1}^{n} \alpha_i (\lambda_i/\lambda_1)^k v_i \right) \cong \lambda_1^k \sum_{1}^{r} \alpha_i v_i$$

(since $(\lambda_i/\lambda_1)^k$ is small for large values of k). *This shows that in this case the power method converges to some vector in the subspace spanned by $v_1, \ldots, v_n$.*

MATCOM Note: Algorithm 9.1 has been implemented in the MATCOM program **POWER-ITERATION**.

Convergence of the Power Method

The rate of convergence of the power method is determined by the ratio $\left|\frac{\lambda_2}{\lambda_1}\right|$. This is seen as follows. Consider

$$\|x_k - \alpha_1 v_1\| = \left\|\alpha_2 \left(\frac{\lambda_2}{\lambda_1}\right)^k v_2 + \cdots + \alpha_n \left(\frac{\lambda_n}{\lambda_1}\right)^k v_n\right\|$$

$$\leq |\alpha_2| \left|\frac{\lambda_2}{\lambda_1}\right|^k \|v_2\| + \cdots + |\alpha_n| \left|\frac{\lambda_n}{\lambda_1}\right|^k \|v_n\|$$

$$\leq \left|\frac{\lambda_2}{\lambda_1}\right|^k (|\alpha_2|\|v_2\| + \cdots + |\alpha_n| \|v_n\|)$$

(since $|\frac{\lambda_i}{\lambda_1}| \leq |\frac{\lambda_2}{\lambda_1}|, i = 3, 4, \ldots, n$). Thus we have

$$\|x_k - \alpha_1 v_1\| \leq \alpha \left|\frac{\lambda_2}{\lambda_1}\right|^k, \quad k = 1, 2, 3, \ldots,$$

where

$$\alpha = (|\alpha_2| \|v_2\| + \cdots + |\alpha_n| \|v_n\|).$$

This shows that the rate at which x_k approaches $\alpha_1 v_1$ depends upon how fast $|\frac{\lambda_2}{\lambda_1}|^k$ goes to zero. The absolute value of the error at each step decreases by the ratio $(\frac{\lambda_2}{\lambda_1})$; that is, *if λ_2 is close to λ_1, then the convergence will be very slow; if this ratio is small, the convergence will be fast.*

The Power Method with a Shift

In some cases, convergence can be significantly improved by using a suitable shift. Thus, if σ is a suitable shift so that $\lambda_1 - \sigma$ is the dominant eigenvalue of $A - \sigma I$, and if the power method is applied to the shifted matrix $A - \sigma I$, then the rate of convergence will be determined by the ratio $|\frac{\lambda_2 - \sigma}{\lambda_1 - \sigma}|$, rather than $|\frac{\lambda_2}{\lambda_1}|$. *(Note that by shifting the matrix A by σ, the eigenvalues get shifted by σ, but the eigenvectors remain unaltered.)*

9.4. Computing Selected Eigenvalues and Eigenvectors

By choosing σ appropriately, in some cases, the ratio $\left|\frac{\lambda_2-\sigma}{\lambda_1-\sigma}\right|$ can be made significantly smaller than $\left|\frac{\lambda_2}{\lambda_1}\right|$, thus yielding the faster convergence. *(Do an example to convince yourself and see also Exercise 9.10.)*

The Inverse Power Method/Inverse Iteration

The following, known as the **inverse iteration method**, is an effective technique for computing an eigenvector when a reasonably good approximation to an eigenvalue is known.

ALGORITHM 9.2. Inverse Iteration.

Inputs: (i) An approximation σ to a real eigenvalue λ such that $|\lambda_i - \sigma| \ll |\lambda_1 - \sigma|$, $i \neq 1$. (ii) Error tolerance ϵ; maximum number of iterations N. (iii) An initial approximation x_0 of the eigenvector.
Output: An approximation x_k to the eigenvector corresponding to σ.

Step 1. Choose x_0.

Step 2. For $k = 1, 2, 3, \ldots$, do
    2.1 Solve $(A - \sigma I)\hat{x}_k = x_{k-1}$.
    2.2 Compute $x_k = \hat{x}_k / \|\hat{x}_k\|_2$.
    2.3 Stop if $\|Ax_k - \sigma x_k\| < \epsilon$ or if $k > N$.
End

Theorem 9.15. *The sequence $\{x_k\}$ converges to the direction of the eigenvector corresponding to λ_1.*

Proof. The eigenvalues of $(A - \sigma I)^{-1}$ are $(\lambda_1 - \sigma)^{-1}$, $(\lambda_2 - \sigma)^{-1}, \ldots, (\lambda_n - \sigma)^{-1}$ and the eigenvectors are the same as those of A. Thus, as in the case of the power method, we can write

$$\hat{x}_k = \frac{c_1}{(\lambda_1 - \sigma)^k} v_1 + \frac{c_2}{(\lambda_2 - \sigma)^k} v_2 + \cdots + \frac{c_n}{(\lambda_n - \sigma)^k} v_n$$

$$= \frac{1}{(\lambda_1 - \sigma)^k} \left[c_1 v_1 + c_2 \left(\frac{\lambda_1 - \sigma}{\lambda_2 - \sigma}\right)^k v_2 + \cdots + c_n \left(\frac{\lambda_1 - \sigma}{\lambda_n - \sigma}\right)^k v_n \right].$$

Since λ_1 is closer to σ than any other eigenvalue, the first term on the right-hand side is the dominating one, and therefore x^k converges to the direction of v_1. It is the direction of v_1 which we are trying to compute. $\square$

Remark. Note that inverse iteration is simply the power method applied to $(A - \sigma I)^{-1}$. That is why it is also known as the *inverse power method*.

An illustration: Let us illustrate the above with $k = 1$. Suppose that $x_0 = c_1 v_1 + c_2 v_2 + \cdots + c_n v_n$. Then

$$\hat{x}_1 = (A - \sigma I)^{-1} x_0 = (\lambda_1 - \sigma)^{-1} c_1 v_1 + (\lambda_2 - \sigma)^{-1} c_2 v_2 + \cdots + (\lambda_n - \sigma)^{-1} c_n v_n.$$

Since λ_1 is closer to σ than any other eigenvalue, the coefficient of the first term in the expansion, namely, $\frac{1}{(\lambda_1-\sigma)}$, is the dominant one (it is the largest). Thus, $\hat{x}_1$ is roughly a multiple of v_1, which is what we desire.

Numerical Stability of the Inverse Iteration

At first sight the inverse iteration procedure seems dangerous, because if σ is near λ_1, then the matrix $(A - \sigma I)$ is obviously ill-conditioned. Consequently, this ill-conditioning might affect the computed approximations of the eigenvector. Fortunately, in practice the ill-conditioning of the matrix $(A - \sigma I)$ is exactly what we want. The error at each iteration grows towards the direction of the eigenvector, and it is the direction of the eigenvector that we are interested in.

Wilkinson (1965, pp. 620–621) has remarked that in practice $\hat{x}_k$ is remarkably close to the solution of $(A - \sigma I + F)x_k = x_{k-1}$, where F is small. For details see Wilkinson (1965, pp. 620–621). "The iterated vectors do indeed converge eventually to the eigenvectors of $A + F$."

Example 9.16. Consider matrix A of Example 9.14. Choose
$$x_0 = (1, 1, 1)^T, \qquad \sigma = 9.$$

k = 1:
$$\hat{x}_1 = (1, 1.5, 2)^T,$$
$$x_1 = \hat{x}_1/\|\hat{x}_1\|_2 = (0.3714, 0.5571, 0.7428)^T.$$

k = 2:
$$\hat{x}_2 = (0.619, 0.8975, 1.1761)^T,$$
$$x_2 = \hat{x}_2/\|\hat{x}_2\|_2 = (0.3860, 0.5597, 0.7334)^T.$$

k = 3:
$$\hat{x}_3 = (0.6176, 0.8974, 1.1772)^T,$$
$$x_3 = \hat{x}_3/\|\hat{x}_3\|_2 = (0.3850, 0.5595, 0.7340)^T.$$

k = 4:
$$\hat{x}_4 = (0.6176, 0.8974, 1.1772)^T,$$
$$x_4 = \hat{x}_4/\|\hat{x}_4\|_2 = (0.3850, 0.5595, 0.7340)^T$$

k = 5:
$$\hat{x}_5 = (0.6177, 0.8974, 1.1772)^T,$$
$$x_5 = \hat{x}_5/\|(\hat{x}_5)\|_2 = (0.3851, 0.5595, 0.7339)^T. \qquad \blacksquare$$

Remark. Note that *scaling is immaterial since we are working towards the direction of the eigenvector.*

Choosing the initial vector x_0. To choose the initial vector x_0 we can run a few iterations of the power method and then switch to the inverse iteration, with the last vector generated by the power method as the initial vector x_0 in the inverse iteration.

9.4. Computing Selected Eigenvalues and Eigenvectors

MATCOM Note: Algorithm 9.2 has been implemented in the MATCOM program **INVITR**.

The Rayleigh Quotient

Theorem 9.17. *Let A be a **symmetric matrix** and let x be a reasonably good approximation to an eigenvector. Then the quotient*

$$R_q = \sigma = \frac{x^T A x}{x^T x}$$

is a good approximation to the eigenvalue λ for which x is the corresponding eigenvector.

Proof. Since A is symmetric there exists a set of orthogonal eigenvectors $v_1, v_2, \ldots, v_n$. Therefore we can write $x = c_1 v_1 + \cdots + c_n v_n$. Assume that v_i, $i = 1, \ldots, n$, are normalized, that is, $v_i^T v_i = 1$. Then, since $A v_i = \lambda_i v_i$, $i = 1, \ldots, n$, and noting that $v_i^T v_j = 0$, $i \neq j$, we have

$$\sigma = \frac{x^T A x}{x^T x} = \frac{(c_1 v_1 + \cdots + c_n v_n)^T A (c_1 v_1 + \cdots + c_n v_n)}{(c_1 v_1 + \cdots + c_n v_n)^T (c_1 v_1 + \cdots + c_n v_n)}$$

$$= \frac{(c_1 v_1 + \cdots + c_n v_n)^T (c_1 \lambda_1 v_1 + \cdots + c_n \lambda_n v_n)}{c_1^2 + c_2^2 + \cdots + c_n^2} = \frac{\lambda_1 c_1^2 + \lambda_2 c_2^2 + \cdots + \lambda_n c_n^2}{c_1^2 + c_2^2 + \cdots + c_n^2}$$

$$= \lambda_1 \left[\frac{1 + \left(\frac{\lambda_2}{\lambda_1}\right)\left(\frac{c_2}{c_1}\right)^2 + \cdots + \left(\frac{\lambda_n}{\lambda_1}\right)\left(\frac{c_n}{c_1}\right)^2}{1 + \left(\frac{c_2}{c_1}\right)^2 + \cdots + \left(\frac{c_n}{c_1}\right)^2} \right].$$

Because of our assumption that x is a good approximation to v_1, c_1 is larger than other c_i, $i = 2, \ldots, n$. Thus, the expression within brackets is close to 1, which means that σ is close to λ_1. □

Definition 9.18. *The quotient $R_q = \frac{x^T A x}{x^T x}$ is called the **Rayleigh quotient**.*[11]

Example 9.19. Let

$$A = \begin{pmatrix} 1 & 2 \\ 2 & 3 \end{pmatrix} \quad \text{and} \quad x = \begin{pmatrix} 1 \\ -0.5 \end{pmatrix}.$$

Then the Rayleigh quotient

$$\sigma = \frac{x^T A x}{x^T x} = -0.2$$

is a good approximation to the eigenvalue -0.2361. ∎

---

[11] John William Strutt (1842–1919), the third Baron Rayleigh, was born in England and studied at Trinity College, Cambridge, and eventually became the chancellor of Cambridge University. His research was mainly mathematical, concerning optics and vibrating systems. He won the Nobel Prize in Physics in 1904.

Note: It can be shown (Exercise 9.14) that for a symmetric matrix A, $\lambda_n \leq R_q \leq \lambda_1$, where λ_n and λ_1 are the smallest and the largest eigenvalue of A, respectively.

Rayleigh Quotient Iteration

The above idea of approximating an eigenvalue of a symmetric matrix can be combined with the inverse iteration procedure (Algorithm 9.2) to compute successive approximations of an eigenvalue and the corresponding eigenvector in an iterative fashion, known as **Rayleigh quotient iteration**, described as follows (see Figure 9.7).

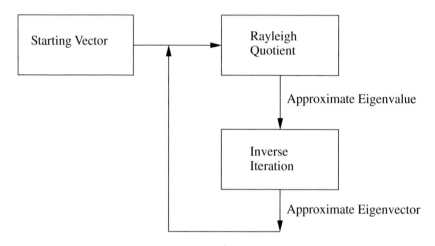

Figure 9.7. *Rayleigh quotient iteration.*

ALGORITHM 9.3. Rayleigh Quotient Iteration.

Inputs: (i) A symmetric matrix A. (ii) Maximum number of iterations N. (iii) An initial approximation x_0 of the eigenvector
Output: An approximate eigenpair.

For $k = 0, 1, 2, \ldots$, do

1. Compute $\sigma_k = x_k^T A x_k / x_k^T x_k$ (Rayleigh quotient).
2. Solve for $\hat{x}_{k+1}$: $(A - \sigma_k I)\hat{x}_{k+1} = x_k$ (inverse iteration).
3. Normalize $x_{k+1} = \hat{x}_{k+1} / \max(|\hat{x}_{k+1}|)$.
4. Stop if the pair (α_k, x_k) is an acceptable eigenvalue-eigenvector pair or if $k > N$.

End

9.4. Computing Selected Eigenvalues and Eigenvectors

Convergence. It can be shown (Wilkinson (1965, p. 630)) that the rate of convergence of Algorithm 9.3 is cubic.

Choice of x_0. As for choosing an initial vector x_0, perhaps the best thing to do is to use the power method itself a few times and then use the last approximation as x_0.

Remark. Rayleigh quotient iteration can also be defined in the nonsymmetric case, where one finds both left and right eigenvectors at each step. We omit the discussion of the nonsymmetric case here and refer the reader to Wilkinson (1965, p. 636). See also Parlett (1974).

Example 9.20. Consider

$$A = \begin{pmatrix} 1 & 2 & 3 \\ 2 & 3 & 4 \\ 3 & 4 & 5 \end{pmatrix}, \quad \text{with } x_0 = \begin{pmatrix} 0.5246 \\ 0.7622 \\ 1.000 \end{pmatrix}.$$

This initial vector x_0 was obtained after 3 iterations of the power method.

$k = 0$:

$$\sigma_0 = x_0^T A x_0 / (x_0^T x_0) = 9.6235, \quad x_1 = \begin{pmatrix} 0.5247 \\ 0.7623 \\ 1.000 \end{pmatrix}.$$

$k = 1$:

$$\sigma_1 = x_1^T A x_1 / (x_1^T x_1) = 9.6235, \quad x_2 = \begin{pmatrix} 1.000 \\ 1.4529 \\ 1.9059 \end{pmatrix}.$$

The normalized eigenvector associated with 9.6255 is

$$\begin{pmatrix} 0.3851 \\ 0.5595 \\ 0.7339 \end{pmatrix}.$$

Note that 0.3851 times x_2 is this eigenvector to three digits. Thus two iterations were sufficient. ∎

MATCOM Note: Algorithm 9.3 has been implemented in the MATCOM program **RAYQOT**.

Computing the Smallest Eigenvalues

It is easy to see that the power method applied to A^{-1} gives us the smallest eigenvalue in magnitude (the least dominant one) of A.

Let A be nonsingular and let the eigenvalues of A be ordered such that

$$|\lambda_1| > |\lambda_2| \geq |\lambda_3| \geq \cdots |\lambda_{n-1}| > |\lambda_n| > 0.$$

Then the eigenvalues of A^{-1} (which are the reciprocals of the eigenvalues of A) are arranged as

$$\left|\frac{1}{\lambda_n}\right| > \left|\frac{1}{\lambda_{n-1}}\right| \geq \left|\frac{1}{\lambda_{n-2}}\right| \geq \cdots \geq \frac{1}{|\lambda_1|} > 0.$$

That is, $\frac{1}{\lambda_n}$ is the dominant eigenvalue of A^{-1}. This suggests that the reciprocal of the smallest eigenvalue can be computed by applying the power method to A^{-1}.

ALGORITHM 9.4. Computing the Smallest Eigenvalue in Magnitude.

Step 1. Apply the power method (Algorithm 9.1) to A^{-1} to compute the dominant eigenvalue of A^{-1}.

Step 2. Take the reciprocal of the eigenvalue obtained in Step 1.

Note: Since the power method is implemented by matrix-vector multiplication only, the inverse of A does not have to be computed explicitly. This is because computing $y = A^{-1}x$, where x is a vector, is equivalent to solving the linear system $Ay = x$.

Example 9.21.

$$A = \begin{pmatrix} 1 & 4 & 5 \\ 2 & 3 & 3 \\ 1 & 1 & 1 \end{pmatrix}.$$

The power method (without shift) applied to A^{-1} with the starting vector $x_0 = (1, -1, 1)^T$ gives the dominant eigenvalue of A^{-1} as $\frac{1}{\lambda_n} = \alpha = 9.5145$. Thus the smallest eigenvalue in magnitude of A is $\frac{1}{\lambda_1} = 0.1051$. (Note that the eigenvalues of A are 6.3850, -1.4901, and 0.1051.) ∎

9.5 Similarity Transformations and Eigenvalue Computations

*A basic idea to numerically compute the eigenvalues of a matrix is to transform the matrix to a "simpler" form by using a **similarity transformation**, from which the eigenvalues can be more easily computed.*

Theorem 9.22. *Two similar matrices have the same eigenvalues.*

Proof. Let A and B be two similar matrices; that is, there exists a nonsingular matrix X such that

$$X^{-1}AX = B.$$

Then

$$\det(B - \lambda I) = \det(X^{-1}AX - \lambda I) = \det(X^{-1}(A - \lambda I)X)$$
$$= \det(X^{-1})\det(X)\det(A - \lambda I) = \det(A - \lambda I).$$

9.5. Similarity Transformations and Eigenvalue Computations

Thus, A and B have the same characteristic polynomial, and therefore the eigenvalues are same. □

Note: *The converse is not true.* Two matrices having the same set of eigenvalues are not necessarily similar. Here is a simple example:

$$A = \begin{pmatrix} 1 & 1 \\ 0 & 1 \end{pmatrix}, \quad B = \begin{pmatrix} 1 & 0 \\ 0 & 1 \end{pmatrix}.$$

A and B have the same eigenvalues, but they cannot be similar.

Some of the "simpler" forms associated with eigenvalue computation that can be obtained via similarity transformations include

- diagonal and block diagonal forms (**Jordan canonical form**),
- Hessenberg form,
- Companion form,
- Triangular form.

The Hessenberg and triangular forms can be achieved via orthogonal transformations *and should be used for eigenvalue computation.* On the other hand, reduction to the *diagonal, block diagonal, and companion forms, in general, require nonorthogonal transformations. The transforming matrices for these forms can be highly ill-conditioned, and therefore these forms should be avoided in eigenvalue computations,* as the following discussions show.

9.5.1 Diagonalization of a Matrix

Definition 9.23. *A matrix A is called **diagonalizable** if $X^{-1}AX$ is a diagonal matrix D. This decomposition is referred to as the **eigenvalue decomposition**.*

We now give a characterization of diagonalizability.

Definition 9.24. *The **algebraic multiplicity** of an eigenvalue λ of A is the number of times it appears as a root of the characteristic equation. An eigenvalue λ is a **simple eigenvalue** if its algebraic multiplicity is 1. The **geometric multiplicity** of λ is the dimension of the nullspace of $A - \lambda I$.*

Example 9.25.

$$A = \begin{pmatrix} 1 & 0 & 0 \\ 0 & 1 & 0 \\ 0 & 0 & 1 \end{pmatrix}, \quad B = \begin{pmatrix} 1 & 1 & 0 \\ 0 & 1 & 1 \\ 0 & 0 & 1 \end{pmatrix}.$$

The algebraic multiplicity of the eigenvalue 1 of both matrices is 3; however, the geometric multiplicity of 1 of matrix A is 3 and that of matrix B is 1. ■

Definition 9.26. *An eigenvalue is called a **defective eigenvalue** if its geometric multiplicity is less than its algebraic multiplicity. A matrix is a **defective matrix** if it has a defective eigenvalue. Otherwise, it is **nondefective**.*

Theorem 9.27. *An $n \times n$ matrix is diagonalizable if and only if it is nondefective.*

Proof. First, suppose that A is diagonalizable. That is, $X^{-1}AX = D$, a diagonal matrix. A diagonal matrix is clearly nondefective. Thus D is nondefective and so is A. Next, suppose that A is nondefective. Since the geometric multiplicity of each eigenvalue is the same as its algebraic multiplicity, matrix A must have n linearly independent eigenvectors. Call them $x_1, \ldots, x_n$. Then $X = (x_1, \ldots, x_n)$ is nonsingular and we have $X^{-1}AX = D$. □

Note: If $X^{-1}AX = D = \mathrm{diag}(\lambda_1, \ldots, \lambda_n)$, then X is the eigenvector matrix.

The above theorem tells us that a matrix is always not diagonalizable. However, the following theorem shows that it is always possible to *block diagonalize* matrix A. A block diagonal matrix A is written as

$$A = \mathrm{diag}(A_1, A_2, \ldots, A_k),$$

where A_i, $i = 1, \ldots, k$, are matrices. A well-known example of a block diagonal matrix is the **Jordan canonical form**.[12]

Theorem 9.28 (Jordan canonical theorem). *If A is $n \times n$, then there exists a nonsingular matrix X such that $X^{-1}AX = \mathrm{diag}(J_1, \ldots, J_k)$, where*

$$J_i = \begin{pmatrix} \lambda_i & 1 & 0 & \cdots & 0 \\ & \ddots & 1 & \ddots & \vdots \\ & & \ddots & \ddots & 0 \\ & & & \ddots & 1 \\ & & & & \lambda_i \end{pmatrix}, \quad i = 1, 2, \ldots, k.$$

If J_i is of order p_i, then $p_1 + p_2 + \cdots + p_k = n$. The matrices J_i are called the Jordan block matrices or simply the Jordan matrices. The number λ_i is an eigenvalue of J_i with multiplicity p_i.

Note: If each $p_i = 1$, then the Jordan matrix J_i is a diagonal matrix.

9.5.2 Numerical Instability of Nonorthogonal Diagonalization

Extreme caution should be taken in using diagonalization or block diagonalization to compute the eigenvalues of a matrix. The following theorem shows that the *conditioning of the transforming matrix X has a significant impact on eigenvalue computation*. A proof of the theorem can be found in Golub and Van Loan (1996, p. 317).

Theorem 9.29.
$$\mathrm{fl}(X^{-1}AX) = X^{-1}AX + E,$$

---

[12] Marie Ennemond Camile Jordan (1838–1922) was a French mathematician known for his many fundamental contributions to mathematics, including complex analysis, linear algebra, mathematical analysis, and group theory. Besides the *Jordan canonical form*, other well-known mathematical terms and results named after him include the *Jordan curve theorem*, the *Jordan measure*, and the *Jordan–Hölder theorem*.

9.5. Similarity Transformations and Eigenvalue Computations

where
$$\|E\|_2 \cong \mu \|X\|_2 \|X^{-1}\|_2.$$

Implication of Theorem 9.29. Thus, if X is ill-conditioned, the computed matrix $X^{-1}AX + E$ will be different from A and the computed eigenvalues will have errors.

> Because of this, it is not advisable to compute the eigenvalues of a matrix A via Jordan canonical form. *Whenever A is close to a defective matrix, the transforming X will be highly ill-conditioned.*

9.5.3 Reduction to Hessenberg Form via Orthogonal Similarity

Theorem 9.30 (Hessenberg reduction theorem). *An arbitrary $n \times n$ matrix can always be transformed into an upper Hessenberg[13] matrix H_u by orthogonal similarity; that is, there exists an orthogonal matrix P such that*
$$PAP^T = H_u.$$

As we will see a little later, the importance of Hessenberg transformation lies in the fact that the *reduction to a Hessenberg form must be performed before applying the QR iteration algorithm to A to compute the eigenvalues.*

Householder's Method

The process of QR factorization using Householder matrices described in Chapter 7 can be easily extended to obtain P and H_u.

> The idea is to reduce the matrix A to an upper Hessenberg matrix H_u by successively premultiplying A with a sequence of Householder matrices followed by postmultiplication with their transposes.

The matrix P in this case is constructed as the product of $(n-2)$ Householder matrices P_1 through P_{n-2}.

- P_1 is constructed to create zeros in the first column of A below the entry $(2, 1)$, resulting in the matrix $P_1 A P_1^T = A^{(1)}$.

- P_2 is determined to create zeros below the entry $(3, 2)$ of the second column of the matrix $A^{(1)}$, resulting in the matrix $P_2 A^{(1)} P_2^T = A^{(2)}$. *The process can be continued.*

---

[13] Karl Hessenberg (1904–1959) was a German engineer whose dissertation *Auflösung Linearer Eigenwertaufgaben mit Hilfe der Hamilton-Cayleyschen Gleichung* (Technische Hochschule, Darmstadt, Germany, 1941) investigated computation of the eigenvalue and eigenvectors of linear operators. The Hessenberg form of a matrix, named after him, appeared later in a paper related to his dissertation. For details visit http://www.Hessenberg.de/karl1.html.

The process consists of $(n-2)$ *steps.* (Note that an $n \times n$ Hessenberg matrix contains at least $\frac{(n-2)(n-1)}{2}$ zeros.)

An illustration: Let $n = 4$. There are only 2 steps.

Step 1. $\begin{pmatrix} \times & \times & \times & \times \\ \times & \times & \times & \times \\ \times & \times & \times & \times \\ \times & \times & \times & \times \end{pmatrix} \xrightarrow{P_1} \begin{pmatrix} \times & \times & \times & \times \\ \times & \times & \times & \times \\ 0 & \times & \times & \times \\ 0 & \times & \times & \times \end{pmatrix} \xrightarrow{P_1^T} \begin{pmatrix} \times & \times & \times & \times \\ \times & \times & \times & \times \\ 0 & \times & \times & \times \\ 0 & \times & \times & \times \end{pmatrix}.$

$\qquad\qquad\qquad A \qquad\qquad\qquad\qquad P_1 A \qquad\qquad\qquad A^{(1)} = P_1 A P_1^T$

Step 2. $\begin{pmatrix} \times & \times & \times & \times \\ \times & \times & \times & \times \\ 0 & \times & \times & \times \\ 0 & \times & \times & \times \end{pmatrix} \xrightarrow{P_2} \begin{pmatrix} \times & \times & \times & \times \\ \times & \times & \times & \times \\ 0 & \times & \times & \times \\ 0 & 0 & \times & \times \end{pmatrix} \xrightarrow{P_2^T} \begin{pmatrix} \times & \times & \times & \times \\ \times & \times & \times & \times \\ 0 & \times & \times & \times \\ 0 & 0 & \times & \times \end{pmatrix}.$

$\qquad\qquad\qquad A^{(1)} \qquad\qquad\qquad\quad P_2 A^{(1)} \qquad\qquad H_u = A^{(2)} = P_2 A^{(1)} P_2^T$

Notes: (i) Each of the matrices P_1 and P_2 is computed in *two substeps* as shown below.

(ii) The zeros created by premultiplication of A by P_1 do not get destroyed by postmultiplication with P_1^T. Similarly for the other steps.

The general case now can be easily written down. *In the following, in order to simplify notation and save computer storage, each of the matrices $A^{(k)}$ will be stored in place of A.*

Step 1. Find a Householder matrix $\widehat{P}_1$ of order $n-1$ such that

$$\widehat{P}_1 \begin{pmatrix} a_{21} \\ a_{31} \\ \vdots \\ a_{n1} \end{pmatrix} = \begin{pmatrix} \times \\ 0 \\ \vdots \\ 0 \end{pmatrix}.$$

Define

$$P_1 = \begin{pmatrix} I_1 & 0 \\ 0 & \widehat{P}_1 \end{pmatrix} \text{ and } \textit{implicitly} \text{ compute } A^{(1)} = P_1 A P_1^T.$$

Then

$$A \equiv A^{(1)} = \begin{pmatrix} \times & \times & \cdots & \times \\ \times & \times & \cdots & \times \\ 0 & a_{32} & \cdots & \times \\ \vdots & \vdots & & \\ 0 & a_{n2} & \cdots & \times \end{pmatrix}.$$

Step 2. Find a Householder matrix $\widehat{P}_2$ of order $(n-2)$ such that

$$\widehat{P}_2 \begin{pmatrix} a_{32} \\ \vdots \\ \vdots \\ a_{n2} \end{pmatrix} = \begin{pmatrix} \times \\ 0 \\ \vdots \\ 0 \end{pmatrix}.$$

9.5. Similarity Transformations and Eigenvalue Computations

Define

$$P_2 = \begin{pmatrix} I_2 & 0 \\ 0 & \widehat{P}_2 \end{pmatrix} \text{ and } \textit{implicitly} \text{ compute } A^{(2)} = P_2 A^{(1)} P_2^T.$$

Then

$$A \equiv A^{(2)} = \begin{pmatrix} \times & \times & \times & \cdots & \times \\ \times & \times & \times & \cdots & \times \\ 0 & \times & \times & \cdots & \times \\ 0 & 0 & \times & \cdots & \times \\ \vdots & \vdots & \vdots & & \\ 0 & 0 & \times & \cdots & \times \end{pmatrix}.$$

The general *Step k* can now easily be written down.

At the end of $(n-2)$ steps, the matrix $A^{(n-2)}$ is an upper Hessenberg matrix H_u.

Obtaining the Orthogonal Transforming Matrix P

Set

$$P = P_{n-2} P_{n-3} \ldots P_2 P_1. \tag{9.21}$$

Then P is *orthogonal* (since it is the product of $(n-2)$ Householder matrices), and it is easy to see that

$$PAP^T = H_u. \tag{9.22}$$

n = 4: $\quad P = P_2 P_1, \; PAP^T = P_2 P_1 A P_1^T P_2^T = P_2 A^{(1)} P_2^T = A^{(2)} = H_u.$

Example 9.31. Let

$$A = \begin{pmatrix} 0 & 1 & 2 \\ 1 & 2 & 3 \\ 1 & 1 & 1 \end{pmatrix}.$$

Since $n = 3$, we have just *one step* to perform.

- Form $\widehat{P}_1 = I_2 - \frac{2 u_1 u_1^T}{u_1^T u_1}$ such that

$$\widehat{P}_1 \begin{pmatrix} 1 \\ 1 \end{pmatrix} = \begin{pmatrix} * \\ 0 \end{pmatrix},$$

$$u_1 = \begin{pmatrix} 1 \\ 1 \end{pmatrix} + \sqrt{2} e_1 = \begin{pmatrix} 1 \\ 1 \end{pmatrix} + \sqrt{2} \begin{pmatrix} 1 \\ 0 \end{pmatrix} = \begin{pmatrix} 1 + \sqrt{2} \\ 1 \end{pmatrix}.$$

So,

$$\widehat{P}_1 = I_2 - 2 \frac{u_1 u_1^T}{u_1^T u_1} \equiv \begin{pmatrix} 1 & 0 \\ 0 & 1 \end{pmatrix} - 0.2929 \begin{pmatrix} 5.8284 & 2.4142 \\ 2.4142 & 1 \end{pmatrix} = \begin{pmatrix} -0.7071 & -0.7071 \\ -0.7071 & 0.7071 \end{pmatrix}.$$

- Form P_1 out of $\widehat{P}_1$ as follows:
$$P_1 = \begin{pmatrix} 1 & 0 & 0 \\ 0 & & \\ 0 & \widehat{P}_1 & \end{pmatrix} = \begin{pmatrix} 1 & 0 & 0 \\ 0 & -0.7071 & -0.7071 \\ 0 & -0.7071 & 0.7071 \end{pmatrix}.$$

- Form the Hessenberg matrix H_u and store it over A:
$$A \equiv A^{(1)} = P_1 A P_1^T = \begin{pmatrix} 0 & -2.1213 & 0.7071 \\ -1.4142 & 3.5000 & -0.5000 \\ 0 & 1.5000 & -0.5000 \end{pmatrix} = H_u. \quad \blacksquare$$

ALGORITHM 9.5. Householder Hessenberg Reduction.

Input: An $n \times n$ matrix A.
Output: An $n \times n$ upper Hessenberg matrix stored over A.

For $k = 1, 2, \ldots, n-2$ do

1. Determine the vector $u_k = (u_{k+1,k}, \ldots, u_{n,k})^T$ defining the Householder matrix $\widehat{P}_k = I_k - 2\frac{u_k u_k^T}{u_k^T u_k}$ of order $(n-k)$ and a scalar σ such that
$$\widehat{P}_k \begin{pmatrix} a_{k+1,k} \\ \vdots \\ \vdots \\ a_{nk} \end{pmatrix} = \begin{pmatrix} \sigma \\ 0 \\ \vdots \\ 0 \end{pmatrix}.$$

2. Store σ over $a_{k+1,k}$:
$$a_{k+1,k} \equiv \sigma.$$

3. Compute $\beta_k = \frac{2}{u_k^T u_k}$ and save u_k and β_k.

4. Update the entries of A in rows $k+1$ through n and columns $k+1$ through n by premultiplication, and then in columns $k+1$ through n and rows 1 to n by postmultiplication by performing the following *multiplications implicitly*. Store them in respective positions of A:
$$A(k+1:n, k:n) = A(k+1:n, k:n) - \beta_k u_k u_k^T A(k+1:n, k:n),$$
$$A(1:n, k+1:n) = A(1:n, k+1:n) - \beta_k A(1:n, k+1:n) u_k u_k^T.$$

End

Note: The algorithm does not explicitly compute the transforming matrix P. However, the latter can be computed out of the Householder vectors u_1 through u_{n-2}. Note that $P = P_{n-2} P_{n-1} \ldots P_2 P_1$, where $P_k = \text{diag}(I_k, I - \beta_k u_k u_k^T)$, $k = 1, 2, \ldots, n-2$.

Example 9.32.
$$A = \begin{pmatrix} 1 & 2 & 5 \\ 3 & 7 & 9 \\ 2 & 5 & 3 \end{pmatrix}.$$

9.5. Similarity Transformations and Eigenvalue Computations

Just one step: $k = 1$.

1. $u_1 = (2.2019, 0.6667)^T$, $\sigma = -3.6056$ (using Algorithm 7.1).

2. $a_{21} \equiv \sigma = -3.6056$.

3. $\beta_1 = \frac{2}{u_1^T u_1} = 0.3779$.

4. *Update A:* $a_{12} = -4.4376$, $a_{13} = 3.0509$; $a_{21} = -3.6056$, $a_{22} = 12.2308$, $a_{23} = -2.8462$; $a_{32} = 1.1538$, $a_{33} = -2.2308$.

So, $A = H_u = P_1 A P_1^T = \begin{pmatrix} 1 & -4.4376 & 3.0509 \\ -3.6056 & 12.2308 & -2.8462 \\ 0 & 1.1538 & -2.2308 \end{pmatrix}$. ∎

Flop-count. Algorithm 9.5 requires $\frac{10}{3}n^3$ flops to compute H_u. *This count does not include the explicit computation of P. P can be stored in factored form. If P is computed explicitly, another $\frac{4}{3}n^3$ flops will be required. However, when n is large, the storage required to form P is prohibitive.*

Round-off property. *The algorithm is stable.* It can be shown (Wilkinson (1965, p. 351)) that *the computed H_u is orthogonally similar to a nearby matrix $A + E$.* Specifically, there exists an orthogonal matrix Q such that $Q^T(A + E)Q = H_u$, with

$$\|E\|_F \leq cn^2 \mu \|A\|_F,$$

where c is a small constant.

MATCOM Note: Algorithm 9.5 has been implemented in the MATCOM program **HOUSHESS**.

Tridiagonal Reduction of a Symmetric Matrix

If the matrix A is symmetric, then from

$$PAP^T = H_u$$

it follows immediately that the upper *Hessenberg matrix H_u is also symmetric and, therefore, is tridiagonal.* Thus, if the algorithm is applied to a symmetric matrix A, the resulting matrix H_u will be a symmetric tridiagonal matrix T. Furthermore, one obviously can take advantage of the symmetry of A to modify the algorithm. For example, a significant savings can be obtained in storage by taking advantage of the symmetry of each $A^{(k)}$.

The symmetric algorithm requires only $\frac{4}{3}n^3$ flops to compute T, compared to $\frac{10}{3}n^3$ flops needed to compute H_u. The round-off property is essentially the same as the nonsymmetric algorithm. *The algorithm is stable.*

Givens Rotations and Reduction to Hessenberg Form

As in the case of QR factorization, the Givens matrices can also be employed to transform an arbitrary $n \times n$ matrix A to an upper Hessenberg matrix H_u by orthogonal similarity: $PAP^T = H_u$. However, to do this, Givens rotations must be constructed in a *certain special manner.* For example, in the first step, Givens rotations $J(2, 3, \theta)$, $J(2, 4, \theta)$, ..., $J(2, n, \theta)$ are successively computed so that with $P_1 = J(2, n, \theta) \cdots J(2, 4, \theta) J(2, 3, \theta)$, the updated matrix $A^{(1)} = P_1 A P_1^T$ has zeros on the first column below the (2, 1) entry. The other steps are similar. We leave the derivation as an exercise (Exercise 9.16). This reduction will require about $\frac{20}{3} n^3$ flops to compute H_u, *almost twice as many as required by the Householder reduction.*

Round-off property. The round-off property is essentially the same as the Householder method. *The method is numerically stable.*

Example 9.33. Consider matrix A of Example 9.31 again.

Step 1. Find c and s such that

$$\begin{pmatrix} c & s \\ -s & c \end{pmatrix} \begin{pmatrix} 1 \\ 1 \end{pmatrix} = \begin{pmatrix} * \\ 0 \end{pmatrix}; \quad c = \frac{1}{\sqrt{2}}, \quad s = \frac{1}{\sqrt{2}}.$$

Step 2.

$$P_1 = J(2, 3, \theta) = \begin{pmatrix} 1 & 0 & 0 \\ 0 & \frac{1}{\sqrt{2}} & \frac{1}{\sqrt{2}} \\ 0 & -\frac{1}{\sqrt{2}} & \frac{1}{\sqrt{2}} \end{pmatrix};$$

$$A \equiv P_1 A P_1^T = \begin{pmatrix} 0 & 2.1213 & 0.7171 \\ 1.4142 & 3.5000 & 0.5000 \\ 0 & -1.5000 & -0.5000 \end{pmatrix}. \quad \blacksquare$$

Observation. Note that the upper Hessenberg matrix obtained here is *essentially the same* as that obtained by Householder's method (Example 9.31). *(They differ only by the signs of the subdiagonal entries.)*

MATCOM Note: The Givens method for Hessenberg reduction has been implemented in the MATCOM program **GIVHESS**.

9.5.4 Uniqueness of Hessenberg Reduction

The above example and the observation made therein brings up the question of uniqueness in Hessenberg reduction. To this end, we state a simplified version of what is known as the **implicit Q theorem**. For a complete statement and proof, see Golub and Van Loan (1996, pp. 346–347).

Theorem 9.34 (implicit Q theorem). *Let P and Q be orthogonal matrices such that $P^T A P = H_1$ and $Q^T A Q = H_2$ are two **unreduced** upper Hessenberg matrices. Suppose that P and Q have the same first columns. Then H_1 and H_2 are essentially the same in the sense that $H_2 = D^{-1} H_1 D$, where*

$$D = \mathrm{diag}(\pm 1, \ldots, \pm 1).$$

9.5. Similarity Transformations and Eigenvalue Computations

Example 9.35. Denote the Hessenberg matrices in Examples 9.31 and 9.33 obtained by the Householder and Givens methods, respectively, by H_1 and H_2.

Using the notation of Theorem 9.34 we have

$$P = P_1^T, \qquad Q^T = J(2, 3, \theta).$$

Both P and Q have the same first columns, namely, the first column of the identity matrix. We verify that $H_2 = D^{-1} H_1 D$, where $D = \text{diag}(1, -1, 1)$. ∎

9.5.5 Eigenvalue Computations Using the Characteristic Polynomial

Why should eigenvalues not be computed via the characteristic polynomial?

Since the eigenvalues of a matrix are the zeros of the characteristic polynomial, it is natural to think of computing the eigenvalues of A by finding the zeros of its characteristic polynomial. However, *this approach is not numerically effective*.

Difficulties with Eigenvalue Computations Using the Characteristic Polynomial

First, the process of explicitly computing the coefficients of the characteristic polynomial may be numerically unstable.

Second, the zeros of the characteristic polynomial may be very sensitive to perturbations of the coefficients of the characteristic polynomial. Thus if the coefficients of the characteristic polynomial are not computed accurately, there will be errors in the computed eigenvalues.

In Chapter 4 we illustrated the sensitivity of the root-finding problem by means of the Wilkinson polynomial and other examples. We will now discuss the difficulty of computing the characteristic polynomial in some detail here.

Computing the characteristic polynomial of a matrix explicitly amounts to transforming the matrix to a block-companion (or Frobenius) form. Every matrix A can be reduced by similarity to $C = \text{diag}(C_1, \ldots, C_k)$, where each C_i is a **companion matrix**. The matrix C is said to be in Frobenius form. If $k = 1$, the matrix A is **nonderogatory**.

Assume that A is nonderogatory and let's see how A can be reduced to a companion matrix by similarity. This can be achieved in two stages.

Reduction of a Matrix to a Companion Matrix

Stage 1: The matrix A is transformed to an upper Hessenberg matrix H by orthogonal similarity using the Householder (Algorithm 9.5) or Givens method.

Stage 2: The transformed unreduced Hessenberg matrix H is further reduced to a companion matrix C by similarity (assuming that H is unreduced).

We have already seen that Stage 1 can be performed in a *numerically stable way*. Consider now Stage 2, that is, the transformation of the unreduced Hessenberg matrix H to a companion matrix C.

Let X be the nonsingular transforming matrix such that $HX = XC$, where

$$C = \begin{pmatrix} 0 & 0 & . & \cdots & 0 & c_1 \\ 1 & 0 & . & \cdots & 0 & c_2 \\ 0 & 1 & 0 & \cdots & 0 & c_3 \\ \vdots & \vdots & \vdots & \ddots & \vdots & \vdots \\ 0 & 0 & . & \cdots & 1 & c_n \end{pmatrix}.$$

If $x_1, x_2, \ldots, x_n$ are the n successive columns of X, then from $HX = XC$, it is easy to see that knowing x_1, one can compute $x_2, \ldots, x_n$ recursively as

$$Hx_i = x_{i+1}, \quad i = 1, \ldots, n-1.$$

Furthermore, if $x_1 = (1, 0, \ldots, 0)^T$, then it is easy to see that matrix X is a lower triangular matrix with $1, h_{21}, \ldots, h_{21}h_{32} \ldots h_{nn-1}$ as the diagonal entries. Thus, X is nonsingular, since $h_{i+1,i} \neq 0$, $i = 1, 2, \ldots, n-1$. However, if *one or more of these subdiagonal entries is small, then clearly X is ill-conditioned.*

> Thus, the first stage, in which A is transformed to a Hessenberg matrix H using the Householder or the Givens method is numerically stable, while the second stage, in which H is further reduced to a companion matrix C, might be highly unstable.

Example 9.36.

$$H = \begin{pmatrix} 1 & 2 & 3 \\ 0.0001 & 1 & 1 \\ 0 & 2 & 3 \end{pmatrix},$$

$$x_1 = (1, 0, 0)^T, \quad x_2 = Hx_1 = (1, 0.0001, 0)^T,$$

$$x_3 = Hx_2 = (1.0002, 0.0002, 0.0002)^T,$$

$$X = \begin{pmatrix} 1 & 1 & 1.0002 \\ 0 & 0.0001 & 0.0002 \\ 0 & 0 & 0.0002 \end{pmatrix}, \quad X^{-1}HX = C = \begin{pmatrix} 0 & 0 & 1 \\ 1 & 0 & -4.9998 \\ 0 & 1 & 5 \end{pmatrix},$$

$$\text{Cond}_2(X) = 3.1326 \times 10^4.$$

(Note that the *existence of a small subdiagonal entry of H*, namely, h_{21}, made the transforming matrix X ill-conditioned.) ∎

Other methods for reduction to companion form. There are also other equivalent methods for reducing H to C. For example, Wilkinson (1965, p. 400) describes a pivoting method for transforming an unreduced Hessenberg matrix H to a companion matrix C using Gaussian elimination, which also *shows that small subdiagonal entries of H can make the method highly unstable*. The subdiagonal entries are used as pivots, and we have seen before that small pivots can be dangerous.

The well-known **LeVerrier's method** (Wilkinson (1965, pp. 434–435)) computes the coefficients of the characteristic polynomial using the traces of the various powers of A.

9.6. Eigenvalue Sensitivity

Here, Wilkinson has shown that *in LeVerrier's method, severe cancellation can take place while computing the coefficients from the traces using Newton's sums.*

Having emphasized the danger of using the Frobenius form in the eigenvalue computations of a matrix, let's point out some remarks of Wilkinson about Frobenius forms of matrices arising in certain applications such as mechanical and electrical systems.

> *Although we have made it clear that we regard the use of the Frobenius form as dangerous, in that it may well be catastrophically worse-conditioned than the original matrix, we have found the program based on its use surprisingly satisfactory in general for matrices arising from damped mechanical or electrical systems. It is common for the corresponding characteristic polynomial to be well-conditioned. When this is true methods based on the use of the explicit characteristic polynomial are both fast and accurate.* [Wilkinson (1965, p. 482)]

Remarks. The above remarks of Wilkinson clearly support a long tradition by engineers of computing the eigenvalues by finding the zeros of the associated companion matrix. However, in general, it is not a good idea.

9.6 Eigenvalue Sensitivity

In the previous two sections we have cautioned the readers about the danger of computing the eigenvalue via Jordan canonical form or the Frobenius form of a matrix. *The danger was mainly the possibility of the transforming matrix X being ill-conditioned.*

In this section we will see now what specific role the condition number of the transforming matrix X, $\text{Cond}(X) = \|X\| \cdot \|X^{-1}\|$, plays in eigenvalue sensitivity.

We start with a well-known theorem by Friedrich L. Bauer and C. T. Fike.

9.6.1 The Bauer–Fike Theorem

Theorem 9.37 (Bauer and Fike (1960)). *Let A be diagonalizable; that is, there exists a nonsingular matrix X such that $X^{-1}AX = D = \text{diag}(\lambda_1, \ldots, \lambda_n)$. Then for an eigenvalue λ of $A + E$, we have*

$$\min |\lambda_i - \lambda| \leq \|X\| \, \|X^{-1}\| \, \|E\|,$$

where $\|\ \|$ is a subordinate matrix norm and $\lambda_1, \lambda_2, \ldots, \lambda_n$ are the eigenvalues of A.

Proof. Consider two cases.

Case 1: $\lambda = \lambda_i$ for some i. The theorem is trivially true.

Case 2: $\lambda \neq \lambda_i$ for any i. Then the diagonal entries of the diagonal matrix $\lambda I - D$ are different from zero. Since the determinant of a matrix is equal to the product of its eigenvalues, the matrix $(\lambda I - D)$ is nonsingular. Now from $(A + E)x = \lambda x$ we have

$$Ex = (\lambda I - A)x = (\lambda I - XDX^{-1})x = X(\lambda I - D)X^{-1}x. \qquad (9.23)$$

Set $X^{-1}x = y$. Then from (9.23) we have, by multiplying the equation by X^{-1} to the left,

$$(\lambda I - D)y = X^{-1}Ex$$

or
$$y = (\lambda I - D)^{-1} X^{-1} E X y \quad \text{(note that } x = Xy\text{)}.$$

Taking a subordinate norm on both sides, we have

$$\|y\| = \|(\lambda I - D)^{-1} X^{-1} E X y\| \le \|(\lambda I - D)^{-1}\| \, \|X^{-1}\| \, \|E\| \, \|X\| \, \|y\|. \tag{9.24}$$

Dividing both sides by $\|y\|$, we get

$$1 \le \|(\lambda I - D)^{-1}\| \, \|X^{-1}\| \, \|E\| \, \|X\|. \tag{9.25}$$

Now for a subordinate norm,

$$\|(\lambda I - D)^{-1}\| = \max_i \frac{1}{|\lambda - \lambda_i|} = \frac{1}{\min_i |\lambda - \lambda_i|}.$$

So, (9.25) becomes

$$1 \le \frac{1}{\min_i |\lambda - \lambda_i|} \|X^{-1}\| \, \|X\| \, \|E\|.$$

or

$$\min_i |\lambda - \lambda_i| \le \|X^{-1}\| \, \|X\| \, \|E\|. \quad \square$$

Implications of the Theorem

The above theorem tells us that if the p-norm condition number of the eigenvectors matrix X, namely,

$$\text{Cond}_p(X) = \|X^{-1}\|_p \, \|X\|_p,$$

is large, then an eigenvalue λ of the perturbed matrix $A + E$ can be significantly different from an eigenvalue λ_i of A. In general, the more ill-conditioned the eigenvector matrix X is, the more ill-conditioned the eigenproblem for A will be.

Remark. In A is not diagonalizable, a similar result also holds. (For details, see Golub and Van Loan (1996, p. 321).)

Example 9.38. Consider the following upper triangular matrix with eigenvalues 1, 2, and 0.9990:

$$A = \begin{pmatrix} 1 & 2 & 3 \\ 0 & 0.9990 & 1 \\ 0 & 0 & 2 \end{pmatrix},$$

$$X = \begin{pmatrix} 1 & -1 & 0.9623 \\ 0 & 0.0005 & 0.1923 \\ 0 & 0 & 0.1925 \end{pmatrix}; \quad X^{-1} A X = \text{diag}(1, 0.9990, 2).$$

Let $E = 10^{-5} \begin{pmatrix} 0 & 0 & 0 \\ 0 & 0 & 0 \\ 1 & 0 & 0 \end{pmatrix}$. The eigenvalues of $A + E$ are $0.9995 \pm 0.0044i$, 2.

Note that these changes in the eigenvalues are due to the relatively large condition number of X, as the following computations show:

$$\text{Cond}_2(X) = 6.8708 \times 10^3 \quad \text{and} \quad \text{Cond}_2(X) \cdot \|E\|_2 = 0.0687. \quad \blacksquare$$

9.6. Eigenvalue Sensitivity

Thus a change of 10^{-5} in (3, 1)th entry of A completely changed the first two eigenvalues. The question that arises, however, is why did the first two eigenvalues of A change and not the third one? The question will be answered in the following section.

9.6.2 Sensitivity of the Individual Eigenvalues

The condition number $\|X\| \|X^{-1}\|$ gives an overall assessment of the changes in eigenvalues with respect to changes in the coefficients of the matrix. However, as we have seen from the examples in Chapter 4, some eigenvalues of A may be more sensitive than others. In fact, some may be very well-conditioned while others are ill-conditioned. Similarly, some eigenvectors may be well-conditioned while others are not.

It is therefore more appropriate to talk about conditioning of the individual eigenvalues, rather than conditioning of the eigenvalue problem. Recall that in Chapter 4 an analysis of the ill-conditioning of the individual eigenvalues of the slightly perturbed Wilkinson matrix was given in terms of the condition numbers of the individual eigenvalues of this matrix. In general, this can be done for any diagonalizable matrix.

Let $X^{-1}AX = \text{diag}(\lambda_1, \ldots, \lambda_n)$. Then the normalized right and left eigenvectors corresponding to an eigenvalue λ_i are given by

$$x_i = \frac{Xe_i}{\|Xe_i\|_2}, \quad y_i = \frac{(X^{-1})^T e_i}{\|(X^{-1})^T e_i\|_2}.$$

Definition 9.39. *The number $\frac{1}{s_i}$, where s_i is defined by*

$$s_i = |y_i^T x_i|,$$

is called the **condition number** *of the eigenvalue λ_i.*

MATCOM and MATLAB Notes: Individual sensitivities of the eigenvalues can be computed using the MATCOM program **SENSEIG**. See also the MATLAB command **condeig**.

A Relationship between s_i and Cond(X)

It is easy to see that the condition numbers s_i and $\text{Cond}_2(X)$ are related. This relationship is derived in the following:

$$s_i = |y_i^T x_i| = \frac{|e_i^T X^{-1} X e_i|}{\|Xe_i\|_2 \|(X^{-1})^T e_i\|_2} = \frac{1}{\|Xe_i\|_2 \|(X^{-1})^T e_i\|_2}.$$

Now

$$\|Xe_i\|_2 \leq \|X\|_2 \|e_i\|_2 = \|X\|_2$$

and

$$\|(X^{-1})^T e_i\|_2 \leq \|(X^{-1})^T\|_2 \|e_i\|_2 = \|(X^{-1})^T\|_2 = \|X^{-1}\|_2.$$

So,

$$\frac{1}{s_i} \leq \|X\|_2 \|X^{-1}\|_2 = \text{Cond}_2(X).$$

Example 9.40. Consider Example 9.38 again:

$$A = \begin{pmatrix} 1 & 2 & 3 \\ 0 & 0.999 & 1 \\ 0 & 0 & 2 \end{pmatrix},$$

$$\frac{1}{s_1} = 2.8305 \times 10^3, \quad \frac{1}{s_2} = 2.8270 \times 10^3, \quad \frac{1}{s_3} = 5.1940,$$

$$\text{Cond}_2(X) = 6.8708 \times 10^3.$$

Thus,

$$\frac{1}{s_i} < \text{Cond}_2(X), \quad i = 1, 2, 3. \quad \blacksquare$$

Remark. Note that the condition numbers of the eigenvalues 1 and 0.9999 are large. That is why they are sensitive to small perturbations.

The Condition Numbers and Linear Dependence of Eigenvectors

Since for a diagonalizable matrix the columns of the matrix X are the eigenvectors of A, $\text{Cond}_2(X)$ gives us an indication of how linearly independent the eigenvectors are:

If $\text{Cond}_2(X)$ *is large, it means that the eigenvectors are nearly dependent.*

Note the almost linear dependence of the first two eigenvectors of the matrix A of Example 9.38. This is because

$$\text{Cond}_2(X) = 6.8708 \times 10^3.$$

The Eigenvalue Sensitivity of a Normal Matrix

A matrix A is called **normal** if $AA^* = A^*A$, where $A^* = (\bar{A})^T$. *A Hermitian matrix is normal. Normal matrices are diagonalizable.* A remarkable property of a normal matrix A is that if X is the transforming matrix that transforms A to a diagonal matrix, then $\text{Cond}_2(X) = 1$.

Thus the following is an immediate consequence of the Bauer–Fike theorem.

Corollary to the Bauer–Fike theorem. *Let A be a normal matrix, and let $\lambda_1, \ldots, \lambda_n$ be the eigenvalues of A. Then for an eigenvalue λ of $A + E$ we have*

$$\min |\lambda_i - \lambda| \leq \|E\|_2.$$

In other words, the eigenvalues of a normal matrix are perfectly well-conditioned.

Remark (the eigenvalue sensitivity of a symmetric matrix). The normal matrices most commonly found in practical applications are symmetric (or *Hermitian*, if complex) matrices. Thus, by the corollary above, the *eigenvalues of a symmetric (or Hermitian) matrix are well-conditioned.*

9.7 Eigenvector Sensitivity

We shall not go into any detail in our discussion on the sensitivity of eigenvectors, but rather just state a theorem (in somewhat crude form) that will highlight the main differences between eigenvalue and eigenvector sensitivities. For an exact statement and proof, see Watkins (2002, pp. 468–472).

Theorem 9.41. *Let ΔA be a very small perturbation of A and let the eigenvalue λ_k of A be perturbed by $\delta\lambda_k$; that is, $\lambda_k + \delta\lambda_k$ is an eigenvalue of $A + \Delta A$. Let $x_k + \delta x_k$ be the eigenvector corresponding to $\lambda_k + \delta\lambda_k$. Then, assuming that the eigenvalues of A are all distinct, we have*

$$x_k + \delta x_k = x_k + \sum_{j \neq k} \frac{\alpha_{jk}}{(\lambda_k - \lambda_j)s_j} x_j + O(\|\Delta A\|^2), \quad \text{where } \alpha_{jk} = y_i^*(\Delta A)x_j.$$

Implications of the theorem. The above theorem tells us that *if A is perturbed by a small amount, then the amount of perturbation an eigenvector x_k experiences is determined by*

1. the condition numbers of all the eigenvalues other than λ_k, and
2. the distance of λ_k from the other eigenvalues.

An immediate consequence of this theorem is that if *there is a multiple eigenvalue or an eigenvalue near another eigenvalue, then there are some ill-conditioned eigenvectors.* This is significant especially for a Hermitian or a symmetric matrix, because we know that *the eigenvalues of such a matrix are all well-conditioned, but the eigenvectors could be ill-conditioned.* If the eigenvalues are well-separated and well-conditioned, then the eigenvectors are well-conditioned.

Example 9.42. Consider the following diagonal matrix:

$$A = \begin{pmatrix} 1 & 0 & 0 \\ 0 & 0.99 & 0 \\ 0 & 0 & 2 \end{pmatrix}.$$

Let

$$A' = A + \Delta A = \begin{pmatrix} 1 & 0.0001 & 0 \\ 0.0001 & 0.99 & 0 \\ 0 & 0 & 2 \end{pmatrix}.$$

The eigenvalues of $A + \Delta A$ are $1, 0.99, 2$. *(No change; since A is symmetric the eigenvalues are well-conditioned.)* However, the eigenvectors of A' are

$$\begin{pmatrix} -1 \\ -0.01 \\ 0 \end{pmatrix}, \begin{pmatrix} 0.01 \\ -1 \\ 0 \end{pmatrix}, \text{ and } \begin{pmatrix} 0 \\ 0 \\ 1 \end{pmatrix},$$

while those of A are

$$\begin{pmatrix} 1 \\ 0 \\ 0 \end{pmatrix}, \begin{pmatrix} 0 \\ 1 \\ 0 \end{pmatrix}, \text{ and } \begin{pmatrix} 0 \\ 0 \\ 1 \end{pmatrix}.$$

Note that the eigenvector corresponding to $\lambda_3 = 2$ has not changed, while the other two eigenvectors have changed; this is due to the proximity of the associated eigenvalues, 1 and 0.99. ∎

9.8 The Real Schur Form and QR Iterations

In the preceding discussions we have seen that *computing eigenvalues of A via reduction of A to the companion or the Jordan canonical form is not numerically effective*. If the transforming matrix is ill-conditioned, then there may be large errors in the computed canonical form, and this in turn will introduce large errors in the eigenvalues.

Therefore, the *lesson is that we should avoid nonorthogonal transformations in eigenvalue or eigenvector computations* and use only orthogonal or unitary transformations, which are perfectly conditioned.

Indeed, if a matrix A is transformed to a matrix B using unitary similarity transformation, then a perturbation in A will result in a perturbation in B of the same magnitude. That is, if
$$B = U^*AU \quad \text{and} \quad U^*(A + \Delta A)U = B + \Delta B,$$
then
$$\|\Delta B\|_2 \approx \|\Delta A\|_2.$$

Example 9.43.

$$A = \begin{pmatrix} 1 & 2 & 3 \\ 3 & 4 & 5 \\ 6 & 7 & 8 \end{pmatrix}, \quad U = \begin{pmatrix} -0.5774 & -0.5774 & -0.5774 \\ -0.5774 & 0.7887 & -0.2113 \\ -0.5774 & -0.2113 & 0.7887 \end{pmatrix},$$

$$B = U^*AU = \begin{pmatrix} 13 & -0.6340 & -2.3660 \\ -0.9019 & 0 & 0 \\ -6.0981 & 0 & 0 \end{pmatrix}.$$

Let
$$\Delta A = 10^{-5} \times I_{3\times 3}.$$
Then
$$A_1 = A + \Delta A = \begin{pmatrix} 1.00001 & 2 & 3 \\ 3 & 4.00001 & 5 \\ 6 & 7 & 8.00001 \end{pmatrix}$$
and
$$B_1 = U^*(A + \Delta A)U = \begin{pmatrix} 13.00001 & -0.633974 & -2.3660 \\ -0.9019 & 0.00001 & 0 \\ -6.0981 & 0 & 0.00001 \end{pmatrix}.$$

So, $\Delta B = B_1 - B = 10^{-5} \times I_{3\times 3}$ and $\|\Delta A\|_2 = \|\Delta B\|_2 = 10^{-5}$. ∎

A perfect canonical form displaying the eigenvalues is a triangular form (the diagonal entries are the eigenvalues). In this context we now recall a classical result due to Schur.[14]

---
[14]Issai Schur (1875–1941), a Lithuanian–German–Israeli mathematician, is well known for his fundamental work on the representation theory of groups, but also worked in number theory, analysis, and linear algebra.

9.8. The Real Schur Form and QR Iterations

Theorem 9.44 (Schur triangularization theorem). *If A is an $n \times n$ matrix, then there exists a unitary matrix U such that*
$$U^*AU = T,$$
where T is a triangular matrix with the eigenvalues $\lambda_1, \lambda_2, \ldots, \lambda_n$ as the diagonal entries.

Proof. We will prove the theorem using induction on n.

If $n = 1$, the theorem is trivially true. Next assume the theorem is true for $n = k - 1$. Then we will show that it is also true for $n = k$.

Let u be a normalized eigenvector of A associated with an eigenvalue λ_1. Define $U_1 = (u, V)$, where V is such that U_1 is unitary. Then $A_1 = U_1^* A U_1 = \begin{pmatrix} \lambda_1 & * \\ 0 & \hat{A} \end{pmatrix}$, where $\hat{A}$ is $(k - 1) \times (k - 1)$. By our hypothesis there exists a unitary matrix V_1 of order $(k-1)$ such that $\hat{T} = V_1^*(\hat{A})V_1$ is triangular. Then, defining $U_2 = \text{diag}(1, V_1)$, we see that U_2 is unitary (because V_1 is so) and
$$U_2^* A_1 U_2 = U_2^* U_1^* A U_1 U_2 = U^* A U = \begin{pmatrix} \lambda_1 & * \\ 0 & \hat{T} \end{pmatrix}$$

Because $\hat{T}$ is triangular, so is U^*AU. Since the eigenvalues of a triangular matrix appear on the diagonal, we are done. $\square$

Since a real matrix can have complex eigenvalues (occurring in complex conjugate pairs), even for a real matrix A, U and T in the Schur theorem above can be complex. However, *we can choose U to be real orthogonal if T is replaced by a quasi-triangular matrix R,* known as the **real Schur form** of A.

Theorem 9.45 (real Schur triangularization theorem). *Let A be an $n \times n$ real matrix. Then there exists an $n \times n$ orthogonal matrix Q such that*
$$Q^T A Q = R = \begin{pmatrix} R_{11} & R_{12} & \cdots & R_{1k} \\ 0 & R_{22} & \cdots & R_{2k} \\ \vdots & & \ddots & \vdots \\ 0 & 0 & \cdots & R_{kk} \end{pmatrix},$$
where each R_{ii} is either a scalar or a 2×2 matrix. The scalar diagonal entries correspond to real eigenvalues and 2×2 matrices on the diagonal correspond to complex conjugate eigenvalues.

Proof. The proof is similar to that of Theorem 9.44. $\square$

Definition 9.46. *The matrix R in Theorem 9.45 is known as the **real Schur form** of A.*

Notes:

- The 2×2 matrices on the diagonal are usually referred to as "**bumps**."

- The columns of Q are called *Schur vectors*. For each $k (1 \leq k \leq n)$, the first k columns of Q form an orthonormal basis for the invariant subspace corresponding to the first k eigenvalues.

Remark. Since the proofs of both the theorems are based on the knowledge of eigenvalues and eigenvectors of matrix A, they cannot be considered to be constructive. They do not help us in computing the eigenvalues and eigenvectors.

We present below a method, known as the QR iteration method, for computing the real Schur form of A. A properly implemented QR method is widely used nowadays for computing the eigenvalues of an arbitrary matrix. As the name suggests, the method is based on the QR factorization and is iterative in nature. The QR iteration method was proposed in algorithmic form by Francis (1961), though its roots can be traced to a work of Rutishauser (1958). The method was also independently discovered by the Russian mathematician Kublanovskaya (1961).

Note: Since the eigenvalues of a matrix A are the n zeros of its characteristic polynomial, and it is well known (proved by Galois more than a century ago) that *the roots of a polynomial equation of degree higher than four cannot be found in a finite number of steps*, any numerical eigenvalue method for an arbitrary matrix has to be iterative in nature.

9.8.1 The Basic QR Iteration

The idea behind the QR iteration method is to iteratively construct a sequence of matrices $\{A_k\}$, starting from $A_0 = A$, such that each A_{i+1} is orthogonally similar to A_i, with an expectation that the sequence will converge to a real Schur matrix from which the eigenvalues of A can be easily extracted. Each matrix in the sequence is constructed by taking QR factorization of the previous matrix and then multiplying the matrices Q and R in reverse order. Specifically, the **basic QR iteration** method is as follows.

ALGORITHM 9.6. Basic QR Iteration Algorithm.

Input: An $n \times n$ matrix A.
Output: A sequence of matrices $\{A_k\}$ containing the eigenvalues of A.

Step 1. Set $A_0 = A$.

Step 2. Compute now a sequence of matrices (A_k) defined as follows:
For $k = 1, 2, \ldots$ do

    2.1. Find the QR factorization of A_{k-1} : $A_{k-1} = Q_{k-1} R_{k-1}$
    (QR factorization).

    2.2. Compute $A_k = R_{k-1} Q_{k-1}$ (reverse multiplication).
End

Eigenvalue property of $\{A_k\}$. The matrices in the sequence $\{A_k\}$ have a very interesting property. *Each matrix in the sequence is orthogonally similar to the previous one and is therefore orthogonally similar to the original matrix.* It is easy to see this. For example,

$$A_1 = R_0 Q_0 = Q_0^T A_0 Q_0 \quad \text{(since } Q_0^T A_0 = R_0\text{)},$$
$$A_2 = R_1 Q_1 = Q_1^T A_1 Q_1.$$

9.8. The Real Schur Form and QR Iterations

Thus A_1 is orthogonally similar to A, and A_2 is orthogonally similar to A_1. Therefore, A_2 is orthogonally similar to A, as the following computation shows:

$$A_2 = Q_1^T A_1 Q_1 = Q_1^T (Q_0^T A_0 Q_0) Q_1 = (Q_0 Q_1)^T A_0 (Q_0 Q_1) = (Q_0 Q_1)^T A Q_0 Q_1.$$

Since each matrix is orthogonally similar to the original matrix A, it has the same eigenvalues as A. Thus, if the sequence $\{A_k\}$ converges to a triangular or quasi-triangular matrix, we will be done. The following result shows that under certain conditions, this indeed happens (see Wilkinson (1965, pp. 518–519)).

A Condition for Convergence

Theorem 9.47 (convergence theorem for basic QR iteration). *Let the eigenvalues $\lambda_1, \ldots, \lambda_n$ be such that $|\lambda_1| > |\lambda_2| > \cdots > |\lambda_n|$, and let the eigenvector matrix X of the left eigenvectors (that is, the right eigenvectors of X^{-1}) be such that its leading principal minors are nonzero. Then $\{A_k\}$ converges to an upper triangular matrix or to the real Schur form.*

In fact, it can be shown that under the above conditions, the first column of A_k approaches a multiple of e_1. Thus, for sufficiently large k we get

$$A_k = \begin{pmatrix} \lambda_1 & u \\ 0 & \overline{A_k} \end{pmatrix}.$$

We can apply the QR iteration again to $\overline{A}_k$ and the process can be continued to show that the sequence converges to an upper triangular matrix.

Example 9.48.

$$A = \begin{pmatrix} 1 & 2 \\ 3 & 4 \end{pmatrix}.$$

The eigenvalues of A are 5.3723 and -0.3723. $|\lambda_1| > |\lambda_2|$.

$k = 0$: $\quad A_0 = A = Q_0 R_0,$

$$Q_0 = \begin{pmatrix} -0.3162 & -0.9487 \\ -0.9487 & 0.3162 \end{pmatrix}, \quad R_0 = \begin{pmatrix} -3.1623 & -4.4272 \\ 0 & -0.6325 \end{pmatrix}.$$

$k = 1$: $\quad A_1 = R_0 Q_0 = \begin{pmatrix} 5.2 & 1.6 \\ .6 & -.2 \end{pmatrix} = Q_1 R_1,$

$$Q_1 = \begin{pmatrix} -0.9934 & -0.1146 \\ -0.1146 & -0.9934 \end{pmatrix}, \quad R_1 = \begin{pmatrix} -5.2345 & -1.5665 \\ 0 & -0.3821 \end{pmatrix}.$$

$k = 2$: $\quad A_2 = R_1 Q_1 = \begin{pmatrix} 5.3796 & -0.9562 \\ 0.0438 & -0.3796 \end{pmatrix} = Q_2 R_2.$

(Note that we have already made some progress towards obtaining the eigenvalues.)

$$Q_2 = \begin{pmatrix} -1 & -0.0082 \\ -0.0081 & 1 \end{pmatrix}, \quad R_2 = \begin{pmatrix} -5.3797 & 0.9593 \\ 0 & -0.3718 \end{pmatrix}.$$

$k = 3$: $\quad A_3 = R_2 Q_2 = \begin{pmatrix} 5.3718 & 1.0030 \\ 0.0030 & -0.3718 \end{pmatrix} = Q_3 R_3,$

$$Q_3 = \begin{pmatrix} 1 & -0.0006 \\ -0.0006 & 1 \end{pmatrix}, \quad R_3 = \begin{pmatrix} -5.3718 & -1.0028 \\ 0 & -0.3723 \end{pmatrix}.$$

$k = 4$: $\quad A_4 = R_3 Q_3 = \begin{pmatrix} 5.3723 & -0.9998 \\ 0.0002 & -0.3723 \end{pmatrix}.$ ∎

MATCOM Note: Algorithm 9.6 has been implemented in the MATCOM program **QRITRB**.

9.8.2 The Hessenberg QR Iteration

The QR iteration method as presented above is not efficient if the matrix A is full and dense. We have seen in Chapter 7 that the QR factorization of such a matrix A requires $O(n^3)$ flops, and thus n QR iterations will require $O(n^4)$ flops, making the method impractical.

Fortunately, something simple can be done: *Reduce the matrix A to a Hessenberg matrix by orthogonal similarity before starting the QR iterations.*

The question now is, *Will the Hessenberg structure be preserved at each iteration step?* The answer is *yes* and provided by the following theorem. Note that the Hessenberg matrix at each iteration has to be *unreduced*. This is not a restriction, because the eigenvalue problem for a reduced Hessenberg matrix can be split into eigenvalue problems of unreduced Hessenberg matrices (Exercise 9.21(c)).

Theorem 9.49. *Let A_k be an unreduced upper Hessenberg matrix and let $A_k = Q_k R_k$ be the QR factorization of A_k. Then $A_{k+1} = R_k Q_k$ is also upper Hessenberg.*

Proof. Suppose Givens rotations are used to factorize A_k into $Q_k R_k$. Then

$$Q_k = J(2, 1, \theta) J(3, 2, \theta) \cdots J(n, n-1, \theta)$$

is also upper Hessenberg.

Again, since R_k is upper triangular and Q_k is upper Hessenberg, $A_{k+1} = R_k Q_k$ is also upper Hessenberg. □

An implication. Since the QR factorization of a Hessenberg matrix requires only $O(n^2)$ flops, the *QR iteration method with the initial reduction of A to a Hessenberg matrix* will be an $O(n^3)$ method.

Example 9.50. **Illustration of invariance of Hessenberg form in Hessenberg QR iteration.**

$$A = A_0 = \begin{pmatrix} 0.2190 & -0.5651 & -0.6418 \\ -0.6805 & 0.1226 & 0.4398 \\ -0.0000 & 0.8872 & 0.8466 \end{pmatrix} \quad \text{(Hessenberg)}.$$

9.8. The Real Schur Form and QR Iterations

k = 1: 1. $Q_0 = \begin{pmatrix} -0.3063 & -0.4676 & 0.8291 \\ 0.9519 & -0.1505 & 0.2668 \\ 0.0000 & 0.8710 & 0.4913 \end{pmatrix}$,

$R_0 = \begin{pmatrix} -0.7149 & 0.2898 & 0.6152 \\ 0.0000 & 1.0186 & 0.9714 \\ -0.0000 & 0 & 0.0011 \end{pmatrix}$.

2. $A_1 = R_0 Q_0 = \begin{pmatrix} 0.4949 & 0.8265 & -0.2132 \\ 0.9697 & 0.6928 & 0.7490 \\ 0.0000 & 0.0010 & 0.0006 \end{pmatrix}$ (Hessenberg).

k = 2: 1. $Q_1 = \begin{pmatrix} 0.4546 & -0.8907 & -0.0021 \\ 0.8907 & 0.4546 & 0.0011 \\ 0.0000 & -0.0023 & 1.0000 \end{pmatrix}$,

$R_1 = \begin{pmatrix} 1.0886 & 0.9928 & 0.5702 \\ 0.0000 & -0.4213 & 0.5303 \\ 0.0000 & -0.0000 & 0.0018 \end{pmatrix}$.

2. $A_2 = R_1 Q_1 = \begin{pmatrix} 1.3792 & -0.5197 & 0.5690 \\ -0.3752 & -0.1927 & 0.5299 \\ -0.0000 & -0.0000 & 0.0018 \end{pmatrix}$ (Hessenberg). ∎

MATCOM Note: The Hessenberg QR iteration algorithm has been implemented in the MATCOM program **QRITRH**.

9.8.3 Convergence of the QR Iterations and the Shift of Origin

Although an initial reduction to a Hessenberg matrix makes the QR iteration algorithm an $O(n^3)$ method, the rate of convergence, that is, *the rate at which a subdiagonal entry approaches zero, can still be very slow if an eigenvalue λ_i is close to the previous one, λ_{i-1}.* This is because the rate is determined by the ratio

$$\left| \frac{\lambda_i}{\lambda_{i-1}} \right|^k.$$

Fortunately, the *rate can be improved substantially by using a shift $\hat{\lambda}_i$, close to the eigenvalue λ_i*, as illustrated by the following example (Ortega and Poole (1981, p. 227)):

Suppose $\lambda_i = 0.99$, $\lambda_{i-1} = 1.1$, and $\hat{\lambda}_i = 1$. Then $\left| \frac{\lambda_i - \hat{\lambda}_i}{\lambda_{i-1} - \hat{\lambda}_i} \right| = 0.1$, while $\left| \frac{\lambda_i}{\lambda_{i-1}} \right| = 0.9$.

This observation tells us that if we apply the QR iteration to the shifted matrix $\hat{H} = H - \hat{\lambda} I$, where $\hat{\lambda}$ is a suitable shift, rather than to the original matrix H, then the rate of convergence will be faster. Of course, once an eigenvalue of $\hat{H}$ is found, the corresponding eigenvalue of H can be computed just by adding the shift back (Exercise 9.9). *At each iteration, the (n, n)th element of the current matrix can be taken as the shift.*

The above procedure is known as **single-shift Hessenberg QR iteration method.** The details of the procedure are left as an exercise for the readers (Exercise 9.30). The process is meaningful if an approximation to a real eigenvalue is desired. In the case of approximating a complex conjugate pair of eigenvalues, the following modified procedure, known as the **double-shift QR iteration**, is to be used *by taking the eigenvalues of the 2×2 trailing principal submatrix on the bottom right-hand corner as the shift parameters.*

9.8.4 The Double-Shift QR Iteration

In the following we describe one iteration step of the double-shift Hessenberg QR iterations.

One Iteration Step of the Double-Shift QR (Complex)

Let the pair of complex conjugate eigenvalues of the 2×2 bottom right-hand corner of the starting Hessenberg matrix H_0 be k_1 and $k_2 = \bar{k}_1$. Then the *first iteration step* of the double-shift QR iteration is given by

$$H_0 - k_1 I = Q_0 R_0, \quad H_1 = R_0 Q_0 + k_1 I,$$
$$H_1 - k_2 I = Q_1 R_1, \quad H_2 = R_1 Q_1 + k_2 I.$$

Avoiding Complex Arithmetic in Double-Shift QR Iteration

Since k_1 and k_2 are complex, the above double-shift QR iteration step will require complex arithmetic for implementation, even though the starting matrix H_0 is real. *However, with a little manipulation complex arithmetic can be avoided.* We will discuss this aspect now.

We will show that matrix H_2 is orthogonally similar to H_0 via a real transforming matrix, and can be formed directly from H_0 without computing H_1.

Step 1. (H_2 *is orthogonally similar to H_0.*)

$$H_2 = R_1 Q_1 + k_2 I = Q_1^*(H_1 - k_2 I)Q_1 + k_2 I = Q_1^*(R_0 Q_0 + (k_1 - k_2)I)Q_1 + k_2 I$$
$$= Q_1^*(Q_0^*(H_0 - k_1 I)Q_0 + (k_1 - k_2)I)Q_1 + k_2 I = Q_1^* Q_0^* H_0 Q_0 Q_1.$$

Thus, $H_2 = (Q_0 Q_1)^* H_0 Q_0 Q_1$, proving that H_2 and H_0 are orthogonally similar.

Step 2. (*The matrix $Q_0 Q_1$ from Step 1 is a real matrix.*) To show this, we define the matrix

$$N = (H_0 - k_2 I)(H_0 - k_1 I).$$

Then we show that (i) N is a real matrix, and (ii) the matrix $Q_0 Q_1$ is the Q matrix of the QR factorization of N.

- (*Matrix N is real.*) $N = (H_0 - k_2 I)(H_0 - k_1 I) = H_0^2 - (k_1 + k_2)H_0 + k_1 k_2 I = H_0^2 - tH_0 + dI$, where $t = k_1 + k_2$ and $d = k_1 k_2$. Thus matrix N is real (since $k_2 = \bar{k}_1$).

- ($Q_0 Q_1$ *is the Q matrix of the QR factorization of N.*) $N = (H_0 - k_2 I)(H_0 - k_1 I) = (H_0 - k_2 I)Q_0 R_0 = Q_0 Q_0^*(H_0 - k_2 I)Q_0 R_0 = Q_0(H_1 - k_2 I)R_0 = Q_0 Q_1 R_1 R_0$.

  So, the matrix $Q_0 Q_1 R_1 R_0$ is the *QR factorization of N*, and the matrix $Q_0 Q_1$ is the Q matrix. Since N is real, so is $Q_0 Q_1$. Combining the result of the above two steps, we see that H_0 is similar to H_2 via a real orthogonal similarity transformation.

9.8. The Real Schur Form and QR Iterations

This allows us to write the *one-step* of double-shift QR iteration in *real arithmetic* as follows.

One Step of Explicit Double-Shift QR Iteration (Real Arithmetic)

- Form the real matrix $N = H_0^2 - tH_0 + dI$.
- Find the QR factorization of N: $N = QR$.
- Form $H_2 = Q^T H_0 Q$.

We will call the above computation **explicit double-shift** QR iteration for reasons to be stated in the next section.

Example 9.51.

$$H = H_0 = \begin{pmatrix} 1 & 2 & 3 \\ 1 & 0 & 1 \\ 0 & -2 & 2 \end{pmatrix}; \quad t = 2, \quad d = 2.$$

$$N = H^2 - tH + dI = \begin{pmatrix} 3 & -8 & 5 \\ -1 & 2 & 3 \\ -2 & 0 & 0 \end{pmatrix}.$$

The Q matrix of the QR factorization of N:

$$Q = \begin{pmatrix} -0.8018 & -0.5470 & -0.2408 \\ 0.2673 & 0.0322 & -0.9631 \\ 0.5345 & -0.8365 & 0.1204 \end{pmatrix},$$

$$H_2 = Q^T H_0 Q = \begin{pmatrix} -0.8571 & 1.1007 & 2.5740 \\ -1.1867 & 3.0455 & -0.8289 \\ 0.0000 & 1.8437 & 0.8116 \end{pmatrix}. \quad \blacksquare$$

MATCOM Note: The explicit single-shift and double-shift QR iterations have been implemented in the MATCOM programs **QRITRSSE** and **QRITRDSE**, respectively.

9.8.5 Implicit QR Iteration

After all this, we note, with utter disappointment, that *the above double-shift (explicit) QR iteration is not practical.* The reason for this is that forming the matrix N itself in Step 2 requires $O(n^3)$ flops. Fortunately, a little trick again allows us to implement the step in $O(n^2)$ flops.

One Iteration of the Implicit Double-Shift QR

1. Compute the first column n_1 of the matrix N.
2. Find a Householder matrix P_0 such that $P_0 n_1$ is a multiple of e_1.
3. Form $H_0' = P_0^T H_0 P_0$.
4. Compute Householder matrices P_1 through P_{n-2} such that if $Z = P_0 P_1 \ldots P_{n-2}$, then $H_2' = Z^T H_0' Z$ is upper Hessenberg and the first column of Q and Z are the same.

By using the implicit Q theorem (Theorem 9.34) *we can then show that matrices* H_2 *of the explicit QR method and* H_2' *of the implicit QR method are both unreduced upper Hessenberg and are essentially the same matrix.*

The above four steps constitute *one iteration* of the double-shift implicit QR.

It now remains to show that (i) *the above implicit computation requires only* $O(n^2)$ *flops* (instead of $O(n^3)$), and (ii) Z and Q have the same first column.

A Close Look at $O(n^2)$ computation of the Double-shift Implicit QR Iteration Step

- The entries of vector n_1, the first column of N, can be explicitly written down:
$$n_1 = (h_{11}^2 + h_{12}h_{21} - th_{11} + d, h_{21}(h_{11} + h_{22} - t), h_{21}h_{32}, 0, \ldots, 0)^T.$$

- Because n_1 has almost three nonzero entries, P_0 has the form $P_0 = \text{diag}(\hat{P}_0, I_{n-3})$, where $\hat{P}_0$ is a 3×3 Householder matrix. Thus the computation of P_0 requires only $O(1)$ flops.

- Because of the above structure P_0 and the matrix H_0 being Hessenberg, the matrix $P_0^T H_0 P_0$ is not a full matrix. *It is a Hessenberg matrix with a bulge.*

  For example, when $n = 6$, we have

$$H_0' = P_0^T H_0 P_0 = \begin{pmatrix} \times & \times & \times & \times & \times & \times \\ \times & \times & \times & \times & \times & \times \\ + & \times & \times & \times & \times & \times \\ + & + & \times & \times & \times & \times \\ 0 & 0 & 0 & \times & \times & \times \\ 0 & 0 & 0 & 0 & \times & \times \end{pmatrix}.$$ (The entries indicated by + form a bulge.)

Bulge-chasing phenomenon. A bulge will be created at each step of the reduction of H_0 to Hessenberg form, and the constructions of Householder matrices P_1 through P_{n-2} amount to chasing these bulges systematically, as shown below with the previous 6×6 case. The entries (3, 1), (4, 1), and (4, 2) form a bulge in H_0'.

1. Create P_1:

$$P_1^T H_0' P_1 = \begin{pmatrix} \times & \times & \times & \times & \times & \times \\ \times & \times & \times & \times & \times & \times \\ 0 & \times & \times & \times & \times & \times \\ 0 & + & \times & \times & \times & \times \\ 0 & + & + & \times & \times & \times \\ 0 & 0 & 0 & 0 & 0 & \times \end{pmatrix}.$$

The entries (3,1) and (4,1) of H_0' are annihilated and the 2×2 bulge has been chased one column down.

2. Create P_2:

$$P_2^T (P_1^T H_0' P_1) P_2 = \begin{pmatrix} \times & \times & \times & \times & \times & \times \\ \times & \times & \times & \times & \times & \times \\ 0 & \times & \times & \times & \times & \times \\ 0 & 0 & \times & \times & \times & \times \\ 0 & 0 & + & \times & \times & \times \\ 0 & 0 & + & + & \times & \times \end{pmatrix}.$$

9.8. The Real Schur Form and QR Iterations

The entries (4, 2) and (5, 2) of the matrix $P_1^T H_0' P_1$ have been annihilated and the 2×2 bulge has been chased further down one column.

3. *Create P_3:*

$$P_3^T (P_2^T P_1^T H_0' P_1 P_2) P_3 = \begin{pmatrix} \times & \times & \times & \times & \times & \times \\ \times & \times & \times & \times & \times & \times \\ 0 & \times & \times & \times & \times & \times \\ 0 & 0 & \times & \times & \times & \times \\ 0 & 0 & 0 & \times & \times & \times \\ 0 & 0 & 0 & + & \times & \times \end{pmatrix}.$$

The entries (5, 3) and (6, 3) of the matrix $P_2^T P_1^T H_0' P_1 P_2$ have been annihilated and the 2×2 bulge has been chased to a 1×1 bulge still further one column down.

4. *Create P_4:*

$$P_4^T (P_3^T P_2^T P_1^T H_0' P_1 P_2 P_3) P_4 = Z^T H_0' Z = H_2' = \begin{pmatrix} \times & \times & \times & \times & \times & \times \\ \times & \times & \times & \times & \times & \times \\ 0 & \times & \times & \times & \times & \times \\ 0 & 0 & \times & \times & \times & \times \\ 0 & 0 & 0 & \times & \times & \times \\ 0 & 0 & 0 & 0 & \times & \times \end{pmatrix}.$$

The last bulge has now been eliminated and the matrix H_2' is Hessenberg again, this time without any bulge and possibly some smaller subdiagonal entries.

In the general case, $(n-2)$ Householder matrices $P_1, \ldots, P_{n-2}$ have to be created and each P_k, $k = 1, 2, \ldots, n-3$, has the form

$$P_k = \begin{pmatrix} I_k & & 0 \\ & \hat{P}_k & \\ 0 & & I_{n-k-3} \end{pmatrix},$$

where $\hat{P}_k$ is a 3×3 Householder matrix. The last Householder matrix P_{n-2} has the form

$$P_{n-2} = \begin{pmatrix} I_{n-2} & 0 \\ 0 & \hat{P}_{n-2} \end{pmatrix}.$$

Taking into consideration the above structures of computations, we see that one step of the double-shift implicit QR iteration requires only $O(n^2)$ flops.

For details of this $O(n^2)$ computations of one iteration of the double-shift implicit QR, see the book by Stewart (1972, pp. 375–378).

To see that the matrices Q and Z have the same first column, observe that (i) $P_k e_1 = e_1$ for $k = 1, \ldots, n-2$, and (ii) P_0 and Q have the same first column. Thus $Z e_1 = P_0 P_1 \ldots P_{n-2} e_1 = P_0 e_1 = Q e_1$.

ALGORITHM 9.7. One Iteration Step of the Double-Shift Implicit QR.

Input: An unreduced upper Hessenberg matrix H.
Output: An unreduced upper Hessenberg matrix $Q^T H Q$, where $Q = P_0 P_1 \ldots P_{n-2}$ is a product of Householder matrices. The matrix $Q^T H Q$ is stored over H.

Step 1. Compute the numbers t and d as follows:
$$t = h_{n-1,n-1} + h_{nn}, \quad d = h_{n-1,n-1} h_{nn} - h_{n,n-1} h_{n-1,n}.$$

Step 2. Compute the first three nonzero entries of the first column of $N = H^2 - tH + dI$:
$$x \equiv n_{11} = h_{11}^2 - t h_{11} + d + h_{12} h_{21},$$
$$y \equiv n_{21} = h_{21}(h_{11} + h_{22} - t),$$
$$z \equiv n_{32} = h_{21} h_{32}.$$

Step 3. Compute the Householder matrices $P_0 P_1 \ldots P_{n-2}$ such that the final matrix is upper Hessenberg:

(a) For $k = 0, 1, 2, \ldots, n-3$ do

(i) Find a 3×3 Householder matrix $\hat{P}_k$ such that
$$\hat{P}_k \begin{pmatrix} x \\ y \\ z \end{pmatrix} = \begin{pmatrix} * \\ 0 \\ 0 \end{pmatrix}.$$

Implicitly form $P_k^T H P_k$, where $P_k = \text{diag}(I_k, \hat{P}_k, I_{n-k-3})$, and store it over H: $H \equiv P_k^T H P_k$.

(ii) Update x, y, and z:
$$x \equiv h_{k+2,k+1}, \; y \equiv h_{k+3,k+1}, \text{ and (if } k < n-3\text{) } z \equiv h_{k+4,k+1}.$$

End

(b) Find a Householder matrix $\hat{P}_{n-2}$ of order 2 such that
$$\hat{P}_{n-2} \begin{pmatrix} x \\ y \end{pmatrix} = \begin{pmatrix} * \\ 0 \end{pmatrix}.$$

Implicitly form $P_{n-2}^T H P_{n-2}$, where $P_{n-2} = \text{diag}(I_{n-2}, \hat{P}_{n-2})$, and store it over $H : H \equiv P_{n-2}^T H P_{n-2}$.

Flop-count. One iteration of the implicit double-shift QR method takes about $10n^2$ flops. If the transforming matrix Q is needed and accumulated, then another $10n^2$ flops will be needed (see Golub and Van Loan (1996, p. 358).

9.8. The Real Schur Form and QR Iterations

Example 9.52. Consider the same matrix H as in Example 9.51.

Step 1. $t = 2$, $d = 2$,

$$H = \begin{pmatrix} 1 & 2 & 3 \\ 1 & 0 & 1 \\ 0 & -2 & 2 \end{pmatrix}.$$

Step 2. $x = n_{11} = 3$, $y = n_{21} = -1$, $z = n_{31} = -2$.

Step 3. $k = 0$:

$$P_0 = I - \frac{2uu^T}{u^T u}, \quad \text{where } u = \begin{pmatrix} 6.7417 \\ -1.000 \\ -2.000 \end{pmatrix},$$

$$P_0 = \begin{pmatrix} -0.8018 & 0.2673 & 0.5345 \\ 0.2673 & 0.9604 & -0.0793 \\ 0.5345 & -0.0793 & 0.8414 \end{pmatrix},$$

$$H \equiv P_0^T H P_0 = \begin{pmatrix} -0.8571 & -2.6248 & -0.9733 \\ 0.0581 & 0.8666 & 1.9505 \\ \boxed{1.1852} & -0.7221 & 2.9906 \end{pmatrix}.$$

Update x and y:

$$x = h_{21} = .0581 \quad y = h_{31} = 1.1852.$$

Find P_1:

$$\hat{P}_1 = \begin{pmatrix} 0.0490 & -0.9988 \\ -0.9988 & 0.0490 \end{pmatrix}, \quad \hat{P}_1 \begin{pmatrix} x \\ y \end{pmatrix} = \begin{pmatrix} -1.1866 \\ 0 \end{pmatrix},$$

$$P_1 = \begin{pmatrix} 1 & 0 & 0 \\ 0 & 0.0490 & -0.9988 \\ 0 & -0.9988 & 0.0490 \end{pmatrix}; \quad H \equiv P_1^T H P_1 = \begin{pmatrix} -0.8571 & 1.1007 & 2.5740 \\ -1.1867 & 3.0455 & -0.8289 \\ 0 & 1.8437 & 0.8116 \end{pmatrix}.$$

Note: The matrix H obtained by the implicit QR is the same as H_2 obtained earlier in Example 9.51 using the explicit QR. ∎

MATCOM Note: Algorithm 9.7 has been implemented in the MATCOM program **QRITRDSI**.

9.8.6 Obtaining the Real Schur Form A

Putting together the results of the preceeding section, we can now formulate the procedure of obtaining the real Schur form as follows:

Step 1. Transform the matrix A to Hessenberg form H.

Step 2. Perform the QR iteration method on the Hessenberg matrix H using implicit double-shift.

Typically, after two to three steps of the doubly-shift implicit QR iteration, one or two (and sometime more) subdiagonal entries from the bottom of the Hessenberg matrix converge to zero.

This then will give us a real or a pair of complex conjugate eigenvalues. Once a real or a pair of complex conjugate eigenvalues is computed, the last row and the last column in the first case, or the last two rows and the last two columns in the second case, can be deleted, and computation of the other eigenvalues can be continued with the submatrix. This process is known as **deflation**.

Note that the eigenvalues of the deflated submatrix are also the eigenvalues of the original matrix. For, suppose immediately before deflation, the matrix has the form

$$H_k = \begin{pmatrix} A' & C' \\ 0 & B' \end{pmatrix},$$

where B' is the 2×2 trailing submatrix or is a 1×1 matrix. Then the characteristic equation of H_k is

$$\det(\lambda I - H_k) = \det(\lambda I - A') \det(\lambda I - B').$$

Thus, the eigenvalues of H_k are the eigenvalues of A' together with those of B'. Since H_k is orthogonally similar to the original matrix A and therefore has the same eigenvalues as A, the eigenvalues of B' are also the eigenvalues of A.

When to Accept a Subdiagonal Entry as Zero

A major decision that we have to make during the iteration procedure is when to accept a subdiagonal entry as zero so that the matrix can be deflated.

Accept a subdiagonal entry $h_{i,i-1}$ to be zero if

$$|h_{i,i-1}| \leq \text{tol} \, (|h_{ii}| + |h_{i-1,i-1}|),$$

where tol is a tolerance greater than the unit round-off (Golub and Van Loan (1996, p. 359)).

Example 9.53. Consider

$$H = \begin{pmatrix} 1 & 2 & 3 \\ 1 & 0 & 1 \\ 0 & -2 & 2 \end{pmatrix}.$$

Iteration	h_{21}
1	-1.1867
2	0.3543
3	0.0129
4	0.0000

9.8. The Real Schur Form and QR Iterations

The real Schur form is

$$\begin{pmatrix} -1.1663 & -1.3326 & -2.0531 \\ 0 & \boxed{1.2384 \quad 1.6659} \\ 0 & \boxed{-1.9409 \quad 2.9279} \end{pmatrix}.$$

The eigenvalues of the 2×2 right-hand lower corner submatrix are $2.0832 \pm 1.5874i$. ∎

Thus the eigenvalues of A are $-1.1663, 2.0832 \pm 1.5874i$

Example 9.54. Consider

$$H = \begin{pmatrix} 0.2190 & -0.0756 & 0.6787 & -0.6391 \\ -0.9615 & 0.9032 & -0.4571 & 0.8804 \\ 0 & -0.3822 & 0.4526 & -0.0641 \\ 0 & 0 & -0.1069 & -0.0252 \end{pmatrix}.$$

Iteration	h_{21}	h_{32}	h_{43}
1	0.3860	-0.5084	-0.0064
2	-0.0672	-0.3773	0.0001
3	0.0089	-0.3673	0
4	-0.0011	-0.3590	0
5	0.0001	-0.3905	0

The real Schur form is

$$H = \begin{pmatrix} 1.4095 & 0.7632 & -0.1996 & 0.8394 \\ 0.0001 & \boxed{0.1922 \quad 0.5792} & 0.0494 \\ 0 & \boxed{-0.3905 \quad 0.0243} & -0.4089 \\ 0 & 0 & 0 & -0.0763 \end{pmatrix}.$$

The eigenvalues of $\begin{pmatrix} 0.1922 & 0.5792 \\ -0.3905 & 0.0243 \end{pmatrix}$ are $0.1082 \pm 0.4681i$.
Thus, the eigenvalues of H are $1.4095, 0.1082 \pm 0.4681i$, and -0.0763. ∎

Flop-count. Since the QR iteration method is an iterative method, it is hard to give an exact flop-count for this method. However, empirical observations have established that it takes about two QR iterations per eigenvalue. *Thus, it will require about $10n^3$ flops to compute all the eigenvalues* (Golub and Van Loan (1996, p. 359)). *If the transforming matrix Q and the final real Schur matrix T are also needed, then the cost will be about $25n^3$ flops.*

Round-off property. *The QR iteration method is stable.* An analysis of the round-off property of the algorithm shows that the computed real Schur form T is orthogonally similar to a nearby matrix $A + E$. Specifically,

$$Q^T(A+E)Q = T,$$

where $Q^T Q = I$ and $\|E\|_F \leq \phi(n)\mu\|A\|_F$. Here $\phi(n)$ is a *slowly growing function of n*. The computed orthogonal matrix $\hat{Q}$ is also almost orthogonal.

Balancing. As in the process of solving a linear system problem, it is advisable to balance the entries of the original matrix A, if they vary widely, before starting the QR iterations.

The balancing is equivalent to transforming the matrix A to $D^{-1}AD$, where the diagonal matrix *D is chosen so that a norm of each row is approximately equal to the norm of the corresponding column.*

In general, *preprocessing the matrix by balancing improves the accuracy of the QR iteration method.* Note that no round-off error is involved in this computation and it takes only $O(n^2)$ flops. The MATLAB command **balance** finds balancing of a matrix. For more on this topic, see Parlett and Reinsch (1969).

9.8.7 The Real Schur Form and Invariant Subspaces

Definition 9.55. *Let S be a subspace of the complex plane $\mathbb{C}^n$. Then S is be called an invariant subspace (with respect to premultiplication by A) if $x \in S$ implies that $Ax \in S$.*

Thus, since $Ax = \lambda x$ for each eigenvalue λ, *each eigenvector is an invariant subspace of dimension 1 associated with the corresponding eigenvalue.*

The real Schur form of A displays information on the invariant subspaces as stated below.

Basis of an Invariant Subspace from the Real Schur Form

Let
$$Q^T A Q = R = \begin{pmatrix} R_{11} & R_{12} \\ 0 & R_{22} \end{pmatrix},$$
and let us assume that R_{11} and R_{22} do not have eigenvalues in common. *Then the first p columns of Q, where p is the order of R_{11}, form a basis for the invariant subspace associated with the eigenvalues of R_{11}.*

Ordering the eigenvalues. In many applications, such as in the solution of algebraic Riccati equations (see Datta (2003), Patel, Laub, and Van Dooren (1994), Petkov et al. (1991), and Van Dooren (1981a, 1981b, 1991)), one needs to compute the orthonormal bases of an invariant subspace associated with a selected number of eigenvalues. Unfortunately, *the real Schur form obtained by QR iteration may not give the eigenvalues in some desired order.* Thus, if the eigenvalues are not in a desired order, one wonders if some extra work can be done to bring them into that order. That this can indeed be done is seen from the following simple discussion. Let A be 2×2.

Let
$$Q_1^T A Q_1 = \begin{pmatrix} \lambda_1 & r_{12} \\ 0 & \lambda_2 \end{pmatrix}, \quad \lambda_1 \neq \lambda_2.$$

If λ_1 and λ_2 are not in the right order, all we need to do to reverse the order is to form a Givens rotation $J(1, 2, \theta)$ such that

$$J(1, 2, \theta) \begin{pmatrix} r_{12} \\ \lambda_2 - \lambda_1 \end{pmatrix} = \begin{pmatrix} * \\ 0 \end{pmatrix}.$$

Then $Q = Q_1 J(1, 2, \theta)^T$ is such that

$$Q^T A Q = \begin{pmatrix} \lambda_2 & r_{12} \\ 0 & \lambda_1 \end{pmatrix}.$$

Example 9.56.

$$A = \begin{pmatrix} 1 & 2 \\ 2 & 3 \end{pmatrix},$$

$$Q_1 = \begin{pmatrix} 0.8507 & 0.5257 \\ -0.5257 & 0.8507 \end{pmatrix}, \quad Q_1^T A Q_1 = \begin{pmatrix} -0.2361 & 0.0000 \\ 0.0000 & 4.2361 \end{pmatrix},$$

$$J(1, 2, \theta) = \begin{pmatrix} 0 & 1 \\ 1 & 0 \end{pmatrix}, \quad J(1, 2, \theta) \begin{pmatrix} 0 \\ 4.4722 \end{pmatrix} = \begin{pmatrix} 4.4722 \\ 0 \end{pmatrix},$$

$$Q = Q_1 J(1, 2, \theta)^T = \begin{pmatrix} 0.5257 & 0.8507 \\ 0.8507 & -0.5257 \end{pmatrix}, \quad Q^T A Q = \begin{pmatrix} 4.2361 & 0.001 \\ 0.001 & -0.2361 \end{pmatrix}. \quad \blacksquare$$

The above simple process can be easily extended to achieve any desired ordering of the eigenvalues in the real Schur form.

The process is quite inexpensive. It requires only $k(12n)$ flops, where k is the number of interchanges required to achieve the desired order. For more on eigenvalue ordering or ordering in real Schur form, see Bai and Demmel (1993b) and Bai and Stewart (1998). The latter provides a Fortran subroutine.

MATLAB Note: The MATLAB commands **ordschur** and **ordeig** are important in the context of ordering the eigenvalue in several specified regions or in some order.

$$[US, TS] = \mathbf{ordschur}\,(U,\ T,\ \text{keyword}),$$

where U and T are the matrices produced by the **schur** command and "keyword" specifies one of the following regions:

lhp – left half plane
rhp – right half plane
udi – interior of the unit disk
udo – exterior of the unit disk

9.9 Computing the Eigenvectors

9.9.1 The Hessenberg-Inverse Iteration

As soon as an eigenvalue λ is computed by QR iteration, we can invoke inverse iteration (Algorithm 9.2) to compute the corresponding eigenvector. However, since A is initially reduced to a Hessenberg matrix H for the QR iteration, it is natural to take advantage of the structure of the Hessenberg matrix H in the solutions of the linear system that need to be solved in the process of inverse iteration. Once an approximate eigenvector of the Hessenberg matrix H is found, the corresponding eigenvector of A can be recovered by an orthogonal matrix multiplication, as follows:

Let y be an eigenvector of H, and let $P^T A P = H$. Then $Hy = \lambda y$ or $P^T A P y = \lambda y$, that is, $APy = \lambda Py$, showing that $x = Py$ is an eigenvector of A.

Thus the **Hessenberg-Inverse iteration** can be stated as follows.

> **ALGORITHM 9.8. The Hessenberg-Inverse Iteration.**
>
> **Inputs:** (i) An $n \times n$ matrix A, (ii) an integer N, maximum number of iterations, (iii) ϵ, a tolerance, and (iv) an initial approximate eigenvector $y^{(0)}$ of the transformed Hessenberg matrix H.
> **Output:** An approximate eigenvector x of A.
>
> **Step 1.** Reduce the matrix A to an upper Hessenberg matrix H: $P^T A P = H$.
>
> **Step 2.** Compute an eigenvalue λ, whose eigenvector x is sought, using the implicit QR iteration.
>
> **Step 3.** Apply the inverse iteration
>   For $k = 1, 2, \ldots$ do
>   3.1. Solve the Hessenberg system $(H - \lambda I)z^{(k)} = y^{(k-1)}$.
>   3.2. Normalize $y^{(k)} = z^{(k)}/\|z_k\|_2$.
>   3.3. Stop if $\|Hy^{(k)} - \lambda y^{(k)}\| < \epsilon$ or if $k > N$.
>   End
>
> **Step 4.** Recover the eigenvector x:
> $$x = Py^{(k)}.$$

9.10 Review and Summary

This chapter has been devoted to the study of the eigenvalue problem, the problem of computing the eigenvalues and eigenvectors of a matrix $Ax = \lambda x$.

Here are the highlights of the chapter.

9.10.1 Applications of the Eigenvalues and Eigenvectors

The eigenvalue problem arises in a wide variety of practical applications. Mathematical models of many of the problems arising in engineering applications are systems of differential and difference equations, and the eigenvalue problem arises mainly in solutions and analysis of stability of these equations. Maintaining the stability of a system is a real concern for engineers. For example, in the study of vibrations of structures, the eigenvalues and eigenvectors are related to the natural frequencies and amplitude of the masses, and if any of the natural frequencies becomes equal or close to a frequency of the imposed periodic force on the structure, resonance occurs. An engineer would like to avoid this situation. In this chapter we have included examples on the *European arms race, buckling of a beam, simulating transient current of an electric circuit, vibration of a building*, and *principal component analysis* in statistics with a reference to a *stock market analysis*. We have attempted just to show how important the eigenvalue problem is in practical applications. These examples are given in Section 9.2.

9.10.2 Localization of Eigenvalues

In several applications, explicit knowledge of the eigenvalues is not required; all that is needed is a knowledge of the distribution of the eigenvalues in a region of the complex

9.10. Review and Summary

plane or estimates of some specific eigenvalues.

- The Geršgorin disk theorems (Theorems 9.6 and 9.9) can be used to obtain a region of the complex plane containing all the eigenvalues or, in some cases, a number of the eigenvalues in a region. *The estimates are, however, very crude.*

- Also, $|\lambda| \leq \|A\|$ (Theorem 9.11). *This result says that the upper bound of any eigenvalue of A can be found by computing its norm.*

  This result plays an important role in convergence analysis of iterative methods for linear systems (Chapter 12).

9.10.3 The Power Method and the Inverse Iteration

There are applications such as *analysis of dynamical systems, vibration analysis of structures, buckling of a beam, and principal component analysis in statistics,* where only the largest or the smallest (in magnitude) eigenvalue or only the first or last few eigenvalues and their corresponding eigenvectors are needed.

The power method (Algorithm 9.1) and the inverse power method (Algorithm 9.2) based on implicit construction of powers of A can be used to compute these eigenvalues and the eigenvectors. The power method is extremely simple to implement and is suitable for large and sparse matrices, but there are certain numerical limitations.

In practice, the power method should be used with a suitable shift. The inverse power method is simply the power method applied to $(A - \sigma I)^{-1}$, where σ is a suitable shift.

It is widely used to compute an eigenvector when a reasonably good approximation to an eigenvalue is known.

9.10.4 The Rayleigh Quotient Iteration

The quotient
$$R_q = \frac{x^T A x}{x^T x},$$
known as the *Rayleigh quotient*, gives an estimate of the eigenvalue λ of a symmetric matrix A for which x is the corresponding eigenvector.

This idea, when combined with the inverse iteration method (Algorithm 9.2), can be used to compute an approximation to an eigenvalue and the corresponding eigenvector. The process is known as the *Rayleigh quotient iteration* (Algorithm 9.3).

9.10.5 Sensitivity of Eigenvalues and Eigenvectors

- *The Bauer–Fike theorem* (Theorem 9.37) tells us that if A is a diagonalizable matrix, then the condition number of the transforming matrix X, $\text{Cond}(X) = \|X\| \, \|X^{-1}\|$, plays the role of the condition number of the eigenvalue problem. If this number is large, then a small change in A can cause significant changes in the eigenvalues.

  Since a symmetric matrix A can be transformed into a diagonal matrix by orthogonal similarity and the condition number of an orthogonal matrix (with respect to the 2-norm) is 1, it immediately follows from the Bauer–Fike theorem that the *eigenvalues of a symmetric matrix are insensitive to small perturbations.*

- If an eigenvalue problem is ill-conditioned, then it *might happen that some eigenvalues are more sensitive than others.* It is thus important to know the sensitivity of the individual eigenvalues. Unfortunately, to measure the sensitivity of an individual eigenvalue, one needs the knowledge of both left and right eigenvectors corresponding to that eigenvalue (Section 9.6.2). The condition number of the simple eigenvalue λ_i is the reciprocal of the number $|y_i^T x_i|$, where x_i and y_i are, respectively, the normalized right and left eigenvectors corresponding to λ_i.

- The sensitivity of an eigenvector x_k corresponding to an eigenvalue λ_k depends upon (i) the condition number of all the eigenvalues other than λ_k, and (ii) the distance of λ_k from the other eigenvalues (Theorem 9.41).

    Thus, if the eigenvalues are well-separated and well-conditioned, then the eigenvectors are well-conditioned. On the other hand, if there is a multiple eigenvalue or there is an eigenvalue close to another eigenvalue, then there are some ill-conditioned eigenvectors. This is especially significant for a symmetric matrix. *The eigenvalues of a symmetric matrix are well-conditioned, but the eigenvectors can be quite ill-conditioned.*

9.10.6 Eigenvalue Computation via the Characteristic Polynomial and the Jordan Canonical Form

A similarity transformation preserves the eigenvalues, and it is well known that a matrix A can be transformed by similarity (Theorem 9.28) to the Jordan canonical form and to the Frobenius form (or a companion form if A is nonderogatory). The eigenvalues of these condensed forms are rather easily computed. The Jordan canonical form displays the eigenvalues explicitly, and with the companion or Frobenius form, the characteristic polynomial of A is trivially computed and then a root-finding method can be applied to the characteristic polynomial to obtain the eigenvalues, which are the zeros of the characteristic polynomial.

However, *computation of eigenvalues via the characteristic polynomial or the Jordan canonical form is not recommended in practice.* Obtaining these forms may require a very ill-conditioned transforming matrix, and the sensitivity of the eigenvalue problem depends upon the condition number of this transforming matrix (Sections 9.5.2 and 9.6.1).

In general, ill-conditioned similarity transformation should be avoided in eigenvalue computation. The use of well-conditioned transforming matrices, such as orthogonal matrices, is necessary.

9.10.7 Hessenberg Transformation

An arbitrary matrix A can always be transformed to a Hessenberg matrix by orthogonal similarity transformation. Two numerically stable methods, Householder's (Algorithm 9.5) and Givens' methods, are described in Section 9.5.3.

9.10.8 The QR Iteration Algorithm

The most widely used algorithm for finding the eigenvalues of a matrix is the QR iteration algorithm.

For a real matrix A, the algorithm iteratively constructs the real Schur form of A by orthogonal similarity. Since the algorithm is based on repeated QR factorizations and each QR factorization of an $n \times n$ full matrix requires $O(n^3)$ flops, the n steps of the QR iteration algorithm, if implemented naively (which we call the *basic QR iteration*), will require $O(n^4)$ flops, making the algorithm impractical.

- Matrix A is, therefore, initially reduced to a Hessenberg matrix H by orthogonal similarity before the start of the QR iteration. The key observations here are (i) the *reduction of A to H has to be made once for all, and* (ii) *the Hessenberg form is preserved at each iteration* (Theorem 9.49).

- The convergence of the Hessenberg QR iteration algorithm, however, can be quite slow in the presence of a near-multiple or a multiple eigenvalue. The convergence can be accelerated by using suitable shifts.

    In practice, double shifts are used. At each iteration, the shifts are the eigenvalues of the 2×2 submatrix at the bottom right-hand corner. Since the eigenvalues of a real matrix can be complex, complex arithmetic is usually required. However, computations can be arranged so that complex arithmetic can be avoided. Also, the eigenvalues of the 2×2 bottom right-hand corner matrix at each iteration do not need to be computed explicitly. The process is known as the *double-shift implicit QR iteration* (Algorithm 9.7).

    With double shifts, the eigenvalues are computed two at a time. Once two eigenvalues are computed, the matrix is deflated, and the process is applied to the deflated matrix.

    The double-shift implicit QR iteration is the most practical algorithm for computing the eigenvalues of a nonsymmetric dense matrix of modest size.

9.10.9 Ordering the Eigenvalues

The eigenvalues appearing in real Schur form obtained by the QR iteration algorithm do not appear in the desired order, although there are some applications which need this. However, with a little extra work, the eigenvalues can be put in the desired order (Section 9.8.7).

9.10.10 Computing the Eigenvectors

Once an approximation to an eigenvalue is obtained for the QR iteration, inverse iteration can be invoked to compute the corresponding eigenvector.

Since the matrix A is initially reduced to a Hessenberg matrix for practical implementation of the QR iteration algorithm, advantage can be taken of the structure of a Hessenberg matrix in computing an eigenvector using inverse iteration (Algorithm 9.8).

9.11 Suggestions for Further Reading

Most books on vibration discuss eigenvalue problems arising in vibration of structures. However, almost all eigenvalue problems here are generalized eigenvalue problems; as a matter of fact, they are symmetric definite problems (see Chapter 11).

For references of well-known books on vibration, see Chapter 11. Those by Inman (2006, 2007) and Thomson (1992) are, in particular, very useful and important books in this area.

For learning more about how the eigenvalue problem arises in other areas of engineering, see the books on numerical methods in engineering by Chapra and Canale (2002) and O'Neil (1991), referenced in Chapter 6. There are other engineering books (too numerous to list here), especially in the areas of electrical, mechanical, civil, and chemical engineering, containing discussions on eigenvalue problems in engineering. The real Schur form of a matrix is an important tool in numerically effective solutions of many important control problems, such as solutions of the Lyapunov, Sylvester, and algebraic Riccati matrix equations (see Datta (2003)).

For some generalizations of the Geršgorin disk theorems, see the paper by Brualdi and Mellendorf (1994). This paper contains results giving a region of the complex plane for each eigenvalue; for a full description of the Geršgorin disk theorems and applications, see the book by Horn and Johnson (1985). The recent book by Varga (2004) has been devoted exclusively to the subject. For the Geršgorin theorem for partitioned matrices, see Johnston (1971).

A nice description of stability theory in dynamic systems is given in the classic book by Luenberger (1979).

For more results on eigenvalue bounds and eigenvalue sensitivity, see Bhatia (2007), Varah (1968b), and Davis and Moler (1978).

For computation of the Jordan canonical form, see the papers by Golub and Wilkinson (1976), Kagström and Ruhe (1980a, 1980b), Demmel (1983), and the recent work of Zeng and Li (2008).

For computation of condition numbers and estimators for eigenvalue problems, see Bai et al. (1993) and Van Loan (1987).

Descriptions of the standard techniques for eigenvalue and eigenvector computations, including the power method, the inverse power method, the Rayleigh quotient iteration method, and the QR iteration method, can be found in all numerical linear algebra books: Golub and Van Loan (1996), Trefethen and Bau (1997), Demmel (1997), Stewart (2001a), Watkins (2002), and Hager (1988). See also the papers by Dongarra et al. (1983, 1992) and Parlett (1965, 1966, 1968).

The papers by Varah (1968a, 1970) and Peters and Wilkinson (1979) are important in the context of eigenvalue computation and inverse iteration. See also an interesting paper by Dhilon (1998) in this context.

Eigenvalue problems studied in this chapter concern computing the eigenvalues of a matrix. On the other hand, *inverse eigenvalue problems* concern constructing a matrix from the knowledge of a partial or complete set of eigenvalues and eigenvectors. Active research on this topic is currently being carried out. An authoritative account on inverse eigenvalue problems can be found in the recent book by Chu and Golub (2005). Another important book on this topic is Gladwell (2004). For an account of a special type of inverse eigenvalue problem arising in control theory, usually known as pole-placement or the *eigenvalue assignment problem*, see the book by Datta (2003, Chapters 10 and 11). See also the thesis by Arnold (1992).

Hessenberg and real Schur matrices arise in a wide variety of applications. For their applications in control theory, see the books by Datta (2003), Patel, Laub, and Van Dooren (1994), and Petkov, Christov, and Konstantinov (1991). The book by Bhaya and Kaszkurewicz (2006) gives an account of control perspectives of numerical algorithms and matrix problems.

Exercises on Chapter 9

(Use MATLAB, whenever needed and appropriate.)

EXERCISES ON SECTION 9.2

9.1 Consider the following model for the vertical vibration of a motor car:

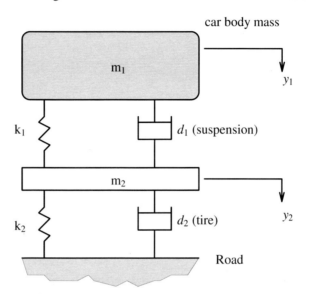

(a) Show that the equation of motion of the car, neglecting the damping constants d_1 of the shock absorber and d_2 of the tire, is given by

$$M\ddot{y} + Ky = 0$$

where $M = \text{diag}(m_1, m_2)$, $K = \begin{pmatrix} k_1 & -k_1 \\ -k_1 & k_1+k_2 \end{pmatrix}$, and $y = \begin{pmatrix} y_1 \\ y_2 \end{pmatrix}$. Determine the stability of motion when $m_1 = m_2 = 1200\,\text{kg}$, $k_1 = k_2 = 300\,\frac{\text{N}}{\text{cm}}$.

(b) Show that the equation of motion when just the damping d_2 of the tire is neglected can be written as

$$M\ddot{y} + D\dot{y} + Ky = 0,$$

where M and K are the same as in part (a), and $D = \begin{pmatrix} d_1 & -d_1 \\ -d_1 & d_1 \end{pmatrix}$.
Investigate the stability of motion in this case when $d_1 = 4500\,\frac{\text{Ns}}{\text{m}}$.
Hints: Show that the system

$$M\ddot{y} + D\dot{y} + Ky = 0$$

is equivalent to the first-order system $\dot{x}(t) = Ax(t)$, where

$$A = \begin{pmatrix} 0 & I \\ -M^{-1}K & -M^{-1}D \end{pmatrix}, \text{ with } x(t) = \begin{pmatrix} y(t) \\ \dot{y}(t) \end{pmatrix}.$$

9.2 Write the solution of the equation $M\ddot{y} + Ky = 0$ with the numerical values of M and K as given in Exercise 9.1(a), using initial conditions $y(0) = 0$ and $\dot{y}(0) = (1, 1, \ldots, 1)^T$.

9.3 Develop an eigenvalue problem for an LC network similar to the case study given in Section 9.2.4, but with only three loops. Compute the natural frequencies. Find the modes and illustrate how the currents oscillate in these modes.

EXERCISES ON SECTION 9.3

9.4 Apply the Geršgorin disk theorems to obtain bounds for the eigenvalues of the following matrices:

(a) $A = \begin{pmatrix} 10 & 1 & 1 \\ 2 & 10 & 1 \\ 2 & 2 & 10 \end{pmatrix}$.

(b) $\begin{pmatrix} 1 & 0 & 0 \\ 2 & 5 & 0 \\ 1 & 1 & 6 \end{pmatrix}$.

(c) $\begin{pmatrix} 2 & -1 & 0 & 0 \\ -1 & 2 & -1 & 0 \\ 0 & -1 & 2 & -1 \\ 0 & 0 & -1 & 2 \end{pmatrix}$.

(d) $\begin{pmatrix} 1 & -1 & 0 & 0 \\ -1 & 2 & -1 & 0 \\ 0 & -1 & 2 & -1 \\ 0 & 0 & -1 & 2 \end{pmatrix}$.

(e) A randomly generated matrix of order 4.

(f) $\begin{pmatrix} 1-i & 0 & 1 \\ 1 & 1 & 1 \\ 0 & 1-i & 1+i \end{pmatrix}$.

9.5 Using a Geršgorin disk theorem, prove that a strictly diagonally dominant matrix is nonsingular.

9.6 Let x be an eigenvector corresponding to a distinct eigenvalue λ in the Geršgorin disk R_k. Prove that $|x_k| > |x_i|$ for $i \neq k$, where $x = (x_1, x_2, \ldots, x_n)^T$.

9.7 Let $A = (a_{ij})$ be an $n \times n$ symmetric matrix. Then using a Geršgorin disk theorem prove that each eigenvalue of A will lie in one of the intervals: $[a_{ij} - r_i, a_{ij} + r_i]$.

Find an interval where all the eigenvalues of A must lie.

EXERCISES ON SECTION 9.4

9.8 Applying the power method and inverse power method, find the dominant eigenvalue and the corresponding eigenvector for each of the matrices in Exercise 9.4.

9.9 Prove that if $\lambda_1, \ldots, \lambda_n$ are the eigenvalues of A and $v_1, \ldots, v_n$ are the corresponding eigenvectors, then $\lambda_1 - \sigma, \ldots, \lambda_n - \sigma$ are the eigenvalues of $A - \sigma I$, and the corresponding eigenvectors are $v_1, \ldots, v_n$.

9.10 Explain the slow rate of convergence of the power method with the following matrices:

(a) $A = \begin{pmatrix} 3 & 2 & 3 \\ 0 & 2.9 & 1 \\ 0 & 0 & 1 \end{pmatrix}$. (b) $A = \begin{pmatrix} 1 & 0 & 0 \\ 1 & 10 & 0 \\ 1 & 1 & 9.8 \end{pmatrix}$.

Choose a suitable shift σ and then apply the shifted power method to each of the matrices and observe the improvement of the rates of convergence.

9.11 (*Orthogonal iteration.*) The following iterative procedure generalizes the power method and is known as the *orthogonal iteration process*. The process can be used to compute p ($p > 1$) largest eigenvalues (in magnitude) and the corresponding eigenvectors.

Let Q_1 be an $n \times p$ matrix with orthonormal columns.
Then

For $k = 2, 3, \ldots$ do
(1) Compute $B_k = AQ_{k-1}$.
(2) Factorize B_k into reduced QR: $B_k = Q_k R_k$.
End

Apply the above method to compute the first two dominant eigenvalues and eigenvectors for each of the matrices in Exercise 9.4.

9.12 (*Inverse orthogonal iteration.*) The following iteration, called the *inverse orthogonal iteration*, generalizes the inverse power method and can be used to compute the p smallest eigenvalues (in magnitude).

Let Q_1 be an $n \times p$ matrix with orthonormal columns.

For $k = 2, 3, \ldots$ do
(1) Solve for B_k: $AB_k = Q_{k-1}$.
(2) Factorize B_k into reduced QR: $B_k = Q_k R_k$.
End

Apply the inverse orthogonal iteration to compute the 2 smallest (least dominant) eigenvalues of each of the matrices in Exercise 9.4.

9.13 Let T be a symmetric tridiagonal matrix. Let the Rayleigh quotient iteration be applied to T with $x_0 = e_n$, then prove that $x_1 = q_n$, where q_n is the last column of Q in $(T - \sigma_0 I) = QR$.

9.14 Prove that for a symmetric matrix A, the Rayleigh quotient lies between the smallest and the largest eigenvalues.

9.15 Compute the smallest eigenvalue of each of the matrices A in Exercise 9.4 by applying the power method to A^{-1}, without explicitly computing A^{-1}.

EXERCISES ON SECTIONS 9.5

9.16 Develop an algorithm, based on the use of Givens rotations, to transform a matrix A to an upper Hessenberg matrix.

9.17 Apply both the Householder and the Givens methods of reduction to the matrix

$$A = \begin{pmatrix} 0 & 1 & 0 & 0 & 0 \\ 0 & 0 & 1 & 0 & 0 \\ 0 & 0 & 0 & 1 & 0 \\ 0 & 0 & 0 & 0 & 1 \\ 1 & 2 & 3 & 4 & 5 \end{pmatrix}$$

to reduce it to a Hessenberg matrix by similarity. Compare the results in the context of the implicit Q theorem (Theorem 9.34).

9.18 (a) Develop an algorithm to transform a symmetric matrix A into a symmetric tridiagonal matrix using Householder transformations, which takes advantages of the symmetry of A.

(b) Given the pair (A, b), develop an algorithm to compute an orthogonal matrix P such that PAP^T is an upper Hessenberg matrix H and Pb is a multiple of e_1. What conditions guarantee that H is unreduced and b is a nonzero multiple of e_1? Construct an example to illustrate the algorithm with a matrix of order 4.

9.19 (a) Show that it requires $\frac{10}{3}n^3$ flops to compute the upper Hessenberg matrix H_u using the Householder method of reduction. (*Hint*: $\sum_{k=1}^{n-2} 4(n-k)^2 + \sum_{k=1}^{n-2} 4n(n-k) \approx \frac{4}{3}n^3 + 2n^3 = \frac{10n^3}{3}$.)

(b) Show that if the transforming matrix P is required explicitly, another $\frac{4}{3}n^3$ flops will be needed.

(c) Work out the corresponding flop-count for reduction to Hessenberg form using Givens rotations.

(d) If A is symmetric, then show that the corresponding count in (a) is $\frac{4n^3}{3}$.

9.20 (a) Given an unreduced upper Hessenberg matrix H, show that matrix X defined by $X = (e_1, He_1, \ldots, H^{n-1}e_1)$ is nonsingular and is such that $X^{-1}HX$ is a companion matrix in upper Hessenberg form.

(b) What are the possible numerical difficulties with the above computations?

(c) Give an example illustrating the numerical difficulties.

9.21 (a) Prove that if R is upper triangular and Q is upper Hessenberg, then RQ is upper Hessenberg.

(b) Given an eigenpair (λ, x) of an upper Hessenberg matrix H, develop an algorithm using Givens rotations to compute an orthogonal matrix P such that

$$P^T H P = \begin{pmatrix} \lambda & \times \\ 0 & \hat{H} \end{pmatrix},$$

where $\hat{H}$ is $(n-1) \times (n-1)$ upper Hessenberg. How are the eigenvalues of H and $\hat{H}$ related? Construct a 4×4 example to illustrate the algorithm.

(c) Prove that the eigenvalue problem of a reduced Hessenberg matrix can be split into eigenvalue problems of unreduced Hessenberg matrices.

EXERCISES ON SECTIONS 9.6 and 9.7

9.22 Construct a simple example to show that an arbitrary similarity transformation can worsen the conditioning of the eigenvalues of the transformed matrix.

9.23 If $B = U^*AU$, where U is unitary and $U^*(A + \Delta A)U = B + \Delta B$, then show that $\|\Delta B\|_2 = \|\Delta A\|_2$.

9.24 (a) Prove that A is normal if there exists an unitary matrix U such that $U^*AU = D$, where D is diagonal.

(b) Prove that a matrix A has a set of n orthonormal eigenvectors if and only if A is normal.

(c) Prove that a normal matrix A is unitary if and only if its eigenvalues are on the unit circle; that is, for each eigenvalue λ of A, we have $|\lambda| = 1$.

(d) How does the real Schur form of a normal matrix look?

(e) Using the eigenvector-sensitivity theorem (Theorem 9.41), show that if A is normal, then the eigenvector x_k corresponding to the eigenvalue λ_k is well-conditioned if λ_k is well-separated from the other eigenvalues.

9.25 Explain both theoretically and experimentally that two of the eigenvectors of the matrix
$$A = \text{diag}(1 + \epsilon, 1 - \epsilon, 2),$$
where ϵ is a very small positive number, are ill-conditioned. However, the eigenvalues are well-conditioned.

9.26 Show that the unitary similarity transformation preserves the condition number of an eigenvalue.

9.27 Prove the Bauer–Fike theorem using Geršgorin's first theorem.

9.28 (a) Given
$$A = \begin{pmatrix} 1 & 1 \\ 0 & 1+\epsilon \end{pmatrix},$$
find the eigenvector matrix X such that $X^{-1}AX$ is diagonal; hence show that the eigenvalues of A are ill-conditioned.

Verify the ill-conditioning of the eigenvalues of A computationally by constructing a small perturbation to A and finding the eigenvalues of the perturbed matrix.

(b) Consider
$$A = \begin{pmatrix} 12 & 11 & 10 & \cdots & 2 & 1 \\ 11 & 11 & 10 & \cdots & 2 & 1 \\ 0 & 10 & 10 & \cdots & 2 & 1 \\ \cdots & \cdots & \cdots & \cdots & 2 & 1 \\ \cdots & \cdots & \cdots & \cdots & \cdots & \cdots \\ 0 & 0 & 0 & 0 & 11 & 1 \end{pmatrix}.$$

Show that the largest eigenvalues of A are well-conditioned while the smallest ones are very ill-conditioned by computing the eigenvalues using MATLAB.

EXERCISES ON SECTION 9.8

9.29 Apply 3 iterations of the single-shift QR iteration to each of the following matrices and observe the convergence or nonconvergence of the subdiagonal entries:

(i) $A = \begin{pmatrix} 1 & 2 & 0 \\ 2 & 3 & 4 \\ 0 & 4 & 1 \end{pmatrix}.$ (ii) $A = \begin{pmatrix} 4 & 5 & 6 \\ 1 & 0 & 1 \\ 0 & -2 & 2 \end{pmatrix}.$

9.30 (*Implicit single-shift QR.*) Consider one step of the single-shift Hessenberg QR iteration:

$$H_k - h_{nn}^{(k)} I = Q_k R_k, \quad H_{k+1} = R_k Q_k + h_{nn}^{(k)} I$$

or (simply)

$$H - \lambda I = QR, \quad \bar{H} = RQ + \lambda I, \quad \lambda \text{ is real.}$$

(a) Prove that the first column of Q is a multiple of the first column of $H - \lambda I$, and therefore contains only two nonzero entries.

(b) Denote the first column of $H - \lambda I$ by $h_1 = (h_{11} - \lambda, h_{21}, 0, \ldots, 0)^T$. Find a Givens rotation P_0 such that $P_0 h_1$ is a multiple of e_1. Show that the first column of P_0 is the same as the first column of Q, except possibly for signs.

(c) Form $H' = P_0^T H P_0$. Find Givens rotations $J_{32}, J_{43}, \ldots, J_{n,n-1}$ such that

$$H_1' = (J_{32} J_{43}, \ldots, J_{n,n-1})^T H'(J_{32}, \ldots, J_{n,n-1})$$

is upper Hessenberg. Show that the matrix

$$\tilde{Q} = P_0 J_{32}, \ldots, J_{n,n-1}$$

has the same first column as P_0 and hence the same first column as Q. Conclude finally from the implicit Q theorem (Theorem 9.34) that the Hessenberg matrix H_1' is *essentially* the same as $\bar{H}$.

Steps (a) to (c) constitute *one step* of the implicit single-shift Hessenberg QR iteration.

Apply one step of the implicit single-shift QR iteration to the symmetric tridiagonal matrix

$$\begin{pmatrix} 2 & -1 & \cdots & 0 \\ -1 & \ddots & \cdots & \\ \vdots & \ddots & \ddots & -1 \\ 0 & \cdots & -1 & 2 \end{pmatrix}.$$

9.31 Construct one step of the explicit double-shift QR iteration (real arithmetic) and one step of the implicit double-shift QR iteration for the matrix

$$A = \begin{pmatrix} 1 & 2 & 3 & 4 \\ 3 & 4 & 5 & 6 \\ 0 & 1 & 0 & 1 \\ 0 & 0 & -2 & 2 \end{pmatrix}$$

and show that the obtained Hessenberg matrices in both cases are (essentially) the same.

9.32 (*LR iteration.*) In analogy with the QR iteration algorithm, develop an LR iteration algorithm, based on LU decomposition of A, making the necessary assumptions.

(1) Set $A = A_1$.
(2) Compute $A_k = L_k R_k$, $A_{k+1} = R_k L_k$, $k = 1, 2, \ldots$.

Why is this algorithm not to be preferred over the QR iteration algorithm?

9.33 Considering the structures of the matrices P_i, $i = 0, 1, \ldots, n-2$, in the implicit double-shift QR iteration step; show that it requires about $10n^2$ flops to implement this step.

9.34 Show that the matrices H_0 and H_2 in the double-shift QR iteration have the same eigenvalues.

9.35 Prove the following:
Let $H = H_0$ be an upper Hessenberg matrix. Generate the sequence $\{H_k\}$:
$$H_k - \mu_k I = Q_k R_k, \quad H_{k+1} = R_k Q_k + \mu_k I.$$
Then
$$\Pi_{i=1}^n (H - \mu_i I) = (Q_1, \ldots, Q_n)(R_n, \ldots, R_1).$$

9.36 (*Deflation using invariant subspace.*) Suppose that we have an $n \times m$ matrix X with independent columns and an $m \times m$ matrix M such that
$$AX = XM.$$
Consider the QR factorization of X: $Q^T X = \binom{R}{0}$. Then show that

(a) $QAQ^T = \begin{pmatrix} A_1 & A_2 \\ 0 & A_3 \end{pmatrix}$;

(b) the eigenvalues of A are those of A_1 and A_3;

(c) the eigenvalues of A_1 are those of M.

MATLAB and MATCOM Programs and Problems on Chapter 9

In Problems M9.1–M9.3(a) (i) x_o stands for the initial vector, (ii) epsilon is the tolerance for convergence, and (iii) n is the order of matrix A.

M9.1 Write a MATLAB program to compute the dominant eigenvalue of a matrix using the power method as follows:
$$[\text{lambda1}] = \mathbf{power}(A, x0, \text{epsilon}, n).$$

(a) Modify the program **power** to incorporate a shift sigma:
$$[\text{lambda1}] = \mathbf{powershift}(A, x0, \text{sigma}, \text{epsilon}, n).$$

(b) Apply **power** and **powershift** to the following matrices and compare the speed of convergence:

(i) $A = $ a randomly generated matrix of order 15.
(ii) $A = $ the Wilkinson bidiagonal matrix of order 20.

M9.2 Using **givhs**(A) from MATCOM and the MATLAB function **hess**(A) on each of the two matrices from Problem M9.1, verify the implicit Q theorem (Theorem 9.34).

M9.3 (a) Write a MATLAB program called **invitr** to implement the inverse iteration algorithm (Algorithm 9.2) as follows:

$$x = \mathbf{invitr}(A, x0, sigma, epsilon, n).$$

(b) Write a MATLAB program called **powersmall** to compute lambdan, the smallest eigenvalue (in magnitude) of a matrix A:

$$lambdan = \mathbf{powersmall}(A, x0, epsilon, n).$$

Test data and experiment:

(a) Consider the symmetric tridiagonal matrix given in (9.5) of order 200 appearing in the buckling problem of Section 9.2.3. Apply **power** to compute the dominant eigenvalue lambda1 by choosing x_0 arbitrarily.

(b) Now compute the smallest eigenvalue in magnitude, **lambdan**, by using

   (i) **powershift** with sigma = lambda1;
   (ii) **powersmall** with the same x_0 as used to implement **power**.

(c) Compare the flop-count of (i) and (ii) in (b).

(d) Find the smallest critical load that will buckle the beam.

(e) Taking sigma = lambdan, find the eigenvector corresponding to the smallest eigenvalue λ_n using **invitr**.

M9.4 (*The purpose of this exercise is to study how the eigenvalues of a matrix A are affected by conditioning of the transforming matrix.*)

(a) For each of the following matrices, construct a matrix X of appropriate order which is upper triangular with all the entries equal to 1 except for a few very small diagonal entries. Then compute the eigenvalues of A and those of $X^{-1}AX$ using MATLAB commands **eig** and **inv**:

   (i) $A = $ an upper triangular matrix of order 200 with several eigenvalues clustered around 1.
   (ii) $A = $ the Wilkinson bidiagonal matrix of order 20.

(b) Repeat part (a) by taking X as a Householder matrix of appropriate order (*note: this X is orthogonal*).

(c) Compare the results of (a) and (b).

M9.5 (a) Compute the eigenvalues of the matrices in Exercise M9.4 using

   (i) MATLAB commands **poly** and **roots**;
   (ii) MATLAB command **eig**.

(b) Compare your results of (i) and (ii) for each matrix.

M9.6 (*The purpose of this exercise is to study the sensitivities of the eigenvalues of some well-known matrices with ill-conditioned eigenvalues.*)

Perform the following on each of the matrices in the test data:

(a) Using the MATLAB command $[V, D] = $ **eig**(A), find the eigenvalues and the matrix of right eigenvectors. Then find the matrix of left eigenvectors W as follows $W = ($**inv**$(V))'$ /**norm(inv**$(V)')$.

(b) Compute $s_i = w_i^T v_i$, $i = 1, \ldots, n$, where w_i and v_i are the ith columns of W and V.

(c) Compute c_i = the condition number of the ith eigenvalue = $1/s_i$, $i = 1, 2, \ldots, n$.

(d) Perturb the $(n, 1)$th entry of A by $\epsilon = 10^{-10}$. Then compute the eigenvalues $\hat{\lambda}_i$, $i = 1, \ldots, n$, of the perturbed matrix using the MATLAB command **eig**.

(e) Make the following table for each matrix.

λ_i	$\hat{\lambda}_i$	$|\lambda_i - \hat{\lambda}_i|$	Cond (V)	c_i

(f) Write your conclusions.

Test data:

(1) $A = $ the Wilkinson bidiagonal matrix of order 20.

(2) $A = $ the transpose of the Wilkinson bidiagonal matrix of order 20.

(3)
$$A = \begin{pmatrix} 12 & 11 & 10 & \cdot & \cdot & 3 & 2 & 1 \\ 11 & 11 & 10 & \cdot & \cdot & 3 & 2 & 1 \\ \cdot & \cdot & \cdot & \cdot & \cdot & \cdot & \cdot & \cdot \\ \cdot & \cdot & \cdot & \cdot & \cdot & \cdot & \cdot & \cdot \\ \cdot & \cdot & \cdot & \cdot & \cdot & 2 & 2 & 1 \\ 0 & 0 & 0 & \cdot & \cdot & \cdot & 1 & 1 \end{pmatrix}.$$

M9.7 Study the sensitivities of the eigenvectors of the following matrices by actually computing the the eigenvectors of the original and perturbed matrices using the MATLAB command $[V, D] = $ **eig**(A), $[V, D] = $ **eig**$(\hat{A})$, where $\hat{A}$ is the matrix obtained from

A by perturbing the $(n, 1)$th entry by $\epsilon = 10^{-5}, 10^{-7}$, and 10^{-10}:

(1)
$$A = \begin{pmatrix} 1 & 0 & 0 & 0 \\ 0 & 0.999 & 0 & 0 \\ 0 & 0 & 0 & 2 \\ 0 & 0 & 0 & 0.0005 \end{pmatrix}.$$

(2) $A = \text{diag}(1, 0.9999, 1, 0.9999, 1)$.

(3) Randomly generated matrices of order 25, 30, 40, and 50.

M9.8 Write a MATLAB program called **qritrb** to implement the basic QR iteration.

(a) $[A] = \textbf{qritrb}(A, num)$, where *num* is the maximum number of iterations.

(b) Modify the program now to implement the Hessenberg QR iteration $[A] = \textbf{qritrh}(A, num)$, where *num* is the number of iterations

(c) Compare the flop-count in (a) and (b).
Use random matrices of order 50, 60, and 100.

M9.9 *(The purpose of this exercise is to verify that explicit and implicit double-shift QR iterations produce essentially the same Hessenberg matrix.)*

(a) Write a MATLAB program to compute one step of the explicit double-shift QR iteration: $[A] = \textbf{qritrdse}(A)$.

(b) Write a MATLAB program to compute one step of the implicit QR iteration with double shift: $[A] = \textbf{qritrdsi}(A)$.

(c) Compare your results of (a) and (b) and conclude that they are essentially the same, using random matrices of order 10, 15, 20, and 50.

M9.10 Write a MATLAB program to deflate the last k rows and k columns of a Hessenberg matrix in the following form: hprime = **deflat**(H, k).

Test your program with a randomly generated matrix with different values of k. Note that for $k = 1$, hprime will be of order $n - 1$, for $k = 2$, hprime will be of order $n - 2$, and so on.

M9.11 Using **qritrdsi** and **deflat**, write a MATLAB program to determine the real Schur form of a Hessenberg matrix A in the following form:

$$[h] = \textbf{rsf}(h, eps),$$

where *eps* is the tolerance.

Test: Generate randomly a 20×20 Hessenberg matrix and make the following table using **rsf**.

Iteration	h_{21}	h_{32}	h_{43}	h_{54}	$\cdots$	$\cdots$	$h_{20,19}$

Note: Some of the above programs are available in MATCOM.

Chapter 10

Numerical Symmetric Eigenvalue Problem and Singular Value Decomposition

Background Material Needed

- Basic properties of eigenvalues and eigenvectors (Theorem 2.7)
- Vector and matrix norms (Section 2.5)
- Singular value decomposition and its properties (Section 7.8)

10.1 Introduction

In this chapter, we mainly describe computational algorithms for two intimately related problems: the *symmetric eigenvalue problem* and *singular value decomposition* (SVD).

The symmetric eigenvalue problem enjoys certain remarkable special properties, and to exploit these properties, specialized methods have been developed. These include the tridiagonal QR iteration, bisection, divide-and-conquer, and Jacobi methods (Section 10.2.2–10.2.5). A brief description of these methods will be given here. In addition, a brief review of the special properties of the symmetric eigenvalue problem will be presented (Section 10.2.1).

The concept of the SVD has been introduced in Chapter 7 and some properties have been described there.

The SVD has a long and fascinating history. The names of at least five classical and celebrated mathematicians—E. Beltrami (1835–1899), C. Jordan (1838–1922), J. Sylvester (1814–1897), E. Schmidt (1876–1959), and H. Weyl (1885–1955)—can be associated with the development of the theory of the SVD. Some details of the contributions of these mathematicians to the SVD can be found in an interesting paper by Stewart (1993b). Also the book by Horn and Johnson (1991) contains a nice history of the SVD.

As we have already seen in Chapter 7, in recent years the SVD has become a computationally viable tool for solving a wide variety of problems arising in many practical applications.

In this chapter, we will give a formal proof of the SVD theorem (Theorem 10.8). There are several proofs of the SVD theorem available in the literature: see Golub and Van Loan (1996, p. 70) and Horn and Johnson (1991). See also Pan and Sigmon (1994). We will, however, give a more traditional and constructive proof that *exhibits the relationship* between the singular values and singular vectors of A with the eigenvalues and eigenvectors of $A^T A$.

We will also discuss the sensitivity of the singular values (Section 10.3.2). *The singular values are insensitive to perturbations*, and this is remarkable.

Finally, we describe the popular *Golub–Kahan–Reinsch* algorithm (Section 10.3.6) and one of its variants, called the *Chan–Lawson–Hanson* algorithm (Section 10.3.7) for computing the SVD.

10.2 Computational Methods for the Symmetric Eigenvalue Problem

In this section, we will first describe some special properties of the symmetric eigenvalue problem which have been exploited in development of several special algorithms for the problem. We shall then describe briefly four such methods:

- The bisection method (Algorithm 10.1).
- The symmetric QR iteration algorithm (Algorithm 10.2).
- The divide-and-conquer method (Algorithm 10.3).
- The Jacobi method (Section 10.2.5).

Except for the Jacobi method, all the other ones compute the eigenvalues by first transforming the symmetric matrix A into the symmetric tridiagonal form.

If the eigenvectors are desired, the *inverse iteration method* (Algorithm 9.8) has to be invoked by replacing the Hessenberg matrix with the symmetric tridiagonal matrix T.

10.2.1 Some Special Properties of the Symmetric Eigenvalue Problem

A. *The real Schur form of a real symmetric matrix is a diagonal matrix.* That is, there exists an orthogonal matrix Q such that
$$Q^T A Q = D = \text{diag}(\lambda_1, \ldots, \lambda_u),$$
where $\lambda_i, i = 1, \ldots, n$, are the eigenvalues of A.

B. *The eigenvalues of a symmetric matrix are real* and *the eigenvectors can be chosen to be orthogonal*.

C. *Minimax characterization (Courant[15]–Fischer minimax theorem).* Let $\lambda_1 \geq \lambda_2 \geq \cdots \geq \lambda_n$ be the eigenvalues of a symmetric matrix A. Then
$$\lambda_i = \min_S \max_{0 \neq x \in S} \frac{x^T A x}{x^T x}, \tag{10.1}$$

---

[15]Richard Courant (1888–1972) was born in Lublinitz, Germany (now Lubliniec, Poland). He studied under such celebrated mathematicians as Hilbert and Minkowski and obtained his doctorate from Göttingen, Germany, in 1910 under Hilbert's supervision. One of the most famous mathematical contributions of Courant is the finite element method. He served as Professor of Mathematics at New York University from 1936 to 1972. The present Courant Institute of Mathematical Sciences at New York University was named after him in 1964.

10.2. Computational Methods for the Symmetric Eigenvalue Problem

where the minimum is taken over all subspaces of dimension $(n - i + 1)$ and the maximum is taken over all nonzero vectors in the subspace S. In particular,

$$\lambda_1 = \lambda_{\max} = \max_{x \neq 0} \frac{x^T A x}{x^T x} \quad \text{and} \quad \lambda_n = \lambda_{\min} = \min_{x \neq 0} \frac{x^T A x}{x^T x}. \tag{10.2}$$

Proof. See Golub and Van Loan (1996, p. 394). □

D. *General perturbation property.* Let A be an $n \times n$ real symmetric matrix. Let $A' = A + E$, where E is a real symmetric perturbation of the matrix A, and let $\lambda_1 \geq \lambda_2 \geq \cdots \geq \lambda_n$ and $\lambda'_1 \geq \lambda'_2 \geq \cdots \geq \lambda'_n$ be the eigenvalues of A and A', respectively. Then it follows from the Bauer–Fike theorem (Theorem 9.37) that

$$\lambda_i - \|E\|_2 \leq \lambda'_i \leq \lambda_i + \|E\|_2, \quad i = 1, 2, \ldots, n. \tag{10.3}$$

This result is remarkable. It says that *the eigenvalues of a real symmetric matrix are well-conditioned*; that is, small changes in the elements of A can cause only small changes in the eigenvalues of A. Specifically, it says that *the eigenvalues of the perturbed matrix A' cannot differ from the eigenvalues of the original matrix A by more than the largest eigenvalue of the perturbed matrix E.* (See also the corollary of the Bauer–Fike theorem given in Section 9.7.2.)

Example 10.1.

$$A = \begin{pmatrix} 1 & 2 & 3 \\ 2 & 3 & 4 \\ 3 & 4 & 6 \end{pmatrix}, \quad E = 10^{-4} \times I_{3 \times 3}.$$

The eigenvalues of A are -0.4203, 0.2336, and 10.1867. The eigenvalues of $A + E$ are -0.4203, 0.2337, and 10.1868. Note that $\|E\|_2 = 10^{-4}$. ∎

E. *Rank-one perturbation property.* In this section we state a theorem that shows how the eigenvalues are shifted if E is a rank-one perturbation matrix. The result plays an important role in the *divide-and-conquer algorithm* (Dongarra and Sorensen (1987)) for the symmetric eigenvalue problem, to be discussed in Section 10.2.4.

Eigenvalues of a Rank-One Perturbed Matrix

Theorem 10.2. *Suppose $B = A + \alpha bb^T$, where A is an $n \times n$ symmetric matrix, α is a scalar, and b is an n-vector. Let $\lambda_1 \geq \lambda_2 \geq \cdots \geq \lambda_n$ be the eigenvalues of A and $\lambda'_1 \geq \cdots \geq \lambda'_n$ be the eigenvalues of B. Then*

$$\begin{aligned} \lambda'_i \in [\lambda_i, \lambda_{i-1}], & \quad i = 2, \ldots, n, \quad \text{if } \alpha \geq 0, \\ \lambda'_i \in [\lambda_{i+1}, \lambda_i], & \quad i = 1, \ldots, n-1, \quad \text{if } \alpha < 0. \end{aligned} \tag{10.4}$$

Proof. See Wilkinson (1965, pp. 97–98). □

Example 10.3.
$$A = \begin{pmatrix} 1 & 2 & 3 \\ 2 & 4 & 5 \\ 3 & 5 & 6 \end{pmatrix}, \quad \alpha = -1, \ b = (1, 2, 3)^T.$$

The eigenvalues of B are $\lambda'_3 = -3.3028$, $\lambda'_2 = 0$, $\lambda'_1 = 0.3028$. The eigenvalues of A are $\lambda_3 = -0.5157$, $\lambda_2 = 0.1709$, $\lambda_1 = 11.3448$. It is easily verified that $\lambda_2 < \lambda'_1 < \lambda_1$ and $\lambda_3 < \lambda'_2 < \lambda_2$. ∎

10.2.2 The Bisection Method for the Symmetric Tridiagonal Matrix

In this section we describe a method for finding the eigenvalues of a symmetric matrix. The method is particularly useful if eigenvalues are required in an interval. In principle, however, it can be used to find all eigenvalues.

First, the symmetric matrix A is transformed into a symmetric tridiagonal matrix T using Householder's method described in Chapter 9; that is, an orthogonal matrix P is constructed such that

$$PAP^T = T = \begin{pmatrix} \alpha_1 & \beta_1 & & & & \\ \beta_1 & \alpha_2 & \beta_2 & & \text{\Large 0} & \\ & \ddots & \ddots & \ddots & & \\ & & \ddots & \ddots & \ddots & \\ & \text{\Large 0} & & \beta_{n-2} & \alpha_{n-1} & \beta_{n-1} \\ & & & & \beta_{n-1} & \alpha_n \end{pmatrix}. \qquad (10.5)$$

A three-term recursion. Let $p_i(\lambda)$ denote the characteristic polynomial of the $i \times i$ principal submatrix of T. Then these polynomials satisfy a three-term recursion:

$$p_i(\lambda) = (\alpha_i - \lambda) p_{i-1}(\lambda) - \beta_{i-1}^2 p_{i-2}(\lambda), \quad i = 2, 3, \ldots, n, \qquad (10.6)$$

with

$$p_0(\lambda) = 1 \quad \text{and} \quad p_1(\lambda) = \alpha_1 - \lambda.$$

Without loss of generality, we may assume that $\beta_i \neq 0$, $i = 1, 2, \ldots, n-1$. Recall that matrix T with this property is called **unreduced**. If a subdiagonal entry of T is zero, then T becomes a block diagonal matrix and its eigenproblem can thus be reduced to that of its submatrices. *The eigenvalues of an unreduced symmetric tridiagonal matrix T are all real and distinct.* Since the characteristic polynomial $P_n(\lambda)$ of T can be easily computed, an immediate idea that comes in one's mind is to apply the well-known **bisection method** for root-finding to $P_n(\lambda)$ to locate an eigenvalue. This will be in contradiction to the warning that we gave in Section 9.5.5 that the eigenvalues should not be computed by finding the zeros of the characteristic polynomial. *The difference here, however, is that the coefficients of the characteristic polynomial do not have to be explicitly computed; all that is needed is to determine the signs of $p_i(\lambda)$, $i = 1, \ldots, n$, for a given number μ.* This is possible by exploiting some additional remarkable spectral properties of the symmetric tridiagonal matrix T, as given in the following theorem.

10.2. Computational Methods for the Symmetric Eigenvalue Problem

Theorem 10.4 (interlacing property). *Let T be an unreduced symmetric tridiagonal matrix. Let the eigenvalues of the kth leading principal minor $T^{(k)}$ of T be denoted by $\lambda_1^{(k)} < \lambda_2^{(k)} < \cdots < \lambda_k^{(k)}$. Then*

$$\lambda_i^{(k+1)} < \lambda_i^{(k)} < \lambda_{i+1}^{(k+1)}, \quad k = 1, 2, \ldots, n-1, \quad i = 1, 2, \ldots, k-1. \tag{10.7}$$

An illustration: Suppose T is a 4×4 unreduced symmetric tridiagonal matrix. Then $T^{(1)}$ has just one eigenvalue, $\lambda_1^{(1)}$; $T^{(2)}$ has two, $\lambda_1^{(2)}, \lambda_2^{(2)}$; $T^{(3)}$ has three, $\lambda_1^{(3)}, \lambda_2^{(3)}$, and $\lambda_3^{(3)}$; and $T^{(4)} = T$ has four, $\lambda_1^{(4)}, \lambda_2^{(4)}, \lambda_3^{(4)}, \lambda_4^{(4)}$. Figure 10.1 shows how the interlacing of these eigenvalues will then look like.

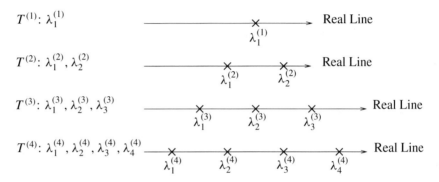

Figure 10.1. *The interlacing property.*

The above results lead to the following remarkable property of the polynomials $p_i(\lambda)$.

Theorem 10.5. *The number of sign agreements between the consecutive terms of the sequence of polynomials $\{p_0(\mu), p_1(\mu), \ldots, p_n(\mu)\}$ equals the number of eigenvalues of T, which are strictly greater than μ.*

Proof. See Wilkinson (1965, pp. 300–301). □

Note: The sequence $\{p_k(\mu)\}$ might contain zeros. In this case the convention is that $p_k(\mu)$ has the opposite sign of $p_{k-1}(\mu)$ if $p_k(\mu) = 0$.

Example 10.6. Let

$$T = \begin{pmatrix} 2 & 1 & 0 \\ 1 & 2 & 1 \\ 0 & 1 & 2 \end{pmatrix}.$$

The characteristic polynomials $\{p_k(\lambda)\}$ are

$$p_0(\lambda) = 1; \quad p_1(\lambda) = 2 - \lambda; \quad p_2(\lambda) = (2-\lambda)^2 - 1;$$
$$p_3(\lambda) = (2-\lambda)p_2(\lambda) - p_1(\lambda) = (2-\lambda)^3 - 2(2-\lambda).$$

Let $\mu = 0$. Then the sequence $\{p_0, p_1(\mu), p_2(\mu), p_3(\mu)\}$ is $\{1, 2, 3, 4\}$. There are three agreements in sign. Thus all the eigenvalues of T are greater than or equal to zero. In fact, since $p_3(0) = 4 \neq 0$, it follows that all the eigenvalues of T are positive.

Let $\mu = 2$. Then the sequence $\{p_0(\mu), p_1(\mu), p_2(\mu), p_3(\mu)\}$ is $\{1, 0, -1, 0\}$. The signs here are $+ - - +$. There is only one agreement in sign confirming that T has one eigenvalue greater than 2.

Verify: The eigenvalues of T are $\{2, 2 + \sqrt{2}, 2 - \sqrt{2}\}$. ∎

Bisection idea. Theorem 10.5 makes it possible to use the well-known **bisection algorithm** to locate a zero of the characteristic polynomial $p_n(\lambda)$ or, in other words, a specific eigenvalue of A.

ALGORITHM 10.1. The Bisection Algorithm for the Symmetric Eigenvalue Problem.

Inputs: An $n \times n$ symmetric tridiagonal matrix T, an integer $m \leq n$, and $\epsilon > 0$.

Output: An approximation to the eigenvalue λ_{n-m+1}, assuming that $\lambda_1 < \lambda_2 < \cdots < \lambda_n$.

Step 1. Find an interval $[s_1, s_2]$ containing λ_{n-m+1}. Since $\lambda_i \leq \|T\|$, initially, we can take $s_1 = -\|T\|_\infty$, $s_2 = \|T\|_\infty$.

Step 2. Compute $s_3 = \dfrac{s_1 + s_2}{2}$.

Step 3. Compute $N(s_3)$ = the number of agreements in sign in the sequence
$$\{1, p_1(s_3), p_2(s_3), \ldots, p_n(s_3)\}.$$
If $N(s_3) < m$, set $s_2 = s_3$; otherwise, set $s_1 = s_3$.

Step 4. Test whether $|s_2 - s_1| < \epsilon$. If so, accept $s_3 = \dfrac{s_1 + s_2}{2}$ as an approximate value of λ_{n-m+1}. Otherwise go to Step 2.

Note: *After k steps, the desired zero is located in an interval of width $\frac{(s_2 - s_1)}{2^k}$.*

Example 10.7. Consider the matrix T in Example 10.6. Suppose we want to approximate $\lambda_1 = 2 - \sqrt{2}$. Then $m = 3$.

Iteration 0. Initially, $s_1 = -4$, $s_2 = 4$, $s_3 = 0$; $N(s_3) = N(0) = 3$. Set $s_1 = s_3$.

Iteration 1. $s_1 = 0$, $s_2 = 4$, $s_3 = \frac{0+4}{2} = 2$; $N(s_3) = N(2) = 2 < 3$. Set $s_2 = s_3$.

Iteration 2. $s_1 = 0$, $s_2 = 2$, $s_3 = \frac{0+2}{2} = 1$; $N(s_3 = 2 < 3$. Set $s_2 = s_3$.

Iteration 3. $s_1 = 0$, $s_2 = 1$, $s_3 = 0.5$; $N(s_3) = 3$. Set $s_1 = s_3$.

The eigenvalue λ_1 is clearly in the interval $[0.5, 1]$, which is, in fact, the case. We can continue our iterations until the length of the interval $|s_2 - s_1| < \epsilon$. ∎

10.2. Computational Methods for the Symmetric Eigenvalue Problem

Flop-count and stability. Once A is transformed into T, it requires about $O(n)$ flops for evaluation of the sequence $\{p_i(\mu)\}$. Thus to *find k eigenvalues, only $O(kn)$ flops will be needed*. A remarkable fact is that *absolute errors in the computed eigenvalues are small*; but the relative errors in the small eigenvalues may be large. If eigenvectors are desired, inverse iteration can be invoked. Computing one eigenvector then requires only $O(n)$ flops, since an $n \times n$ tridiagonal system can be solved using $O(n)$ flops (Chapter 6). Thus *in principle, all the eigenvalues and eigenvectors can be computed in $O(n^2)$ flops by this method*, once the symmetric matrix A has been transformed to the symmetric tridiagonal matrix T. However, the *method is best used to find a selected number of eigenvalues, the eigenvalues in an interval or a prescribed number of eigenvalues to the left or right of a given eigenvalue*.

10.2.3 The Symmetric QR Iteration Method

To apply the QR iteration method of Chapter 9 to a symmetric tridiagonal matrix, we note that if the *starting matrix is a symmetric tridiagonal matrix T, then so is each matrix T_k in the sequence*

$$T_k - \mu_k I = Q_k R_k,$$

and, furthermore, we need only $O(n)$ flops to generate each T_k (note that the QR factorization of a symmetric tridiagonal matrix requires only $O(n)$ flops). Thus, the tridiagonal symmetric QR iteration is an $O(n^2)$ algorithm.

Also, since the eigenvalues of a symmetric matrix are all real and the real Schur form of a symmetric matrix is a diagonal rather than a triangular matrix, the double-shift strategy discussed for the general eigenvalue problem in Chapter 9 is not needed in this case. However, in this case a popular shift, known as the Wilkinson shift, defined below, is normally used.

Instead of taking the (n, n)th entry at every iteration as the shift, the eigenvalue of the trailing 2×2 matrix that is closer to the (n, n)th entry is usually chosen as the shift. This is known as the **Wilkinson shift.** Thus if a trailing 2×2 submatrix of T_k is given by

$$\begin{pmatrix} t_{n-1,n-1}^{(k)} & t_{n,n-1}^{(k)} \\ t_{n,n-1}^{(k)} & t_{nn}^{(k)} \end{pmatrix},$$

then the Wilkinson shift is

$$\mu = t_{nn}^{(k)} + r - \operatorname{sign}(r)\sqrt{r^2 + \left(t_{n,n-1}^{(k)}\right)^2}, \tag{10.8}$$

where

$$r = \frac{(t_{n-1,n-1}^{(k)} - t_{nn}^{(k)})}{2}. \tag{10.9}$$

Remark. It is possible to compute T_{k+1} from T_k without explicitly forming the matrix $T_k - \mu_k I$. This is known as the *implicit symmetric QR algorithm*. For details, see Golub and Van Loan (1996, pp. 420–421). See also Exercise 9.30 in Chapter 9.

ALGORITHM 10.2. Symmetric QR Iteration with the Wilkinson Shift.

Input: A symmetric matrix A.
Output: The approximate eigenvalues of A.

Phase I. Transform A into a symmetric tridiagonal matrix T using orthogonal similarity transformations:
$$PAP^T = T.$$

Phase II. Apply *single-shift QR iteration* to T with the *Wilkinson shift* μ:

Set $T = T_1$.
    For $k = 1, 2, \ldots$ do until convergence

    Find a real shift μ

        1. $T_k - \mu I = Q_k R_k$ (*shifted QR factorization*).
        2. $T_{k+1} = R_k Q_k + \mu I$ (*reverse multiplication with shift added*).

End

Convergence of the Symmetric QR Iteration with the Wilkinson Shift

The QR algorithm with the Wilkinson shift always converges. The rate of convergence is cubic for most matrices; in the worst case it is at least quadratic.

Flop-count.

- Transformation to T: $\frac{4}{3}n^3$.

- Eigenvalue computations: $O(n^2)$. (Note that the QR factorization of a tridiagonal matrix requires only $O(n)$ flops (Chapter 6)).

- All the eigenvectors of T: a little over $6n^3$ on average.

Round-off error property. As in the general nonsymmetric case, the *symmetric* QR with implicit shift is *stable*. It can be shown that, given a symmetric matrix A, the symmetric QR algorithm with implicit shift generates an orthogonal matrix Q and a diagonal matrix D such that
$$Q^T A Q = D + E,$$
where
$$\|E\|_F \approx \mu \, \phi(n) \|A\|_F.$$

$\phi(n)$ is a slowly growing function of n.

Accuracy of the computed eigenvalues. Each computed eigenvalue $\hat{\lambda}_i$ satisfies the inequality

$$|\lambda_i - \hat{\lambda}_i| \leq \phi(n)\mu\|A\|_2.$$

Thus, the *absolute error in each computed eigenvalue is small.*

10.2.4 The Divide-and-Conquer Method

As the title suggests, this method is based on the divide-and-conquer principle. *The algorithm first divides a given symmetric tridiagonal eigenvalue problem into two smaller subproblems, and then combines the solutions of the subproblems to recover (conquer) the solution of the original problem.* The method was originally suggested by Cuppen (1981).

The method can be used to compute all the eigenvalues and the corresponding eigenvectors of a symmetric matrix, and *it is faster than the symmetric QR iteration method* just described. We present here a very brief sketch of the method.

Suppose that the symmetric matrix A has been transformed to a symmetric tridiagonal matrix T by an orthogonal similarity. Let

$$T = \begin{pmatrix} a_1 & b_1 & & 0 \\ b_1 & \ddots & \ddots & \\ & \ddots & \ddots & b_{n-1} \\ 0 & & b_{n-1} & a_n \end{pmatrix}. \tag{10.10}$$

Define

$$T_1 = \begin{pmatrix} a_1 & b_1 & & 0 \\ b_1 & \ddots & \ddots & \\ & \ddots & a_{k-1} & b_{k-1} \\ 0 & & b_{k-1} & a_k - b_k \end{pmatrix}, \tag{10.11}$$

$$T_2 = \begin{pmatrix} a_{k+1} - b_k & b_{k+1} & & 0 \\ b_{k+1} & \ddots & \ddots & \\ & \ddots & \ddots & b_{n-1} \\ 0 & & b_{n-1} & a_n \end{pmatrix}. \tag{10.12}$$

Then

$$T = \begin{pmatrix} T_1 & 0 \\ 0 & T_2 \end{pmatrix} + b_k v v^T,$$

where

$$v = (0, 0, \ldots, 1, 1, 0, \ldots, 0)^T,$$

Since T_1 and T_2 are symmetric tridiagonal, we can find orthogonal matrices Q_1 and Q_2 such that

$$T_1 = Q_1 D_1 Q_1^T \quad \text{and} \quad T_2 = Q_2 D_2 Q_2^T, \quad \text{where } D_1 \text{ and } D_2 \text{ are diagonal matrices.}$$

Then

$$T = \begin{pmatrix} Q_1 & 0 \\ 0 & Q_2 \end{pmatrix} \left[\begin{pmatrix} D_1 & 0 \\ 0 & D_2 \end{pmatrix} + b_k u u^T \right] \begin{pmatrix} Q_1^T & 0 \\ 0 & Q_2^T \end{pmatrix},$$

where

$$u = \begin{pmatrix} Q_1^T & 0 \\ 0 & Q_2^T \end{pmatrix} v.$$

Therefore, the eigenvalues of T are the same as those of

$$\hat{D} = D + b_k u u^T = D + \rho u u^T, \tag{10.13}$$

where D is given by

$$D = \begin{pmatrix} D_1 & 0 \\ 0 & D_2 \end{pmatrix},$$

and $\rho = b_k$. We therefore concentrate now on how to obtain the eigenvalues and eigenvectors of the rank-one perturbed diagonal matrix $\hat{D} = D + \rho u u^T$.

Assume without any loss of generality that $\|u\|_2 = 1$ and $\rho = b_k \neq 0$. Let $D = \text{diag}(d_1, d_2, \ldots, d_n)$. Assume that $d_1 < d_2 < d_3 < \cdots < d_n$ and *none of the components of the vector u is zero*.

In fact, a *zero component of u is a blessing in disguise*. We can show (Exercise 10.7) that, in this case, we get an eigenvalue and eigenvector pair free. Also, if k eigenvalues of D are equal, then the problem can be deflated by deleting $(k-1)$ rows and columns (Exercise 10.8).

Let (λ, q) be an eigenpair of $\hat{D}$. Then we show that

(i) λ is a root of the equation

$$1 + \rho u^T (D - \lambda I)^{-1} u = 0 \tag{10.14}$$

and

(ii)

$$q = (D - \lambda I)^{-1} u \tag{10.15}$$

is an eigenvector of $(D + \rho u u^T)$ corresponding to λ.

To show (i), we note that since (λ, q) is an eigenpair of $\hat{D}$ we must have

$$(D + \rho u u^T) q = \lambda q \quad \text{for some } q \neq 0,$$

10.2. Computational Methods for the Symmetric Eigenvalue Problem

that is,
$$(D - \lambda I)q = -\rho(u^T q)u.$$

Now our assumptions that $\rho \neq 0$, that $d_1 < d_2 < d_3 < \cdots < d_n$, and that none of the components of u is zero imply that (a) $(D - \lambda I)$ is nonsingular, and (b) $u^T q$ is nonzero (Exercise 10.10).

Multiplying by $(D - \lambda I)^{-1}$ we have
$$q = -\rho(u^T q)(D - \lambda I)^{-1} u. \tag{10.16}$$

Multiplying both sides of (10.16) by u^T and dividing by the nonzero scalar $u^T q$, we have
$$1 + \rho u^T (D - \lambda I)^{-1} u = 0. \tag{10.17}$$

To show (ii), we note that
$$\begin{aligned}(D + \rho u u^T)(D - \lambda I)^{-1} u &= (D - \lambda I + \lambda I + \rho u u^T)(D - \lambda I)^{-1} u \\ &= u + \lambda (D - \lambda I)^{-1} u + u \rho u^T (D - \lambda I)^{-1} u \\ &= u + \lambda (D - \lambda I)^{-1} u + u(-1) \quad \text{(using 10.14)} \\ &= u + \lambda (D - \lambda I)^{-1} u - u = \lambda (D - \lambda I)^{-1} u.\end{aligned}$$

Locating the roots of (10.14). Note that
$$1 + \rho u^T (D - \lambda I)^{-1} u = 0$$
can be written in terms of the components u_i of u as follows:
$$f(\lambda) = 1 + \rho u^T (D - \lambda I)^{-1} u = 1 + \rho \sum_{j=1}^{n} \frac{u_j^2}{d_j - \lambda} = 0. \tag{10.18}$$

This equation is usually known as the **secular equation.**

Again, because d_i's are all distinct and none of the components of u is zero, we can show (Exercises 10.11) that $f(\lambda) = 0$ has precisely n roots, one in each of the intervals (d_j, d_{j+1}), $j = 1, 2, \ldots, n - 1$, and one to the right of d_n if $\rho > 0$ or one to the left of d_1 if $\rho < 0$.

For example if $u = (0.7, 0.8, 0.9, 1)^T$, $\rho = \frac{1}{2}$, and $D = \text{diag}(1, 2, 3, 4)$, the graph of the secular equation will look like the graph in Figure 10.2.

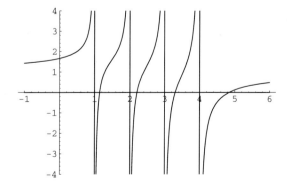

Figure 10.2. *An illustrative graph of the secular equation.*

Obtaining the eigenvalues and eigenvectors. Knowing that the roots of $f(\lambda) = 0$ are located in such specified intervals, we can then apply the bisection method or Newton's method to find these roots in each of these intervals. Once a root is found, the corresponding eigenvector can be obtained from (10.15).

However, a more stable way of computing the vector u is as follows (see Demmel (1997, pp. 224–225)):

Let $d_n < \lambda_n < \cdots < d_{i+1} < \lambda_i \cdots < d_1 < \lambda_1$.

- Compute the ith component u_i as

$$|u_i| = \left[\frac{\Pi_{j=1}^n (\lambda_j - d_i)}{\Pi_{j=1, j \neq i}^n (d_j - d_i)} \right]^{\frac{1}{2}}. \quad (10.19)$$

- Compute the eigenvectors of $D + \rho u u^T$ using (10.15).

ALGORITHM 10.3. Divide-and-Conquer Method.

Input: A symmetric tridiagonal matrix T as given in (10.10).
Output: Approximate eigenvalues and eigenvectors of T.

Step 1. Form

$$T = \begin{pmatrix} T_1 & 0 \\ 0 & T_2 \end{pmatrix} + b_k v v^T,$$

where T_1 and T_2 are as given by (10.11) and (10.12).

Step 2. Find orthogonal matrices Q_1 and Q_2 such that $Q_1^T T_1 Q_1 = D_1$ and $Q_2^T T_2 Q_2 = D_2$, where D_1 and D_2 are diagonal matrices.

Step 3. Form $D = \text{diag}(D_1, D_2) = \text{diag}(d_1 \ldots d_n)$ and $u = \text{diag}(Q_1^T, Q_2^T) v$.

Step 4. Find the eigenvalues of T by solving the secular equation (10.18),

$$f(\lambda) = 1 + \rho \sum_{j=1}^n \frac{u_j^2}{d_j - \lambda} = 0,$$

and obtain the eigenvectors of $D + \rho u u^T$ from (10.15) with u computed by (10.19).

Step 5. Recover the eigenvectors of T: If Q' is the eigenvector matrix of $D + \rho u u^T$, then the eigenvector matrix of T is given by

$$\begin{pmatrix} Q_1 & 0 \\ 0 & Q_2 \end{pmatrix} Q'.$$

Flop-count. Assuming that one Newton iteration step costs about $O(n)$ flops, *the algorithm will require only $O(n^2)$ flops for all the n eigenvalues*. The cost of computing each eigenvector from a computed eigenvalue is also $O(n)$ flops.

10.2. Computational Methods for the Symmetric Eigenvalue Problem

Remarks. (i) Bunch, Nielsen, and Sorensen (1978) proposed a method for solving $f(\lambda) = 0$ using rational function approximations. Their method converges quadratically. For more on this method and implementational details, see Bunch, Nielsen, and Sorensen (1978), Dongarra and Sorensen (1987), and Sorensen and Tang (1991).

(ii) The divide-and-conquer method is naturally parallel.

Note that, because T_1 and T_2 are both symmetric tridiagonal, the eigenvalue problem for each of these matrices can further be decomposed into two subproblems, resulting in four subproblems. These four subproblems again can be decomposed into eight smaller problems and the process can be continued for as long as desired (possibly until the problem sizes become 1×1 or 2×2). Since each of these subproblems is independent, the original problem can be divided into many independent subproblems. See Dongarra and Sorensen (1987) for parallel implementational aspects of this method. In fact, the divide-and-conquer algorithm was originally targeted as a parallel scheme, but it turns out to be faster than the symmetric QR algorithm if properly implemented (see Gu and Eisenstat (1995a)).

10.2.5 The Jacobi Method

One of the classical methods for computing the eigenvalues of a symmetric matrix is the method introduced by C. J. Jacobi[16] in 1846. Since a symmetric matrix A can be diagonalized by orthogonal similarity, the idea is to create orthogonal matrices $J_0, J_1, \ldots, J_{k-1}$ such that the sequence $\{A_k\}$ defined by $A_0 = A$,

$$A_k = J_{k-1}^T A_{k-1} J_{k-1}^T, \quad k = 1, 2, \ldots,$$

approaches a diagonal matrix for large k.

In Jacobi's method, the orthogonal matrices are nothing but Givens rotations; but they were originally invented by Jacobi. These matrices are created successively to make one pair of off-diagonal entries zeros, one pair at a time. Recall from Chapter 7 that each of these rotations is uniquely determined by two numbers, c and s. *Formulas for $c = \cos\theta$ and $s = \sin\theta$ used for QR factorization in Chapter 7 have to be modified here.*

It can be easily verified that (i, j) and (j, i) entries of the matrix $A_{r+1} = J_r^T A_r J_r$ can be made zeros simultaneously in the Jacobi method: if $A_r = (a_{ij}^{(r)})$, then this will happen if $J_i(i, j, c, s)$ is constructed with c and s defined by

$$c = \frac{1}{\sqrt{1+t^2}}, \quad s = ct,$$

where

$$t = \frac{\text{sign}(\tau)}{(|\tau| + \sqrt{1+\tau^2})}, \quad \tau = \frac{a_{ii}^{(r)} - a_{jj}^{(r)}}{2a_{ij}^{(r)}}.$$

Unfortunately, the zeros created at an earlier step get destroyed by subsequent steps. However, as in the QR iteration algorithm for eigenvalue computations, the nonzero entries decreases steadily as the iteration proceeds.

---

[16]Carl Jakob Jacobi (1804–1851) was a German mathematician. He wrote the classic treatise on *elliptic functions*, studied *Jacobi theta functions*, proved *Fermat's polygonal number theorem*, put the *determinant* in its modern form, found the *Jacobi integral*, and did much to develop the *Hamilton–Jacobi theory*

Choosing the off-diagonal entries for zeroing. In the classical Jacobi scheme, the indices i and j are chosen so that the entry a_{ij} is the largest off-diagonal entry in magnitude at each step.

It can be shown (Exercise 10.12) that the sum of the squares of the off-diagonal entries, denoted by off$^2(A)$,

$$\text{off}^2(A) = \sum_{i=1}^{n} \sum_{\substack{j \neq n \\ j=1}}^{n} a_{ij}^2,$$

decreases at least by the factor $1 - \frac{2}{n(n-1)}$ at each step; that is,

$$\text{off}^2(A) - a_{ij}^2 \leq \frac{2}{n(n-1)} \text{off}^2(A).$$

Thus the classical *Jacobi scheme converges at least linearly;* in practice the convergence is *actually quadratic.* Note that implementation of the scheme involves an $O(n^2)$ search for the largest entry at each step.

In practice, a scheme called the *cyclic Jacobi* scheme is used, in which the off-diagonal entries are annihilated in the rowwise order $(1, 2), (1, 3), \ldots, (1, n); (2, 3), (2, 4), \ldots, (2, n)$; and so on.

This scheme is faster since it does not require off-diagonal search, and is more accurate. *The rate of convergence is also ultimately quadratic* (Wilkinson (1965, p. 270)). The details can be found in Demmel (1997), Golub and Van Loan (1996), and Parlett (1998).

10.2.6 Comparison of the Symmetric Eigenvalue Methods

- **Tridiagonal QR iteration.** The QR iteration algorithm applied to an $n \times n$ symmetric tridiagonal matrix requires only $O(n^2)$ flops to compute all the eigenvalues. However, finding all the eigenvectors requires another $6n^3$ flops approximately.

- **Divide-and-conquer method.** Like the QR iteration algorithm, the divide-and-conquer algorithm also requires about $O(n^2)$ flops to compute all the eigenvalues of a symmetric tridiagonal matrix.

    However, if all the eigenvectors are also desired, this algorithm is more efficient than the QR iteration because it can be shown that the flop-count for all eigenvectors is about $\frac{4n^3}{3}$, compared to $6n^3$ needed by the QR iteration algorithm. There are several other faster implementations of this popular algorithm (see Demmel (1997) and Gu and Eisenstat (1995a) for details).

- **Bisection method.** This method needs only $O(nk)$ flops if k number of eigenvalues are required. If the eigenvalues are well-separated, the cost of computing the eigenvectors via *inverse iteration* is also $O(nk)$. Thus, in principle it takes only $O(n^2)$ flops to compute all the eigenvalues and eigenvectors of a symmetric tridiagonal matrix, making the method much faster than both the QR and divide-and-conquer methods. In the worst case, when several eigenvalues are clustered together, the cost becomes $O(n^3)$ for inverse iteration, and, furthermore, the accuracy of the computed eigenvectors is not guaranteed. However, there has been some progress in obtaining the eigenvectors more accurately with not much more than $O(n)$ flops per eigenvector (Demmel (1997)).

10.3. The Singular Value Decomposition and Its Computation

The method is best if only a few eigenvalues or those in an interval are desired.

- **The Jacobi method.** The method also requires $O(n^3)$ flops to compute all the eigenvalues and eigenvectors of a symmetric matrix A (note that the method does not require tridiagonalization). However, *it is in general much slower than the other methods. The method sometimes, however, computes the eigenvalues with a relative high accuracy.*

Conclusion:

- The divide-and-conquer method is the fastest algorithm for symmetric matrices if all the eigenvalues and eigenvectors are desired.

- The symmetric QR algorithm with the Wilkinson shift is the fastest practical algorithm for finding all the eigenvalues of small-order symmetric matrices.

- The bisection method may be used to compute a small number of eigenvalues of a symmetric matrix or a number of eigenvalues in a specified interval.

10.3 The Singular Value Decomposition and Its Computation

We remind the reader of the statement of the SVD theorem from Chapter 7.

Theorem 10.8 (SVD theorem). *Let $A \in \mathbb{R}^{m \times n}$. Then there exist orthogonal matrices $U \in \mathbb{R}^{m \times m}$ and $V \in \mathbb{R}^{n \times n}$ such that*

$$A = U \Sigma V^T, \tag{10.20}$$

where Σ is an $m \times n$ "diagonal" matrix. The diagonal entries of Σ are all nonnegative and can be arranged in nonincreasing order.

Proof. Denote the eigenvalues of the symmetric positive semidefinite matrix $A^T A$, which are nonnegative, by $\lambda_1 = \sigma_1^2, \lambda_2 = \sigma_2^2, \ldots, \lambda_n = \sigma_n^2$ and the corresponding eigenvectors by $v_1, \ldots, v_n$. Let $\sigma_1 \geq \sigma_2 \geq \cdots \geq \sigma_r > 0$ and $\sigma_{r+1} = \sigma_n = 0$. Set $V_1 = (v_1, v_2, \ldots, v_r)$, $V_2 = (v_{r+1} \ldots v_n)$, and $V = (V_1, V_2)$. Then V is an $n \times n$ **orthogonal matrix**. Also, since $\{v_1, v_2, \ldots, v_n\}$ forms an **orthonormal set of eigenvectors** of $A^T A$, we have

$$v_i^T A^T A v_i = \sigma_i^2 \quad \text{and} \quad v_i^T A^T A v_j = 0, \quad i \neq j. \tag{10.21}$$

Define now a set of vectors $\{u_i\}$ by

$$u_i = \frac{1}{\sigma_i} A v_i, \quad i = 1, \ldots, r. \tag{10.22}$$

The u_i's, $i = 1, \ldots, r$, then form an orthonormal set, because

$$u_i^T u_j = \frac{1}{\sigma_i} (A v_i)^T \frac{1}{\sigma_j} (A v_j) = \frac{1}{\sigma_i \sigma_j} (v_i^T A^T A v_j) = \begin{cases} 0 & \text{when } i \neq j, \\ 1 & \text{when } i = j. \end{cases} \tag{10.23}$$

Set $U_1 = (u_1, \ldots, u_r)$, and choose $U_2 = (u_{r+1}, \ldots, u_m)$ such that $u_j^T A = 0$, $j = r+1, \ldots, m$. Then the set $\{u_1, \ldots, u_r, u_{r+1}, \ldots, u_m\}$ forms an orthonormal basis of the m-space $\mathbb{C}^m$.

Now set $U = (U_1, U_2)$. Then U and V are orthonormal, and using (10.22), we obtain

$$U^T A V = \begin{pmatrix} u_1^T \\ u_2^T \\ \vdots \\ u_m^T \end{pmatrix} A(v_1, \ldots, v_n) = \begin{pmatrix} \frac{1}{\sigma_1} v_1^T A^T \\ \frac{1}{\sigma_2} v_2^T A^T \\ \vdots \\ \frac{1}{\sigma_r} v_r^T A^T \\ u_{r+1}^T \\ \vdots \\ u_m^T \end{pmatrix} A(v_1, \ldots, v_n) \qquad (10.24)$$

$$= \begin{pmatrix} \frac{1}{\sigma_1} \cdot \sigma_1^2 & 0 & 0 & \cdots & 0 & 0 \\ 0 & \frac{1}{\sigma_2} \cdot \sigma_2^2 & 0 & \cdots & 0 & 0 \\ \vdots & \ddots & \ddots & \cdots & \vdots & \\ 0 & \cdots & \frac{1}{\sigma_r} \cdot \sigma_r^2 & 0 & \cdots & 0 \\ 0 & \cdots & & 0 & \ddots & 0 \\ 0 & \cdots & \cdots & \cdots & \cdots & 0 \end{pmatrix} = \Sigma \quad \text{(using (10.21))}. \quad \square \qquad (10.25)$$

10.3.1 The Relationship between the Singular Values and the Eigenvalues

The above proof of the SVD theorem reveals the following interesting relationship between the singular values and singular vectors of A with the eigenvalues and eigenvectors of $A^T A$:

- The singular values of A are nonnegative square roots of the eigenvalues of $A^T A$.
- The right singular vectors are the eigenvectors of $A^T A$.
- Furthermore, the SVD of A is related to the eigendecomposition of $A^T A$ as
$$V^T A^T A V = \Sigma^T \Sigma.$$

Theorem 10.9. *The nonzero singular values of an $m \times n$ ($m \geq n$) matrix A are positive eigenvalues of the matrix*
$$C = \begin{pmatrix} 0_{m \times m} & A \\ A^T & 0_{n \times n} \end{pmatrix}.$$

Proof. Let
$$A = U \Sigma V^T \text{ be the SVD of } A.$$
Partition
$$U = \begin{pmatrix} U_1 & U_2 \\ m \times n & m \times (m-n) \end{pmatrix}, \qquad \Sigma = \begin{pmatrix} \Sigma_{1(n \times n)} \\ 0_{(m-n) \times n} \end{pmatrix}.$$

10.3. The Singular Value Decomposition and Its Computation

Define
$$P = \begin{pmatrix} \tilde{U}_1 & -\tilde{U}_1 & U_2 \\ \tilde{V} & \tilde{V} & 0_{n \times (m-n)} \end{pmatrix},$$
where
$$\tilde{U}_1 = \frac{1}{\sqrt{2}} U_1 \quad \text{and} \quad \tilde{V} = \frac{1}{\sqrt{2}} V.$$
Then it is easy to verify that
$$P^T C P = \begin{pmatrix} \Sigma_1 & 0 & 0 \\ 0 & -\Sigma_1 & 0 \\ 0 & 0 & 0 \end{pmatrix},$$
which shows that the nonzero eigenvalues of C are $\sigma_1, \ldots, \sigma_k, -\sigma_1, \ldots, -\sigma_k$, where σ_1 through σ_k are the nonzero singular values of A. □

10.3.2 Sensitivity of the Singular Values

We stated a remarkable property of the singular values in Chapter 7: *the singular values of a matrix are well-conditioned.* We now give a proof of this result.

Theorem 10.10 (perturbation theorem for singular values). *Let A and $B = A + E$ be two $m \times n$ matrices ($m \geq n$). Let σ_i, $i = 1, \ldots, n$, and $\tilde{\sigma}_i$, $i = 1, \ldots, n$, be, respectively, the singular values of A and $A + E$, appearing in decreasing order. Then $|\tilde{\sigma}_i - \sigma_i| \leq \|E\|_2$ for each i.*

Proof. Define
$$\tilde{A} = \begin{pmatrix} 0 & A \\ A^T & 0 \end{pmatrix}.$$

By Theorem 10.9, we have that the nonzero eigenvalues of $\tilde{A}$ are $\sigma_1, \ldots, \sigma_k$, $-\sigma_1, \ldots, -\sigma_k$, where σ_1 through σ_k are the nonzero singular values of A. The remaining eigenvalues of $\tilde{A}$ are, of course, zero. Define now $\tilde{B} = \begin{pmatrix} 0 & B \\ B^T & 0 \end{pmatrix}$, $\tilde{E} = \begin{pmatrix} 0 & E \\ E^T & 0 \end{pmatrix}$. Then $\tilde{B} - \tilde{A} = \tilde{E}$.

The eigenvalues of $\tilde{B}$ and $\tilde{E}$ are related, respectively, to the singular values of B and E in the same way the eigenvalues of $\tilde{A}$ are related to the singular values of A. The result now follows immediately by applying the Bauer–Fike theorem (Theorem 9.37) to $\tilde{B}$, $\tilde{A}$, and $\tilde{E}$. □

10.3.3 Computing the Variance-Covariance Matrix with SVD

One can easily compute the variance-covariance matrix $(A^T A)^{-1}$ using the QR factorization of A. We note here that this matrix can also be computed immediately once the singular values and the left singular vectors of A have been computed.

Computing $(A^T A)^{-1}$ using the SVD. Let
$$A = U \Sigma V^T$$

be the SVD of A. Then the entries of the matrix $C = (A^T A)^{-1} = (c_{ij})$ are given by $c_{ij} = \sum_{k=1}^{n} \frac{v_{ik} v_{jk}}{\sigma_k^2}$, where $n = \text{rank}(A)$. (Note that $C = (A^T A)^{-1} = V \Sigma^{-2} V^T$.)

Example 10.11.

$$A = \begin{pmatrix} 1 & 2 \\ 2 & 3 \\ 3 & 4 \end{pmatrix}.$$

$$c_{11} = \frac{v_{11}^2}{\sigma_1^2} + \frac{v_{12}^2}{\sigma_2^2} = 4.833, \quad c_{12} = \frac{v_{11} v_{21}}{\sigma_1^2} + \frac{v_{12} v_{22}}{\sigma_2^2} = -3.3333,$$

$$c_{22} = \frac{v_{21}^2}{\sigma_1^2} + \frac{v_{22}^2}{\sigma_2^2} = 2.3333. \quad \blacksquare$$

MATCOM Note: The MATCOM program **COVSVD** computes the variance-covariance matrix using the SVD of A.

10.3.4 Computing the Pseudoinverse with SVD

In Chapter 8 we have seen that when A is an $m \times n$ ($m \geq n$) matrix having full rank, the pseudoinverse of A is given by $A^\dagger = (A^T A)^{-1} A^T$. A formal definition of the pseudoinverse of any matrix A (whether it has full rank or not) can be given as follows: The pseudoinverse is also known as the *Moore–Penrose inverse*.

Four properties of the pseudoinverse. The pseudoinverse of an $m \times n$ matrix A is an $n \times m$ matrix X satisfying the following properties:

1. $AXA = A$.
2. $XAX = X$.
3. $(AX)^T = AX$.
4. $(XA)^T = XA$.

The pseudoinverse of a matrix always exists and is unique. We now show that the SVD provides a nice expression for the pseudoinverse.

Let $A = U \Sigma V^T$ be the SVD of A; then it is easy to verify that the matrix

$$A^\dagger = V \Sigma^\dagger U^T, \quad \text{where } \Sigma^\dagger = \text{diag}\left(\frac{1}{\sigma_j}\right) \quad (10.26)$$

$$\left(\text{if } \sigma_j = 0, \text{ use } \frac{1}{\sigma_j} = 0\right),$$

satisfies all the four properties above and therefore is the pseudoinverse of A. Note that this expression for the pseudoinverse coincides with A^{-1} when A is nonsingular, because

$$A^{-1} = (A^T A)^{-1} A^T = (V \Sigma^T U^T U \Sigma V^T)^{-1} V \Sigma^T U^T \quad (10.27)$$
$$= V \Sigma^{-1} (\Sigma^T)^{-1} V^T V \Sigma^T U^T = V \Sigma^{-1} U^T.$$

(Note that in this case $\Sigma^\dagger = \Sigma^{-1}$.)

The process for computing the pseudoinverse $A^\dagger$ of A using the SVD of A can be summarized as follows.

10.3. The Singular Value Decomposition and Its Computation

ALGORITHM 10.4. Computing the Pseudoinverse Using the SVD.

Input: An $m \times n$ matrix A.
Output: $A^\dagger$, the pseudoinverse of A.

Step 1. Find the SVD of A:
$$A = U\Sigma V^T.$$

Step 2. Compute
$$\Sigma^\dagger = \text{diag}\begin{pmatrix} \frac{1}{\sigma_1} & & & & \\ & \frac{1}{\sigma_2} & & & \\ & & \ddots & & 0 \\ & & & \frac{1}{\sigma_r} & \\ \hline & 0 & & & 0 \end{pmatrix},$$

where $\sigma_1, \ldots, \sigma_r$ are the r nonzero singular values of A.

Step 3. Compute $A^\dagger = V\Sigma^\dagger U^T$.

Example 10.12. Find the pseudoinverse of

$$A = \begin{pmatrix} 0 & 0 & -1 \\ -1 & 0 & 0 \\ 0 & -1 & 0 \end{pmatrix} \begin{pmatrix} 1 & 0 & 0 \\ 0 & 2 & 0 \\ 0 & 0 & 0 \end{pmatrix} \begin{pmatrix} \frac{1}{3} & -\frac{6}{9} & -\frac{6}{9} \\ -\frac{6}{9} & \frac{1}{3} & -\frac{6}{9} \\ -\frac{6}{9} & -\frac{6}{9} & \frac{1}{3} \end{pmatrix},$$

$$A = U\Sigma V^T, \qquad A^\dagger = V\Sigma^\dagger U^T,$$

$$\Sigma^\dagger = \begin{pmatrix} 1 & 0 & 0 \\ 0 & \frac{1}{2} & 0 \\ 0 & 0 & 0 \end{pmatrix}, \quad U^T = \begin{pmatrix} 0 & -1 & 0 \\ 0 & 0 & -1 \\ -1 & 0 & 0 \end{pmatrix}, \quad V^T = V.$$

Thus

$$A^\dagger = \begin{pmatrix} \frac{1}{3} & -\frac{6}{9} & -\frac{6}{9} \\ -\frac{6}{9} & \frac{1}{3} & -\frac{6}{9} \\ -\frac{6}{9} & -\frac{6}{9} & \frac{1}{3} \end{pmatrix} \begin{pmatrix} 1 & 0 & 0 \\ 0 & \frac{1}{2} & 0 \\ 0 & 0 & 0 \end{pmatrix} \begin{pmatrix} 0 & -1 & 0 \\ 0 & 0 & -1 \\ -1 & 0 & 0 \end{pmatrix} = \begin{pmatrix} 0 & -\frac{1}{3} & \frac{1}{3} \\ 0 & \frac{2}{3} & -\frac{1}{6} \\ 0 & \frac{2}{3} & \frac{1}{3} \end{pmatrix}. \blacksquare$$

10.3.5 Computing the SVD

Since the singular values of a matrix A are just the nonnegative square roots of the eigenvalues of the symmetric matrix $A^T A$, it is natural to think of computing the singular values of A by finding the eigenvalues of the symmetric matrix $A^T A$. However, *this is not a numerically effective process because*, as we have seen in Chapter 7, *some vital information may be lost due to round-off error in the process of computing $A^T A$*.

The following simple example illustrates the phenomenon.

Example 10.13.
$$A = \begin{pmatrix} 1.0001 & 1.0000 \\ 1.0000 & 1.0001 \end{pmatrix}. \tag{10.28}$$

The singular values of A are 2.0010 and 0.0001.
   Now
$$A^T A = \begin{pmatrix} 2.0002 & 2.0002 \\ 2.0002 & 2.0002 \end{pmatrix}$$

(to four significant digits). The eigenvalues of $A^T A$ are 0 and 4.0004. Thus the singular values of A will be computed as 0, 2.0001 (in 4-digit arithmetic), whereas the actual singular values are 0.0001 and 2.0010. ∎

10.3.6 The Golub–Kahan–Reinsch Algorithm

For more than three decades, the following algorithm, called the Golub–Kahan–Reinsch algorithm (Golub and Kahan (1965), Golub and Reinsch (1970)), has been a standard algorithm for SVD computation. There have been some recent developments and recent algorithms such as the *zero-shift QR algorithm*, the *differential QD algorithm*, and the *divide-and-conquer algorithm* (see later in this section for further remarks on these algorithms). We describe here the Golub–Kahan–Reinsch algorithm. The algorithm comes into two phases, as illustrated in Figure 10.3.

Phase 1 (direct).
$$A \longrightarrow B = \begin{pmatrix} \searrow & 0 \\ & \searrow \\ 0 & \end{pmatrix}.$$
Bidiagonal

Phase 2 (iterative).
$$B \longrightarrow \Sigma = \begin{pmatrix} \searrow & 0 \\ 0 & \searrow \end{pmatrix}.$$
Diagonal

Phase 1 (*bidiagonalization*). The $m \times n$ matrix $A(m \geq n)$ is transformed into an upper bidiagonal matrix by orthogonal equivalence; that is, the matrices $U_0 \in \mathbb{R}^{m \times m}$ and $V_0 \in \mathbb{R}^{n \times n}$ are created such that

$$U_0^T A V_0 = \begin{pmatrix} B \\ 0 \end{pmatrix}, \tag{10.29}$$

where B is an $n \times n$ bidiagonal matrix given by

$$\begin{pmatrix} b_{11} & b_{12} & & 0 \\ 0 & \ddots & \ddots & \vdots \\ \vdots & \ddots & \ddots & b_{n-1,n} \\ 0 & 0 & 0 & b_{nn} \end{pmatrix}.$$

10.3. The Singular Value Decomposition and Its Computation

Phase 2 (reduction to diagonal form). The bidiagonal matrix B is further reduced by orthogonal equivalence to a diagonal matrix Σ; that is, the orthogonal matrices U_1 and V_1 are created such that

$$U_1^T B V_1 = \Sigma = \mathrm{diag}(\sigma_1, \ldots, \sigma_n). \tag{10.30}$$

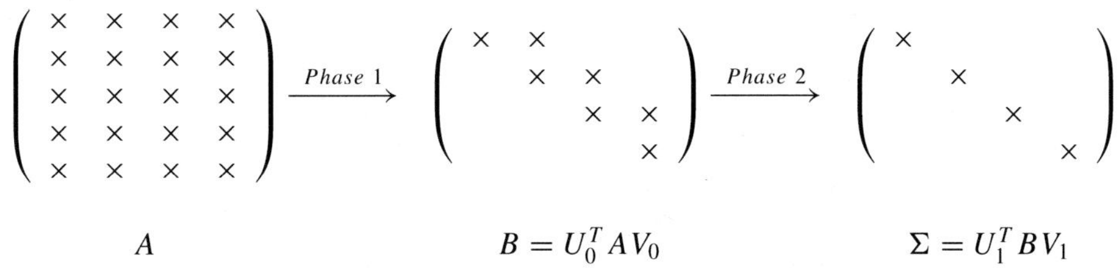

$$\begin{pmatrix} \times & \times & \times & \times \\ \times & \times & \times & \times \\ \times & \times & \times & \times \\ \times & \times & \times & \times \\ \times & \times & \times & \times \end{pmatrix} \xrightarrow{Phase\ 1} \begin{pmatrix} \times & \times & & \\ & \times & \times & \\ & & \times & \times \\ & & & \times \end{pmatrix} \xrightarrow{Phase\ 2} \begin{pmatrix} \times & & & \\ & \times & & \\ & & \times & \\ & & & \times \end{pmatrix}$$

$$A \qquad\qquad\qquad B = U_0^T A V_0 \qquad\qquad \Sigma = U_1^T B V_1$$

Figure 10.3. *Illustration of the two-phase procedure.*

Obtaining the SVD of A from Phases 1 and 2

- Σ — the matrix of the singular values.

- The singular vector matrices U and V are given by $U = U_0 U_1$, $V = V_0 V_1$.

Remark. In the numerical linear algebra literature, Phase 1 is known as the **Golub–Kahan** bidiagonal procedure, and Phase 2 is known as the **Golub–Reinsch** algorithm. We will call the combined two-stage procedure the **Golub–Kahan–Reinsch method**.

High relative accuracy of the singular values of bidiagonal matrices. The following result due to Demmel and Kahan (1990) *shows that the singular values of a bidiagonal matrix can be computed with very high accuracy.*

Theorem 10.14. *Let $B = (b_{ij})$ be an $n \times n$ bidiagonal matrix. Let $\Delta B = (\delta b_{ij})$ also be bidiagonal. Suppose that $b_{ii} + \delta b_{ii} = \alpha_{2i-1} b_{ii}$ and $\delta b_{i,i+1} + b_{i,i+1} = \alpha_{2i} b_{i,i+1}$, $\alpha_j \neq 0$.*
Let $\overline{\alpha} = \Pi_{i=1}^{2n-1} \max(|\alpha_i|, |\alpha_i^{-1}|)$. Let $\sigma_1 \geq \cdots \geq \sigma_n$ be the singular values of B and let $\sigma_1' \geq \cdots \geq \sigma_n'$ be the singular values of $B + \Delta B$. Then

$$\frac{\sigma_i}{\overline{\alpha}} \leq \sigma_i' \leq \overline{\alpha} \sigma_i, \qquad i = 1, \ldots, n. \tag{10.31}$$

Phase 1. Reduction to Bidiagonal Form. The matrices U_0 and V_0 in Phase 1 are constructed as the product of Householder matrices as follows:

$$U_0 = U_{01} U_{02} \cdots U_{0n} \tag{10.32}$$

and

$$V_0 = V_{01} V_{02} \cdots V_{0,n-2}. \tag{10.33}$$

Let's illustrate the bidiagonalization process with $m = 5$ and $n = 4$. "*" *indicates the entry to be zeroed.*

Step 1. Apply a Householder matrix U_{01} to the left of A to create zeros in positions (2, 1) through (5, 1), then apply another Householder matrix V_{01} to the right to create zeros in (1, 3) and (1, 4) positions of A.

$$\begin{pmatrix} \times & \times & \times & \times \\ * & \times & \times & \times \\ * & \times & \times & \times \\ * & \times & \times & \times \\ * & \times & \times & \times \end{pmatrix} \xrightarrow{U_{01}} \begin{pmatrix} \times & \times & * & * \\ 0 & \times & \times & \times \\ 0 & \times & \times & \times \\ 0 & \times & \times & \times \\ 0 & \times & \times & \times \end{pmatrix} \xrightarrow{V_{01}} \begin{pmatrix} \times & \times & 0 & 0 \\ 0 & \times & \times & \times \\ 0 & \times & \times & \times \\ 0 & \times & \times & \times \\ 0 & \times & \times & \times \end{pmatrix} = A^{(1)}.$$

$\qquad\qquad A \qquad\qquad\qquad\qquad U_{01}A \qquad\qquad\qquad U_{01}AV_{01}$

Step 2. Apply a Householder matrix U_{02} to the left of $A^{(1)}$ and apply another Householder matrix V_{02} to the right to create zeros in the places indicated by "*".

$$\begin{pmatrix} \times & \times & 0 & 0 \\ 0 & \times & \times & \times \\ 0 & * & \times & \times \\ 0 & * & \times & \times \\ 0 & * & \times & \times \end{pmatrix} \xrightarrow{U_{02}} \begin{pmatrix} \times & \times & 0 & 0 \\ 0 & \times & \times & * \\ 0 & 0 & \times & \times \\ 0 & 0 & \times & \times \\ 0 & 0 & \times & \times \end{pmatrix} \xrightarrow{V_{02}} \begin{pmatrix} \times & \times & 0 & 0 \\ 0 & \times & \times & 0 \\ 0 & 0 & \times & \times \\ 0 & 0 & \times & \times \\ 0 & 0 & \times & \times \end{pmatrix} = A^{(2)}.$$

$\qquad\quad A^{(1)} \qquad\qquad\qquad\quad U_{02}A^{(1)} \qquad\qquad\quad U_{02}A^{(1)}V_{02}$

Step 3. Apply a Householder matrix U_{03} to the left of $A^{(2)}$ to create zeros in the position indicated by "*".

$$\begin{pmatrix} \times & \times & 0 & 0 \\ 0 & \times & \times & 0 \\ 0 & 0 & \times & \times \\ 0 & 0 & * & \times \\ 0 & 0 & * & \times \end{pmatrix} \xrightarrow{U_{03}} \begin{pmatrix} \times & \times & 0 & 0 \\ 0 & \times & \times & 0 \\ 0 & 0 & \times & \times \\ 0 & 0 & 0 & \times \\ 0 & 0 & 0 & \times \end{pmatrix} = A^{(3)}.$$

$\qquad\quad A^{(2)} \qquad\qquad\qquad\quad U_{03}A^{(2)}$

Step 4. Apply the Householder matrix U_{04} to the left of $A^{(3)}$ to create zeros in the positions indicated by "*".

$$\begin{pmatrix} \times & \times & 0 & 0 \\ 0 & \times & \times & 0 \\ 0 & 0 & \times & \times \\ 0 & 0 & 0 & \times \\ 0 & 0 & 0 & * \end{pmatrix} \xrightarrow{U_{04}} \begin{pmatrix} \times & \times & 0 & 0 \\ 0 & \times & \times & 0 \\ 0 & 0 & \times & \times \\ 0 & 0 & 0 & \times \\ 0 & 0 & 0 & 0 \end{pmatrix} = A^{(4)} = \begin{pmatrix} B \\ 0 \end{pmatrix} \text{ (bidiagonal).}$$

$\qquad\quad A^{(3)} \qquad\qquad\qquad\quad U_{04}A^{(3)}$

General step for an $n \times n$ matrix. In general, at the kth step, U_{0k} is constructed to create zeros in the kth column, while V_{0k} introduces zero in the kth row in appropriate places.

10.3. The Singular Value Decomposition and Its Computation

Computational Notes.

- In Step 2, one can work with the 4×3 submatrix of $A^{(1)}$ indicated by the box. Call it $\hat{A}_1$. Thus, orthogonal matrices $\hat{U}_{02}$ and $\hat{V}_{02}$ may be constructed such that

$$\hat{U}_{02} \hat{A}_1 \hat{V}_{02} = \begin{pmatrix} \times & \times & 0 \\ 0 & \times & \times \\ 0 & \times & \times \\ 0 & \times & \times \end{pmatrix}.$$

Then U_{02} and V_{02} are constructed from $\hat{U}_{02}$ and $\hat{V}_{02}$ in the usual way by embedding them into the identity matrices of appropriate orders.

- This holds similarly for the other steps.

- *Construction of a Householder matrix V for zeroing entries in a row vector.* At every step, V matrices are created to zeros in a row vector. We now show how to do this. Let $x^T = (x_1, x_2, \ldots, x_n)$ be a row vector. Suppose we want to construct a Householder matrix V such that $x^T V = (\times, 0, \ldots, 0)$. To do this, just construct V such that

$$V^T x = \begin{pmatrix} \times \\ 0 \\ \vdots \\ 0 \end{pmatrix}.$$

Then $x^T V = (\times, 0, \ldots, 0)$.

Flop-count. The above process will require approximately $4mn^2 - \frac{4n^3}{3}$ flops. If the matrices U_0 and V_0 are also explicitly needed, then their accumulations will require $4m^2n - \frac{4n^3}{3}$ and $\frac{4n^3}{3}$ flops, respectively.

Example 10.15.

$$A = \begin{pmatrix} 1 & 2 & 3 \\ 3 & 4 & 5 \\ 6 & 7 & 8 \end{pmatrix}.$$

Step 1.

$$U_{01} = \begin{pmatrix} -0.1474 & -0.4423 & -0.8847 \\ -0.4423 & 0.8295 & -0.3410 \\ -0.8847 & -0.3410 & 0.3180 \end{pmatrix}, \quad U_{01}A = \begin{pmatrix} -6.7823 & -8.2567 & -9.7312 \\ 0 & 0.0461 & 0.0923 \\ 0 & -0.9077 & -1.8154 \end{pmatrix},$$

$$V_{01} = \begin{pmatrix} 1 & 0 & 0 \\ 0 & -0.6470 & -0.7625 \\ 0 & -0.7625 & 0.6470 \end{pmatrix}, \quad U_{01}AV_{01} = \begin{pmatrix} -6.7823 & 12.7620 & 0 \\ 0 & -1.0002 & 0.0245 \\ 0 & 1.9716 & -0.4824 \end{pmatrix} = A^{(1)}.$$

Step 2.

$$U_{02} = \begin{pmatrix} 1 & 0 & 0 \\ 0 & -0.0508 & 0.9987 \\ 0 & 0.9987 & 0.0508 \end{pmatrix},$$

$$B = U_{02}A^{(1)} = U_{02}U_{01}AV_{01} = \begin{pmatrix} -6.7823 & 12.7620 & 0 \\ 0 & 1.9741 & -0.4830 \\ 0 & 0 & 0 \end{pmatrix} = A^{(2)}.$$

Note that from the above expression of B, it immediately follows that zero is a singular value of A. ∎

MATCOM Note: The above bidiagonalization process has been implemented in the MATCOM program **BIDIAG**.

Phase 2. Finding the SVD of the Bidiagonal Matrix. The process is a variant of the QR iteration. Starting from the $n \times n$ bidiagonal matrix B obtained in Phase 1, it successively constructs a sequence of bidiagonal matrices $\{B_k\}$ such that each B_i has possibly smaller off-diagonal entries than the previous one. *The ith iteration is equivalent to applying the implicit symmetric QR algorithm, described in Chapter 9, with the Wilkinson shift to the symmetric tridiagonal matrix $B_i^T B_i$ without, of course, forming the product $B_i^T B_i$ explicitly.* The effective tridiagonal matrices are assumed to be **unreduced** (note that the implicit symmetric QR works with unreduced matrices); otherwise we would work with decoupled SVD problems. For example, if $b_{k,k+1} = 0$, then B can be written as the direct sum of two bidiagonal matrices B_1 and B_2 and $\sigma(B) = \sigma(B_1) \cup \sigma(B_2)$.

The process has guaranteed convergence, and the rate of convergence is quite fast. The details of the process can be found in Golub and Van Loan (1996, pp. 452–456). We outline the process briefly in the following.

In the following *just one iteration step of the method is described.* To simplify the notation, let's write

$$B = \begin{pmatrix} \alpha_1 & \beta_2 & & \\ & \ddots & \ddots & \\ & & \ddots & \beta_n \\ & & & \alpha_n \end{pmatrix}. \tag{10.34}$$

Wilkinson shift $\sigma =$ the eigenvalue λ of the 2×2 right-hand corner submatrix of $B^T B$,

$$\begin{pmatrix} \alpha_{n-1}^2 + \beta_{n-1}^2 & \alpha_{n-1}\beta_n \\ \beta_n \alpha_{n-1} & \alpha_n^2 + \beta_n^2 \end{pmatrix}, \tag{10.35}$$

which is closer to $\alpha_n^2 + \beta_n^2$.

Step 1. Form the Givens rotation J_1, such that $J_1(\alpha_1^2 - \sigma, \alpha_1\beta_2, 0, \ldots, 0)^T = (\times, 0, \ldots, 0)^T$. This is done in two steps as follows.

1.1. Compute a Givens rotation J_1' such that $J_1' \begin{pmatrix} \alpha_1^2 - \sigma \\ \alpha_1 \beta_2 \end{pmatrix} = \begin{pmatrix} \times \\ 0 \end{pmatrix}$.

1.2. Form $J_1 = \begin{pmatrix} J_1' & 0 \\ 0 & I_{n-2} \end{pmatrix}$.

10.3. The Singular Value Decomposition and Its Computation

Step 2. Apply J_1 to the right of B and overwrite BJ_1 with B.

$$B \equiv BJ_1 = \begin{pmatrix} \times & \times & & & & \\ + & \ddots & \ddots & & & \\ & & \ddots & \ddots & & \\ & & & \ddots & \times & \\ & & & & \times & \end{pmatrix}, \qquad (10.36)$$

where $+$ indicates a fill-in. *(The fill-in is at the $(2, 1)$ position.)*
The idea is now to chase the nonzero entry "+" down the subdiagonal to the end of the matrix by applying the Givens rotations in an appropriate order, as indicated by the following.

Step 3. Form the Givens rotation J_2 such that the fill-in at the $(2, 1)$ position is eliminated.

$$B \equiv J_2 B = \begin{pmatrix} \times & \times & + & & & \\ & \times & \ddots & & & \\ & & \ddots & \ddots & & \\ & & & \ddots & \times & \\ & & & & \times & \end{pmatrix}.$$

(The fill-in is at the $(1, 3)$ position.)

Step 4. Form the Givens rotation J_3 to eliminate the fill-in of the $(1, 3)$ position.

$$B \equiv BJ_3 = \begin{pmatrix} \times & \times & & & & \\ & \times & \ddots & & & \\ & + & \ddots & \ddots & & \\ & & & \ddots & \times & \\ & & & & \times & \end{pmatrix}.$$

(The fill-in is at $(3, 2)$ position.)

Step 5. Form J_4 to eliminate the fill-in of the $(3, 2)$ position.

$$B \equiv J_4 B = \begin{pmatrix} \times & \times & & & & \\ & \times & \times & + & & \\ & & \ddots & \ddots & & \\ & & & \ddots & \times & \\ & & & & \times & \end{pmatrix}.$$

(The fill-in is at $(2, 4)$ position.)

The process is continued. *The general process is now clear.* The entries $(2, 1)$, $(3, 2)$, $(4, 3)$, etc., are annihilated by a premultiplication, whereas the entries $(1, 3)$, $(2, 4)$, $(3, 5)$, etc., are annihilated by a postmultiplication.

At the end of one iteration we will have a new bidiagonal matrix $\overline{B}$ orthogonally equivalent to the original bidiagonal matrix B:

$$\overline{B} = (J_{2n-2} \cdots J_4 J_2) B (J_1 J_3 \cdots J_{2n-3}).$$

Example 10.16.

$$B = \begin{pmatrix} 1 & 2 & 0 \\ 0 & 2 & 3 \\ 0 & 0 & 1 \end{pmatrix}.$$

Step 1. The Wilkinson shift $\sigma = 15.0828$:

$$J_1 = \begin{pmatrix} -0.9901 & 0.1406 & 0 \\ -0.1406 & -0.9901 & 0 \\ 0 & 0 & 1 \end{pmatrix}.$$

Step 2.

$$B \equiv B J_1 = \begin{pmatrix} -1.2713 & -1.8395 & 0 \\ -0.2812 & -1.9801 & 3 \\ 0 & 0 & 1 \end{pmatrix}.$$

(The fill-in is at the (2, 1) position.)

Step 3. Form

$$J_2 = \begin{pmatrix} -0.9764 & -0.2160 & 0 \\ 0.2160 & -0.9764 & 0 \\ 0 & 0 & 1 \end{pmatrix},$$

$$B \equiv J_2 B J_1 = \begin{pmatrix} 1.3029 & 2.2238 & -0.6480 \\ 0 & 1.5361 & -2.9292 \\ 0 & 0 & 1 \end{pmatrix}.$$

(The fill-in is at the (1, 3) position.)

Step 4. Form

$$J_3 = \begin{pmatrix} 1 & 0 & 0 \\ 0 & 0.9601 & 0.2797 \\ 0 & -0.2797 & 0.9601 \end{pmatrix},$$

$$B \equiv J_2 B J_1 J_3 = \begin{pmatrix} 1.3029 & 2.3163 & 0 \\ 0 & 2.2942 & -2.3827 \\ 0 & -0.2797 & 0.9601 \end{pmatrix}.$$

(The fill-in is at the (3, 2) position.)

Step 5. Form

$$J_4 = \begin{pmatrix} 1 & 0 & 0 \\ 0 & 0.9926 & -0.1210 \\ 0 & 0.1210 & 0.9926 \end{pmatrix},$$

$$B \equiv J_4 J_2 B J_1 J_3 = \begin{pmatrix} 1.3020 & 2.3163 & 0 \\ 0 & 2.3112 & -2.4812 \\ 0 & 0 & 0.6646 \end{pmatrix}. \blacksquare$$

10.3. The Singular Value Decomposition and Its Computation

Stopping criterion. The algorithm typically requires a few iterations before the off-diagonal entry β_n becomes negligible. A criterion for off-diagonal negligibility follows.

Criterion for neglecting an off-diagonal entry (Golub and Van Loan (1996, p. 455)). Accept an off-diagonal β_i to be zero if

$$|\beta_i| \leq \epsilon(|\alpha_i| + |\alpha_{i-1}|).$$

Accept a diagonal entry α_i to be zero if

$$|\alpha_i| \leq \epsilon \|B\|,$$

where ϵ is a small multiple of the machine precision μ.

Flop-count. The cost of the two-phase SVD method is determined by the cost of Phase 1. Phase 2 is iterative and is quite cheap. The *estimated* flop-count is $4m^2n + 8mn^2 + 9n^3 (m \geq n)$. *This count includes the cost of U, Σ, and V.* There are applications (e.g., least squares) where all three matrices are not explicitly required. A nice table of different flop-counts of the Golub–Kahan–Reinsch SVD and the triangular SVD (to be described in the next section) for different requirements of U, Σ, and V appears in Golub and Van Loan (1996, p. 254).

Computing of Σ alone will cost about $4mn^2 - \frac{4n^3}{3}$ flops by the Golub–Kahan–Reinsch algorithm.

Round-off property. It can be shown that the computed SVD, $\hat{U}\hat{\Sigma}(\hat{V})^T$, produced by the Golub–Kahan–Reinsch algorithm, is nearly the exact SVD of $A + E$, that is,

$$A + E \approx (\hat{U} + \delta\hat{U})\hat{\Sigma}(\hat{V} + \delta\hat{V}),$$

where $\hat{U} + \delta\hat{U}$ and $\hat{V} + \delta\hat{V}$ are orthogonal. Specifically,

$$(\|E\|_2 \big| \|A\|_2) \leq p(m,n)\mu, \quad \|\delta\hat{U}\| \leq p(m,n)\mu,$$
$$\|\delta\hat{V}\| \leq p(m,n)\mu,$$

and $p(m, n)$ is a slowly growing function of m and n.

Entrywise errors of the singular values. Furthermore, let $\bar{\sigma}_i$ be a computed singular value. Then

$$|\sigma_i - \bar{\sigma}_i| \leq \mu p(n)\|A\|_2 = \mu p(n)\sigma_{\max},$$

where $p(n)$ is a slowly growing function of n.

The result says that the *computed singular values cannot differ from the true singular values by an amount larger than* $\delta = \mu p(n)\sigma_{\max}$.

Thus, the singular values which are not much smaller than $\sigma_{\max}$ will be computed by the algorithm quite accurately.

10.3.7 The Chan SVD Algorithm

The Golub–Kahan–Reinsch procedure can be made faster sometimes if matrix A is triangularized first by QR factorization and then the procedure is applied to the upper triangular matrix R. The idea was mentioned in Lawson and Hanson in their celebrated 1974 book *Solving Least-Squares Problems* and later fully analyzed by Chan (1982a). The triangular SVD, to be called the Chan SVD, can be described as follows.

Step 1. Find the QR factorization of A:

$$Q^T A = \begin{pmatrix} R \\ 0 \end{pmatrix}. \tag{10.37}$$

Step 2. Find the SVD of R using the Golub–Kahan–Reinsch algorithm:

$$R = X \Sigma Y^T. \tag{10.38}$$

Step 3. Compute the singular values and singular vectors of A.
The singular values of A are just the singular values of R. The singular vector matrices U and V are given by

$$U = Q\,\mathrm{diag}\,(X, I_{m-n}), \quad V = Y. \tag{10.39}$$

Flop-count. The triangular SVD (the Chan SVD) requires about $4m^2n + 22n^3$ flops to compute Σ, U, and V, compared to the $4m^2n + 8mn^2 + 9n^3$ flops required by the Golub–Kahan–Reinsch SVD algorithm. *Clearly, there will be savings with the triangular-SVD when* $m \geq \frac{5n}{3}$. Note that in this case one needs to bidiagonalize an upper triangular matrix rather than a full matrix.

Example 10.17.

$$A = \begin{pmatrix} 1 & 2 \\ 2 & 3 \\ 4 & 5 \end{pmatrix}.$$

Step 1. *The QR factorization of A:*

$$Q = \begin{pmatrix} -0.2182 & -0.8165 & -0.5345 \\ -0.4364 & -0.4082 & 0.8018 \\ -0.8729 & 0.4082 & -0.2673 \end{pmatrix}, \quad R = \begin{pmatrix} -4.5826 & -6.1101 \\ 0 & -0.8165 \end{pmatrix}.$$

Step 2. *The SVD of R:*

$$R = X\Sigma Y^T, \quad X = \begin{pmatrix} -0.9963 & 0.0856 \\ -0.0856 & -0.9963 \end{pmatrix},$$

$$Y = \begin{pmatrix} 0.5956 & -0.8033 \\ 0.8033 & 0.5956 \end{pmatrix}, \quad \Sigma = \begin{pmatrix} 7.6656 & 0 \\ 0 & 0.4881 \end{pmatrix}.$$

The singular value decomposition of $A = U\Sigma V^T$. The singular values of A are 7.6656, 0.4881.
$$U = \begin{pmatrix} 0.2873 & 0.7948 & -0.5345 \\ 0.4698 & 0.3694 & 0.8018 \\ 0.8347 & 0.4814 & -0.2673 \end{pmatrix}, \quad V = Y. \quad \blacksquare$$

Flop-count for the least-squares problem using the SVD and other methods. In view of two SVD algorithms just described, let's have another close look at the flop-count of *different approaches for least-squares solutions* of $Ax = b$, where A is $m \times n$ ($m \geq n$).

- Using the Golub–Khan–Reinsch SVD: $4mn^2 + 8n^3$.

- Using the Chan SVD: $2mn^2 + 11n^3$.

- Using normal equations: $mn^2 + \dfrac{n^3}{3}$.

- Using Householder QR: $2mn^2 - \dfrac{2n^3}{3}$.

- Using modified Gram–Schmidt (MGS): $2mn^2$.

(See Golub and Van Loan (1996, p. 263) for a comprehensive list.)

Recent developments. The other notable works on SVD computation include:

- **Zero-shift QR iteration** (Demmel and Kahan (1990)). Demmel and Kahan show that, using zero-shift, the tiny singular values and the singular vectors can be found (almost) as accurately as the data permits.

- **Differential QD algorithm for large matrices** (Fernando and Parlett (1994)). This is a variation of the QR iteration algorithm for finding *all the singular values with high relative accuracy. It is the fastest algorithm* now for computing all the singular values of a bidiagonal matrix.

- **Divide-and-conquer algorithm** (Jessup and Sorensen (1994), Gu and Eisenstat (1995b), etc.). *A fast algorithm, but does not guarantee that the tiny singular values will be computed with high relative accuracy.*

- **Jacobi method.** *For some classes of matrices the Jacobi method computes the singular values and singular vectors to high relative accuracy* by implicitly forming the matrix BB^T or B^TB. This method is not discussed here; see Demmel (1997). See also the recent papers of Drmač and Veselić (2008a, 2008b).

10.4 Generalized SVD

The SVD theorem (Theorem 10.8) can be generalized for a pair of matrices A and B, and this generalized SVD is useful in certain applications such as *constrained least squares problems* (Golub and Van Loan (1996, pp. 586–587)).

The generalization was first obtained by Van Loan (1976). We state only the theorem here without proof. For a proof, see Golub and Van Loan (1996, p. 466).

Theorem 10.18 (generalized SVD theorem). *Let A and B be, respectively, real matrices of order $m \times n$ and $p \times n$ ($m \geq n$). Then there exist orthogonal matrices $U \in \mathbb{R}^{m \times m}$ and $V \in \mathbb{R}^{p \times p}$ and an $n \times n$ nonsingular matrix W such that*

$$U^T A W = C = \mathrm{diag}(c_1, \ldots, c_n), \quad c_i \geq 0,$$

and

$$V^T B W = D = \mathrm{diag}(d_1, \ldots, d_q), \quad d_i \geq 0,$$

where $q = \min(p, n)$ and $d_1 \geq \cdots \geq d_r > d_{r+1} = \cdots = d_q = 0$, $r = \mathrm{rank}(B)$.

The elements $(\frac{c_1}{d_1}, \frac{c_2}{d_2}, \ldots, \frac{c_r}{d_r})$ are called the generalized singular values of A and B.

10.5 Review and Summary

The two closely related topics, the *symmetric eigenvalue problem* and the *SVD*, are discussed in this chapter. Emphasis here is on computations of the eigenvalues and singular values.

10.5.1 The Symmetric Eigenvalue Computation

Here we have described

- the bisection method (Algorithm 10.1);

- the QR iteration method with Wilkinson shift (Algorithm 10.2);

- the divide-and-conquer method (Algorithm 10.3);

- The Jacobi method (Section 10.2.5).

10.5.2 The SVD

- **Existence and uniqueness of the SVD.** The SVD of a matrix A always exists (Theorem 10.8).
  $$A = U \Sigma V^T.$$
  The singular values (the diagonal entries of Σ) are unique, but the singular vectors are not.

- **Relationship of the singular values and singular vectors with the eigenvalues.** The singular values of A are the nonnegative square roots of the eigenvalues of $A^T A$. See Theorem 10.9 for another interesting relationship.

- **Sensitivity of the singular values.** The singular values are insensitive to small perturbations (Theorem 10.10).

- **Computing the SVD.** The most widely used approach for computing the SVD of A is the Golub–Kahan–Reinsch algorithm (Section 10.3.6). This algorithm works in two phases. In Phase 1, the matrix A is reduced to a bidiagonal matrix by orthogonal equivalence, and in Phase 2, the bidiagonal matrix is further reduced to a diagonal matrix by orthogonal similarity using implicit QR iteration with Wilkinson shift. Unfortunately, very tiny singular values may not be computed with very high relative accuracy by this method. A modification of this method, known as the *zero-shift QR iteration* or the *QR iteration with a zero shift* has been proposed by *Demmel and Kahan* in 1990. The Demmel–Kahan method computes all the singular values and singular vectors with high relative accuracy for small-order (about 25) matrices. The dqds algorithm of Fernando and Parlett (1994), however, computes all the singular values most accurately.

10.6 Suggestions for Further Reading

A book specialized to the symmetric eigenvalue problem is Parlett (1998); a fair amount of discussion on this problem also appears in most contemporary numerical linear algebra books, including Demmel (1997), Golub and Van Loan (1996), Trefethen and Bau (1997), and Watkins (2002). For more on the interlacing property, see Hill and Parlett (1992). A book devoted to perturbation analysis, including that of the symmetric eigenvalue problem, is by Stewart and Sun (1990). Many important results, including some classical ones, can be found in this book. See also Horn and Johnson (1985) and Stewart (1991) for perturbation analysis of the SVD. For some other important papers on SVD and symmetric eigenvalue computations, see Demmel and Kahan (1990), Gu and Eisenstat (1995b), Fernando and Parlett (1994), Parlett (1995), Demmel et al. (1999), Demmel and Veselić (1992), Drmač and Veselić (2008a, 2008b), Bai (1988), Charlier et al. (1988), and Bai and Demmel (1993a).

For more on the generalized SVD and its variations, see Van Loan (1976), Paige and Saunders (1981), Kagström (1985), De Moor and Van Dooren (1992), De Moor and Zha (1991), Stewart (1983), Paige (1986), De Moor (1991, 1992), De Moor and Golub (1991), and Paige and Van Dooren (1986). For perturbation analysis of the generalized SVD, see Sun (1983) and Stewart (1991).

Exercises on Chapter 10

EXERCISES ON SECTION 10.2

10.1 (a) Develop the symmetric tridiagonal QR iteration algorithm in detail using the implicit symmetric QR step with Wilkinson shift.

(b) Apply your algorithm to compute all the eigenvalues of a symmetric matrix of order 20 generated randomly: $A = \text{rand}(20)$, $A \equiv A + A^T$.

10.2 (a) Let $A = \begin{pmatrix} 1 & -1 & 0 & 0 \\ -1 & 2 & -1 & 0 \\ 0 & -1 & 2 & -1 \\ 0 & 0 & -1 & 2 \end{pmatrix}$.

Without computing the eigenvalues show that $|\lambda| < 4$ for each eigenvalue λ of A. Show that there are exactly two eigenvalues greater than 2 and two less than 2.

Apply the bisection algorithm (Algorithm 10.1) to compute the eigenvalue close to 2.

(b) Apply the inverse Hessenberg iteration algorithm to A to compute the eigenvector associated with the eigenvalue close to 2.

(c) Compute the eigenvalues of A by applying the symmetric QR iteration with Wilkinson shift (Algorithm 10.2).

10.3 (a) Prove that the eigenvalues of an unreduced real symmetric tridiagonal matrix are real and distinct.

(b) Prove that if λ is an eigenvalue of multiplicity k of an unreduced symmetric tridiagonal matrix T, then at least $(k-1)$ subdiagonal entries of T must be zero.

10.4 (a) Develop a QR-type algorithm to compute the eigenvalues of a symmetric positive definite matrix A, based upon the Cholesky decomposition (Section 6.12.3).

(b) Test your algorithm with matrix A of Exercise 10.2.

10.5 Let
$$A = \begin{pmatrix} \alpha & \beta \\ \beta & \gamma \end{pmatrix}$$

Prove that the eigenvalue δ of A closest to γ is given by
$$w = \gamma - \frac{\text{sign}(r)\beta^2}{|r| + \sqrt{r^2 + \beta^2}},$$
where $r = (\alpha - \gamma)/2$. Explain why this formula is better than the one given by (10.8)–(10.9).

10.6 Let $A = A_1 + iA_2$ be a Hermitian matrix. Then prove that
$$B = \begin{pmatrix} A_1 & -A_2 \\ A_2 & A_1 \end{pmatrix}$$
is symmetric. How are the eigenvalues and eigenvectors of A related to those of B?

10.7 Prove that if the ith component u_i of u in (10.13) is zero, then

(a) the ith column of $\hat{D}$ in (10.13) is $d_i e_i$;

(b) the ith row of $\hat{D}$ is $d_i e_i^T$;

(c) d_i is an eigenvalue of $\hat{D}$ with the associated eigenvector e_i.

10.8 Prove that if k eigenvalues of D are equal, then the eigenvalue problem of $\hat{D}$ in (10.13) can be deflated by deleting $(k-1)$ rows and columns.

10.9 Assume that

(a) $\rho \neq 0$;

(b) the eigenvalues of D are arranged in the order $d_1 < d_2 < \cdots < d_n$; and

(c) none of the components of u is zero.

Then prove that each interval (d_i, d_{i+1}) contains exactly one eigenvalue of $\hat{D} = D + \rho u u^T$.

10.10 Prove that if the assumptions of Exercise 10.9 hold, then $D - \lambda I$ is nonsingular and $u^T q \neq 0$.

10.11 By drawing the graphs of the secular equation or otherwise, prove that

(a) $d_1 < \lambda_1 < d_2 < \lambda_2 < \cdots < \lambda_n < d_n + \rho$ if $\rho > 0$;

(b) $d_1 + \rho < \lambda_1 < d_2 < \lambda_2 < \cdots < d_{n-1} < \lambda_n < d_n$ if $\rho < 0$.

10.12 Show that in the classical Jacobi scheme, the sum of the squares of the off-diagonal entries decreases by at least the factor of $1 - \frac{2}{n(n-1)}$ at each step.

EXERCISES ON SECTION 10.3

10.13 (a) Derive Theorem 10.10 without using Theorem 10.9.

(b) Given
$$A = \begin{pmatrix} 1 & 2 \\ 3 & 4 \\ 5 & 6 \end{pmatrix},$$
find the singular values σ_1 and σ_2 of A by computing the eigenvalues of $A^T A$. Then find the orthogonal matrix P such that
$$P^T S P = \text{diag}(\sigma_1, \sigma_2, -\sigma_1, -\sigma_2, 0),$$
where $S = \begin{pmatrix} 0_{3 \times 3} & A \\ A^T & 0_{2 \times 2} \end{pmatrix}$.

10.14 Using the constructive proof of Theorem 10.8, find the SVD of the following matrices:

(i) $A = \begin{pmatrix} 1 & 2 \\ 3 & 4 \\ 5 & 6 \end{pmatrix}$,

(ii) $A = \begin{pmatrix} 1 & 2 & 3 \end{pmatrix}$,

(iii) $A = \begin{pmatrix} 1 \\ 1 \\ 1 \end{pmatrix}$,

(iv) $A = \text{diag}(1, 0, 2, 0, -5)$, $A = \begin{pmatrix} 1 & 1 \\ \epsilon & 0 \\ 0 & \epsilon \end{pmatrix}$, where $\epsilon = 10^{-5}$.

10.15 Prove that the singular values of a symmetric positive definite matrix are the same as its eigenvalues. How are the singular vectors and eigenvectors related?

10.16 Let

$$D = \begin{pmatrix} \sigma_1 & & & & & 0 \\ & \ddots & & & & \\ & & \sigma_r & & & \\ & & & 0 & & \\ & & & & \ddots & \\ 0 & & & & & 0 \end{pmatrix}, \quad \sigma_i > 0, \; i = 1, \ldots, r.$$

Then show that

$$D^\dagger = \begin{pmatrix} \frac{1}{\sigma_1} & & & & & 0 \\ & \ddots & & & & \\ & & \frac{1}{\sigma_r} & & & \\ & & & 0 & & \\ & & & & \ddots & \\ 0 & & & & & 0 \end{pmatrix}.$$

10.17 Verify that the matrix $A^\dagger = V \Sigma^\dagger U^T$, where $\Sigma^\dagger = \text{diag}(\frac{1}{\sigma_j})$ (with the convention that if $\sigma_j = 0$, we use $\frac{1}{\sigma_j} = 0$), is the pseudoinverse of A. (Check all four conditions for the definition of the pseudoinverse.)

10.18 For any nonzero matrix A, show that

(a) $AA^\dagger v = v$ for any vector v in $R(A)$;

(b) $A^\dagger x = 0$ for any x in $N(A^T)$;

(c) $(A^T)^\dagger = (A^\dagger)^T$;

(d) $(A^\dagger)^\dagger = A$.

10.19 Let A be an $m \times n$ matrix. Show the following:

(a) If A has full column rank, then
$$A^\dagger = (A^T A)^{-1} A^T.$$

(b) If A has full row rank, then
$$A^\dagger = A^T (A A^T)^{-1}.$$

10.20 From the SVD of A, compute the SVDs of the projection matrices $P_1 = A^\dagger A$, $P_2 = I - A^\dagger A$, $P_3 = AA^\dagger$, and $P_4 = I - AA^\dagger$. Also verify that each of these is a projection matrix.

10.21 (a) Let B be an upper bidiagonal matrix having a multiple singular value. Then prove that B must have a zero either on its diagonal or superdiagonal.

(b) Prove that if the entries of both diagonals of a bidiagonal matrix are all nonzero, then its singular values are distinct.

Exercises on Chapter 10

10.22 Consider the family of bidiagonal matrices

$$B(\eta) = \begin{pmatrix} 1-\eta & \beta(1+\eta) & & & \\ & \ddots & \ddots & & \\ & & \ddots & \ddots & \\ & & & \ddots & \beta(1+\eta) \\ & & & & 1-\eta \end{pmatrix},$$

$\beta \gg 1$. It can be shown (Demmel and Kahan (1990)) that the smallest singular value of $B(\eta)$ is approximately $\beta^{1-n}(1-(2n-1)\eta)$.

Taking $\beta = 10^6$, and using $\eta = 0$, verify the above result.

10.23 Develop a procedure to upper bidiagonalize an $n \times n$ tridiagonal matrix using Givens rotations.

10.24 Based on discussions in Section 10.3.6, develop an algorithmic procedure to implement Phases 1 and 2 of the SVD computation.

10.25 Develop the Jacobi algorithm for computing the SVD.

10.26 Prove that flop-count for an overdetermine least-squares problem using the Chan SVD scheme is about $2mn^2 + 2n^3$.

MATLAB Programs and Problems on Chapter 10

M10.1 (a) Write a MATLAB program called **polysymtri** to compute the characteristic polynomial $p_n(\lambda)$ of an unreduced symmetric tridiagonal matrix T, based on the recursion in Section 10.2.2:

[valpoly] = **polysymtri**(T, lambda).

(b) Using **polysymtri**, write a MATLAB program called **signagree** that finds the number of eigenvalues of T greater than a given real number μ, based on Theorem 10.5:

[number] = **signagree**(T, meu).

(c) Using **polysmtri** and **signagree**, implement the *bisection algorithm* (Algorithm 10.1):

[lambda] = **bisection**(T, m, n).

Compute λ_{n-m+1} for $m = 1, 2, 3, \ldots$, using *bisection*, and then compare your results with those obtained by using **eig(T)**.

Test data:

$A =$ the symmetric tridiagonal matrix arising in the *buckling problem* in Section 8.3.2, with $n = 200$.

M10.2 (*The purpose of this exercise is to study the sensitivities (insensitivities) of the singular values of a matrix.*)

Using the MATLAB commands **svd** and **norm**, verify the inequalities in Theorem 10.10.

Test data:

(i)
$$A = \begin{pmatrix} 1 & 1 & 1 & 1 \\ 0 & 0.99 & 1 & 1 \\ 0 & 0 & 0.99 & 1 \\ 0 & 0 & 0 & 0.99 \end{pmatrix}.$$

(ii) A = the Wilkinson bidiagonal matrix of order 20.

In each case, construct a suitable E so that $(A + E)$ differs from A in the $(n, 1)$th element only by an $\epsilon = 10^{-5}$.
(*Note that the eigenvalues of both matrices are ill-conditioned.*)

M10.3 Let A = **rand** (10, 3), and X = **pinv**(A). Verify that X satisfies all four conditions of the pseudoinverse using MATLAB: $AXA = X$, $XAX = X$, $(AX)^T = AX$, $(XA)^T = XA$.

M10.4 Write a MATLAB program called **chansvd** to implement the Chan SVD algorithm described in Section 10.3.7, using the MATLAB commands **qr** and **svd**:

$$[U, S, V] = \textbf{chansvd}(A).$$

Run your program with a randomly generated 500×40 matrix A = **rand** (50, 40) and compare the flop-count and elapsed time with those obtained by using **svd**(A).

M10.5 Write a MATLAB program called **bidiag** to bidiagonalize a matrix A using the sketch of the procedure give in (Section 10.3.6).

(a) [B] = **bidiag**(A, tol),
where B is a bidiagonal matrix and **tol** is the tolerance.
Test your program using A = **rand**$(15, 10)$.

(b) Use **bidiag** to write a MATLAB program to compute the singular values of the bidiagonal matrix B.
Test your program by using a randomly generated matrix of order 100×10.

M10.6 Write MATLAB programs to compute the singular values of a matrix A in the following two ways: (a) by calling the standard SVD routine from MATLAB, and (b) by explicitly forming $A^T A$ and then computing its eigenvalues.

Test matrix: an upper triangular matrix of order 100 with 0.0001 on the main diagonal and 1 everywhere above the main diagonal.

Plot your results.

Chapter 11

Generalized and Quadratic Eigenvalue Problems

Background Material Needed

- The Householder and Givens methods to create zeros in a vector and the corresponding QR factorization algorithms (Algorithms 7.1, 7.2, 7.4, and 7.5)
- The Cholesky factorization algorithm (Algorithm 6.8)
- The methods for symmetric eigenvalue problem (Algorithms 10.1, 10.2, and 10.3)
- The inverse iteration algorithm (Algorithm 9.2)
- The Rayleigh quotient algorithm (Algorithm 9.3)

11.1 Introduction

In this chapter we consider the generalized eigenvalue problem for a matrix pair (A, B) defined as follows.

Statement of the Generalized Eigenvalue Problem

Given $n \times n$ matrices A and B, find n scalars λ and nonzero vectors x such that $Ax = \lambda B x$.

Note that the standard eigenvalue problem for matrix A considered in Chapter 9 is a special case of this problem (take $B = I$).

Definition 11.1. *The matrix $A - \lambda B$ is called a **matrix pencil**. This pencil is conveniently denoted by (A, B). It is very often referred to as the **pair** (A, B).*

The pencil $A - \lambda B$ is **singular** if for all λ, $\det(A - \lambda B) = 0$. Otherwise, the pencil is **regular**. We will assume throughout the whole chapter that the pencil is regular.

Example 11.2. The pencil (A, B) defined by $A = \begin{pmatrix} 1 & 0 \\ 0 & 0 \end{pmatrix}$, $B = \begin{pmatrix} 0 & 1 \\ 0 & 0 \end{pmatrix}$ is a singular pencil, since $\det(A - \lambda B) = 0$ for all λ. ∎

Definition 11.3. *The scalars $\lambda \in \mathbb{C}$ such that $\det(A - \lambda B) = 0$ are called the **eigenvalues** of the pencil (A, B). A nonzero vector x is a **right eigenvector** of the pencil (A, B) if*

$$Ax = \lambda Bx.$$

*The vector $y \neq 0$ is a **left eigenvector** if*

$$y^*A = \lambda y^*B.$$

*The polynomial $\det(A - \lambda B)$ is called the **characteristic polynomial of the pencil (A, B)**. The eigenvalues of the pencil (A, B) are the zeros of the characteristic polynomial.*

The finite and infinite eigenvalues of a regular pencil. If B is nonsingular, then a regular pencil $A - \lambda B$ of order n has n eigenvalues. An eigenvalue λ of the pair (A, B) is also an eigenvalue of $B^{-1}A$. If B is singular, then the characteristic polynomial will have degree less than n. In this case, there will be less than n finite eigenvalues and the missing eigenvalues will be set to ∞. Thus, *if the degree of $\det(A - \lambda B)$ is $r(< n)$, then there will be r finite and $n - r$ infinite eigenvalues.*

Example 11.4.

$$A = \begin{pmatrix} 1 & 0 \\ 0 & 0 \end{pmatrix}, \quad B = \begin{pmatrix} 0 & 0 \\ 0 & 1 \end{pmatrix}.$$

The degree of the characteristic polynomial is 1. The eigenvalues of this regular pencil $A - \lambda B$ are 0 and ∞. ∎

A note on the use of the word "pencil." "The rather strange use of the word 'pencil' comes from optics and geometry. An aggregate of (light) rays converging to a point does suggest the sharp end of a pencil and, by a natural extension, the term came to be used for any **one parameter** family of curves, spaces, matrices, or other mathematical objects." (Parlett (1998, p. 339)).

If A and B are real symmetric matrices and, furthermore, if B is positive definite, then the generalized eigenvalue problem $Ax = \lambda Bx$ is called the **symmetric definite generalized eigenvalue problem.**

This chapter is devoted to the study of the generalized eigenvalue problem with particular attention to the symmetric definite problem. The chapter is organized in the following manner.

In Section 11.3 we present a result that shows how the generalized eigenvalues and eigenvectors can be extracted once the matrices A and B are reduced to generalized Schur or generalized real Schur forms.

In Section 11.4 we describe the QZ algorithm (Algorithm 11.2) for the generalized eigenvalue problem. *It is a natural generalization of the QR iteration algorithm described in Chapter 9.*

In Section 11.5 we show how to compute a generalized eigenvector when an approximation of a generalized eigenvalue is known using the inverse iteration (Algorithm 11.3).

Sections 11.6–11.8 are devoted to the study of the symmetric definite generalized eigenvalue problems. Several case studies on the problems arising in vibration of structures are presented. A popular algorithm widely used in engineering practice—namely, the

11.3. Generalized Schur and Real Schur Decompositions

simultaneous diagonalization algorithm (Algorithm 11.5), is described and several engineering applications of this technique are discussed. The *generalized Rayleigh quotient iteration* for a symmetric definite pencil is presented in Algorithm 11.6.

Finally, in Section 11.9, we include a brief discussion of the *quadratic eigenvalue problem*.

11.2 Eigenvalue-Eigenvector Properties of Equivalent Pencils

Definition 11.5. *If X and Y are nonsingular matrices, then the pencil (A, B) and (Y^*AX, Y^*BX) is called **equivalent** to (A, B).*

The following are easily proven properties of two equivalent pencils (Exercise 11.3):

- The eigenvalues of two equivalent pencils $A - \lambda B$ and $Y^*AX - \lambda Y^*BX$ are the same.
- If x is a right eigenvector of $A - \lambda B$, then $X^{-1}x$ is a right eigenvector of $Y^*AX - \lambda Y^*BX$.
- If y is a left eigenvector of $A - \lambda B$, then $Y^{-1}y$ is a left eigenvector of $Y^*AX - \lambda Y^*BX$.

Thus, *in order to compute the eigenvalues of $A - \lambda B$, we will seek orthogonal matrices to transform the pair (A, B) into an equivalent pair from which the eigenvalues can be more easily computed.* Also, once the eigenvectors of the transformed pencil are computed, the *eigenvectors of the original pencil can be recovered from those of the transformed pencil by appropriate matrix multiplications as shown above*.

11.3 Generalized Schur and Real Schur Decompositions

Fortunately, analogous to the Schur decomposition of matrix A, there exists the generalized Schur decomposition of the pair (A, B) of the matrix pencil $A - \lambda B$.

- **Schur decomposition of A.** There exists a unitary matrix U such that $U^*AU = T$, an upper triangular matrix.

- **Generalized Schur decomposition of (A, B).** There exist unitary matrices U_1 and U_2 such that $U_1^*AU_2$ and $U_2^*BU_2$ are upper triangular:

$$U_1^*AU_2 = T_1 = \begin{pmatrix} t_{11} & & & \\ 0 & \ddots & & * \\ \vdots & & \ddots & \\ 0 & & & t_{nn} \end{pmatrix}, \qquad (11.1)$$

$$U_1^*BU_2 = T_2 = \begin{pmatrix} t'_{11} & & & \\ 0 & \ddots & & * \\ \vdots & & \ddots & \\ 0 & & & t'_{nn} \end{pmatrix}. \qquad (11.2)$$

The finite eigenvalues λ_i, $i = 1, \ldots, n$, of the regular pencil $(A - \lambda B)$ are then given by $\lambda_i = t_{ii}/t'_{ii}$, $t'_{ii} \neq 0$. By convention, the eigenvalues corresponding to the zero diagonal entries of T_2 are ∞.

- **Generalized real Schur decomposition.** Analogous to the real Schur decomposition of a single matrix A, there also exists the generalized real Schur decomposition of (A, B). In the case where both A and B are real, the matrices U_1 and U_2 can be chosen to be orthogonal. That is, when A and B are both real, there exist orthogonal matrices Q and Z such that

$$Q^T A Z = R, \quad \text{an upper real Schur matrix,}$$
$$Q^T B Z = T, \quad \text{an upper triangular matrix.} \tag{11.3}$$

The pair (R, T) is said to be the **generalized real Schur form** of (A, B).

The eigenvalues of (A, B) can then be extracted from R and T as follows:

(i) The 1×1 diagonal blocks of (R, T) contain the real eigenvalues of (A, B)

(ii) The 2×2 diagonal blocks of (R, T) contain the pairs of complex conjugate eigenvalues.

For example, if

$$R = \begin{pmatrix} 0 & 1 & 1 & 2 \\ -1 & 0 & 1 & 1 \\ 0 & 0 & 3 & 2 \\ 0 & 0 & 0 & 3 \end{pmatrix} \quad \text{and} \quad T = \begin{pmatrix} 1 & 0 & 0 & 0 \\ 0 & 1 & 3 & 3 \\ 0 & 0 & 2 & 2 \\ 0 & 0 & 0 & -1 \end{pmatrix},$$

then the real eigenvalues are $\frac{3}{2}$ and $-\frac{3}{1}$, and the two pairs of complex conjugate eigenvalues are the eigenvalues of the pair $\left(\begin{pmatrix} 0 & 1 \\ -1 & 0 \end{pmatrix}, \begin{pmatrix} 1 & 0 \\ 0 & 1 \end{pmatrix} \right)$, which are i and $-i$.

11.4 The QZ Algorithm

A standard algorithm for finding the generalized real Schur form of the pair (A, B) is the *QZ iteration algorithm*, developed by Moler and Stewart (1973). *It is a natural analogue of the QR iteration for computing the eigenvalues of A.*

Like the QR iteration algorithm, the QZ algorithm also comes in two stages:

QR Iteration for the matrix A:

Stage I. $A \xrightarrow{P} P^T A P = H$, upper Hessenberg

Stage II. $H \xrightarrow{Q} Q^T H Q = T$, real Schur

QZ Iteration for the pair (A,B):

Stage I. $A \xrightarrow{Q', Z'} Q'^T A Z' = A'$, upper Hessenberg

$B \xrightarrow{Q', Z'} (Q')^T B Z' = B'$, upper triangular

Stage II. $A' \xrightarrow{Q, Z} Q^T A' Z = R$, upper real Schur

$B' \xrightarrow{Q, Z} Q^T B' Z = T$, upper triangular

Stage I is *direct* and Stage II is *iterative*. The Stage II is achieved by applying *implicit QR iteration algorithm* to the matrix $B^{-1}A$ without explicitly forming the matrix.

11.4. The QZ Algorithm

A Note of Caution. If B is ill-conditioned, and $C = B^{-1}A$ is explicitly computed, then C will not be computed accurately and thus the computed eigenvalues of the pencil $A - \lambda B$ will be inaccurate. *This is true even if the eigenvalues themselves are well-conditioned.*

11.4.1 Stage I: Reduction to Hessenberg Triangular Form

Let A and B be two $n \times n$ matrices. Then:

Step 1. Triangularize the matrix B by QR factorization. That is, find an orthogonal matrix U such that
$$B \equiv U^T B \text{ is an upper triangular matrix.}$$
Form
$$A \equiv U^T A \text{ (in general, } A \text{ will be full).}$$

Step 2. Now reduce A obtained in Step 1 to upper Hessenberg form while preserving the triangular structure of B.

This step is achieved as follows for the case $n = 4$ ("+" indicates a fill-in).

2.1. Apply a Givens transformation $Q_{34} = J(3, 4, \theta)$ in the (3, 4) plane to the left of A to make the entry (4, 1) of A zero and then update B.

$$A = \begin{pmatrix} \times & \times & \times & \times \\ \times & \times & \times & \times \\ \times & \times & \times & \times \\ \times & \times & \times & \times \end{pmatrix} \xrightarrow{Q_{34}} \begin{pmatrix} \times & \times & \times & \times \\ \times & \times & \times & \times \\ \times & \times & \times & \times \\ 0 & \times & \times & \times \end{pmatrix}; \quad B \xrightarrow{Q_{34}} \begin{pmatrix} \times & \times & \times & \times \\ 0 & \times & \times & \times \\ 0 & 0 & \times & \times \\ 0 & 0 & + & \times \end{pmatrix}.$$

$$A \equiv Q_{34}A \qquad\qquad B \equiv Q_{34}B$$

2.2. Apply the Givens rotation Z_{34} to the right of B to make the (4, 3) entry of B zero and then update A.

$$B = \begin{pmatrix} \times & \times & \times & \times \\ 0 & \times & \times & \times \\ 0 & 0 & \times & \times \\ 0 & 0 & + & \times \end{pmatrix} \xrightarrow{Z_{34}} \begin{pmatrix} \times & \times & \times & \times \\ 0 & \times & \times & \times \\ 0 & 0 & \times & \times \\ 0 & 0 & 0 & \times \end{pmatrix}; \quad A \xrightarrow{Z_{34}} \begin{pmatrix} \times & \times & \times & \times \\ \times & \times & \times & \times \\ \times & \times & \times & \times \\ 0 & \times & \times & \times \end{pmatrix}.$$

$$B \equiv BZ_{34} \qquad\qquad A \equiv AZ_{34}$$
$$\qquad\qquad\qquad\qquad\qquad \textit{(no fill-in)}$$

2.3. Apply the Givens rotation Q_{23} to the left of A to make the (3, 1) entry zero and then update B.

$$A = \begin{pmatrix} \times & \times & \times & \times \\ \times & \times & \times & \times \\ \times & \times & \times & \times \\ 0 & \times & \times & \times \end{pmatrix} \xrightarrow{Q_{23}} \begin{pmatrix} \times & \times & \times & \times \\ \times & \times & \times & \times \\ 0 & \times & \times & \times \\ 0 & \times & \times & \times \end{pmatrix}; \quad B \xrightarrow{Q_{23}} \begin{pmatrix} \times & \times & \times & \times \\ 0 & \times & \times & \times \\ 0 & + & \times & \times \\ 0 & 0 & 0 & \times \end{pmatrix}.$$

$$A \equiv Q_{23}A \qquad\qquad B \equiv Q_{23}B$$
$$\qquad\qquad\qquad\qquad\qquad \textit{(fill-in at the}$$
$$\qquad\qquad\qquad\qquad\qquad \textit{(3, 2) entry)}$$

2.4. Apply the Givens rotation Z_{23} to the right of B to make the $(3, 2)$ entry zero and then update A.

$$B = \begin{pmatrix} \times & \times & \times & \times \\ 0 & \times & \times & \times \\ 0 & + & \times & \times \\ 0 & 0 & 0 & \times \end{pmatrix} \xrightarrow{Z_{23}} \begin{pmatrix} \times & \times & \times & \times \\ 0 & \times & \times & \times \\ 0 & 0 & \times & \times \\ 0 & 0 & 0 & \times \end{pmatrix}; \quad A \xrightarrow{Z_{23}} \begin{pmatrix} \times & \times & \times & \times \\ \times & \times & \times & \times \\ 0 & \times & \times & \times \\ 0 & \times & \times & \times \end{pmatrix}.$$

$$B \equiv BZ_{23} \qquad\qquad AZ_{23}$$
$$\text{(no fill-in)}$$

Note: At this point, B is upper triangular and A is Hessenberg in its first column.

2.5. Apply the Givens rotation Q_{12} to the left of A to make the entry $(4, 2)$ zero and then update B.

$$A = \begin{pmatrix} \times & \times & \times & \times \\ \times & \times & \times & \times \\ 0 & \times & \times & \times \\ 0 & \times & \times & \times \end{pmatrix} \xrightarrow{Q_{12}} \begin{pmatrix} \times & \times & \times & \times \\ \times & \times & \times & \times \\ 0 & \times & \times & \times \\ 0 & 0 & \times & \times \end{pmatrix}; \quad B \xrightarrow{Q_{12}} \begin{pmatrix} \times & \times & \times & \times \\ + & \times & \times & \times \\ 0 & 0 & \times & \times \\ 0 & 0 & 0 & \times \end{pmatrix}.$$

$$A \equiv Q_{12}A \qquad\qquad B \equiv Q_{12}B$$
$$\text{(fill-in at the}$$
$$(2, 1) \text{ entry)}$$

2.6. Apply the Givens rotation Z_{12} to the right of B to make the $(2, 1)$ entry zero and then update A.

$$B = \begin{pmatrix} \times & \times & \times & \times \\ + & \times & \times & \times \\ 0 & 0 & \times & \times \\ 0 & 0 & 0 & \times \end{pmatrix} \xrightarrow{Z_{12}} \begin{pmatrix} \times & \times & \times & \times \\ 0 & \times & \times & \times \\ 0 & 0 & \times & \times \\ 0 & 0 & 0 & \times \end{pmatrix}; \quad A \xrightarrow{Z_{12}} \begin{pmatrix} \times & \times & \times & \times \\ \times & \times & \times & \times \\ 0 & \times & \times & \times \\ 0 & 0 & \times & \times \end{pmatrix}.$$

$$B \equiv BZ_{12} \text{ (triangular)} \qquad\qquad A \equiv AZ_{12}$$
$$\text{(upper Hessenberg)}$$

General case. The process is similar. For each a_{ij} to be zeroed, two Givens rotations are used: one, applied to the left of A for zeroing an entry of A, and the other, applied to the right of B to recover the B's triangularity.

Flop-count. The process requires about $8n^3$ flops. If Q' and Z' are accumulated and are explicitly required, then it will additionally require about $4n^3$ and $3n^3$ flops, respectively.

Example 11.6.

$$A = \begin{pmatrix} 1 & 2 & 3 \\ 1 & 3 & 4 \\ 1 & 3 & 3 \end{pmatrix}, \quad B = \begin{pmatrix} 1 & 1 & 1 \\ 0 & 1 & 2 \\ 0 & 0 & 2 \end{pmatrix}.$$

Since B is already an upper triangular matrix, *Step 1 is skipped.* ∎

11.4. The QZ Algorithm

Step 2. Reduce A to upper Hessenberg while retaining upper triangular structure of B.

Step 2.1. *Form Q_{23} to make a_{31} zero and update B:*

$$Q_{23} = \begin{pmatrix} 1 & 0 & 0 \\ 0 & 0.7071 & -0.7071 \\ 0 & 0.7071 & 0.7071 \end{pmatrix},$$

$$A \equiv A^{(1)} = Q_{23}A = \begin{pmatrix} 1 & 2 & 3 \\ 1.4142 & 4.2426 & 4.9497 \\ 0 & 0 & -0.7071 \end{pmatrix},$$

$$B \equiv B^{(1)} = Q_{23}B = \begin{pmatrix} 1 & 1 & 1 \\ 0 & 0.7071 & 2.8284 \\ 0 & -0.7071 & 0 \end{pmatrix} \quad \text{(fill-in at the (3, 2) entry)}.$$

Step 2.2. *Form Z_{23} to make b_{32} zero and update A:*

$$Z_{23} = \begin{pmatrix} 1 & 0 & 0 \\ 0 & 0 & -1 \\ 0 & 1 & 0 \end{pmatrix},$$

$$B \equiv B^{(1)}Z_{23} = Q_{23}BZ_{23} = \begin{pmatrix} 1 & 1 & -1 \\ 0 & 2.8284 & -0.7071 \\ 0 & 0 & 0.7071 \end{pmatrix},$$

$$A \equiv A^{(1)}Z_{23} = Q_{23}AZ_{23} = \begin{pmatrix} 1 & 3 & -2 \\ 1.4142 & 4.9497 & -4.2426 \\ 0 & -0.7071 & 0 \end{pmatrix}.$$

A is in upper Hessenberg and B is in upper triangular form.

MATCOM Note: The algorithmic process of reduction of (A, B) to a Hessenberg triangular pair has been implemented in the MATCOM function **HESSTRI**.

11.4.2 Stage II: Reduction to the Generalized Real Schur Form

From Stage I, we have

$$A \equiv Q'^T A Z' = \quad \text{upper Hessenberg}$$
$$\text{(assume it is } \textit{unreduced}\text{)}$$
$$B = Q'^T B Z' = \quad \text{upper triangular.}$$

The basic idea now is to apply an implicit QR step to AB^{-1} without ever completely forming this matrix explicitly.

One Iteration of the QZ Algorithm

Step 1. Compute the first column of $N = (C - \alpha_1 I)(C - \alpha_2 I)$, where $C = AB^{-1}$, and α_1 and α_2 are suitably chosen shifts.

Step 2. Find a Householder matrix Q_1, such that $Q_1 N e_1$ is a multiple of e_1.

Step 3. Form $Q_1 A$ and $Q_1 B$.

Step 4. Transform simultaneously $Q_1 A$ to an upper Hessenberg matrix A_1 and $Q_1 B$ to an upper triangular matrix B_1; that is, find orthogonal matrices Q and Z such that $Q(Q_1 A)Z = A_1$ (upper Hessenberg), $Q(Q_1 B)Z = B_1$ (upper triangular).

Using the implicit Q theorem (Theorem 9.34), we can then show that the *matrix* $A_1 B_1^{-1}$ *is essentially the same as what we would have obtained by applying an implicit QR iteration step directly to* AB^{-1}.

Application of a few QZ steps sequentially will then yield a real Schur matrix $R = Q^T A Z$ and an upper triangular $T = Q^T B Z$. The generalized eigenvalues now can be computed from the real Schur triangular pair (R, T), as shown in Section 11.3.

Implementation of Step 1: Computation of the First Column of N

The real bottleneck in implementing the whole algorithm is in computing the first column of $(C - \alpha_1 I)(C - \alpha_2 I)$ without forming $C = AB^{-1}$ explicitly. Fortunately, this can be done. First, note that because A is upper Hessenberg and B is upper triangular, this first column of N contains at most three nonzero entries in the first three places:

$$n_1 = N e_1 = (C - \alpha_1 I)(C - \alpha_2 I) e_1 = (x, y, z, 0, \ldots, 0)^T.$$

To compute x, y, and z all we need to know is the first two columns of C, which can be obtained just by inverting the 2×2 leading principal submatrix of B^{-1}; *the whole B^{-1} does not need to be computed*. Thus, if c_1 and c_2 are the first two columns of $C = AB^{-1}$, then

$$(c_1, c_2) = (a_1, a_2) \begin{pmatrix} b_{11} & b_{12} \\ 0 & b_{22} \end{pmatrix}^{-1},$$

where a_i, $i = 1, 2$, are the first two columns of A. Note that c_1 has at most two nonzero entries and c_2 has at most three.

Let $c_1 = (c_{11}, c_{21}, 0, \ldots, 0)^T$ and $c_2 = (c_{12}, c_{22}, c_{32}, 0, \ldots, 0)^T$.

Then it is easy to see that

$$\begin{pmatrix} x \\ y \\ z \end{pmatrix} = \begin{pmatrix} (c_{11} - \alpha_1)(c_{11} - \alpha_2) + c_{12} c_{21} \\ c_{21}(c_{11} - \alpha_2) + c_{21}(c_{22} - \alpha_1) \\ c_{21} c_{32} \end{pmatrix}.$$

Example 11.7. Let

$$A = \begin{pmatrix} 1 & 1 & 1 & 1 \\ 2 & 1 & 4 & 1 \\ 0 & 1 & 1 & 1 \\ 0 & 0 & 1 & 1 \end{pmatrix}, \quad B = \begin{pmatrix} 1 & 2 & 3 & 4 \\ 0 & 1 & 1 & 1 \\ 0 & 0 & 1 & 2 \\ 0 & 0 & 0 & 3 \end{pmatrix}.$$

11.4. The QZ Algorithm

The 2×2 leading principal submatrix of $B = \begin{pmatrix} 1 & 2 \\ 0 & 1 \end{pmatrix}$:

$$c_1 = (1, 2, 0, 0)^T, \quad c_2 = (-1, -3, -1, 0)^T.$$

Choose $\alpha_1 = 1$, $\alpha_2 = 1$; then $x = -2$, $y = -8$, $z = 2$. ∎

Implementation of Step 2. Since the first column n_1 of $N = (C - \alpha_1 I)(C - \alpha_2 I)$ has at most three nonzero entries, the Householder matrix Q_1 that transforms n_1 into a multiple of e_1 has the form

$$Q_1 = \begin{pmatrix} \hat{Q}_1 & 0 \\ 0 & I_{n-3} \end{pmatrix},$$

where $\hat{Q}_1$ is a 3×3 Householder matrix.

Implementation of Step 4: Computation of A_1 and B_1. The matrices $Q_1^T A$ and $Q_1^T B$ now have the following structures, as illustrated with $n = 6$. (In the following "+" denotes a possible fill-in.)

$$A \equiv Q_1 A = \begin{pmatrix} \times & \times & \times & \times & \times & \times \\ \times & \times & \times & \times & \times & \times \\ + & \times & \times & \times & \times & \times \\ 0 & 0 & \times & \times & \times & \times \\ 0 & 0 & 0 & \times & \times & \times \\ 0 & 0 & 0 & 0 & \times & \times \end{pmatrix}, \quad B \equiv Q_1 B = \begin{pmatrix} \times & \times & \times & \times & \times & \times \\ + & \times & \times & \times & \times & \times \\ + & + & \times & \times & \times & \times \\ 0 & 0 & 0 & \times & \times & \times \\ 0 & 0 & 0 & 0 & \times & \times \\ 0 & 0 & 0 & 0 & 0 & \times \end{pmatrix}.$$

That is, both the Hessenberg form of A and the triangular form of B are now lost in that there is now fill-in at the $(3, 1)$ position of A and the $(2, 1)$, $(3, 1)$, and $(3, 2)$ positions of B. *The job now at hand is to cleverly chase away these unwanted nonzero entries to restore the original Hessenberg form of A and the triangular form of B.* This is done iteratively as shown below.

- Apply a Householder matrix Z_1 to the right of B to eliminate the $(3, 1)$ and $(3, 2)$ entries, followed by another Householder matrix Z_2 to the right to eliminate the $(2, 1)$ entry.

$$B \equiv \begin{pmatrix} \times & \times & \times & \times & \times & \times \\ + & \times & \times & \times & \times & \times \\ + & + & \times & \times & \times & \times \\ 0 & 0 & 0 & \times & \times & \times \\ 0 & 0 & 0 & 0 & \times & \times \\ 0 & 0 & 0 & 0 & 0 & \times \end{pmatrix} \xrightarrow{Z_1}$$

$$\begin{pmatrix} \times & \times & \times & \times & \times & \times \\ + & \times & \times & \times & \times & \times \\ 0 & 0 & \times & \times & \times & \times \\ 0 & 0 & 0 & \times & \times & \times \\ 0 & 0 & 0 & 0 & \times & \times \\ 0 & 0 & 0 & 0 & 0 & \times \end{pmatrix} \xrightarrow{Z_2} \begin{pmatrix} \times & \times & \times & \times & \times & \times \\ 0 & \times & \times & \times & \times & \times \\ 0 & 0 & \times & \times & \times & \times \\ 0 & 0 & 0 & \times & \times & \times \\ 0 & 0 & 0 & 0 & \times & \times \\ 0 & 0 & 0 & 0 & 0 & \times \end{pmatrix}.$$

$$BZ_1 \qquad\qquad BZ_1 Z_2$$

Update A:

$$A = \begin{pmatrix} \times & \times & \times & \times & \times & \times \\ \times & \times & \times & \times & \times & \times \\ + & \times & \times & \times & \times & \times \\ 0 & 0 & \times & \times & \times & \times \\ 0 & 0 & 0 & \times & \times & \times \\ 0 & 0 & 0 & 0 & \times & \times \end{pmatrix} \xrightarrow{Z_1 Z_2} \begin{pmatrix} \times & \times & \times & \times & \times & \times \\ \times & \times & \times & \times & \times & \times \\ + & \times & \times & \times & \times & \times \\ + & + & \times & \times & \times & \times \\ 0 & 0 & 0 & \times & \times & \times \\ 0 & 0 & 0 & 0 & \times & \times \end{pmatrix}.$$

$$A \equiv A Z_1 Z_2$$

(Updating A created two additional fill-ins. We now have unwanted zeros in the $(3, 1)$, $(4, 1)$, and $(4, 2)$ positions of A.)

- Apply a Householder matrix Q_2 to the left of A to eliminate the $(3, 1)$ and $(4, 1)$ entries.

$$A = \begin{pmatrix} \times & \times & \times & \times & \times & \times \\ \times & \times & \times & \times & \times & \times \\ + & \times & \times & \times & \times & \times \\ + & + & \times & \times & \times & \times \\ 0 & 0 & 0 & \times & \times & \times \\ 0 & 0 & 0 & 0 & \times & \times \end{pmatrix} \xrightarrow{Q_2} \begin{pmatrix} \times & \times & \times & \times & \times & \times \\ \times & \times & \times & \times & \times & \times \\ 0 & \times & \times & \times & \times & \times \\ 0 & + & \times & \times & \times & \times \\ 0 & 0 & 0 & \times & \times & \times \\ 0 & 0 & 0 & 0 & \times & \times \end{pmatrix}.$$

$$A \equiv Q_2 A$$

Update B:

$$B = \begin{pmatrix} \times & \times & \times & \times & \times & \times \\ 0 & \times & \times & \times & \times & \times \\ 0 & 0 & \times & \times & \times & \times \\ 0 & 0 & 0 & \times & \times & \times \\ 0 & 0 & 0 & 0 & \times & \times \\ 0 & 0 & 0 & 0 & 0 & \times \end{pmatrix} \xrightarrow{Q_2} \begin{pmatrix} \times & \times & \times & \times & \times & \times \\ 0 & \times & \times & \times & \times & \times \\ 0 & + & \times & \times & \times & \times \\ 0 & + & + & \times & \times & \times \\ 0 & 0 & 0 & 0 & \times & \times \\ 0 & 0 & 0 & 0 & 0 & \times \end{pmatrix}.$$

$$B \equiv Q_2 B$$

At this point, the submatrices of the current A and B enclosed by the boxes have the same structure as that of the original matrices $Q_1 A$ and $Q_1 B$. The problem is now deflated. So, we can now work with these submatrices and the process can be continued until the Hessenberg triangular structure of the pair (A, B) is restored.

In view of the above discussion, let's now summarize one iteration step of the QZ algorithm.

11.4. The QZ Algorithm

ALGORITHM 11.1. One Iteration Step of the QZ Algorithm.

Inputs: (i) $A \in \mathbb{R}^{n \times n}$, an unreduced upper Hessenberg matrix. (ii) $B \in \mathbb{R}^{n \times n}$, an upper triangular matrix
Output: The orthogonal matrices Q and Z such that $A_1 = Q^T A Z$ is upper Hessenberg and $B_1 = Q^T B Z$ is upper triangular.

1. Choose the shifts α_1 and α_2.

2. Compute the first column of $N = (C - \alpha_1 I)(C - \alpha_2 I)$, where $C = AB^{-1}$, without explicitly forming B^{-1}: Let (c_1, c_2) be the first two columns of C. Then $(c_1, c_2) = (a_1, a_2) \begin{pmatrix} b_{11} & b_{12} \\ 0 & b_{22} \end{pmatrix}^{-1}$. The three nonzero entries of the first column of N are given by

$$x = (c_{11} - a_1)(c_{11} - a_2) + c_{12}c_{21},$$
$$y = c_{21}(c_{11} - a_2) + c_{21}(c_{22} - a_1),$$
$$z = c_{21}c_{32}.$$

The first column of $N = n_1 = (x, y, z, 0, \ldots, 0)^T$.

3. Find a Householder matrix Q_1 such that

$$Q_1 n_1 = \begin{pmatrix} * \\ 0 \\ \vdots \\ 0 \end{pmatrix}.$$

4. Form $Q_1 A$ and $Q_1 B$.

5. Transform the matrices $Q_1 A$ and $Q_1 B$, respectively, into an upper Hessenberg matrix A_1 and an upper triangular matrix B_1 by orthogonal equivalence in the way shown previously, creating orthogonal matrices Q_2 through Q_{n-1} and Z_1 through Z_{n-1}.

Obtaining the transforming matrices. The transforming matrices Q and Z are obtained as follows.
The matrix Q:
$$Q = Q_1 Q_2 \cdots Q_{n-1}.$$
The matrix Z:
$$Z = Z_1 Z_2 \cdots Z_{n-1}.$$
Note that Q has the same first row as Q_1.

Flop-count. One QZ iteration step requires about $22n^2$ flops. If Q and Z are to be accumulated, then an additional $8n^2$ and $13n^2$ flops, respectively, will be required.

Choosing the shifts. The double shifts α_1 and α_2 at a QZ step can be taken as the eigenvalues of the lower 2×2 submatrix of $C = AB^{-1}$. The 2×2 lower submatrix of C can be computed without explicitly forming B^{-1} (Exercise 11.5).

> **ALGORITHM 11.2. The Complete QZ Algorithm.**
>
> **Inputs:** (i) $A \in \mathbb{R}^{n \times n}$. (ii) $B \in \mathbb{R}^{n \times n}$.
> **Outputs:** (i) R, real Schur form of A. (ii) T, an upper triangular matrix. The pair (R, T) contains the eigenvalues of (A, B).
>
> **Step 1.** Transform (A, B) into a Hessenberg triangular pair by orthogonal equivalence *(assume that A is unreduced)*:
>
> $$A \equiv Q^T A Z, \quad \text{upper Hessenberg,}$$
> $$B \equiv Q^T B Z, \quad \text{upper triangular.}$$
>
> **Step 2.** Iterate with Algorithm 11.1 to produce $\{A_k\}$ and $\{B_k\}$, choosing the shifts for each iteration as described above.
>
> **Step 3.** Monitor the convergence of the sequences $\{A_k\}$ and $\{B_k\}$:
>
> $$\{A_k\} \to R, \quad \text{real Schur,}$$
> $$\{B_k\} \to T, \quad \text{upper triangular.}$$

Remark. In a computational setting, it will be necessary to monitor the subdiagonal entries of A and the diagonal entries of B in each iteration step to see if a decoupling is possible. *The same criterion for deflation as used for the QR iteration algorithm in Chapter 9 can be used.* For details, see Golub and Van Loan (1996).

Flop-count. Algorithm 11.2 requires about $30n^3$ flops. The formation of Q and Z, if required, needs, respectively, another $16n^3$ and $20n^3$ flops *(from experience it is known that about two QZ steps per eigenvalue are needed)*.

Round-off properties. The QZ iteration algorithm is as *stable* as the QR iteration algorithm. It can be shown that the computed $\hat{R}$ and $\hat{S}$ satisfy

$$Q_0^T (A + E) Z_0 = \hat{R} \quad \text{and} \quad Q_0^T (B + F) Z_0 = \hat{S},$$

where Q_0 and Z_0 are orthogonal and

$$\|E\| \cong \mu \|A\| \quad \text{and} \quad \|F\| \cong \mu \|B\|,$$

where μ is the machine precision.

MATLAB Note: The MATLAB program **qz** finds the QZ factorization for generalized eigenvalues. $[R, T, Q, Z, V] = $ **qz**(A, B) produces complex upper triangular matrices R and T such that $QAZ = R$ and $QBZ = T$, and the matrix V contains the generalized eigenvectors. The MATLAB program **ordqz** reorders the eigenvalues in QZ factorization. Several options of reordering are available.

11.5 Computations of Generalized Eigenvectors

Once an approximate generalized eigenvalue λ is computed, the corresponding generalized eigenvector v can be computed using the inverse iteration as before.

11.6. The Symmetric Positive Definite Generalized Eigenvalue Problem

---

ALGORITHM 11.3. Computation of a Generalized Eigenvector.

Input: $A \in \mathbb{R}^{n \times n}$, $B \in \mathbb{R}^{n \times n}$, and approximate eigenvalue λ of the pencil $A - \lambda B$.
Output: An approximate eigenvector corresponding to λ.

Step 1. Choose an initial eigenvector v_0.
Step 2. For $k = 1, 2, \ldots$ do

  2.1. Solve $(A - \lambda B)\hat{v}_k = B v_{k-1}$.

  2.2. $v_k = \hat{v}_k / \|\hat{v}_k\|_2$.

End

A Remark on Solving $(A - \lambda B)v_k = B v_{k-1}$. In solving $(A - \lambda B)v_k = B v_{k-1}$, *substantial savings can be made by exploiting the Hessenberg triangular structure to which the pair (A, B) is reduced at Stage I of the QZ algorithm.* Note that for a given λ, the matrix $A - \lambda B$ is a Hessenberg matrix. Thus, at *each iteration only a Hessenberg system needs to be solved.* Note that when B is nonsingular, this is equivalent to solving the system with $B^{-1}A$ and Algorithm 11.3 becomes identical to Algorithm 9.2

Example 11.8.

$$A = 10^9 \begin{pmatrix} 3 & -1.5 & 0 \\ -1.5 & 3 & -1.5 \\ 0 & -1.5 & 1.5 \end{pmatrix}, \quad B = \begin{pmatrix} 2 & 0 & 0 \\ 0 & 3 & 0 \\ 0 & 0 & 4 \end{pmatrix}.$$

$\lambda_1 =$ a generalized eigenvalue of $(A - \lambda B) = 1.9508$, $v_0 = (1, 1, 1)^T$.

$k = 1$. Solve for v_1: $(A - \lambda_1 B)\hat{v}_1 = B v_0$.

$$v_1 = \hat{v}_1 / \|\hat{v}_1\|_2 = (0.8507, -0.5114, 0.1217)^T. \quad \blacksquare$$

MATCOM Note: Algorithm 11.3 has been implemented in the MATCOM function **INVITRGN**.

11.6 The Symmetric Positive Definite Generalized Eigenvalue Problem

In this section, we study the *symmetric definite generalized eigenvalue problem $Ax = \lambda Bx$.*
The problem routinely arises in vibration analysis of structures (Inman (2006)).

11.6.1 Eigenvalues and Eigenvectors of Symmetric Definite Pencil

We start with an important (but not surprising) property of the symmetric definite pencil.

Theorem 11.9. *The symmetric definite pencil $(A - \lambda B)$ has real eigenvalues and linearly independent eigenvectors.*

Proof. Since B is symmetric positive definite, it admits the Cholesky decomposition $B = LL^T$. So, from $Ax = \lambda Bx$ we have $Ax = \lambda LL^T x$. So,

$$L^{-1}A(L^T)^{-1}L^T x = \lambda L^T x,$$

or

$$Cy = \lambda y, \qquad \text{where } y = L^T x.$$

The matrix $C = L^{-1}A(L^T)^{-1}$ is **symmetric**; therefore λ is real. The assertion about the eigenvectors is obvious, since a symmetric matrix has a set of n independent eigenvectors. □

An interval containing the eigenvalues of a symmetric definite pencil. The eigenvalues of the symmetric definite pencil $A - \lambda B$ lie in the interval $[-\|B^{-1}A\|, \|B^{-1}A\|]$. (Exercise 11.14).

11.6.2 Conditioning of the Eigenvalues of the Symmetric Definite Pencil

If x is an eigenvector of the symmetric definite pencil (A, B) corresponding to the eigenvalue λ, then the number

$$\nu = \frac{\|x\|_2}{\sqrt{(x^*Ax)^2 + (x^*Bx)^2}}$$

is a *condition number* for the eigenvalue λ (Stewart and Sun (1990)). *(Compare this with the definition of the condition number of a simple eigenvalue of a matrix given in Chapter 9.)*

Two important consequences

- In contrast with the eigenvalues of a symmetric matrix, the eigenvalues of a definite pair (A, B) are not necessarily well-conditioned.

  For example, the eigenvalue 1 of the pair $\left(\begin{smallmatrix}1 & 0 \\ 0 & 0.002\end{smallmatrix}\right), \left(\begin{smallmatrix}1 & 0 \\ 0 & 0.001\end{smallmatrix}\right)$ is well-conditioned (with a perturbation of order 10^{-4}) but the eigenvalue 2 is ill-conditioned.

- Even though the eigenvalues of a definite pair may be ill-conditioned, the degree of ill-conditioning can be bounded.

For further discussions on conditioning of the generalized eigenvalue problem, see Stewart and Sun (1990) and Golub and Van Loan (1996, p.378). See also Tisseur (2000), Higham and Higham (1998), and Tisseur and Meerbergen (2001).

11.6.3 The QZ Method for the Symmetric Definite Pencil

The QZ algorithm described in the previous section for the regular pencil $A - \lambda B$ can, of course, be applied to a symmetric definite pencil. However, *the drawback here is that both*

11.6.4 The Cholesky QR Algorithm for the Symmetric Definite Pencil

> **ALGORITHM 11.4. The Cholesky QR Algorithm for the Symmetric Definite Pencil.**
>
> **Input:** (i) $A \in \mathbb{R}^{n \times n}$, symmetric, and (ii) $B \in \mathbb{R}^{n \times n}$, symmetric positive definite.
> **Output:** The eigenvalues and eigenvectors of the definite pencil $A - \lambda B$.
>
> **Step 1.** Find the Cholesky factorization of B: $B = LL^T$.
>
> **Step 2.** Form $C = L^{-1}A(L^T)^{-1}$ by taking advantage of the symmetry of A.
>
> **Step 3.** Compute the eigenvalues λ_i and the eigenvectors y_i, $i = 1, \ldots, n$, of the symmetric matrix C using the QR iteration with single shift, specialized for symmetric matrices. *(The eigenvalues of the pencil $A - \lambda B \equiv$ the eigenvalues of C.)*
>
> **Step 4.** Compute the *generalized eigenvectors* x_i of the pencil $A - \lambda B$ by solving $L^T x_i = y_i$, $i = 1, \ldots, n$.

Stability of the Cholesky QR Algorithm

When B is well-conditioned, there is nothing objectionable about the algorithm. However, if B is ill-conditioned or nearly singular, then so is L^{-1}, and then matrix C cannot be computed accurately. Therefore, in this case, the eigenvalues and the eigenvectors will be inaccurate.

Specifically, it can be shown (Golub and Van Loan (1996)) that a computed eigenvalue $\tilde{\lambda}$ obtained by the algorithm is the exact eigenvalue of the matrix

$$(L^{-1}A(L^T)^{-1} + E), \quad \text{where} \quad \|E\|_2 \approx \mu \|A\|_2 \|B^{-1}\|_2.$$

Thus, ill-conditioning of B will severely affect the computed eigenvalues, even if they are themselves well-conditioned.

Another disadvantage of this algorithm is the loss of the sparsity—*matrix C is in general full, even though A and B may be sparse*. Since many problems in practical application are large and sparse, the algorithm will not be able to take advantage of the sparsity of the problem in a computational setting. For an analysis of this method with iterative refinement, see Davies, Higham, and Tisseur (2001). The best known algorithm in the 1990s for the sparse symmetric definite generalized eigenvalue problem, which has been incorporated in some well-known structural engineering software packages, is by Grimes, Lewis, and Simon (1994), which is based on the block Lanczos method.

MATLAB and MATCOM Note: The MATLAB function **eig** $(A, B, \text{'chol'})$ has implemented Algorithm 11.4. The algorithm has also been implemented in the MATCOM function **CHOLQR**.

As we will see in the next section, in many practical applications only a few of the smallest generalized eigenvalues are of interest. *These smallest eigenvalues sometimes can be computed reasonably accurately, even when B is ill-conditioned,* by using the ordered RSF of B in which the eigenvalues are ordered from the smallest to the largest. We leave this an exercise for readers (Exercise 11.11).

11.6.5 Diagonalization of the Symmetric Definite Pencil: Simultaneous Diagonalization of A and B

The Cholesky QR iteration algorithm (Algorithm 11.4) of the symmetric definite pencil gives us a method for finding a nonsingular matrix P that transforms A and B simultaneously to diagonal forms by congruence. This can be seen as follows:

Let Q be an orthogonal matrix such that
$$Q^T C Q = \text{diag}(c_1, \ldots, c_n).$$
Set $P = (L^{-1})^T Q$. Then
$$P^T A P = Q^T L^{-1} A (L^{-1})^T Q = Q^T C Q = \text{diag}(c_1, c_2, \ldots, c_n)$$
and
$$P^T B P = Q^T L^{-1} B (L^{-1})^T Q = Q^T L^{-1} L L^T (L^{-1})^T Q = I \quad \text{(note that } B = L L^T\text{)}.$$

ALGORITHM 11.5. Simultaneous Diagonalization of a Symmetric Definite Pencil.

Input: A symmetric definite pair (A, B); $A = A^T$, $B = B^T > 0$.
Output: A nonsingular matrix P such that $P^T B P = I$ and $P^T A P$ is a diagonal matrix.

Step 1. Compute the Cholesky factorization of B: $B = LL^T$.

Step 2. Form $C = L^{-1} A (L^T)^{-1}$ by taking advantage of the symmetry of A (C is symmetric).

Step 3. Applying the symmetric QR iteration algorithm to C, find an orthogonal matrix Q such that $Q^T C Q$ is a diagonal matrix.

Step 4. Form $P = (L^{-1})^T Q$.

Flop-count. Algorithm 11.5 requires about $14n^3$ flops.

Example 11.10. Consider
$$A = \begin{pmatrix} 1 & 2 & 3 \\ 2 & 3 & 4 \\ 3 & 4 & 5 \end{pmatrix}, \quad B = \begin{pmatrix} 10 & 1 & 1 \\ 1 & 10 & 1 \\ 1 & 1 & 10 \end{pmatrix}.$$
A is symmetric and B is symmetric positive definite.

11.6. The Symmetric Positive Definite Generalized Eigenvalue Problem

Step 1. *The Cholesky decomposition of* $B = LL^T$:

$$L = \begin{pmatrix} 3.1623 & 0 & 0 \\ 0.3162 & 3.1464 & 0 \\ 0.3162 & 0.2860 & 3.1334 \end{pmatrix}.$$

Step 2. *Form* $C = L^{-1}A(L^{-1})^T$:

$$C = \begin{pmatrix} 0.1000 & 0.1910 & 0.2752 \\ 0.1910 & 0.2636 & 0.3320 \\ 0.2752 & 0.3320 & 0.3864 \end{pmatrix}.$$

Step 3. *Find an orthogonal* Q *such that* $Q^T C Q = \mathrm{diag}(c_1, \ldots, c_n)$:

$$Q = \begin{pmatrix} 0.4220 & -0.8197 & -0.3873 \\ 0.5684 & -0.0936 & 0.8174 \\ 0.7063 & 0.5651 & -0.4262 \end{pmatrix}.$$

Step 4. *Form*

$$P = (L^{-1})^T Q = \begin{pmatrix} 0.09409 & -0.2726 & -0.1361 \\ 0.1601 & -0.0462 & 0.2722 \\ 0.2254 & 0.1803 & -0.1361 \end{pmatrix}.$$

Step 5. *Verify*

$$P^T A P = \mathrm{diag}(0.8179, -0.0679, 0) \quad \text{and} \quad P^T B P = \mathrm{diag}(1, 1, 1). \blacksquare$$

MATCOM Note: *Algorithm* 11.5 has been implemented in the MATCOM function **SIMDIAG**.

Frequencies, Modes, and Modal Matrix

As we will see a little later, in vibration engineering, a frequently arising eigenvalue problem is the symmetric definite generalized eigenvalue problem of the form

$$Kx = \lambda M x.$$

The matrices M and K are called, respectively, *mass* and *stiffness* matrices. The eigenvalues of this problem are related to the *natural frequencies*, and "the size and sign of each element of an eigenvector determines the shape of the vibration at any instant of time." The eigenvectors are, therefore, referred to as *mode shapes* or simply as *modes*.

"The language of modes, mode shapes, and natural frequencies form the basis for discussing vibration phenomena of complex systems. An entire industry has been formed around the concept of modes" (Inman (2006)).

The diagonal matrix P that simultaneously diagonalize M and K (see Algorithm 11.5) is called the *modal matrix*, and the columns of matrix P are called *normal modes*.

Orthogonality of the eigenvectors. Note that $P = (p_1, p_2, \ldots, p_n)$ is an eigenvector matrix and it is easy to see that

$$p_i^T B p_i = 1, \quad i = 1, \ldots, n; \quad p_i^T B p_j = 0, \quad i \neq j,$$

and

$$p_i^T A p_i = c_i, \quad i = 1, \ldots, n; \quad p_i^T A p_j = 0, \quad i \neq j.$$

11.6.6 Generalized Rayleigh Quotient

The Rayleigh quotient iteration defined for a symmetric matrix A in Chapter 9 can easily be generalized to the symmetric definite pair (A, B).

Definition 11.11. *The number*

$$\lambda = \frac{x^T A x}{x^T B x} \quad (\|x\|_2 = 1),$$

is called the **generalized Rayleigh quotient**.

Significance of the generalized Rayleigh quotient. It can be shown that the generalized Rayleigh quotient as defined above has the following property: *the generalized Rayleigh quotient λ minimizes*

$$f(\lambda) = \|Ax - \lambda Bx\|_B,$$

where $\|\cdot\|_B$ is defined by $\|z\|_B^2 = z^T B^{-1} z$ (Exercise 11.15).

It can be used to compute approximations to generalized eigenvalues λ_k and eigenvectors x_k for the **symmetric definite generalized eigenvalue problem**, as shown in the following algorithm, which is a natural generalization of Algorithm 9.3.

ALGORITHM 11.6. Generalized Rayleigh Quotient Iteration.

Input: A symmetric definite pair (A, B); $A = A^T, B = B^T > 0$.
Output: An approximate eigenpair of (A, B).

Step 0. Choose x_0 such that $\|x_0\| = 1$.
Step 1. For $k = 0, 1, \ldots$ do until convergence
    1.1. Compute $\lambda_k = \frac{x_k^T A x_k}{x_k^T B x_k}$ (generalized Rayleigh quotient).
    1.2. Solve for $\hat{x}_{k+1}$: $(A - \lambda_k B)\hat{x}_{k+1} = Bx_k$ (generalized eigenvector).
    1.3. Normalize $\hat{x}_{k+1}$: $x_{k+1} = \frac{\hat{x}_{k+1}}{\|\hat{x}_{k+1}\|_2}$ (normalized generalized eigenvectors).
End

11.7. SPD Generalized Eigenvalue Problems in Vibrations of Structures

MATCOM Note: Algorithm 11.6 has been implemented in the MATCOM function **GENRAYQT**.

11.7 Symmetric Definite Generalized Eigenvalue Problems Arising in Vibrations of Structures

In this section, we present some case studies on the symmetric definite generalized eigenvalue problem arising in vibration analysis of buildings, airplanes, and others.

11.7.1 Vibration of a Building: A Case Study

Consider a four-story reinforced concrete building as shown in Figure 11.1. The floors and roofs, which are fairly rigid, are represented by lumped masses m_1 to m_4 having a horizontal motion caused by shear deformation of columns, and k_1 to k_4 are equivalent spring constants of columns that act as springs in parallel.

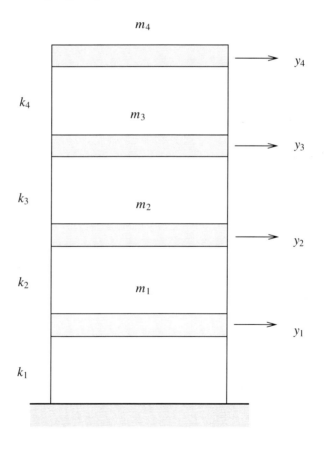

Figure 11.1. *Schematic of four-story building.*

We would like to study the configuration when the building is vibrating in its first two modes (corresponding to two smallest eigenvalues).

Formulation of the generalized eigenvalue problem. To find these modes, first we formulate the problem as a **symmetric definite generalized eigenvalue problem** in terms of mass and stiffness matrices. $Kx = \lambda Mx$ as follows: The equations of motion of the system are $M\ddot{y} + Ky = 0$, where $y = (y_1, y_2, y_3, y_4)^T$, $M = \text{diag}(m_1, m_2, m_3, m_4)$, and

$$K = \begin{pmatrix} k_1 + k_2 & -k_2 & 0 & 0 \\ -k_2 & k_2 + k_3 & -k_3 & 0 \\ 0 & -k_3 & k_3 + k_4 & -k_4 \\ 0 & 0 & -k_4 & k_4 \end{pmatrix}.$$

Assuming harmonic motion, we can write $y_k = x_k e^{i\omega t}$, $k = 1, 2, 3, 4$, where x_k is the **amplitude** of the mass m_k and ω denotes the **natural frequency**. Now, substituting these expressions for y_1, y_2, y_3, and y_4 into the equations of motion, and noting that $\ddot{y}_k = -\omega^2 x_k e^{i\omega t}$, $k = 1, 2, 3, 4$, we obtain the generalized eigenvalue problem $Kx = \lambda Mx$.

Since M and K are both symmetric and M is positive definite, this is a **symmetric positive definite generalized eigenvalue problem**.

Take $m_1 = 5 \times 10^7$, $m_2 = 4 \times 10^7$, $m_3 = 3 \times 10^7$, $m_4 = 2 \times 10^7$, and $k_1 = 10 \times 10^{14}$, $k_2 = 8 \times 10^{14}$, $k_3 = 6 \times 10^{14}$, $k_4 = 4 \times 10^{14}$. We then have

$$K = 10^{14} \begin{pmatrix} 1.8 & -8 & 0 & 0 \\ -8 & 14 & -6 & 0 \\ 0 & -6 & 10 & -4 \\ 0 & 0 & -4 & 4 \end{pmatrix} \quad \text{and} \quad M = 10^7 \times \text{diag}(5, 4, 3, 2).$$

Solution of the eigenvalue problem $Kx = \lambda Mx$ using the Cholesky QR algorithm (Algorithm 11.4). The eigenvalues are $10^7\{6.1432, 1.8516, 0.3435, 4.0950\}$. The eigenvectors corresponding to the two smallest eigenvalues $10^7(0.3435)$ and $10^7(1.8516)$ are

$$10^{-3} \begin{pmatrix} 0.0370 \\ 0.0753 \\ 0.1091 \\ 0.1318 \end{pmatrix}, \quad 10^{-3} \begin{pmatrix} -0.0785 \\ -0.0858 \\ 0.0104 \\ 0.1403 \end{pmatrix}.$$

The first two modes of vibration corresponding to these two smallest eigenvalues are shown in Figures 11.2 and 11.3, respectively.

11.7.2 Forced Harmonic Vibration: Phenomenon of Resonance

In the previous example, we considered vibration of a system without any external force. Consider now a system with two degrees of freedom with different masses, but having the same stiffness coefficients, excited by a harmonic force $F_1 \sin \omega t$, as shown in Figure 11.4. *This example will explain how resonance can occur in structure.*

Then the equations of motion of the system are

$$\begin{aligned} m_1 \ddot{y}_1 &= -k(y_1 - y_2) - ky_1 + F_1 \sin \omega t, \\ m_2 \ddot{y}_2 &= k(y_1 - y_2) - ky_2. \end{aligned} \tag{11.4}$$

11.7. SPD Generalized Eigenvalue Problems in Vibrations of Structures

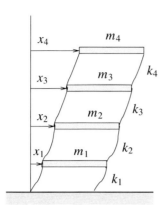

Figure 11.2. *First mode of vibration of four-story building.*

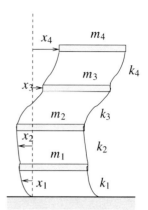

Figure 11.3. *Second mode of vibration of four-story building.*

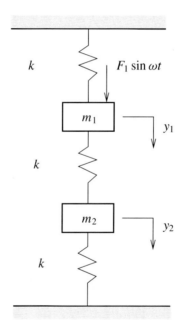

Figure 11.4. *Forced vibration of a two-degrees-of-freedom system.*

Assuming the solution $\begin{pmatrix} y_1 \\ y_2 \end{pmatrix} = \begin{pmatrix} x_1 \\ x_2 \end{pmatrix} \sin \omega t$, and substituting this into the equation (11.4), we get

$$(2k - m_1\omega^2)x_1 - kx_2 = F_1,$$
$$-kx_1 + (2k - m_2\omega^2)x_2 = 0. \tag{11.5}$$

The solution of (11.5) can be written as

$$x_1 = \frac{(2k - m_2\omega^2)F_1}{m_1 m_2(\omega_1^2 - \omega^2)(\omega_2^2 - \omega^2)}, \tag{11.6}$$

$$x_2 = \frac{kF_1}{m_1 m_2(\omega_1^2 - \omega^2)(\omega_2^2 - \omega^2)}, \tag{11.7}$$

where ω_1 and ω_2 are the modal frequencies. For the special case when $m_1 = m_2 = m$, ω_1 and ω_2 are given by $\omega_1 = \sqrt{\frac{k}{m}}$ and $\omega_2 = \sqrt{3\frac{k}{m}}$.

$$x_1 = \frac{(2k - m\omega^2)F_1}{m^2(\omega_1^2 - \omega^2)(\omega_2^2 - \omega^2)},$$

$$x_2 = \frac{kF_1}{m^2(\omega_1^2 - \omega^2)(\omega_2^2 - \omega^2)}. \tag{11.8}$$

From above, it follows immediately that *whenever ω is equal to or close to ω_1 or ω_2, the amplitude becomes arbitrarily large, signaling the occurrence of **resonance**.* Note that in this case, the denominator is zero or close to it.

In other words, when the frequency of the imposed periodic force becomes equal to or nearly equal to one of the natural frequencies of a system, resonance results, a situation which is quite alarming for some applications.

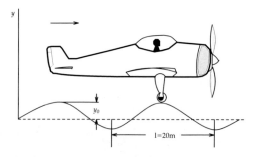

Figure 11.5. *Airplane landing on a runway.*

Example 11.12. Illustration of resonance. Consider the landing of an airplane on a rough runway. The fuselage and engine are assumed to have a combined mass m_1. The wings are modeled by lumped masses m_2 and m_3, and stiffness k_2 and k_3; k_1 represents the combined stiffness of the landing gear and tires. The mass of the wheels and landing gear is assumed negligible compared to that of the fuselage and the wings.

The runway is modeled by a sinusoidal curve as shown in Figure 11.5. Let the contour be described by $y = y_0 \sin \omega t$, and let the airplane be subjected to a forcing input of $f_1 \sin \omega t$, where $f_1 = k_1 y_0$. Let y_1, y_2, and y_3 represent motion relative to the ground.

11.8. Decoupling and Model Reduction

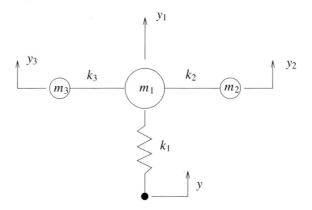

Figure 11.6. *Model of the airplane.*

So, the equations of motion for the three-degrees-of-freedom system so described are given by

$$\begin{pmatrix} m_1 & 0 & 0 \\ 0 & m_2 & 0 \\ 0 & 0 & m_3 \end{pmatrix} \begin{pmatrix} \ddot{y}_1 \\ \ddot{y}_2 \\ \ddot{y}_3 \end{pmatrix} + \begin{pmatrix} k_1+k_2+k_3 & -k_2 & -k_3 \\ -k_2 & k_2 & 0 \\ -k_3 & 0 & k_3 \end{pmatrix} \begin{pmatrix} y_1 \\ y_2 \\ y_3 \end{pmatrix} = \begin{pmatrix} f_1 \sin \omega t \\ 0 \\ 0 \end{pmatrix}.$$

The airplane is shown schematically in Figure 11.6.

Let

$$k_1 = 1.7 \times 10^5 \text{ N/m}; \quad k_3 = k_2 = 6 \times 10^5 \text{ N/m},$$
$$m_1 = 1300 \text{kg}; \quad m_2 = m_3 = 300 \text{kg}.$$

The natural frequencies obtained by solving the *generalized eigenvalue problem* $Kx = \lambda Mx = \omega^2 Mx$ are given by

$$\omega_1 = 9.39 \text{ rad/sec}, \quad \omega_2 = 44.72 \text{ rad/sec}, \quad \omega_3 = 54.46 \text{ rad/sec}.$$

The forcing frequency ω is related to the landing velocity v by $v = \omega \ell / 2\pi$. So if $\omega = \omega_1$, then

$$v = \frac{\omega_1 \ell}{2\pi} = \frac{9.39 \times 20 \text{m/sec}}{2\pi} = 29.8 \text{m/sec} = 107 \text{km/hr}.$$

Thus, if *the landing velocity is 107 km/hr, or close to it, then there is danger of excitation at or near the resonance.* ∎

11.8 Applications of Symmetric Positive Definite Generalized Eigenvalue Problem to Decoupling and Model Reduction

In this section we will mention a few more engineering applications of the generalized eigenvalue problem. These include (i) decoupling of a second-order system of differential equations, and (ii) model reduction.

11.8.1 Decoupling of a Second-Order System

Case 1. The Undamped System

As we have seen before, vibration problems in structural analysis may be modeled by a homogeneous system of second-order differential equations of the form

$$M\ddot{y} + Ky = 0, \tag{11.9}$$

where $y = (y_1, y_2, \ldots, y_n)^T$ and $\ddot{y} = \frac{d^2 y}{dt^2}$.

The matrices M and K are, as usual, the *mass* and *stiffness matrices*. Assuming that these matrices are symmetric and M is positive definite, we will now show how the simultaneous diagonalization technique described earlier (Algorithm 11.5) can be employed to solve this system of second-order differential equations.

The *idea is to decouple the system into n uncoupled equations so that each of these uncoupled equations can be solved using a standard technique.* Let P be the **modal matrix** such that

$$P^T M P = I, \quad P^T K P = \Lambda = \text{diag}(\omega_1^2, \ldots, \omega_n^2). \tag{11.10}$$

Let $y = Pz$, so the homogeneous system $M\ddot{y} + Ky = 0$ becomes

$$MP\ddot{z} + KPz = 0.$$

Next premultiplying the above equation by P^T, we have

$$P^T M P \ddot{z} + P^T K P z = 0$$

or

$$\ddot{z} + \Lambda z = 0. \tag{11.11}$$

Denoting $z = (z_1, z_2, \ldots, z_n)^T$ we see that (11.11) is a set of n uncoupled equations:

$$\ddot{z}_i + \omega_i^2 z_i = 0, \quad i = 1, 2, \ldots, n.$$

The solution of the original system (11.9) now can be obtained by solving these n uncoupled equations using standard techniques and then recovering the original solution y from

$$y = Pz.$$

Thus, if the solutions of the decoupled system (11.11) are given by

$$z_i = A_i \cos \omega_i t + B_i \sin \omega_i t, \quad i = 1, 2, \cdots, n,$$

then the solutions of the original system (11.9) are

$$\begin{pmatrix} y_1 \\ y_2 \\ \vdots \\ y_n \end{pmatrix} = P \begin{bmatrix} A_1 \cos \omega_1 t + B_1 \sin \omega_1 t \\ A_2 \cos \omega_2 t + B_2 \sin \omega_2 t \\ \vdots \\ A_n \cos \omega_n t + B_n \sin \omega_n t \end{bmatrix}. \tag{11.12}$$

The constants A_i and B_i are to be determined from the initial conditions. For example,

$$y_i|_{t=0} = \text{displacement at time } t = 0, \quad \dot{y}_i|_{t=0} = \text{initial velocity}.$$

Example 11.13. We will illustrate the decoupling technique with the following mass-spring example (see Figure 11.7).

11.8. Decoupling and Model Reduction

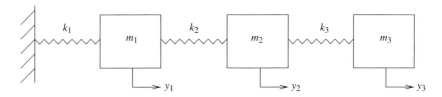

Figure 11.7. *Three-degrees-of-freedom spring-mass system: Decoupling.*

The equations of motion are $M\ddot{y} + ky = 0$, where $y = y_1, y_2, y_3)^T$, $M = \text{diag}(m_1, m_2, m_3)$ and

$$K = \begin{pmatrix} k_1 + k_2 & -k_2 & 0 \\ -k_2 & k_2 + k_3 & -k_3 \\ 0 & -k_3 & k_3 \end{pmatrix}.$$

Take $m_1 = 2 \times 10^4 kg$, $m_2 = 3 \times 10^4 kg$, $m_3 = 4 \times 10^4 kg$, and $k_1 = k_2 = k_3 = 10^9 \times 1.5 N/M$. Then the natural frequencies $\{\omega_1, \omega_2, \omega_3\} = 10^2(4.4168, 2.8951, 0.9273)$.

Suppose that the system, when released from rest at $t = 0$, is subjected to a displacement.

We would like to find the undamped **time response** of the system. The initial conditions are

$$y = (y_1, y_2, y_3)^T = (1, 2, 3)^T \quad \text{and} \quad \dot{y} = (0, 0, 0)^T.$$

Since the initial velocities are zeros, we obtain

$$\dot{y}_i = PB_i\omega_i = 0, \quad i = 1, 2, 3.$$

These equations give $B_1 = B_2 = B_3 = 0$.

Again, at $t = 0$, we have from (11.12)

$$\begin{pmatrix} y_1 \\ y_2 \\ y_3 \end{pmatrix} = P \begin{pmatrix} A_1 \\ A_2 \\ A_3 \end{pmatrix} = \begin{pmatrix} 1 \\ 2 \\ 3 \end{pmatrix}.$$

The modal matrix P *(obtained by using Algorithm* 11.5*)* corresponding to the natural frequencies is given by

$$P = \begin{pmatrix} 0.0056 & -0.0040 & 0.0017 \\ -0.0034 & -0.0035 & 0.0031 \\ 0.0008 & 0.0028 & 0.0040 \end{pmatrix}.$$

The solution of the linear system

$$P \begin{pmatrix} A_1 \\ A_2 \\ A_3 \end{pmatrix} = \begin{pmatrix} 1 \\ 2 \\ 3 \end{pmatrix}$$

is

$$A_1 = 6.1816, \quad A_2 = 50.6264, \quad A_3 = 705.2650.$$

Substituting these values of A_1, A_2, A_3 and the values of $\omega_1, \omega_2,$ and ω_3 obtained earlier, we get

$$\begin{pmatrix} y_1 \\ y_2 \\ y_3 \end{pmatrix} = P \begin{pmatrix} z_1 \\ z_2 \\ z_3 \end{pmatrix} = P \begin{pmatrix} A_1 \cos \omega_1 t \\ A_2 \cos \omega_2 t \\ A_3 \cos \omega_3 t \end{pmatrix}.$$

The values of $y_1, y_2,$ and y_3 give the undamped time response of the systems subject to the given initial conditions. ∎

Case 2. The Damped Systems

Some damping, such as that due to air resistance, fluid and solid friction, etc., is present in all structures. Let us now consider damped homogeneous systems.

Let D be the damping matrix. Then the equations of motion of the damped system become

$$M\ddot{y} + D\dot{y} + Ky = 0. \tag{11.13}$$

Assume that D is a linear combination of M and K; that is,

$$D = \alpha M + \beta K, \tag{11.14}$$

where α and β are constants. Damping of this type is called **proportional** or **Rayleigh damping**. Let P be the modal matrix. Then we have

$$P^T D P = \alpha P^T M P + \beta P^T K P = \alpha I + \beta \Lambda.$$

Let $z = P^T y$. Then the above homogeneous damped equations are transformed to n uncoupled equations:

$$\ddot{z}_i + (\alpha + \beta \omega_i^2)\dot{z}_i + \omega_i^2 z_i = 0, \quad i = 1, 2, \ldots, n. \tag{11.15}$$

In engineering practice it is customary to assume **modal damping**, that is, α and β are chosen so that

$$\alpha + \beta \omega_i^2 = 2\zeta_i \omega_i.$$

The number ζ_i is called the **modal damping ratio** of the ith mode. The quantity ζ_i is usually taken as a small number between 0 and 1. The most common values are $0 \leq \zeta \leq 0.05$. (See Inman (2007).) However, in some applications, such as in the design of flexible structures, $\{\zeta_i\}$ are taken to be as low as 0.005. On the other hand, for an automobile shock absorber, a value as high as $\zeta = 0.5$ is possible.

Assuming modal damping, the decoupled equations (11.15) become

$$\ddot{z}_i + 2\zeta_i \omega_i \dot{z}_i + \omega_i^2 z_i = 0, \quad i = 1, 2, \ldots, n.$$

The solutions of these equations are given by

$$z_i = e^{-\zeta_i \omega_i t}\left(A_i \cos \omega_i \sqrt{1 - \zeta_i^2} t + B_i \sin \omega_i \sqrt{1 - \zeta_i^2} t\right), \quad i = 1, 2, \ldots, n,$$

where the constants A_i and B_i are to be determined from the given initial equations.

The original system can now be solved by solving these n uncoupled equations separately and then recovering the original solution y from $y = Pz$.

11.8. Decoupling and Model Reduction

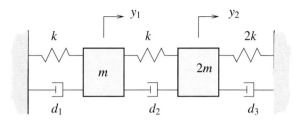

Figure 11.8. *Two-degrees-of-freedom spring-mass system: Nonproportional damping.*

Remark. We have just seen that if damping is proportional, then the system can be decoupled. Unfortunately, however, the *concept of proportional damping is more of theoretical interest rather than practical.* Systems can always be constructed whose damping cannot be proportional. We cite a simple example below.

Example 11.14. Nonproportional damping. Consider the following system with two degrees of freedom (as shown in Figure 11.8):

The equations of motion of the system are developed by considering a free body diagram for each mass.

For mass m:

Thus, the equation of motion for the mass m is
$$-d_1 \dot{y}_1 + d_2(\dot{y}_2 - \dot{y}_1) - k y_1 + k(y_2 - y_1) = m \ddot{y}_1.$$
Similarly, the equation of motion for the mass $2m$ is
$$-d_2(\dot{y}_2 - \dot{y}_1) - d_3 \dot{y}_2 - k(y_2 - y_1) - 2k y_2 = 2m \ddot{y}_2.$$
Thus, for the whole system we have
$$\begin{pmatrix} m & 0 \\ 0 & 2m \end{pmatrix} \begin{pmatrix} \ddot{y}_1 \\ \ddot{y}_2 \end{pmatrix} + \begin{pmatrix} d_1 + d_2 & -d_2 \\ -d_2 & d_2 + d_3 \end{pmatrix} \begin{pmatrix} \dot{y}_1 \\ \dot{y}_2 \end{pmatrix} + \begin{pmatrix} 2k & -k \\ -k & 3k \end{pmatrix} \begin{pmatrix} y_1 \\ y_2 \end{pmatrix} = \begin{pmatrix} 0 \\ 0 \end{pmatrix}.$$

Now let's take $k = 2, m = 5, d_1 = 2, d_2 = 4, d_3 = 1$.

If the relation $D = \alpha M + \beta K$ were satisfied, then there would exist α and β satisfying the equations
$$6 = 5\alpha + 4\beta, \tag{11.16}$$
$$-4 = -2\beta, \tag{11.17}$$
$$5 = 10\alpha + 6\beta. \tag{11.18}$$

However, the above equations cannot all be satisfied with any set of values of α and β. This is a case of *nonproportional damping.* Proportional damping is not possible in this case. ∎

Damped Systems under Force Excitation

When a damped system is subject to an external force F, the equations of motion are given by

$$M\ddot{y} + D\dot{y} + Ky = F(t) = \begin{pmatrix} F_1(t) \\ F_2(t) \\ \vdots \\ F_n(t) \end{pmatrix}. \quad (11.19)$$

Assuming that M is symmetric positive definite K is symmetric, and that damping is proportional, it is easy to see from our previous discussion that the above equations can be decoupled using simultaneous diagonalization.

Let $P = (p_{ij})$ be the **modal matrix**. Then the uncoupled equations will be given by

$$\ddot{z}_i + 2\zeta_i\omega_i\dot{z}_i + \omega_i^2 z_i = p_{1i}F_1 + p_{2i}F_2 + \cdots + p_{ni}F_n$$

or

$$\ddot{z}_i + 2\zeta_i\omega_i\dot{z}_i + \omega_i^2 z_i = E_i(t), \quad (11.20)$$

where $E_i(t) = \sum_{j=1}^n p_{ji}F_j$, $i = 1, 2, \ldots, n$.

The function $E_i(t)$ is called the **exciting** or **forcing function** of the ith mode. If each force F_i is written as

$$F_i = f_i s(t),$$

then

$$E_i(t) = s(t) \sum_{j=i}^n p_{ji} f_j.$$

Definition 11.15. *The expression*

$$\sum_{j=1}^n p_{ji} f_i$$

*is called the **mode participation factor** for the ith mode.*

Once the uncoupled equations (11.20) are solved for z_i, the solutions of the original equations are given by

$$y = Pz.$$

Remark. The solutions of the uncoupled equations

$$\ddot{z}_i + 2\zeta_i\omega_i\dot{z}_i + \omega_i^2 z_i = E_i(t)$$

depend upon the nature of the force $F(t)$. For example, when the force is a shock-type force, such as an earthquake, one is normally interested in maximum responses. The maximum values of $z_1, z_2, \ldots, z_n$ can be obtained from the responses of a single equation of one degree of freedom.

11.8.2 The Reduction of a Large Model

Many applications give rise to a *very large* system of second-order differential equations:

$$M\ddot{y} + D\dot{y} + Ky = F(t). \tag{11.21}$$

For example, the **large space structure** is a distributed parameter system. It is therefore infinite-dimensional in theory. A finite element generated model can have many degrees of freedom (e.g., several million). Naturally, the solution of a large system will lead to a solution of a very large generalized eigenvalue problem. *Unfortunately, effective numerical techniques for computing generalized eigenvalues and eigenvectors of a large generalized eigenvalue problem are not very well developed.* State-of-the-art computational techniques can compute only a few extremal eigenvalues of a large pair (A, B) (see Bai et al. (2000)). It is, therefore, natural to think of solving a vibration problem by constructing a **reduced-order model** with the help of a few eigenvalues and eigenvectors which are feasible to compute. Such a thought is based on an assumption that, in many instances, *the response of the structure depends mainly on the first few eigenvalues (lower frequencies). Usually the higher modes do not get excited.*

We will now show how the *computations can be simplified by using only the knowledge of a few eigenvalues and eigenvectors.*

Suppose that, under the usual assumption that M and K are symmetric and of order n and that M is positive definite, we were able to compute only the first few normal modes, perhaps m of them where $m \ll n$. Let the matrix of these normal modes be $P_{n \times m}$. Then from Algorithm 11.5, we have

$$P^T M P = I_{m \times m};$$
$$P^T K P = \Lambda_{m \times m} = \text{diag}(\omega_1^2, \ldots, \omega_m^2).$$

Setting $y = Pz$ and assuming that the *damping is proportional to mass or stiffness*, the system of n differential equations (11.21) then reduces to m equations:

$$\ddot{z}_i + 2\zeta_i \omega_i \dot{z}_i + \omega_i^2 z_i = E_i(t), \quad i = 1, 2, \ldots, m,$$

where E_i is the ith coordinate of the vector $P^T F$. Once this small number of equations is solved, the displacement of any masses under the external force can be computed from

$$y_i = Pz_i, \quad i = 1, \ldots, n.$$

Sometimes only the maximum value of the displacement is of interest.

Several vibration groups in industry and the military use the following approximation to obtain the maximum value of y_i (see Thomson (1992)):

$$|y_i|_{\max} = |p_1 z_{1(\max)}| + \sqrt{\sum_{j=2}^{m} |p_j z_{j(\max)}|^2},$$

where p_i is the ith column of P evaluated at z_i.

11.8.3 A Case Study on the Potential Damage of a Building Due to an Earthquake

Suppose we are interested in finding the absolute maximum responses of the four-story building considered in the example of Section 11.7.1, when the building is subjected to a strong earthquake (see Figure 11.9), using some known responses of the building due to a previous earthquake. *We will use only the first two modes (the modes corresponding to the two lowest frequencies) in our calculations.*

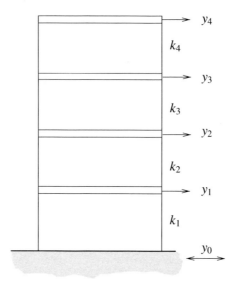

Figure 11.9. *Building subject to an earthquake.*

Let y_0 denote the displacement of the moving support. The uncoupled normal mode equations in modal form in this case can be written as

$$\ddot{z}_i + 2\zeta_i\omega_i\dot{z}_i + \omega_i^2 z_i = -E_i\ddot{y}_0, \quad i = 1, 2,$$

where

$\ddot{y}_0$ = absolute acceleration of the moving support,

$$E_i = \sum_{j=1}^{4} p_{ji}m_j = \text{mode participation factor of the chosen mode } p_i$$

due to support existence.

Here p_{ji} are the coordinates of the participating mode P_i, that is,

$$P = (p_1, p_2) = \left(\begin{pmatrix} p_{11} \\ p_{21} \\ p_{31} \\ p_{41} \end{pmatrix}, \begin{pmatrix} p_{12} \\ p_{22} \\ p_{32} \\ p_{42} \end{pmatrix} \right),$$

where p_1 and p_2 are the two chosen participating modes.

Let R_1 and R_2 denote the maximum relative responses of $z_{1(\max)}$ and $z_{2(\max)}$ obtained from a previous experience. Then we can take

$$z_{1(\max)} = E_1 R_1, \quad z_{2(\max)} = E_2 R_2.$$

11.8. Decoupling and Model Reduction

This observation immediately gives

$$\begin{pmatrix} y_1 \\ y_2 \\ y_3 \\ y_4 \end{pmatrix}_{max} = E_1 R_1 \begin{pmatrix} p_{11} \\ p_{21} \\ p_{31} \\ p_{41} \end{pmatrix} + E_2 R_2 \begin{pmatrix} p_{12} \\ p_{22} \\ p_{32} \\ p_{42} \end{pmatrix}.$$

Using now the data of the example in Section 11.7.1 (vibration of the four-story building), we have

$$E_1 = m_1 p_{11} + m_2 p_{21} + m_3 p_{31} + m_4 p_{41} = 1.0772 \times 10^4,$$
$$E_2 = m_1 p_{12} + m_2 p_{22} + m_3 p_{32} + m_4 p_{42} = -4.2417 \times 10^3.$$

Assume that $R_1 = 1.5$ inches, $R_2 = 0.25$ inches. We then obtain

$$\begin{pmatrix} y_1 \\ y_2 \\ y_3 \\ y_4 \end{pmatrix}_{max} = 10^{-3} \times 10^4 \, (1.0772)(1.5) \begin{pmatrix} 0.0370 \\ 0.0753 \\ 0.1091 \\ 0.1318 \end{pmatrix} + 10^{-3} \times 10^3 (-4.2417)(0.25)$$

$$\times \begin{pmatrix} -0.0785 \\ 0.0858 \\ 0.0104 \\ 0.1403 \end{pmatrix} = \begin{pmatrix} 0.5979 \\ 1.2169 \\ 1.7636 \\ 2.1293 \end{pmatrix} + \begin{pmatrix} 0.0833 \\ 0.0910 \\ -0.0110 \\ -0.1488 \end{pmatrix} = \begin{pmatrix} 0.6812 \text{ in.} \\ 1.3079 \text{ in.} \\ 1.7526 \text{ in.} \\ 1.9805 \text{ in.} \end{pmatrix}.$$

Thus, the *absolute maximum displacement* (relative to the moving support) *of the first floor* is 0.6812 in., that *of the second floor* is 1.3079 in., etc. (see Figure 11.10).

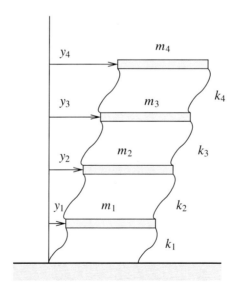

Figure 11.10. *Absolute maximum displacement.*

Note: The contribution to the second participating mode to responses is small in comparison with the contribution of the first mode.

The *absolute maximum relative displacements* are obtained by adding the terms using their absolute values:

$$\begin{pmatrix} y_1 \\ y_2 \\ y_3 \\ y_4 \end{pmatrix}_{\text{(abs. max.)}} = \begin{pmatrix} 0.6812 \text{ in.} \\ 1.3079 \text{ in.} \\ 1.7747 \text{ in.} \\ 2.2781 \text{ in.} \end{pmatrix}.$$

Average maximum relative displacement of the masses. *The absolute maximum relative displacement of the masses provides us with an upper bound for the largest relative displacements the masses can have, and thus help us to choose design parameters.* Another practice for such a measure in engineering literature has been to use the **root sum square** of the same terms, giving the "average" maximum relative displacement values:

$$(y_i)_{\text{average max.}} = \sqrt{(E_1 R_1 p_{i1})^2 + (E_2 R_2 p_{i2})^2 + \cdots + (E_k R_k p_{ik})^2}.$$

For the above example, $k = 2$ and the average maximum relative displacements are given by

$$(y_1)_{\text{average max.}} = \sqrt{(E_1 R_1 p_{11})^2 + (E_2 R_2 p_{12})^2} = 0.9975 \text{ inches},$$

$$(y_2)_{\text{average max.}} = \sqrt{(E_1 R_1 p_{21})^2 + (E_2 R_2 p_{22})^2} = 1.5610 \text{ inches},$$

and so on.

11.9 The Quadratic Eigenvalue Problem

In this section, we discuss a more general eigenvalue problem, called the **quadratic eigenvalue problem** (QEP):

$$(\lambda^2 M + \lambda D + K)x = 0, \qquad (11.22)$$

where M, D, and K are $n \times n$ matrices. The scalars λ are the *eigenvalues*, and the vectors x are the *right eigenvectors*.

The *left eigenvectors* y are given by

$$y^*(\lambda^2 M + \lambda D + K) = 0. \qquad (11.23)$$

The matrix $P_2(\lambda) = \lambda^2 M + \lambda D + K$ is called the *quadratic matrix pencil*.

The pencil is called **regular** if $\det(P_2(\lambda))$ is not identically zero for all values of λ; otherwise it is called **singular**. *Unless otherwise stated, we will assume that the pencil is regular.*

- When M is *nonsingular*, the pencil is regular and has $2n$ *finite* eigenvalues. In fact, it is easy to see these are the $2n$ eigenvalues of the $2n \times 2n$ matrix

$$A = \begin{pmatrix} 0 & I \\ -M^{-1}K & -M^{-1}D \end{pmatrix}, \qquad (11.24)$$

where I is an $n \times n$ identity matrix. An eigenvector u of A corresponding to the eigenvalue λ is of the form $u = \begin{pmatrix} x \\ \lambda x \end{pmatrix}$. Thus *an eigenvector x of $P_2(\lambda)$ is just the vector of the first n components of u.*

11.9. The Quadratic Eigenvalue Problem

- When M is singular, the degree of $\det(P_2(\lambda))$ is $r < 2n$. In this case $P_2(\lambda)$ has r *finite eigenvalues* and the remaining $(2n - r)$ eigenvalues are *infinite eigenvalues*.

Consider the pencil $P_2(\lambda)$ with $M = \mathrm{diag}(1, 0)$, $K = \begin{pmatrix} 4 & -2 \\ -2 & 6 \end{pmatrix}$, and $D = \begin{pmatrix} 1 & 1 \\ 1 & 1 \end{pmatrix}$. Then $\det(P_2(\lambda)) = \lambda^3 + 6\lambda^2 + 14\lambda + 20$ is a polynomial of degree 3. Therefore, $P_2(\lambda)$ has three finite eigenvalues and one infinite. Readers are invited to verify this using MATLAB command **polyeig**.

- The *algebraic multiplicity* of an eigenvalue λ is the order α of the corresponding zero in $\det P_2(\lambda)$. The *geometric multiplicity* of λ is the dimension of $\mathrm{Ker}(P_2(\lambda))$. An eigenvalue is *semisimple* if its algebraic multiplicity is the same as the geometric multiplicity. A *defective* eigenvalue is an eigenvalue that is not semisimple. *An eigenvalue of multiplicity $k > n$ is necessarily defective.*

11.9.1 Orthogonality Relations of the Eigenvectors of Quadratic Matrix Pencil

- Recall that the eigenvectors of a symmetric matrix can be chosen to be orthogonal.

- The eigenvector matrix Φ of the symmetric positive definite generalized eigenvalue problem $K - \lambda M$ can be chosen such that $\Phi^T M \Phi = I$ (Section 11.6.5).

It is natural to wonder if such relations hold for a symmetric positive definite quadratic pencil as well. To this end, the following result on the orthogonality of the eigenvectors of a symmetric positive definite quadratic pencil has been proved by Datta, Elhay, and Ram (1997).

Theorem 11.16 (orthogonality of the eigenvectors of quadratic pencil). *Let $P(\lambda) = \lambda^2 M + \lambda D + K$, where $M = M^T > 0$, $D = D^T$, and $K = K^T$. Assume that the eigenvalues $\lambda_1, \ldots, \lambda_n$ are all distinct and different from zero. Let $\Lambda = \mathrm{diag}(\lambda_1, \ldots, \lambda_{2n})$ be the eigenvalue matrix and let $X = (x_1, \ldots, x_n)$ be the corresponding matrix of eigenvectors. Then there exist diagonal matrices D_1, D_2, and D_3 such that*

$$\Lambda X^T M X \Lambda - X^T K X = D_1, \tag{11.25}$$

$$\Lambda X^T D X \Lambda + \Lambda X^T K X + X^T K X \Lambda = D_2, \tag{11.26}$$

$$\Lambda X^T M X + X^T M X \Lambda + X^T D X = D_3. \tag{11.27}$$

Furthermore

$$D_1 = D_3 \Lambda, \quad D_2 = -D_1 \Lambda, \quad D_2 = -D_3 \Lambda^2. \tag{11.28}$$

Proof. By definition, the pair (X, Λ) must satisfy the $n \times 2n$ system of equations (called the *eigendecomposition* of the pencil $P(\lambda) = \lambda^2 M + \lambda D + K$):

$$M X \Lambda^2 + D X \Lambda + K X = 0. \tag{11.29}$$

Isolating the term in D, we have from above
$$-DX\Lambda = MX\Lambda^2 + KX.$$
Multiplying this on the left by ΛX^T gives
$$-\Lambda X^T DX\Lambda = \Lambda X^T MX\Lambda^2 + \Lambda X^T KX.$$
Taking the transpose gives
$$-\Lambda X^T DX\Lambda = \Lambda^2 X^T MX\Lambda + X^T KX\Lambda.$$
Now, subtracting the latter from the former gives, upon rearrangement,
$$\Lambda X^T MX\Lambda^2 - X^T KX\Lambda = \Lambda^2 X^T MX\Lambda - \Lambda X^T KX$$
or
$$(\Lambda X^T MX\Lambda - X^T KX)\Lambda = \Lambda(\Lambda X^T MX\Lambda - X^T KX). \tag{11.30}$$
Thus, the matrix $\Lambda X^T MX\Lambda - X^T KX$, which we denote by D_1, must be diagonal since it commutes with a diagonal matrix, the diagonal entries of which are distinct. We thus have the *first orthogonality relation:*
$$\Lambda X^T MX\Lambda - X^T KX = D_1.$$
Similarly, isolating the term in M of the eigendecomposition equation (11.29) we obtain
$$-MX\Lambda^2 = DX\Lambda + KX,$$
and multiplying this on the left by $\Lambda^2 X^T$ gives
$$-\Lambda^2 X^T MX\Lambda^2 = \Lambda^2 X^T DX\Lambda + \Lambda^2 X^T KX.$$
Taking the transpose, we have
$$-\Lambda^2 X^T MX\Lambda^2 = \Lambda X^T DX\Lambda^2 + X^T KX\Lambda^2.$$
Subtracting the last equation from the previous one and adding $\Lambda X^T KX\Lambda$ to both sides gives, after some rearrangement,
$$\Lambda(\Lambda X^T DX\Lambda + \Lambda X^T KX + X^T KX\Lambda) = (\Lambda X^T DX\Lambda + \Lambda X^T KX + X^T KX\Lambda)\Lambda.$$
Again, this commutativity property implies, since Λ has distinct diagonal entries, that
$$\Lambda X^T DX\Lambda + \Lambda X^T KX + X^T KX\Lambda = D_2$$
is a diagonal matrix. This is the *second orthogonality relation.*

The first and second orthogonality relations together easily imply the *third orthogonality relation:*
$$\Lambda X^T MX + X^T MX\Lambda + X^T DX = D_3.$$
To prove (11.28), we multiply the last equation on the right by Λ, giving
$$\Lambda X^T MX\Lambda + X^T MX\Lambda^2 + X^T DX\Lambda = D_3\Lambda,$$
which, using the eigendecomposition equation (11.29) of the quadratic pencil, becomes
$$\Lambda X^T MX\Lambda + X^T(-KX) = D_3\Lambda.$$
So, from the first orthogonality relation we see that
$$D_1 = D_3\Lambda. \tag{11.31}$$

Next, using again the eigendecomposition equation (11.29), we rewrite the second orthogonality relation as

$$D_2 = \Lambda X^T(DX\Lambda + KX) + X^T KX\Lambda$$
$$= \Lambda X^T(-MX\Lambda^2) + X^T KX\Lambda = (-\Lambda X^T MX\Lambda + X^T KX)\Lambda.$$

By the first orthogonality relation we then have $D_2 = -D_1\Lambda$.

Finally, from $D_1 = D_3\Lambda$ and $D_2 = -D_1\Lambda$ we have $D_2 = -D_3\Lambda^2$.

We remind the reader that matrix and vector transposition here does not mean conjugation for complex quantities. A real-valued representation of the relations in Theorem 11.16 have been recently obtained in Datta et al. (2009) in the context of finite element model updation (FEMUP). For more on FEMUP, see the authoritative book by Friswell and Mottershead (1995). □

11.9.2 Applications of the Quadratic Eigenvalue Problem

The QEP arises in a wide variety of applications, which include

- vibration of structures
- vibro-acoustic systems
- fluid dynamics
- electric circuit simulation
- signal processing
- microelectronic mechanical systems

A brief account of how a QEP arises in these applications can be found in the recent survey by Tisseur and Meerbergen (2001). The end-users of these applications frequently do not see the QEP as such; these problems are routinely formulated in terms of the standard or generalized eigenvalue problems, because, as we will see here, that is how these problems are usually solved.

Vibration Analysis and the Quadratic Eigenvalue Problem

We have seen in Section 11.7 that vibrating structures are modeled by a system of matrix second-order differential equations of the form

$$M\ddot{x}(t) + D\dot{x}(t) + Kx(t) = F(t). \tag{11.32}$$

By using separation of variables,

$$x(t) = ue^{\lambda t},$$

where u a constant vector, we obtain the quadratic eigenvalue problem

$$P_2(\lambda_k)u_k = 0, \quad k = 1, \ldots, 2n,$$

where $P_2(\lambda) = \lambda^2 M + \lambda D + K$.

Proportional damping. Under the assumption that (i) $M = M^T > 0$, $K = K^T \geq 0$, and (ii) **proportional damping**, we showed in Section 11.8.1 that (11.13) can be decoupled into independent equations which can be solved separately. This decoupling was possible, because under those assumptions, M and K (and therefore D) were simultaneously diagonalized and the modal matrix P was obtained from the solution of the *symmetric positive definite generalized eigenvalue problem*: $Kx = \lambda Mx$. Thus, the QEP did not explicitly appear there.

Nonproportional damping. We have also seen in Example 11.14 that there are systems for which proportional damping is impossible or does not make much sense. *The gyroscopic systems corresponding to spinning structure are other examples of such nonproportional damped systems.* Mathematical models of gyroscopic systems are of the form

$$M\ddot{x}(t) + (D + G)\dot{x}(t) + Kx(t) = F(t),$$

where M, D, and K are the same as before and G is a skew-symmetric matrix: $G = -G^T$.

Decoupling of such systems would be possible if the coefficient matrices were simultaneously diagonalized. *However, this is, unfortunately, not possible in general.* Indeed, it has been shown in Williams and Laub (1992) that *under general damping, these matrices cannot even be simultaneously triangularized.* It can be shown (see Inman (2006)) that the most general condition for simultaneous diagonalization of the mass, stiffness, and damping matrices is

$$DM^{-1}K = KM^{-1}D.$$

The details are left as an exercise (Exercise 11.27). *(Note that Rayleigh damping is a special case of this property.)*

Computing Frequencies and Damping Ratios

Let $\lambda_k = \alpha_k + i\beta_k$ be an eigenvalue of the quadratic eigenvalue problem; that is, α_k and β_k are, respectively, the real and imaginary parts of the complex eigenvalue λ_k. Then the **natural frequency** ω_k corresponding to this eigenvalue is given by

$$\omega_k = \sqrt{\alpha_k^2 + \beta_k^2}, \quad k = 1, 2, \ldots, n. \tag{11.33}$$

Note that the eigenvalues λ_k occur in complex conjugate pairs.

The **modal damping ratio** is given by

$$\zeta_k = \frac{-\alpha_k}{\sqrt{\alpha_k^2 + \beta_k^2}}. \tag{11.34}$$

11.9.3 Numerical Methods for the Quadratic Eigenvalue Problem

A natural way to solve the QEP: $(\lambda^2 M + \lambda D + K)x = 0$ is to transform the problem into a generalized eigenvalue problem (GEP) of the form

$$Az = \lambda Bz,$$

11.9. The Quadratic Eigenvalue Problem

where A and B are $2n \times 2n$ *matrices* (see below for the general forms of A and B). This is known as **linearization** of the QEP.

Finding the Eigenvalues and Eigenvectors of the QEP from GEP

- The eigenvalues of the regular QEP are the same as those of an associated GEP.
- The right eigenvectors x of the QEP and the right eigenvectors z of the associated GEP are related by $z = \binom{x}{\lambda x}$.

Thus, an eigenvector x of the QEP consists of the first n components of the corresponding eigenvector z of the transformed GEP (note that z has $2n$ components).

The transformation of the QEP into the GEP is not unique. Here are some of the commonly used forms of A and B.

First Companion Form:

$$A = \begin{pmatrix} 0 & W \\ -K & -D \end{pmatrix}, \quad B = \begin{pmatrix} W & 0 \\ 0 & M \end{pmatrix}. \tag{11.35}$$

Second Companion Form:

$$A = \begin{pmatrix} -K & 0 \\ 0 & W \end{pmatrix}, \quad B = \begin{pmatrix} D & M \\ W & 0 \end{pmatrix}, \tag{11.36}$$

where W is *an arbitrary nonsingular matrix* both in (11.35) *and* (11.36).
Note that

$$\det(A - \lambda B) = \det(W) \cdot \det(\lambda^2 M + \lambda D + K).$$

Thus, the eigenvalues of $\lambda^2 M + \lambda D + K$ are the same as those of $A - \lambda B$, and the eigenvectors are related, as shown above.

Special cases of the first companion form.

- $W = I$. Then

$$L_1 : A = \begin{pmatrix} 0 & I \\ -K & -D \end{pmatrix}, \quad B = \begin{pmatrix} I & 0 \\ 0 & M \end{pmatrix}. \tag{11.37}$$

- Assume that $M = M^T$, $K = K^T > 0$, $D = D^T$, and $W = -K$. Then we have the *symmetric linearization*

$$L_2 : A = \begin{pmatrix} 0 & -K \\ -K & -D \end{pmatrix}, \quad B = \begin{pmatrix} -K & 0 \\ 0 & M \end{pmatrix}. \tag{11.38}$$

Computing the eigenvalues and eigenvectors of the QEP from GEP. Once the QEP is transformed into a linearized GEP, the QZ iteration now can be applied to compute the eigenvalues of the transformed linearized GEP $Az = \lambda Bz$ to obtain the eigenvalues of the QEP. The eigenvectors of $A - \lambda B$ are computed by using the generalized inverse iteration and the eigenvectors of the quadratic pencil then can be extracted from the eigenvectors of $A - \lambda B$ as shown above. See Section 12.9 for more on computing the eigenvalues and eigenvectors of the QEP.

Accuracy and stability. *The accuracy of a computed eigenvalue depends upon the choice of linearization.* For details, see Tisseur and Meerbergen (2001) and Tisseur (2000). Furthermore, *although the QZ algorithm is stable for the GEP, it is not stable for the QEP* in the sense that it cannot exploit the special structure of the problem.

Linearization of the symmetric positive-definite QEP. In case the QEP is symmetric positive definite, the natural choice between L_1 and L_2 is probably L_2, which preserves the symmetry. Unfortunately, *it cannot preserve definiteness.* Even though A and B are symmetric, they may be indefinite. *Other symmetric choices are also possible.* See Parlett and Chen (1990) in this context.

MATCOM Note: The symmetric linearization L_2 of the QEP has been implemented in the MATCOM function **QUADEIG2**.

MATLAB Note: The MATLAB function **polyeig** (K, D, M) solves the QEP $(\lambda^2 M + \lambda D + K)x = 0$. It uses a companion realization and the QZ algorithm.

11.10 Review and Summary

This chapter has been devoted to the study of the most commonly arising eigenvalue problem in engineering, namely, the generalized eigenvalue problem involving two matrices: $Ax = \lambda Bx$. We now review and summarize the most important results.

11.10.1 Existence Results

There exist Schur and real Schur analogues (Section 11.3) of the ordinary eigenvalue problem for the generalized eigenvalue problem as well. Once the pair (A, B) is transformed into the generalized real Schur form, the eigenvalues can be easily extracted.

11.10.2 The QZ Algorithm

The most widely used algorithm for the generalized eigenvalue problem is the QZ algorithm (Section 11.4), which constructs the generalized real Schur form of (A, B). The algorithm comes in two stages. In Stage I, the pair (A, B) is reduced to a Hessenberg triangular pair. In Stage II, the *Hessenberg triangular* pair is further reduced to the *generalized real Schur form* by applying the implicit QR iteration to AB^{-1}. The matrix B^{-1} is never formed explicitly.

11.10.3 The Generalized Symmetric Eigenvalue Problem

If we neglect the damping forces, then all eigenvalue problems arising in structural and vibration engineering can be cast in the form $Kx = \lambda Mx$, where M is symmetric positive definite, and K is symmetric positive semidefinite. This is called the *symmetric definite generalized eigenvalue problem.* Because of the importance of this problem, it has been studied in some depth here.

11.10. Review and Summary

Several case studies from vibration engineering have been presented in Section 11.7 to show how this problem arises in important practical applications. These include

(i) *vibration of a free spring-mass system,*

(ii) *vibration of a building,*

(iii) *forced harmonic vibration of a spring-mass system.*

The *natural frequencies* and *amplitudes* of a vibrating system are related, respectively, to the generalized eigenvalues and eigenvectors. *If the frequency of the imposed periodic force becomes equal or nearly equal to one of the natural frequencies of the system, then resonance occurs, and the situation is quite alarming.*

The fall of the Tacoma Bridge in the state of Washington and of Broughton Bridge in England are possibly related to such a phenomenon. (See Chapter 9.)

Here are some of the methods and applications of the generalized eigenvalue problem:

- The QZ method can, of course, be used to solve a symmetric definite generalized eigenvalue problem. However, *both symmetry and definiteness will be lost in general.*

- The *Cholesky QR algorithm* (Algorithm 11.4) computes the eigenvalues of the symmetric definite pencil $A - \lambda B$ by transforming the problem into a symmetric problem using the Cholesky decomposition of B. The accuracy obtained by this algorithm can be severely impaired if the matrix B is ill-conditioned.

  The accuracy can be sometimes improved *by constructing an ordered real Schur form of B rather than its Cholesky factorization, in which the eigenvalues are computed from the smallest to the largest* (Exercise 11.11).

- *Simultaneous diagonalization and applications* (Sections 11.6–11.8) The Cholesky QR algorithm (Algorithm 11.4) for the symmetric definite problem $Ax = \lambda Bx$ basically constructs a nonsingular matrix P that transforms A and B simultaneously into diagonal matrices by congruence: $P^T A P = $ a diagonal matrix, and $P^T B P = I$. This is called *simultaneous diagonalization* of A and B. In vibration and other engineering applications, this decomposition is called *modal decomposition* and the matrix P is called a *modal matrix*. The technique of simultaneous diagonalization is a very useful technique in engineering practice (Algorithm 11.5). Its applications include

  (i) *decoupling of a second-order system of differential equations*
  $$M\ddot{y} + D\dot{y} + ky = 0$$
  to n independent equations
  $$\ddot{z}_i + (\alpha + \beta\omega_i^2)\dot{z}_i + \omega_i^2 z_i = 0, \quad i = 1, 2, \ldots, n,$$
  where $D = \alpha M + \beta K$, and

  (ii) reduction of a very large system of second-order systems to a reduced-order model.

Decoupling and model reduction are certainly very useful approaches for handling a large second-order system of differential equations. *Unfortunately, simultaneous diagonalization technique is not practical for large and sparse problems.* On the other hand, many practical problems, such as the *design of large space structures, power* systems, etc., give rise to very large and sparse symmetric definite eigenvalue problems. Furthermore, simultaneous diagonalization technique destroys the sparsity, bandedness, and other exploitable properties of the matrices M, D, and K. The most practical large problems are sparse, and maintaining sparsity is a major concern for algorithm developers in economizing computer storage requirements.

11.10.4 The Quadratic Eigenvalue Problem

The QEP arises in a wide variety of practical applications, including *vibration analysis and design, vibroacoustic systems, fluid mechanics, processing, and control theory*. In most of these applications, the end users, however, do not see the associated eigenvalue problem as the QEP. These problems are routinely formulated as the standard or generalized eigenvalue problem of the form $Ax = \lambda Bx$, where A and B are $2n \times 2n$ matrices. This is because that is how a QEP is generally solved. There exist, however, some projection methods that work directly on the QEP and compute a few extremal or selected eigenvalues and eigenvectors. The *Jacobi–Davidson method* is an example of such a method. We will discuss it briefly in Chapter 12.

11.11 Suggestions for Further Reading

Almost all books in vibration and structural engineering discuss implicitly or explicitly how generalized eigenvalue problems arise in these applications. Some well-known books in the literature of vibration include Inman (2006, 2007) and Thomson (1992).

The QZ iteration algorithm has been discussed in detail in the books by Golub and Van Loan (1996), Stewart (1973), and Demmel (1997). *(The original paper of Moler and Stewart (1973) is worth reading in this context.)* For further reading on simultaneous diagonalization techniques, see Golub and Van Loan (1996) and the references therein. See Demmel and Kagström (1993) for generalized Schur decomposition of a pencil.

For results on perturbation analysis and sensitivity of the generalized eigenvalue problem, see, for instance, Stewart and Sun (1990), Boley (1990), Stewart (1978, 1979), and Tisseur (2000). For Geršgorin theory of the generalized eigenvalue problem, see Stewart (1975). See Ward (1981) for balancing of the generalized eigenvalue problem.

For applications of the symmetric definite generalized eigenvalue problem to earthquake engineering, see the book by Okamoto (1984).

A technique more efficient than the Cholesky QR iteration method for computing the generalized eigenvalues of a symmetric definite pencil for banded matrices has been proposed by Crawford (1973). See also the papers by Wang and Zhao (1991), Kaufman (1993), Davies, Higham, and Tisseur (2001), and Erxiong (1990).

Williams and Laub (1992) have considered the simultaneous triangularizations of matrices M, D, and K of the second-order system $M\ddot{y} + D\dot{y} + Ky = 0$.

Chapter 15 of the book by Parlett (1998) is a rich source of information on symmetric generalized eigenvalue problems. A delightful survey on theory, methods, and applications of the QEP has been given by Tisseur and Meerbergen (2001). This paper also contains a bibliography rich on the subject. See also an earlier survey on the QEP by Sleijpen, van der Vorst, and van Gijzen (1996). Several papers on conditioning, backward errors, scaling, and other aspects of solving quadratic and other polynomial eigenproblems have been written by N. Higham, Tisseur, and their collaborators in recent years.

These include Tisseur (2000), Higham, Li, and Tisseur (2007), Higham, Mackey, and Tisseur (2006, 2009), Higham et al. (2008), and Higham and Tisseur (2003). Visit the home pages of N. Higham (http://www.maths.manchester.ac.uk/~higham) and F. Tisseur (http://www.maths.manchester.ac.uk/~ftisseur) for more recent papers. See also an earlier paper by Langer et al. (1992) for perturbation analysis.

Other recent papers of interests on computational nonlinear eigenvalue problems include Mackey et al. (2006), Mehrmann and Voss (2004), and Hwang et al. (2003). For a collection of nonlinear eigenvalue problems, see Betcke et al. (2008). For the results on orthogonality of the eigenvectors of the symmetric definite quadratic pencil and their uses in *partial quadratic eigenvalue* and *eigenstructure assignments,* see Datta and Sarkissian (2001), Datta, Elhay, and Ram (1997, 2000), Datta, Ram, and Sarkissian (2002), Brahma and Datta (2009), and Bai, Datta, and Wong (2009). A classical book on the QEP is Lancaster (1966). The book by Gohberg, Lancaster, and Rodman (1982) is a rich source of knowledge in theoretical aspects of polynomial eigenvalue problems.

Several variations of the quadratic orthogonality relations in Theorem 11.16 appear in the dissertations of J. Carvalho and D. R. Sarkissian (available from the author's website, *www.math.niu.edu/~dattab*) and in Datta and Sarkissian (2001).

Exercises on Chapter 11

EXERCISES ON SECTION 11.2 and 11.3

11.1 Compute the eigenvalues and eigenvectors of the following pairs (A, B) by finding the zeros of $\det(A - \lambda B)$:

(i) $A = \begin{pmatrix} 2 & 1 \\ 0 & 1 \end{pmatrix}$, $B = \begin{pmatrix} 1 & 0 \\ 0 & 0 \end{pmatrix}$; (ii) $A = \begin{pmatrix} 1 & 1 \\ 0 & 1 \end{pmatrix}$, $B = \begin{pmatrix} 1 & 0 \\ 0 & 0 \end{pmatrix}$;

(iii) $A = \begin{pmatrix} 1 & 2 \\ 0 & 4 \end{pmatrix}$, $B = \begin{pmatrix} 0 & 1 \\ 0 & 0 \end{pmatrix}$.

11.2 Show that when A and B have a common null vector, the generalized characteristic polynomial is identically zero.

11.3 Let A and B be $n \times n$ matrices. Then prove that (i) $\det(A - \lambda B)$ is a polynomial of degree at most n, (ii) the degree of $\det(A - \lambda B)$ is equal to n if and only if B is nonsingular, and (iii) the eigenvalues of two orthogonally equivalent pencils are the same. How are the eigenvectors related?

EXERCISES ON SECTIONS 11.4–11.6

11.4 Show that the matrix Q_1 in the initial QZ step can be computed just by inverting the 2×2 leading principal submatrix of the triangular matrix B.

11.5 Show that the shifts α_1 and α_2 in a QZ step, which are the eigenvalues of the lower 2×2 principal submatrix of $C = AB^{-1}$, can be computed without forming the complete B^{-1}. (*Hint*: Computation depends only on the lower right 3×3 submatrix of B^{-1}.)

11.6 Using the implicit Q theorem, prove that matrix Q in the QZ step has the same first row as Q_1.

11.7 Work out the flop-count for

(a) Hessenberg triangular reduction with and without accumulations of the transforming matrices;

(b) one step of the QZ iteration;

(c) reduction of (A, B) to the generalized Schur form.

11.8 Consider the Hessenberg triangular reduction of (A, B) with B singular. Show that in this case, if the dimension of the null space AB is k, then the Hessenberg triangular structure takes the form

$$A = \begin{pmatrix} A_{11} & A_{12} \\ 0 & A_{22} \end{pmatrix}, \quad B = \begin{pmatrix} 0 & B_{12} \\ 0 & B_{22} \end{pmatrix},$$

where A_{11} is a $k \times k$ upper triangular matrix, A_{22} is upper Hessenberg, and B_{22} is $(n-k) \times (n-k)$ upper triangular and nonsingular. How does this help in the reduction process of the generalized Schur form?

11.9 Verify the statement on the conditioning of the eigenvalues of the matrix pair

$$\left(\begin{pmatrix} 1 & 0 \\ 0 & 0.002 \end{pmatrix}, \begin{pmatrix} 1 & 0 \\ 0 & 0.001 \end{pmatrix} \right)$$

given in Section 11.6.2.

11.10 Work out the flop-count for the Cholesky QR algorithm (Algorithm 11.4) of the symmetric definite pencil.

11.11 Develop an algorithm for computing the smallest eigenvalues of the symmetric definite pencil $A - \lambda B$ by using an ordered real Schur form of B and then construct an example to show that this algorithm yields better accuracy than Algorithm 11.4.

11.12 (*Generalized orthogonal iteration.*) Consider the following iterative algorithm known as the *generalized orthogonal iteration*:

Step 1. Choose an $n \times m$ orthonormal matrix Q_0 such that $Q_0^T Q_0 = I_{m \times m}$.

Step 2. For $k = 1, 2, \ldots$ do

    2.1 Solve for Z_k: $BZ_k = AQ_{k-1}$.

    2.2 Find QR factorization of Z_k: $Z_k = Q_k R_k$.

Apply the above algorithm to the pair (A, B) given by

$$A = \begin{pmatrix} 1 & 1 & 1 & 1 \\ 1 & 2 & 3 & 4 \\ 1 & 3 & 4 & 5 \\ 1 & 4 & 5 & 6 \end{pmatrix}, \quad B = \begin{pmatrix} 10 & 1 & 1 & 1 \\ 1 & 10 & 1 & 1 \\ 1 & 1 & 10 & 1 \\ 1 & 1 & 1 & 10 \end{pmatrix}.$$

11.13 Given

$$A = \begin{pmatrix} 1 & 1 & 1 \\ 1 & 1 & 1 \\ 1 & 1 & 1 \end{pmatrix}, \quad B = \begin{pmatrix} 10 & 1 & 0 \\ 1 & 10 & 1 \\ 0 & 1 & 10 \end{pmatrix};$$

$$A = \begin{pmatrix} 10 & 1 & 1 \\ 1 & 10 & 1 \\ 1 & 1 & 1 \end{pmatrix}, \quad B = \begin{pmatrix} 1 & \frac{1}{2} & \frac{1}{3} \\ \frac{1}{2} & \frac{1}{3} & \frac{1}{4} \\ \frac{1}{3} & \frac{1}{4} & \frac{1}{5} \end{pmatrix}.$$

Find the eigenvalues and eigenvectors for each of the above pairs using

(a) the QZ algorithm followed by inverse iteration (Algorithms 11.2 and 11.3);

(b) the generalized Rayleigh quotient iteration algorithm (Algorithm 11.6);

(c) techniques of simultaneous diagonalization (Algorithm 11.5).

11.14 Prove that the eigenvalues of a symmetric definite pencil (A, B) lies in the interval $[-\|B^{-1}A\|, \|B^{-1}A\|]$.

11.15 Show that the generalized Rayleigh quotient λ minimizes $f(\lambda) = \|Ax - \lambda Bx\|_B$.

PROBLEMS ON SECTIONS 11.7 AND 11.8

11.16 Suppose that a bridge trestle has a natural frequency of 5 Hz (known from an earlier test). Suppose that it deflects about 2mm at midspan under a vehicle of 90000 kg. What are the natural frequencies of the bridge and the vehicle?

11.17 For the equation of motion $m\ddot{y} + ky = F_1 e^{i\omega t}$, find the *amplitude* and *determine the situation which can give rise to resonance*.

11.18 Consider the spring-mass problem shown in Figure 11.11.

(a) Determine the equations of motion.

(b) Set up the generalized eigenvalue problem in the form $Kx = \lambda Mx$, and then determine the natural frequencies and modes of vibration.

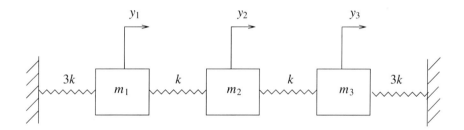

Figure 11.11. *A three-degrees-of-freedom spring-mass system: Generalized eigenvalue problem.*

11.19 Consider the four story building as depicted by Figure 11.1.

Given

$m_1 = 1.0 \times 10^5$ kg, $\quad m_2 = 0.8 \times 10^5$ kg, $\quad m_3 = 0.5 \times 10^5$ kg, $\quad m_4 = 0.6 \times 10^5$ kg,
$k_1 = 15 \times 10^8$ N/m, $\quad k_2 = 12 \times 10^8$ N/m, $\quad k_3 = 15 \times 10^8$ N/m, $\quad k_4 = 10 \times 10^8$ N/m,

find the *maximum amplitude of each floor* for a horizontal displacement of 3mm with a period of 0.25 seconds, assuming zero initial conditions.

430   Chapter 11. Generalized and Quadratic Eigenvalue Problems

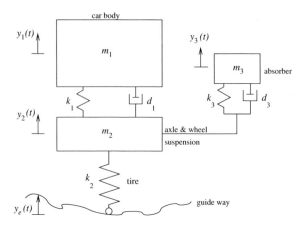

Figure 11.12. *Automobile suspension with vibration absorber.*

11.20 Consider the diagram of an automobile suspension with vibration absorber shown in Figure11.12 (taken from the book *Linear Vibrations* by P. C. Muller and W. O. Schiehlen (1985, p. 226)).

Given
$$m_1 = 1200\text{kg}, \quad m_2 = 80\text{kg}, \quad m_3 = 20\text{kg},$$
$$k_1 = 300 \text{ N/cm}, \quad k_2 = 3200 \text{ N/cm}, \quad k_3 = 600 \text{ N/cm},$$

find the *responses* with different damping values of the absorber $d_3 = 0, 300, 600,$ *and* 1000 Ns/m (the response of a system is measured by the amplitude ratios), assuming that $d_1 = 0$ and that the road profile is described by a sine wave (similar to Example 11.12).

PROBLEMS ON SECTION 11.9

11.21 Find the eigenvalues and eigenvectors of each of the quadratic pencils with the matrices

(i) $M = I_{3\times 3}$, $K = \begin{pmatrix} 2 & -1 & 0 \\ -1 & 2 & -1 \\ 0 & -1 & 2 \end{pmatrix}$, $D = \begin{pmatrix} 1 & 1 & 1 \\ 1 & 1 & 1 \\ 1 & 1 & 1 \end{pmatrix}$;

(ii) $M = \begin{pmatrix} 1 & 1 & 1 \\ 1 & 1 & 1 \\ 1 & 1 & 1 \end{pmatrix}$, $K = \begin{pmatrix} 10 & 1 & 1 \\ 1 & 10 & 1 \\ 1 & 1 & 10 \end{pmatrix}$, $D = 2M + 3K$;

(iii) $M = I_{3\times 3}$, $K = \begin{pmatrix} 1 & 1 & 1 \\ 1 & 1 & 1 \\ 1 & 1 & 1 \end{pmatrix}$, $D = K$;

(iv) $M = \begin{pmatrix} 1 & 2 & 3 \\ 4 & 5 & 6 \\ 7 & 8 & 9 \end{pmatrix}$, $K = \begin{pmatrix} 1 & 1 & 1 \\ 1 & 1 & 1 \\ 1 & 1 & 1 \end{pmatrix}$, $D = 0$;

(v) $M = I_{3\times 3}$, $K = \begin{pmatrix} 1 & 1 & 1 \\ 1 & 1 & 1 \\ 1 & 1 & 1 \end{pmatrix}$, $D = 0$.

11.22 Show that the first and second companion forms (11.35) and (11.36) are linearizations of the quadratic matrix polynomial $P_2(\lambda)$.

11.23 Let (λ, x) be an eigenpair of $(\lambda^2 M + \lambda D + K)x = 0$. Let $\lambda = \alpha + i\beta$. Then prove the following (Datta and Rincón (1993)):

(a) $\alpha = -\frac{|\lambda|^2 D_x}{|\lambda|^2 M_x + K_x}$, where D_x, M_x, and K_x stand for the expressions $x^T D x$, $x^T M x$, and $x^T K x$, respectively.

(b) Using (a), or otherwise, prove that (i) if M, K, and D are positive definite, then $\text{Re}(\lambda) < 0$; (ii) if $M > 0$, $K > 0$, and $D = 0$, then the eigenvalues are purely imaginary; and (iii) if $M > 0$ and $D \leq 0$, $K \geq 0$, then $\text{Re}(\lambda) \geq 0$.

11.24 (a) Construct an example to show that if a regular quadratic pencil $P_2(\lambda)$ has $2n$ distinct eigenvalues, then there exists a set of linearly independent eigenvectors.

(b) Show that an infinite eigenvalue of a quadratic matrix pencil $P_2(\lambda)$ corresponds to a zero eigenvalue of the reverse polynomial
$$\lambda^2 P_2(\lambda^{-1}) = \lambda^2 K + \lambda D + M.$$
Construct an example to illustrate this.

11.25 In finding frequency responses of a vibrating system, one has to solve the linear systems of the form
$$(\omega^2 M + \omega D + K)x = b\omega$$
for many different values of ω. Develop an efficient computational method to solve such systems based on a linearization $A - \omega B$ of the quadratic matrix polynomial $P_2(\omega) = \omega^2 M + \omega D + K$, and using the Schur decomposition of the $2n \times 2n$ linearized form.

11.26 Develop a procedure to solve the damped second-order system
$$M\ddot{x}(t) + D\dot{x}(t) + Kx(t) = f(t)$$
when damping is nonproportional, based on a first-order linearization and assuming that the eigenvalues are all distinct.

11.27 Prove that a necessary and sufficient condition for simultaneous diagonalization of the mass, stiffness, and damping matrices is that the following commutativity relation is satisfied: $DM^{-1}K = KM^{-1}D$.

MATLAB Programs and Problems on Chapter 9

M11.1 Write a MATLAB program, called **hesstri**, to reduce a pair of matrices (A, B) to a Hessenberg triangular pair:
$$[H, T] = \textbf{hesstri}(A, B).$$
Test your program by randomly generating A and B, each of order up to 100.

M11.2 (a) Write a MATLAB program, called **qzitri**, to implement one iteration step of the QZ iteration algorithm (Algorithm 11.1):

$$[A1, B1] = \mathbf{qzitri}\,(A, B),$$

where (A, B) is a Hessenberg triangular pair.

(b) Now apply one step of **qritrdsi** (double-shift implicit QR iteration from Chapter 9) to $C = B^{-1}A$:

$$[C] = \mathbf{qritrdsi}\,(C).$$

(c) Compute $D = A1 * B1^{-1}$, where $(A1, B1)$ is the Hessenberg triangular pair obtained in step (a).

(d) Compare C and D to verify that they are essentially the same.

Test data:
$A = 50 \times 50$ randomly generated unreduced upper Hessenberg matrix.
$B = 50 \times 50$ upper triangular matrix with all entries equal to 1, except five diagonal entries each equal to 10^{-5}.

M11.3 Write a MATLAB program, called **invitrgn**, to implement Algorithm 11.3 by reducing the pair (A, B) first to a Hessenberg triangular pair:

$$[u] = \mathbf{invitrgn}(A, B, \lambda, V_0).$$

Test your program using randomly generated matrices A and B, each of order 50, and then compare the result with that obtained by running the MATLAB command $[U, D] = \mathbf{eig}(A, B)$.

M11.4 *(The purpose of this exercise is to compare the accuracy of different ways of finding the generalized eigenvalues and eigenvectors of the symmetric definite pencil $K - \lambda M$.)*

(a) Use the MATCOM program **CHOLQR** and the MATLAB command **eig**(K, M) to compute the eigenpairs (V_1, D_1) and (V_2, D_2), respectively, of a symmetric positive definite pencil $A - \lambda B$.

(b) Run **eig(inv**$(M) * K)$ from MATLAB to compute $(V3, D3)$: $[V3, D3] = \mathbf{eig}\,(\mathbf{inv}(M) * K)$.

(c) Compare the results obtained in two different ways above.

Test data:

$$M = \mathrm{diag}(m, m, \ldots, m)_{200 \times 200};\ m = 100.$$

$$K = \begin{pmatrix} k & -k & 0 & 0 & \cdots & 0 \\ -k & 2k & -k & 0 & \cdots & 0 \\ \vdots & \ddots & \ddots & \ddots & & \vdots \\ \vdots & & \ddots & \ddots & \ddots & -k \\ 0 & \cdots & 0 & & -k & 2k \end{pmatrix}_{200 \times 200},\ k = 10^8.$$

M11.5 Using **chol, inv, eig** from MATLAB, write a MATLAB program, called **simdiag**, to simultaneously diagonalize a pair of matrices (M, K), where M is symmetric positive definite and K is symmetric (Algorithm 11.5): $[I, D] =$ **simdiag** (M, K).

Test data: Use M and K from the case study on the vibration of a building in Section 11.7.1.

M11.6 (*The purpose of this exercise is to compare different approaches for solving the symmetric positive definite quadratic eigenvalue problem* $(\lambda^2 M + \lambda D + K)x = 0$.)

(i) Form the matrix $A = \begin{pmatrix} 0 & I \\ -M^{-1}K & -M^{-1}D \end{pmatrix}$ by using the MATLAB commands **inv, eig,** and **zeros**. Then compute the eigenvalues and eigenvectors of A using the MATLAB command $[V_1, D_1] =$ **eig** (A).

(ii) Use the program **cholqr** to compute the eigenvalues

$$[V2, D2] = \textbf{cholqr}\,(A, B),$$

where $A = \begin{pmatrix} 0 & -K \\ -K & -D \end{pmatrix}$, $B = \begin{pmatrix} -K & 0 \\ 0 & M \end{pmatrix}$.

(iii) Use the MATLAB command $[V3, D3] =$ **eig** (A, B), where matrices A and B are given as in part (ii).

(iv) Use the MATLAB command $[V_4, D_4] =$ **polyeig**(M, K, D).
Compare the results of the above four approaches with respect to accuracy and flop-count. Use $\frac{\|MV_iD_i^2 + D_iV_iD_i + KV_i\|}{\|V_i\|}$ as a measure of accuracy.

Test data: Use the same M and K as in Problem M11.4 and with $D = 10^{-5} * M$.

M11.7 Find the natural frequencies and modal damping ratios of a quadratic pencil with the matrices $M, K,$ and D as given by the test data of Problem M.11.6 (equations (11.33) and (11.34)).

M11.8 Perform an experiment to compare the relative errors of the largest and the smallest eigenvalues (in magnitude) of the QEP with different linearizations, as stated in Section 11.9.3, using the test data of Problem M11.6.

Chapter 12

Iterative Methods for Large and Sparse Problems: An Overview

12.1 Introduction

The direct methods based on triangularization of matrix A becomes prohibitive in terms of computer time and storage if matrix A is quite large. On the other hand, there are practical situations, such as the discretization of partial differential equations, where the matrix size can be as large as several hundred thousand or even more. For such problems, direct methods for linear systems such as the Gaussian elimination and QR factorization methods become impractical. Furthermore, most large problems are sparse. A *sparse matrix is roughly defined as a matrix with a few nonzero entries and a large number of zero entries.* Unfortunately, the *sparsity gets lost to a considerable extent during the triangularization procedure*, so that at the end we have to deal with a very large matrix with too many nonzero entries, and storage becomes a crucial issue. For such problems, it is advisable to use a class of methods called *iterative methods* that never alter matrix A and require the storage of only a few vectors of length n at a time. These methods, unlike the direct methods, do not produce an exact solution, but rather aim at iteratively improving solutions at each iteration. These methods then allow a user to stop as soon as a certain stopping criterion is satisfied.

In this chapter, we will first study the classical iterative methods for linear systems, such as the *Jacobi, Gauss–Seidel, and successive overrelaxation* (SOR) methods, and then discuss *Krylov subspace methods*, both for *large and sparse linear systems* and *eigenvalue problems*.

Much research has been done in recent years on Krylov subspace methods and the research on this topic is still at developing stage. We only give a very brief overview of these methods and refer the reader to the specialized books and papers in this area.

Our discussions on Krylov subspace methods include (i) the conjugate gradient (CG) method for symmetric positive definite systems, (ii) the generalized minimal residual (GMRES) method, (iii) the bi-conjugate gradient (Bi-CG) method, and (iv) the quasi-minimal residual (QMR) methods for nonsymmetric systems. A brief discussion on *preconditioning techniques* is also included. We also include a brief discussion on Krylov subspace methods to compute extreme eigenvalues of large and sparse matrices.

12.2 The Jacobi, Gauss–Seidel, and SOR Methods

The basic idea behind these methods is to first write the system $Ax = b$ in an equivalent form,
$$x = Bx + d, \tag{12.1}$$
and then, starting with an initial approximation $x^{(1)}$ of the solution vector x, to generate a sequence of approximations $\{x^{(k)}\}$, iteratively defined by
$$x^{(k+1)} = Bx^{(k)} + d, \quad k = 1, 2, \ldots, \tag{12.2}$$
with a hope that under certain mild conditions the sequence $\{x^{(k)}\}$ converges to the solution as $k \to \infty$.

To solve the linear system $Ax = b$ iteratively using this idea, we therefore need to know

(a) how to write the system $Ax = b$ in the form (12.1), and
(b) how $x^{(1)}$ should be chosen so that the iteration (12.2) converges to the limit or under what sort of assumptions the iteration converges to the limit with any arbitrary choice of $x^{(1)}$.

There are three well-known classical iterative methods: *Jacobi, Gauss–Seidel,* and *successive overrelaxation* (SOR). These three methods differ in way matrix B and vector d are computed.

Computations of $x^{(k+1)}$ from $x^{(k)}$ by each of these methods in (12.2) are shown below. Let $x^{(k)} = (x_1^{(k)}, \ldots, x_n^{(k)})^T$.

Jacobi iteration.

$$x_i^{(k+1)} = \frac{1}{a_{ii}} \left(b_i - \sum_{\substack{j=1 \\ i \neq j}}^{n} a_{ij} x_j^{(k)} \right), \quad i = 1, 2, \ldots. \tag{12.3}$$

Matrix form of the Jacobi iteration. Write $A = D + L + U$, where $D = \text{diag}(a_{11}, \ldots, a_{nn})$ (diagonal of A),

$$L = \begin{pmatrix} 0 & 0 & \cdots & 0 \\ a_{21} & 0 & \cdots & 0 \\ \vdots & \ddots & \ddots & \vdots \\ a_{n1} & \cdots & a_{n,n-1} & 0 \end{pmatrix} \quad \text{(lower triangular with zeros on the diagonal)} \tag{12.4}$$

and

$$U = \begin{pmatrix} 0 & a_{12} & \cdots & \cdots & a_{1n} \\ 0 & 0 & a_{23} & \ddots & a_{2n} \\ \vdots & \ddots & \ddots & \ddots & \vdots \\ \vdots & \ddots & \ddots & \ddots & a_{n-1,n} \\ 0 & 0 & 0 & 0 & 0 \end{pmatrix} \quad \text{(upper triangular with zeros on the diagonal)}. \tag{12.5}$$

12.2. The Jacobi, Gauss–Seidel, and SOR Methods

Define $B_J = -D^{-1}(L+U), d_J = D^{-1}b$. Then the Jacobi iterations (12.3) can be written as
$$x^{(k+1)} = B_J x^{(k)} + d_J$$

Gauss–Seidel iteration.

$$x_i^{(k+1)} = \frac{1}{a_{ii}} \left(b_i - \sum_{j=1}^{i-1} a_{ij} x_j^{(k+1)} - \sum_{j=i+1}^{n} a_{ij} x_j^{(k)} \right), \quad i = 1, 2, \ldots. \quad (12.6)$$

The idea is to use each new component, as soon as it is available, in the computation of the next component. This is not done in the Jacobi method.

Matrix form of Gauss–Seidel iteration. Define $B_{GS} = -(D+L)^{-1}U$, $d_{GS} = (D+L)^{-1}b$. Then
$$x^{(k+1)} = B_{GS} x^{(k)} + d_{GS}.$$

SOR iteration.

$$x_i^{(k+1)} = \frac{\omega}{a_{ii}} \left(b_i - \sum_{j=1}^{i-1} a_{ij} x_j^{(k+1)} - \sum_{j=i+1}^{n} a_{ij} x_j^{(k)} \right) + (1-\omega) x_i^k, \quad i = 1, 2, \ldots. \quad (12.7)$$

Matrix form of SOR. Define $B_{SOR} = (D + \omega L)^{-1}[(1-\omega)D - \omega U]$ and $d_{SOR} = \omega(D+\omega L)^{-1}b$. Then the above iteration can be written as
$$x^{(k+1)} = B_{SOR} x^{(k)} + d_{SOR}.$$

Notes: (i) If $\omega = 1$, then the *SOR method becomes identical to the Gauss–Seidel method*.

(ii) If $\omega > 1$, then in computing the $(k+1)$th iteration, more weight is placed on the most current value than when $\omega < 1$, with the hope that convergence will be faster. The number ω is called the **relaxation factor.**

Example 12.1. We apply the Jacobi, Gauss–Seidel, and SOR methods to

$$\begin{pmatrix} 5 & 1 & 1 \\ 1 & 5 & 1 \\ 1 & 1 & 5 \end{pmatrix} \begin{pmatrix} x_1 \\ x_2 \\ x_3 \end{pmatrix} = \begin{pmatrix} 7 \\ 7 \\ 7 \end{pmatrix}$$

with $x^{(1)} = (0, 0, 0)^T$. Note that the exact solution is $x = (1, 1, 1)^T$. ∎

The results are displayed in Table 12.1.

Comment: The matrix A above is *strictly row diagonally dominant and positive definite*. The Jacobi and Gauss–Seidel methods converge for *strictly row* diagonally dominant matrices with an arbitrary initial approximation, and the SOR converges for a symmetric positive definite matrix if $0 < \omega < 2$. See discussions in the next section.

Stopping criteria for iteration (12.2). It is natural to wonder when iteration (12.2) can be terminated. Let $\epsilon > 0$ be the tolerance. Let $\|\cdot\|$ be a subordinate norm.

Table 12.1. *A few iterations of the Jacobi, Gauss–Seidel, and SOR methods for Example* 12.1.

Method	$x^{(2)}$	$x^{(3)}$	$x^{(4)}$	$x^{(5)}$	$x^{(6)}$	$x^{(7)}$	$x^{(8)}$
Jacobi	1.4	0.84	1.0640	0.9744	1.0102	0.9959	1.0016
	1.4	0.84	1.0640	0.9749	1.0102	0.9959	1.0016
	1.4	0.84	1.0640	0.9744	1.0102	0.9959	1.0016
Gauss–Seidel	1.400	0.9968	0.9964	0.9996	1.0000		
	1.1200	1.0214	1.0014	1.0000	1.0000		
	0.8960	0.9964	1.0014	1.0010	1.0000		
SOR	1.6800	0.8047	1.0266	1.0022	0.9979	0.9979	1.0000
($\omega = 1.2$)	1.2768	0.9986	0.9811	1.062	0.9991	0.9991	1.0001
	0.9704	1.0531	0.9875	1.0006	1.0070	1.0007	1.0001

There are several criteria that can be used (see Higham (2002, pp. 335–337)). The one that is the most convenient and widely used in existing code is

$$\text{Relative residual norm:} \quad \frac{\|b - Ax^{(k)}\|}{\|b\|} \leq \epsilon.$$

See also Arioli, Duff, and Ruiz (1992) and Barrett et al. (1994, pp. 57–63). A user might want to stop if the number of iterations exceeds the maximum number of iterations permitted to perform.

12.2.1 Convergence of the Jacobi, Gauss–Seidel, and SOR Methods

It is often hard to make a good guess as to the initial approximation $x^{(1)}$. Thus, it will be nice to have conditions that will guarantee the convergence of iteration (12.2) for any arbitrary choice of $x^{(1)}$.

In the following we derive such a condition.

Theorem 12.2 (iteration convergence theorem). *The iteration*

$$x^{(k+1)} = Bx^{(k)} + d$$

converges to a limit with an arbitrary choice of the initial approximation $x^{(1)}$ *if and only if matrix* $B^k \to 0$ *as* $k \to \infty$, *that is, B is a **convergent matrix**.*

Proof. From

$$x = Bx + d$$

and

$$x^{(k+1)} = Bx^{(k)} + d,$$

we have

$$x - x^{(k+1)} = B(x - x^{(k)}). \tag{12.8}$$

Since this is true for any value of k, we can write

$$x - x^{(k)} = B(x - x^{(k-1)}). \tag{12.9}$$

12.2. The Jacobi, Gauss–Seidel, and SOR Methods

Substituting (12.9) in (12.8), we have
$$x - x^{(k+1)} = B^2(x - x^{(k-1)}). \tag{12.10}$$
Continuing this process k times we can write
$$x - x^{(k+1)} = B^k(x - x^{(1)}).$$
This shows that $\{x^{(k)}\}$ converges to the solution x for any arbitrary choice of $x^{(1)}$ if and only if $B^k \to 0$ as $k \to \infty$. □

Convergence in terms of spectral radius and matrix norm. Using the Jordan canonical theorem (Theorem 9.28) it can be shown that *B is a convergent matrix if and only if the spectral radius of B, $\rho(B)$, is less than* 1. Now $\rho(B) = \max\{|\lambda_i|, \ i = 1, \ldots, n\}$, where λ_1 through λ_n are the eigenvalues of B. Since $|\lambda_i| \leq \|B\|$ for each i (see Theorem 9.11); in particular, $\rho(B) \leq \|B\|$. Thus, if $\|B\| < 1$, then the convergence is guaranteed. Computationally, it is lot easier to check than finding the spectral radius. Unfortunately, *the converse of this fact is not true.*

In the following theorem, we combine the result of Theorem 12.2 with the observation just made.

Theorem 12.3 (conditions for convergence of iteration (12.2)). *A necessary and sufficient condition for the convergence of iteration* (12.2), *for any arbitrary choice of $x^{(1)}$, is that $\rho(B) < 1$. A sufficient condition is that $\|B\| < 1$ for a subordinate matrix norm.*

We now apply the above result to identify classes of matrices for which the Jacobi and/or Gauss–Seidel methods converge for any choice of the initial approximation $x^{(1)}$.

The Jacobi and Gauss–Seidel Methods for Diagonally Dominant Matrices

Corollary 12.4. *If A is strictly row diagonally dominant, then the Jacobi method converges for any arbitrary choice of the initial approximation $x^{(1)}$.*

Proof. Since $A = (a_{ij})$ is strictly row diagonally dominant, we have by definition
$$|a_{ii}| > \sum_{\substack{j=1 \\ i \neq j}}^{n} |a_{ij}|, \quad i = 1, 2, \ldots, n. \tag{12.11}$$
The Jacobi iteration matrix B_J can be written as
$$B_J = \begin{pmatrix} 0 & -\dfrac{a_{12}}{a_{11}} & \cdots & \cdots & -\dfrac{a_{1n}}{a_{11}} \\ -\dfrac{a_{21}}{a_{22}} & 0 & -\dfrac{a_{23}}{a_{22}} & \cdots & -\dfrac{a_{2n}}{a_{22}} \\ \vdots & & \ddots & \ddots & \vdots \\ \vdots & & & \ddots & -\dfrac{a_{n-1,n}}{a_{n-1,n-1}} \\ -\dfrac{a_{n1}}{a_{nn}} & \cdots & \cdots & -\dfrac{a_{n,n-1}}{a_{nn}} & 0 \end{pmatrix}.$$
From (12.11) we therefore have that the absolute row sum of B_J (that is, the row sum taking absolute values) of each row is less than 1, which means $\|B_J\|_\infty < 1$. Thus by Theorem 12.3, we have Corollary 12.4. □

Corollary 12.5. *If A is a strictly row diagonally dominant matrix, then the Gauss–Seidel method converges for any arbitrary choice of $x^{(1)}$.*

Proof. Let λ be an eigenvalue and let $u = (u_1, \ldots, u_n)^T$ be the corresponding eigenvector of the Gauss–Seidel iteration matrix B_{GS}. Then we will show that $\rho(B_{GS}) < 1$. From the expression of B_{GS} given in Section 12.2, we have

$$-Uu = (D + L)\lambda u$$

or

$$-\sum_{j=i+1}^{n} a_{ij} u_j = \lambda \sum_{j=1}^{i} a_{ij} u_j, \quad i = 1, 2, \ldots, n,$$

which can be rewritten as

$$\lambda a_{ii} u_i = -\lambda \sum_{j=1}^{i-1} a_{ij} u_j - \sum_{j=i+1}^{n} a_{ij} u_j, \quad i = 1, 2, \ldots, n.$$

Let u_k be the largest component (having magnitude 1) of the vector u. Then from the above equation, we have

$$|\lambda||a_{kk}| \leq |\lambda| \sum_{j=1}^{k-1} |a_{kj}| + \sum_{j=k+1}^{n} |a_{kj}| \tag{12.12}$$

or

$$|\lambda| \leq \frac{\sum_{j=k+1}^{n} |a_{kj}|}{(|a_{kk}| - \sum_{j=1}^{k-1} |a_{kj}|)}. \tag{12.13}$$

Since A is strictly row diagonally dominant, $|a_{kk}| - \sum_{j=1}^{k-1} |a_{kj}| > \sum_{j=k+1}^{n} |a_{kj}|$. Thus from (12.13), we then conclude that $|\lambda| < 1$, that is, $\rho(B_{GS}) < 1$.
From Theorem 12.3, we now have Corollary 12.5. □

Remark. It is usually true that the greater the diagonal dominance of A, the faster the convergence of the Jacobi method. However, there are simple counterexamples that show that this does not always happen.

The following simple 2×2 example in support of this statement appears in Golub and Van Loan (1996). The example was provided by Richard S. Varga.

$$A_1 = \begin{pmatrix} 1 & -\frac{1}{2} \\ -\frac{1}{2} & 1 \end{pmatrix}, \quad A_2 = \begin{pmatrix} 1 & -\frac{3}{4} \\ -\frac{1}{2} & 1 \end{pmatrix}.$$

It is easy to verify that $\rho(B_J)$ of A_1 is greater than $\rho(B_J)$ of A_2. Readers are invited to try the Jacobi method with A_1 and A_2 to verify the statement.

The Gauss–Seidel Method for a Symmetric Positive Definite Matrix

We show that the Gauss–Seidel method converges, with an arbitrary choice of $x^{(1)}$, for a symmetric positive definite matrix.

Theorem 12.6. *Let A be a symmetric positive definite matrix. Then the Gauss–Seidel method converges for any arbitrary choice of the initial approximation $x^{(1)}$.*

12.2. The Jacobi, Gauss–Seidel, and SOR Methods

Proof. Because A is symmetric, we have $A = L + D + L^T$, where L is as defined in (12.4) and $D = \text{diag}(a_{11}, \ldots, a_{nn})$.

Thus $B_{\text{GS}} = -(D + L)^{-1} L^T$. We will now show that $\rho(B_{\text{GS}}) < 1$.

Let $-\lambda$ be an eigenvalue of B_{GS} and u be the corresponding eigenvector. Then

$$(D + L)^{-1} L^T u = \lambda u.$$

Multiplying the last equation to the left of both sides first by $(D + L)$ and then by u^*, we get

$$u^* L^T u = \lambda u^* (D + L) u$$

or

$$u^* A u - u^* (L + D) u = \lambda u^* (L + D) u \quad \text{(since } A = L + D + L^T\text{)}$$

or

$$u^* A u = (1 + \lambda) u^* (L + D) u. \tag{12.14}$$

Taking the conjugate transpose on both sides, we have

$$u^* A u = (1 + \bar{\lambda}) u^* (L^T + D^T) u. \tag{12.15}$$

Adding (12.14) and (12.15), we obtain

$$\left(\frac{1}{(1+\lambda)} + \frac{1}{(1+\bar{\lambda})} \right) u^* A u = u^* (L + D) u + u^* (L^T + D^T) u$$

$$= u^* (L + D + L^T + D^T) u = u^* (A + D^T) u = u^* (A + D) u > u^* A u.$$

(Note that since A is positive definite, so is D and, therefore, $u^* D u > 0$.)

Dividing both sides of the last equation by $u^* A u (> 0)$ we have

$$\left(\frac{1}{(1+\lambda)} + \frac{1}{(1+\bar{\lambda})} \right) > 1$$

or

$$\frac{(2 + \lambda + \bar{\lambda})}{(1+\lambda)(1+\bar{\lambda})} > 1. \tag{12.16}$$

Let $\lambda = \alpha + i\beta$. Then $\bar{\lambda} = \alpha - i\beta$. From (12.16) we then have

$$\frac{2(1+\alpha)}{(1+\alpha)^2 + \beta^2} > 1,$$

from which it follows that $\alpha^2 + \beta^2 < 1$. That is, $\rho(B_{\text{GS}}) < 1$, since $|\lambda| = \sqrt{\alpha^2 + \beta^2}$. □

Rates of Convergence and a Comparison between the Gauss–Seidel and Jacobi Methods

We have just seen that for strict row diagonally dominant matrices both the Jacobi and the Gauss–Seidel methods converge for an arbitrary $x^{(1)}$. The question as to whether this is true for some other matrices as well naturally arises. *Also, when both methods converge, another question arises: Which one converges faster?*

From our discussion in the last section we know that it is the iteration matrix B that plays a crucial role in the convergence of an iterative method. More specifically, recall from the proof of Theorem 12.2 that e_{k+1} = error at the $(k+1)$th step = $x - x^{k+1}$, and e_1 = initial error = $x - x^{(1)}$ are related by

$$\|e_{k+1}\| \leq \|B^k\| \|e_1\|, \quad k = 1, 2, 3, \ldots.$$

Thus, $\|B^k\|$ gives us an upper bound of the ratio of the error between the $(k+1)$th step and the initial error.

Definition 12.7. *If $\|B^k\| < 1$, then the quantity*

$$-\frac{\ln \|B^k\|}{k}$$

is called the average rate of convergence for k iterations, and the quantity

$$-\ln \rho(B)$$

is called the **asymptotic rate of convergence**.

If the asymptotic rate of convergence of one iterative method is greater than that of the other and both methods are known to converge, then the one with the larger asymptotic rate of convergence converges asymptotically faster than the other.

The following results on the rate of convergence of the Jacobi and Gauss–Seidel methods (Varga (2000)) can be proved.

- If $0 < \rho(B_J) < 1$, then the asymptotic rate of convergence of the Gauss–Seidel method is larger than that of the Jacobi method.

- If matrix A has all its diagonal entries positive and off-diagonal entries nonnegative, then

  (i) the Jacobi and the Gauss–Seidel methods either both converge or both diverge;

  (ii) when both methods converge, the Gauss–Seidel method converges faster than the Jacobi method.

Remarks. Note that in (ii) above we are talking about the *asymptotic* rate of convergence, not the *average* rate of convergence.

Unfortunately, in the general case no such statements about the convergence and the asymptotic rates of convergence of two iterative methods can be made. In fact, there are examples where one method converges but the other diverges. *However, when both the Gauss–Seidel and the Jacobi methods converge, because of the lower storage requirement and the faster rates of convergence, the Gauss–Seidel method should be preferred over the Jacobi method.*

Convergence of the SOR Method: Choice of ω in the SOR Iteration

It is natural to wonder what is the range of ω for which the SOR iteration converges and what is the optimal choice of ω. To this end, we first prove the following important result due to Kahan (1958).

Theorem 12.8 (Kahan). *For the SOR iteration to converge for every initial approximation $x^{(1)}$, ω must lie in the interval $(0, 2)$.*

12.2. The Jacobi, Gauss–Seidel, and SOR Methods

Proof. Recall that SOR iteration matrix B_{SOR} is given by
$$B_{\text{SOR}} = (D + \omega L)^{-1}[(1 - \omega)D - \omega U],$$
where $A = L + D + U$.

The matrix $(D + \omega L)^{-1}$ is a lower triangular matrix with $\frac{1}{a_{ii}}$, $i = 1, \ldots, n$, as the diagonal entries, and the matrix $(1-\omega)D - \omega U$ is an upper triangular matrix with $(1-\omega)a_{ii}$, $i = 1, \ldots, n$, as the diagonal entries. So, $\det(B_{\text{SOR}}) = (1 - \omega)^n$.

Since the determinant of a matrix is equal to the product of its eigenvalues, we conclude that
$$\rho(B_{\text{SOR}}) \geq |1 - \omega|,$$
where $\rho(B_{\text{SOR}})$ is the spectral radius of the matrix B_{SOR}.

Since by Theorem 12.3, $\rho(B_{\text{SOR}})$ has to be less than 1, we conclude that ω must lie in the interval $(0,2)$. ☐

The next theorem, known as the Ostrowski–Reich theorem, shows that the above *condition is also sufficient in case matrix A is symmetric and positive definite.*

The theorem was proved by Reich for the Gauss–Seidel iteration ($\omega = 1$) in 1949 and subsequently extended by Ostrowski[17] in 1954 for the SOR method.

Theorem 12.9 (Ostrowski). *Let A be a symmetric positive definite matrix and let $0 < \omega < 2$. Then the SOR method will converge for any arbitrary choice of $x^{(1)}$.*

There are certain classes of matrices, such as *consistently ordered* and *2-cyclic matrices* with nonzero diagonal entries (see Young (1971, 1972), and Varga (2000)), for which there exists an optimal choice of ω. *The block diagonal matrices with nonsingular diagonal blocks are such matrices.* Note that the block tridiagonal matrix A arising in the discretization of Poisson's equation, encountered in Chapter 6, belongs to this class. In this case, the optimal choice for ω, denoted by ω_{opt}, can be shown to be
$$\omega_{\text{opt}} = \frac{2}{1 + \sqrt{1 - \rho(B_J)^2}}$$
and $\rho(B_{SOR}) = \omega_{\text{opt}} - 1$.

Furthermore, for consistently ordered 2-cyclic matrices, *if the Jacobi method converges, so does the Gauss–Seidel method, and the Gauss–Seidel method converges twice as fast as the Jacobi method.*

Example 12.10.
$$A = \begin{pmatrix} 4 & -1 & 0 & -1 & 0 & 0 \\ -1 & 4 & -1 & 0 & -1 & 0 \\ 0 & -1 & 4 & 0 & 0 & -1 \\ -1 & 0 & 0 & 4 & -1 & 0 \\ 0 & -1 & 0 & -1 & 4 & -1 \\ 0 & 0 & -1 & 0 & -1 & 4 \end{pmatrix}, \quad b = \begin{pmatrix} 1 \\ 0 \\ 0 \\ 0 \\ 0 \\ 0 \end{pmatrix}.$$

---

[17]Alexander Ostrowski (1893–1986) was born in Kiev, Ukraine, and lived in several countries, including the United States, the United Kingdom, Germany, and Switzerland. He studied under such celebrated mathematicians as Hilbert, Klein, and Landau and solved the famous Hilbert's Eighteenth Problem. He made profound contributions in several areas of mathematics, including determinants, matrix theory, algebraic equations, differential equations, number theory, geometry, topology, and numerical analysis. Some details of his contributions can be found in *http://www-history.mcs.st-andrews.ac.uk/References/Ostrowski.html*.

The eigenvalues of B_J are $0.1036, 0.2500, -0.1036, -0.2500, 0.6036, -0.6036$.
$$\rho(B_J) = 0.6036, \qquad \rho(B_{GS}) = 0.3643,$$
$$\omega_{opt} = \frac{2}{1 + \sqrt{1 - (0.6036)^2}} = 1.1128.$$

It took five iterations for the SOR method with ω_{opt} to converge to the exact solution (up to four significant figures), starting with $x_{SOR}^{(1)} = (0, 0, \ldots, 0)^T$, and
$$x_{SOR}^{(5)} = (0.2948, 0.0932, 0.0282, 0.0861, 0.0497, 0.0195)^T.$$

With the same starting vector $x^{(1)}$, the Gauss–Seidel method required nine iterations. (Try it!) Also eighteen iterations is required by the Jacobi method. ∎

MATCOM Note: The Jacobi, Gauss–Seidel, and SOR methods have been implemented, respectively, in the MATCOM programs **jacobi, gaused,** and **sucov.**

Table 12.2 summarizes properties of the three methods discussed above.

Symmetric Successive Overrelaxation (SSOR) Method

There exists a symmetric version of the SOR method which is derived by combining the SOR scheme with the **backward SOR scheme.** The SSOR matrix and the corresponding vector can be written as
$$B_{SSOR} = (D + \omega U)^{-1}(-\omega L + (1 - \omega)D)(D + \omega L)^{-1}(-\omega U + (1 - \omega)D),$$
$$d_{SSOR} = \omega(D + \omega U)^{-1}(I + [-\omega L + (1 - \omega)D](D + \omega L)^{-1})b.$$

The reader is invited to develop the complete algorithm (Exercise 12.12).

Table 12.2. *Comparison of some classical iterative methods.*

Method	Properties
Jacobi	Easy to use. Convergence with an arbitrary initial guess is guaranteed if A is *strictly row diagonally dominant*.
Gauss–Seidel	Typically converges faster than the Jacobi method. Convergence with an arbitrary initial guess is guaranteed for the *strictly row diagonally dominant* and *symmetric positive definite matrices*.
Successive overrelaxation (SOR)	When $\omega > 1$, convergence is typically faster than Gauss–Seidel. The speed of convergence depends upon ω. Guaranteed convergence for a *symmetric positive definite matrix*. Optimal ω is available for certain classes of matrices, such as the *block symmetric positive definite matrices arising in Poisson's equation*.

12.3 Krylov Subspace Methods for Linear Systems: Lanczos, Arnoldi, GMRES, Conjugate Gradient, and QMR

Given an $n \times n$ matrix A and an n-vector x, the sequence $\{x, Ax, \ldots, A^{n-1}x\}$ is the called a **Krylov sequence** and the matrix $(x, Ax, \ldots A^{n-1}x)$ is called the **Krylov matrix,** denoted by $K_n(A, x)$, after A.N. Krylov.[18]

The subspace $\mathcal{K}_m(A, x) = \text{span}\{x, Ax, \ldots, A^{m-1}x\}$ is called the **Krylov subspace** of dimension m, assuming that the vectors are independent. Two basic Krylov subspace methods are the **Lanczos** and **Arnoldi methods**. Historically, the Lanczos method, developed by C. Lanczos (Lanczos (1950)), was the first of its type. When A is symmetric, both methods become identical if the starting vectors are the same.

The Arnoldi and Lanczos methods are used as *projection methods* in numerical linear algebra to solve *large-scale* matrix computational problems which are typically sparse (see Bai et al. (2000), Saad (1992, 2003), and Stewart (2001a)).

A large problem is typically projected onto a Krylov subspace of dimension m ($m \ll n$). Then the projected problem is solved using a standard technique, and finally an approximate solution of the original problem is retrieved from the solution of the smaller projected problem.

We will describe the basic Arnoldi (Algorithm 12.1) and Lanczos (Algorithms 12.4 and 12.6) methods first and then discuss some of their well-known applications to linear system and eigensolutions of large and sparse matrices. Our discussions will include the following:

- Arnoldi-based full orthogonalization (Algorithm 12.2) and GMRES (Algorithm 12.3) methods for nonsymmetric linear systems

- Nonsymmetric Lanczos-based bi-conjugate gradient (Algorithm 12.7) (Bi-CG) and QMR methods for nonsymmetric linear systems (Section 12.3.11).

- Symmetric Lanczos-based conjugate gradient (CG) (Algorithm 12.5) and preconditioned conjugate gradient (Algorithm 12.9) methods for symmetric positive definite systems.

- The Lanczos method for the symmetric eigenvalue problem (Section 12.6.3).

- An explicitly restarted Arnoldi algorithm (Algorithm 12.10) for the nonsymmetric eigenvalue problem.

- The implicit Arnoldi method for the nonsymmetric eigenvalue problem (Algorithm 12.11).

There are some drawbacks to these methods. Both the Arnoldi and Lanczos methods might encounter "breakdowns" or "near breakdowns" during the process of orthogonalization. In

---

[18]Alexei Nikolaevich Krylov (1863–1945) was a Russian maritime engineer. His celebrated paper "On the numerical solution of the equation by which the technical questions frequencies of small oscillations of material systems are determined" [*Izvestija Akad. Nauk SSSR, Otdel. Mat. i Estest. Nauk,* 7(4) (1931), 491–539 (in Russian)] forms the basis of the so-called *Krylov subspace methods*. (Consult *http://www.navy.ru/history/b-krylov.htm* (in Russian).)

case of the Arnoldi and symmetric Lanczos methods, such a "breakdown" is considered to be "happy breakdown" for the eigenvalue problem because in such a case one obtains an exact eigenvalue. With the nonsymmetric Lanczos method, however, this is not always true.

12.3.1 The Basic Arnoldi Method

(See Arnoldi (1951).) Given $A \in \mathbb{R}^{n \times n}$, a nonzero vector v, and an integer $m \leq n$, the idea is to create the set of $(m+1)$ *orthonormal vectors* $\{v_1, \ldots, v_{m+1}\}$ and an $(m+1) \times m$ Hessenberg matrix $\tilde{H}_m$ such that if $V_m = (v_1, \ldots, v_m)$ and $V_{m+1} = (v_1, \ldots, v_m, v_{m+1})$, then

$$AV_m = V_{m+1}\tilde{H}_m \quad (12.17)$$

or

$$A(v_1, v_2, \ldots, v_m) = (v_1, v_2, \ldots, v_m, v_{m+1}) \begin{pmatrix} h_{11} & \cdots & h_{1m} \\ h_{21} & & \vdots \\ & \ddots & \\ & & h_{m+1,m} \end{pmatrix}.$$

From above, it follows that if we set $v_1 = v/\|v\|_2$, then we can compute, at step k, the vector v_{k+1} and the kth column of $\tilde{H}_m$ by comparing the entries of the kth column of both sides, as shown below.

Step 1. Compute v_2 and the entries of the first column of $\tilde{H}_m$. By comparing the entries of the first column of both sides of the above equation, we have

$$Av_1 = h_{11}v_1 + h_{21}v_2.$$

Multiplying by v_1^T on both sides, we get

$$v_1^T A v_1 = h_{11} \quad \text{(since } v_1^T v_1 = 1 \text{ and } v_1^T v_2 = 0\text{)}.$$

Next set

$$\hat{v} = Av_1 - h_{11}v_1$$

and take $h_{21} = \|\hat{v}\|_2$. Then set

$$v_2 = \hat{v}/h_{21} = \hat{v}\big|\|\hat{v}\|_2.$$

The other steps are analogous.

Step k. At step k, the entries of the kth column of the matrix $\tilde{H}_m$ together with v_{k+1} can be generated from the $(k+1)$-*term recursive relation*

$$Av_k = h_{1k}v_1 + \cdots + h_{kk}v_k + h_{k+1,k}v_{k+1} \quad (12.18)$$

and making use of the fact that $\{v_1, v_2, \ldots, v_k\}$ are orthonormal.

Algorithmically, $h_{k,k}$ will not be computed from the relation $h_{k,k} = v_k^T A v_k$ (see Algorithm 12.1).

Thus, the above process is simply the modified Gram–Schmidt process for generating a set of orthonormal vectors. We will now write the process algorithmically in the following. There exists a *Householder version* of the algorithm due to Walker (1988), which is *more numerically stable* but more expensive as well.

12.3. Krylov Subspace Methods for Linear Systems

ALGORITHM 12.1. The Arnoldi Method (Modified Gram–Schmidt Version).

Inputs: (i) A, an $n \times n$ matrix, (ii) v, an n-vector, and (iii) m, a positive integer less than or equal to n.

Outputs: (i) A set of $(m+1)$ orthonormal vectors $\{v_1, v_2, \ldots, v_{m+1}\}$. (ii) A $(m+1) \times m$ Hessenberg matrix $\tilde{H}_m = (h_{ij})$.

Step 0. Normalize the vector v to obtain v_1: $v_1 = \dfrac{v}{\|v\|_2}$.

Step 1. For $k = 1, 2, \ldots, m$ do
   $\hat{v} = Av_k$
   For $j = 1, 2, \ldots, k$ do
      $h_{j,k} = v_j^T \hat{v}$
      $\hat{v} = \hat{v} - h_{j,k} v_j$
   End
   $h_{k+1,k} = \|\hat{v}\|_2$. If $h_{k+1,k} = 0$, stop.
   $v_{k+1} = \hat{v}/h_{k+1,k}$
End

Some Important Relations obtained from Algorithm 12.1.

Here we summarize some important relations which can be easily derived from the Arnoldi[19] method.

Let
$$\begin{aligned} V_m &= (v_1, v_2, \ldots, v_m), \\ V_{m+1} &= (v_1, v_2, \ldots, v_{m+1}), \\ H_m &= \text{the } m \times m \text{ Hessenberg matrix obtained by} \\ &\quad \text{deleting the } (m+1)\text{th row of the matrix } \tilde{H}_m. \end{aligned}$$

I. *(Arnoldi factorization.)* The relation (12.17) can be written as

$$AV_m - V_m H_m = h_{m+1,m}(0, 0, \ldots, 0, v_{m+1}) \tag{12.19}$$

or

$$AV_m = V_m H_m + h_{m+1,m} v_{m+1} e_m^T \tag{12.20}$$

or

$$AV_m = V_m H_m + f_m e_m^T, \qquad \text{where } f_m = h_{m+1,m} v_{m+1}. \tag{12.21}$$

The above factorization is called the **Arnoldi factorization** and can be represented as shown in Figure 12.1.

---

[19]Walter Edwin Arnoldi (1917–1995), an American engineer/scientist, was born in New York and educated at Stevens Institute of Technology and Harvard University. He was employed as an analytic engineer by Hamilton Standard Division of United Aircraft Corporation from 1939 to 1977. His paper "The principle of minimized iterations in the solution of the eigenvalue problem" (Arnoldi (1951)) is perhaps one of the most cited papers in numerical linear algebra. An article about Arnoldi can be found in *American Men & Women of Science* [18th ed. Vol. 7, 1993]. See also NA Digest, Monday, March 4, 1996, Vol. 96, issue 09 (http://www.netlib.org/na-digest-html/96/v96n09.html).

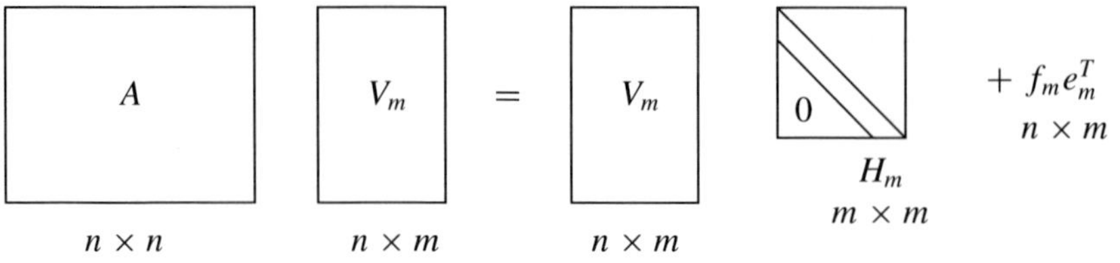

Figure 12.1. *Arnoldi factorization.*

II. *(QR factorization of the Arnoldi–Krylov matrix.)* The matrix V_m is such that
$$K_m = V_m R_m, \tag{12.22}$$
where K_m is the Krylov matrix $(b, Ab, \ldots, A^{m-1}b)$. That is, V_m is the Q matrix of the reduced QR factorization of the Krylov matrix K_m.

III. From $AV_m = V_{m+1}\tilde{H}_m$, it follows that
$$V_m^T A V_m = H_m. \tag{12.23}$$

IV. Finally, it can be established that each $v_i = p_{i-1}(A)v_1$, where p_{i-1} is a polynomial of degree $i - 1$.

Breakdown of the Arnoldi method. *The algorithm breaks down at step j if $\hat{v}$ at that step is a zero vector.* It can be shown that this will happen if and only if the degree of the minimal polynomial of v_1 is exactly j; that is, it is a combination of j eigenvectors. As indicated earlier, this is a *happy breakdown* for the eigenvalue problem. In this case, the subspace $K_j(A, v_1)$ is invariant and the approximate eigenvalues and eigenvectors are exact.

Restarted Arnoldi methods. The storage and computational costs of the Arnoldi method increase substantially as m increases. Note that for *m steps of the process, approximately $O(m^2 n)$ flops and $(mn + \frac{m^2}{2})n$ storage locations are required.* To overcome this difficulty, the Arnoldi method is usually restarted with a different starting vector (keeping m fixed) or by changing m dynamically by introducing a fixed variable m_1 (a small integer) such that the accuracy of the method is checked after every m_1 iterations of the Arnoldi method. Such variations of the Arnoldi method are usually called **restarted Arnoldi methods** and will be described in what follows in the context of applications of Arnoldi methods for solving linear systems. There are, however, other types of restarted Arnoldi methods (Saad (2003)).

12.3.2 Solving $Ax = b$ Using the Arnoldi Method

The Arnoldi method, described in the previous section, can be conveniently used to solve the *large and sparse linear* system $Ax = b$. We will describe two methods: a **Galerkin method** and a **minimal residual method.** The basic idea behind these methods is the same: both are **projection methods** and work as follows.

- Guess an initial approximation x_0 and compute the residual: $r_0 = b - Ax_0$.

- Find a correction vector z_m by solving an m-dimensional problem ($m \ll n$) such that
$$A(x_0 + z_m) = Ax_m \equiv b.$$

12.3. Krylov Subspace Methods for Linear Systems

How to Determine z_m. To determine z_m, the m steps of the Arnoldi method, starting with $v_1 = r_0/\|r_0\|_2$, can be run to generate the matrices H_m, V_m, and $\tilde{H}_m$, and then z_m is sought in the form
$$z_m = V_m y_m$$
for some $y_m \in \mathbb{R}^m$.

The two methods differ in the way the vector z_m is computed.

Let the Arnoldi method be started with $v_1 = r_0/\|r_0\|_2$, where $r_0 = b - Ax_0$ and x_0 is the initial solution. Then the following hold.

- In the *Galerkin method*, the residual vector $r_m = b - Ax_m$ is required to be orthogonal to $\mathcal{K}_m(A, r_0)$. This is equivalent to finding y_m by *solving an $m \times m$ Hessenberg system.* Specifically, the following $m \times m$ Hessenberg system is solved (see the discussion below):
$$H_m y_m = \|r_0\|_2 e_1. \tag{12.24}$$

- In the *generalized minimal residual* (GMRES) method, it is required that r_m be minimized. This is equivalent to finding y_m by solving an m-dimensional least-squares problem (see Theorem 12.12 below). Specifically, y_m minimizes
$$J(y) = \|e_1\|r_0\|_2 - \tilde{H}_m y\|_2. \tag{12.25}$$

The Galerkin Method

In the **Galerkin** (also know as the **Ritz–Galerkin**) method, it is required that *the residual vector $r_m = b - Ax_m$ be orthogonal to $\mathcal{K}_m(A, r_0)$.* Since $r_m = b - A(x_0 + z_m) = r_0 - AV_m y_m$, this condition gives
$$V_m^T(r_0 - AV_m y_m) = 0.$$
That is,
$$V_m^T AV_m y_m = V_m^T r_0.$$

Noting that $v_1 = r_0/\|r_0\|_2$, we can simplify the right-hand side as follows:
$$V_m^T r_0 = \begin{pmatrix} v_1^T \\ v_2^T \\ \vdots \\ v_m^T \end{pmatrix} r_0 = \begin{pmatrix} \frac{r_0^T r_0}{\|r_0\|_2} \\ v_2^T \|r_0\|_2 v_1 \\ \vdots \\ v_m^T \|r_0\|_2 v_1 \end{pmatrix} = \begin{pmatrix} \|r_0\|_2 \\ 0 \\ \vdots \\ 0 \end{pmatrix} = \|r_0\|_2 e_1.$$

(Note that $v_i^T v_1 = 0$, $i = 2, \ldots, m$.) Thus the *projected $m \times m$ system to be solved*, $V_m^T AV_m y_m = V_m^T r_0$, reduces to
$$H_m y_m = \|r_0\|_2 e_1,$$
which is an $m \times m$ Hessenberg system with the above special right-hand side.

This projected problem is now solved to obtain y_m; then the correction vector $z_m = V_m y_m$ is computed, and finally an improved solution vector $x_m = x_0 + z_m$ is obtained. The method is also known as the **full orthogonalization method** (FOM).

Restarting. If the corresponding residual vector $r_m = b - Ax_m$ is not small enough, then the process is restarted, setting $x_0 \equiv x_m$ and $r_0 \equiv r_m$. Other types of restarting methods, such as restarting with an increased value of m, are also possible. However, we will consider here restarting methods only of the first type.

Computing the residual. Since the residual r_m at every iteration (after the fixed m steps of the Arnoldi method) needs to be computed and checked for smallness to see if the method needs to be restarted, it is desirable that this can be computed cheaply. It turns out *this indeed can be done with information available only at the end of m steps of the Arnoldi method, and, in fact, this can be computed even before the next updating, as seen from the following.*

The residual vector r_m and its norm are given by the following theorem (Exercise 12.15).

Theorem 12.11 (residual by FOM). *Let x_0 be an initial approximation and let r_0 be the corresponding residual. If the starting vector in the Arnoldi method is taken as $v_1 = r_0/\|r_0\|_2$, then the residual vector r_m of the approximate solution x_m computed by the FOM and its norm are, respectively, given by*

$$r_m = b - Ax_m = -h_{m+1,m} e_m^T y_m v_{m+1} \tag{12.26}$$

and

$$\|r_m\|_2 = h_{m+1,m} |e_m^T y_m|. \tag{12.27}$$

ALGORITHM 12.2. An Explicitly Restarted Arnoldi Algorithm (FOM) for $Ax = b$ (Galerkin type).

Inputs: (i) A, an $n \times n$ matrix, (ii) m, a positive integer less than n, (iii) ϵ, the tolerance (>0), and (iv) x_0, an initial approximation.

Output: An approximate solution x_m such that the associated residual vector $r_m = b - Ax_m$ is orthogonal to $\mathcal{K}_m(A, r_0)$.

Step 0. Compute $r_0 = b - Ax_0$.
Step 1. Run m steps of the Arnoldi algorithm (Algorithm 12.1) to generate the matrices V_m, H_m, and $h_{m+1,m}$ using $v_1 = r_0/\|r_0\|_2$.
Step 2. Solve the $m \times m$ system $H_m y_m = \|r_0\|_2 e_1$
Step 3. Compute the correction vector z_m: $z_m = V_m y_m$.
Step 4. Compute the new solution vector: $x_m = x_0 + z_m$.
Step 5. Compute the norm of the new residual vector, $\|r_m\|_2 = h_{m+1,m}|e_m^T y_m|$. Stop if $\|r_m\| < \epsilon$ and accept x_m as the approximate solution.
Step 6. Compute $r_m = -h_{m+1,m} e_m^T y_m v_{m+1}$ and set $x_0 \equiv x_m$ and $r_0 \equiv r_m$. Return to Step 1.

12.3. Krylov Subspace Methods for Linear Systems

Numerical Experiment with Algorithm 12.2

Algorithm 12.2 was run with a sparse matrix of order 99 and fixed $m = 4$. Using MATLAB notation, we write

$$A = \text{diag}((1:99)) + \text{diag}(1, 98) + \text{diag}(1, -98),$$

which is a sparse matrix with entries $1, \ldots, 99$ on the diagonal and 1 on the top right and bottom left corners, with a suitably chosen right-hand side vector b.

Choose the initial approximation $x_0 = \begin{pmatrix} 0.1 & 0.1 & \ldots & \ldots & \ldots & 0.1 & 0.1 \end{pmatrix}^T$.

The norms of the residual vectors $b - Ax_i$, $i = 0, 1, \ldots, 20$, are shown in Table 12.3. The table shows that the restarted Arnoldi algorithm for $Ax = b$ converges as the number of iterations i increases.

Table 12.3. *Residual norms by FOM.*

i	$\|b - Ax_i\|_2$	i	$\|b - Ax_i\|_2$	i	$\|b - Ax_i\|_2$
0	515.875101	7	0.489985	14	0.055578
1	17.950331	8	0.337757	15	0.043757
2	4.739720	9	0.263640	16	0.030658
3	2.455344	10	0.183782	17	0.024151
4	1.380971	11	0.144290	18	0.016929
5	0.971805	12	0.100907	19	0.013341
6	0.641555	13	0.079375	20	0.009355

12.3.3 The GMRES Method for Solving $Ax = b$

We now present the other method, the minimized residual method, called the **generalized minimal residual method** (GMRES), developed by Saad and Schultz (1986).

Recall that for this method one seeks an approximate solution x_m of the form $x_0 + V_m y_m$ such that the norm of the residual vector $r_m = b - Ax_m$ is minimized by choosing y_m appropriately. Let e_1 be the *first unit $(m+1)$th vector*: $(1, 0, \ldots, 0)^T$.

Theorem 12.12. *Minimization of the residual norm is equivalent to solving the $m \times m$ least-squares problem: Minimize $J(y) = \| \|r_0\|_2 e_1 - \tilde{H}_m y\|_2$, where $r_0 = b - Ax_0$ and $\tilde{H}_m$ is the $(m+1) \times m$ Hessenberg matrix obtained by applying m steps of the Arnoldi method, starting with $v_1 = r/\|r_0\|_2$.*

Proof.

$$\begin{aligned} r_m &= b - Ax_m = b - A(x_0 + V_m y_m) \\ &= r_0 - AV_m y_m = r_0 - V_{m+1} \tilde{H}_m y_m \quad \text{(note that } AV_m = V_{m+1} \tilde{H}_m\text{)} \\ &= V_{m+1}(e_1 \|r_0\|_2 - \tilde{H}_m y_m) \quad \text{(note that } r_0 = v_1/\|r_0\|_2 = V_{m+1} e_1 \|r_0\|_2\text{)}. \end{aligned}$$

Since V_{m+1} has orthonormal columns, we obtain

$$\|r_m\|_2 = \| e_1 \|r_0\|_2 - \tilde{H}_m y_m\|_2.$$

Thus $\|r_m\|$ will be minimized if y_m is chosen so that

$$J(y) = \| \, \|r_0\|_2 e_1 - \tilde{H}_m y \|_2$$

is minimized over all $y \in \mathbb{R}^m$. □

The above least-squares problem can be solved using the QR factorization method described earlier (see Chapter 8). Once y_m is so obtained, the correction vector $z_m = V_m y_m$ and the new improved solution vector $x_m = x_0 + z_m$ can easily be obtained.

Restarting. If $\|r_m\|_2$ is not small enough, the process can be restarted by setting $x_0 \equiv x_m$ and $r_0 \equiv r_m$, the residual of the approximate solution x_m. But *for this restarting process to be practical, the residual and its norm have to be computed cheaply.* We now show how this can be done.

Computing the Residual and Its Norm Cheaply from Least-Squares Solution

Let the least-squares problem, minimize $\| \, \|r_0\|_2 e_1 - \tilde{H}_m y \|_2$, be solved using QR factorization method with Givens rotations. *Since this is a Hessenberg least-squares problem, the Givens method is ideal and a natural choice.*

Let $Q_m \tilde{H}_m = \tilde{R}_m$ (QR factorization of $\tilde{H}_m$), where $Q_m = J_m J_{m-1} \ldots J_1$; J_i is the Givens rotation in the ith and $(i+1)$th planes. Define

$$\tilde{g}_m = Q_m(\|r_0\|_2 e_1) = \begin{pmatrix} \gamma_1 \\ \gamma_2 \\ \vdots \\ \gamma_{m+1} \end{pmatrix}, \quad (12.28)$$

$R_m = $ the $m \times m$ matrix obtained from the $(m+1) \times m$ matrix (12.29)
$\tilde{R}_m$ by deleting the last row,

$g_m = $ the m-dimensional vector obtained from the $(m+1)$- (12.30)
dimensional vector $\tilde{g}_m$ by deleting its last component.

Then the following can be shown (Saad (2003, p. 169)).

- The vector y_m that minimizes $\| \, \|r_0\|_2 e_1 - \tilde{H}_m y_m \|$ is given by

$$R_m y_m = g_m. \quad (12.31)$$

- The residual vector r_m and its norm can be computed, respectively, as

$$r_m = b - A x_m = V_{m+1} Q_m^T (\gamma_{m+1} e_{m+1}) \quad (12.32)$$

and

$$\|r_m\|_2 = |\gamma_{m+1}|. \quad (12.33)$$

12.3. Krylov Subspace Methods for Linear Systems

ALGORITHM 12.3. An Explicitly Restarted GMRES Method for $Ax = b$.

Inputs: (i) A, an $n \times n$ matrix, (ii) m, a positive integer less than n, (iii) ϵ, the tolerance (> 0), and (iv) x_0, an initial approximation.

Output: An approximate solution x_m such that the associated residual vector $r_m = b - Ax_m$ satisfies $\|r_m\|_2 < \epsilon$.

Step 0. Compute $r_0 = b - Ax_0$.

Step 1. Run m steps of the Arnoldi method (Algorithm 12.1) to generate the $(m+1) \times m$ Hessenberg matrix $\tilde{H}_m$ and the orthonormal matrix V_{m+1}, using $v_1 = r_0/\|r_0\|_2$.

Step 2. Find the vector y_m such that the function
$$J(y) = \| \, \|r_0\|_2 e_1 - \tilde{H}_m y\|_2$$
is minimized over all vector $y \in \mathbb{R}^m$; $e_1 = (1, 0, \ldots, 0)^T \in \mathbb{R}^{m+1}$ by solving the $m \times m$ upper triangular system
$$R_m y_m = g_m,$$
where R_m and g_m are defined by (12.29)–(12.30).

Step 3. Compute the correction vector $z_m = V_m y_m$.

Step 4. Compute the new approximate solution $x_m = x_0 + z_m$.

Step 5. Compute the new residual norm $\|r_m\|_2 = |\gamma_{m+1}|$, where γ_{m+1} is given by (12.28). If $\|r_m\|_2 < \epsilon$, then stop and accept x_m as the approximate solution.

Step 6. Compute the new residual $r_m = b - Ax_m = V_{m+1}Q_m^T(\gamma_{m+1}e_{m+1})$. Set $x_0 \equiv x_m$ and $r_0 \equiv r_m$ and return to Step 1.

Remark. Since the residual at any substep j can be computed without computing the update x_j, one can stop early as soon as the residual norm is small enough. The readers is invited to develop this variation of the GMRES algorithm.

Breakdown of the GMRES method. Clearly, the GMRES method breaks down when the Arnoldi algorithm (Algorithm 12.1) stops at step j. If this happens, then the residual vector is zero; that is, the solution obtained at this step by GMRES is *exact*. The converse is also true. The following can be proved (see Saad (2003, p. 171)):

- If A is nonsingular, then the GMRES method breaks down at step j if and only if the approximate solution x_j is exact.

Convergence of the GMRES Method

The global convergence of the method has been proved only in the case when A is positive definite; that is, $\frac{A+A^T}{2}$ is symmetric positive definite. Two consequences of this result:

- If A is positive definite, then the GMRES converges for any $m \geq 1$.
- If A is not positive definite, then the GMRES may stagnate.

There exist results on the upper bound of the residual norm obtained after m steps of the GMRES method in case the matrix A is diagonalizable. These results involve $\text{Cond}_2(X)$ of the transforming matrix X that diagonalizes A and are thus useful only when this quantity is known in advance. For details, we refer the readers to Saad (2003, pp. 205–227). We present one such result here without proof.

An Error Bound. Let $X^{-1}AX = \text{diag}(\lambda_1, \ldots, \lambda_n)$. Then it can be shown (Greenbaum (1997, p. 54), Saad (2003, p. 206)) that

$$\|r_m\|_2 / \|r_0\|_2 \leq \text{Cond}_2(X) \min_{p_m} \max_{i=1,2,\ldots,n} |p_m(\lambda_i)|,$$

where the minimum is taken over all polynomials $p_m(x)$ of degree less than or equal to m with $p_m(0) = 1$.

Remarks.

- If A is normal, the above error bound is sharp. In this case, if the eigenvalues are clustered around a single point away from the origin, then there will be rapid convergence.

- In case A is not normal but has a reasonably well-conditioned X, the distribution of eigenvalues of A essentially determines the convergence behavior of GMRES.

- In general, however, it is not true that the convergence can be determined from the eigenvalue distribution alone. For example, eigenvalues clustered around 1 is not necessarily a favorable distribution for convergence in case A is a nonnormal matrix.

Choosing m. Unfortunately, there is no definite guideline for choosing m. If it is "too small," then there could be very slow convergence or no convergence at all. If m is "too large," then the storage and computational costs are prohibitive.

12.3.4 Solving Shifted Linear Systems Using the Arnoldi Method

An important observation is that the Arnoldi basis $\{v_1, \ldots, v_m\}$ *is invariant under a diagonal shift σ of A:* if we were to use $A - \sigma I$ instead of A, we would obtain the same sequence $\{v_1, \ldots, v_m\}$. This is because the Krylov subspace $K_m(A, v_1)$ is the same as $K_m(A - \sigma I, v_1)$. Note that from (12.20) we have

$$(A - \sigma I)V_m = V_m(H_m - \sigma I) + h_{m+1,m} v_{m+1} e_m^T, \qquad (12.34)$$

which means that if we run m steps of the Arnoldi method with the matrix $A - \sigma I$, we will obtain the same matrix V_m, but matrix H_m will have its diagonal shifted by σI.

A consequence of this is that for solving several linear systems of the form

$$(A - \mu_i I)x_i = b, \quad i = 1, 2, \ldots,$$

we might use the same information V_m and H_m, generated only once, to solve these systems approximately. See Datta and Saad (1991) and Saad (1987).

12.3.5 The Symmetric Lanczos Algorithm

In case matrix A is symmetric, the Arnoldi algorithm (Algorithm 12.1) becomes what is well known as the **symmetric Lanczos algorithm**, named after C. Lanczos (1952).[20]

In this case, the Hessenberg matrix H_m reduces to the *symmetric tridiagonal matrix T_m*, written as

$$T_m = \begin{pmatrix} \alpha_1 & \beta_1 & \cdots & 0 \\ \beta_1 & \alpha_2 & \ddots & \\ \vdots & \ddots & \ddots & \beta_{m-1} \\ 0 & & \beta_{m-1} & \alpha_m \end{pmatrix}.$$

The $(k+1)$th term recurrence (12.18) in the Arnoldi method now reduces to the well-known **three-term recurrence**:

$$Av_j = \alpha_j v_j + \beta_{j-1} v_{j-1}, +\beta_j v_{j+1}, \quad j = 1, 2, \ldots, m \quad (12.35)$$

(where we assume that $\beta_0 v_0 = 0$).

ALGORITHM 12.4. The Symmetric Lanczos Algorithm.

Inputs: (i) A symmetric $A \in \mathbb{R}^{n \times n}$, (ii) a vector v, and (iii) a positive integer $m < n$.
Outputs: (i) A set of orthonormal vectors $\{v_1, v_2, \ldots, v_{m+1}\}$, and (ii) the entries α_j and β_j of the *symmetric tridiagonal matrix T_m*.
Step 1. Set $v_0 = 0$, $\beta_0 = 0$, $v_1 = v/\|v\|_2$.
Step 2. For $j = 1, \ldots, m$ do
  2.1 $\hat{v}_{j+1} = Av_j - \beta_{j-1} v_{j-1}$.
  2.2 $\alpha_j = v_j^T \hat{v}_{j+1}$.
  2.3 $\hat{v}_{j+1} \equiv \hat{v}_{j+1} - \alpha_j v_j$.
  2.4 $\beta_j = \|\hat{v}_{j+1}\|_2$. If $\beta_j = 0$, stop.
  2.5 $v_{j+1} = \frac{\hat{v}_{j+1}}{\beta_j}$.
    End

The relations (12.17), (12.22), and (12.23), stated for the Arnoldi method, specialize to the case of the symmetric Lanczos method, respectively, as follows. Define $V_j = (v_1, v_2, \ldots, v_j)$.

I. $AV_m = V_{m+1} \tilde{T}_m$ (*symmetric Lanczos factorization*).

II. $K_m = V_m R_m$ (*QR factorization of the symmetric Lanczos–Krylov matrix*).

III. $V_m^T A V_m = T_m$.

---

[20]Cornelius Lanczos (1893–1974), a Hungarian physicist, after receiving his doctorate degree from Budapest Technical University in 1921, moved to Germany, where he worked as an assistant to Albert Einstein during the years 1928–1929. He worked as a Professor of Physics at Purdue University, as a scientist at Boeing and the National Bureau of Standards in the United States, and as Professor of Physics at the Dublin Institute for Advanced Study in Ireland. His paper "An iteration method for the solution of eigenvalue problems of linear differential and integral operations" (Lanczos (1950)) forms the basis of the much-studied *Lanczos methods*.

Loss of orthogonality. The Lanczos algorithm clearly breaks down when any of the β_j equal 0, which is a blessing in disguise for eigenvalue problems. We immediately obtain an invariant subspace. This, however, seldom happens in practice. In general, in the process there is always some loss of orthogonality, which takes place as soon as one eigenvalue converges.

In such cases, procedures such as *Lanczos with complete orthogonalization*, which produces the Lanczos vectors that are orthogonal to working precision, or *Lanczos with selective orthogonalization*, which is used to enforce orthogonality only in selective vectors, are used when needed. For details, we refer the reader to the well-known books on this subject by Parlett (1998) and Cullum and Willoughby (1985). Several papers of Paige (1970, 1971, 1976, 1980), whose pioneering work in the early 1970s rejuvenated the interests of the researchers in this area, are also very useful references. Procedures for complete and selective orthogonalization have been described in some detail in Golub and Van Loan (1996) and Parlett and Scott (1979).

MATCOM Note: Algorithm 12.4 has been implemented in the MATCOM program **LANSYM**.

Solving the Symmetric System Using the Lanczos Method

The symmetric Lanczos algorithm for solving the symmetric system $Ax = b$ can be used exactly in the same way as in the case of the Arnoldi method. In this case, the approximate solutions will be given by

$$x_m = x_0 + V_m y_m, \text{ where } y_m = T_m^{-1}(\|r_0\|_2 e_1). \tag{12.36}$$

12.3.6 The Conjugate Gradient Method

The method was originally devised by Hestenes and Stiefel (1952) and today is widely used to solve *large and sparse symmetric positive definite systems. It is direct in theory, but iterative in practice*. The method is a Krylov subspace method. In fact, *this method can be derived from the symmetric Lanczos method* (Exercise 12.20). We will, however, present this method as an optimization method and then display its connection with a Krylov subspace method. Our derivation is based on the following well-known result.

Theorem 12.13. *Let $A \in \mathbb{R}^{n \times n}$ be symmetric positive definite and let $b \in \mathbb{R}^{n \times 1}$. Define the quadratic function*

$$\phi(z) = \frac{1}{2} z^T A z - z^T b.$$

Then the minimizer z of $\phi(z)$ is the solution of $Ax = b$.

Proof.

$$\phi(z) = \frac{1}{2} z^T A z - z^T A x = \frac{1}{2} z^T A z - z^T A x + \frac{1}{2} x^T A x - \frac{1}{2} x^T A x \tag{12.37}$$

$$= \frac{1}{2}(z-x)^T A(z-x) - \frac{1}{2} x^T A x \quad \text{(note that } z^T A x = x^T A z\text{)}.$$

Since $-\frac{1}{2} x^T A x$ is a constant here, $\phi(z)$ will be minimized if $z = x$. $\square$

12.3. Krylov Subspace Methods for Linear Systems

There is a large number of iterative methods in the literature of optimization for solving this minimization problem (see Luenberger (1973) and Nocedal and Wright (2006)). In these iterative methods the successive approximations x_k are computed recursively:

$$x_{k+1} = x_k + \alpha_k p_k, \tag{12.38}$$

where the vectors $\{p_k\}$ are called the **direction vectors** and the scalars α_k are chosen to minimize $\phi(p)$ in the direction of p_k; that is, α_k is chosen to minimize the function $\phi_\alpha(x_k + \alpha p_k)$. Let $r_k = b - Ax_k$. It will now be shown how α_k and p_k are chosen in the conjugate gradient method.

The algorithm will be developed using the following facts (Exercise 12.18):

(i) The residual vectors $\{r_k\}$ are orthogonal: $r_k^T r_j = 0$ $(k > j)$.

(ii) The direction vectors are A-conjugate: $p_k^T A p_j = 0$ $(k > j)$.

Determining α_k. From (12.38), it follows that the residual vectors $\{r_k\}$ must satisfy the recurrence

$$r_{k+1} = r_k - \alpha_k A p_k. \tag{12.39}$$

Since $\{r_k\}$ are orthogonal, we have $r_k^T r_{k+1} = 0$; that is,

$$r_k^T (r_k - \alpha_k A p_k) = 0,$$

which gives

$$\alpha_k = \frac{r_k^T r_k}{r_k^T A p_k}.$$

Again, the direction vectors $\{p_k\}$ are updated using the residuals as

$$p_{k+1} = r_{k+1} + \beta_k p_k, \tag{12.40}$$

from which it follows that

$$r_k^T A p_k = (p_k - \beta_{k-1} p_{k-1})^T A p_k = p_k^T A p_k$$

(since $A p_k$ is orthogonal to p_{k-1}).

Thus, we have

$$\alpha_k = -\frac{r_k^T r_k}{p_k^T A p_k} = \frac{\|r_k\|_2^2}{p_k^T A p_k}. \tag{12.41}$$

Determining β_k. Since p_{k+1} is orthogonal to $A p_k$, we obtain from (12.40)

$$\beta_k = \frac{-(A p_k)^T r_{k+1}}{(A p_k)^T p_k}.$$

Again, from (12.39), we have $A p_k = -\frac{1}{\alpha_k}(r_{k+1} - r_k)$.

Thus, substituting the value of α_k from (12.41) and noting that $r_k^T r_{k+1} = 0$, we get

$$\beta_k = \frac{1}{\alpha_k} \frac{((r_{k+1} - r_k)^T r_{k+1})}{p_k^T A p_k} = \frac{r_{k+1}^T r_{k+1}}{r_k^T r_k}$$
$$= \frac{\|r_{k+1}\|_2^2}{\|r_k\|_2^2}. \tag{12.42}$$

ALGORITHM 12.5. The Classical Conjugate Gradient Algorithm (CG Algorithm).

Inputs: $A \in \mathbb{R}^{n \times n}$, symmetric positive definite; $b \in \mathbb{R}^{n \times 1}$.

Output: An approximate solution x of $Ax = b$

Step 1. Choose an initial approximation x_0 and a tolerance ϵ. Set $p_0 = r_0 = b - Ax_0$.

Step 2. For $i = 0, 1, 2, 3, \ldots$ do
    2.1 $w = Ap_i$.
    2.2 Compute the step length: $\alpha_i = \|r_i\|_2^2 / p_i^T w$.
    2.3 Update the iterates: $x_{i+1} = x_i + \alpha_i p_i$.
    2.4 Update the residuals: $r_{i+1} = r_i - \alpha_i w$.
    2.5 Test for convergence: If $\|r_{i+1}\|_2^2 \geq \epsilon$, continue.
    2.6 Compute $\beta_i = \frac{\|r_{i+1}\|_2^2}{\|r_i\|_2^2}$.
    2.7 Update the direction vectors: $p_{i+1} = r_{i+1} + \beta_i p_i$.
End

Example 12.14.

$$A = \begin{pmatrix} 5 & 1 & 1 \\ 1 & 5 & 1 \\ 1 & 1 & 5 \end{pmatrix}, \qquad b = \begin{pmatrix} 7 \\ 7 \\ 7 \end{pmatrix},$$

$$x^{(0)} = (0, 0, 0)^T, \qquad p_0 = r_0 = b - Ax^{(0)} = \begin{pmatrix} 7 \\ 7 \\ 7 \end{pmatrix}.$$

$i = 0$:

$$w = Ap_0 = \begin{pmatrix} 49 \\ 49 \\ 49 \end{pmatrix}, \qquad \alpha_0 = \frac{\|r_0\|_2^2}{p_0^T \omega} = 0.1429,$$

$$x_1 = x_0 + \alpha_0 p_0 = \begin{pmatrix} 1.0003 \\ 1.0003 \\ 1.0003 \end{pmatrix}, \qquad r_1 = r_0 - \alpha_0 w = \begin{pmatrix} -0.0021 \\ -0.0021 \\ -0.0021 \end{pmatrix},$$

$$\beta_0 = 9 \times 10^{-8}, \qquad p_1 = r_1 + \beta_0 p_0 = \begin{pmatrix} -0.0021 \\ -0.0021 \\ -0.0021 \end{pmatrix}.$$

$i = 1$:

$$w = Ap_1 = \begin{pmatrix} -0.0147 \\ -0.0147 \\ -0.0147 \end{pmatrix}, \quad \alpha_1 = 0.1429, \quad x_2 = x_1 + \alpha_1 p_1 = \begin{pmatrix} 1.0000 \\ 1.0000 \\ 1.0000 \end{pmatrix}. \quad \blacksquare$$

MATCOM Note: Algorithm 12.5 has been implemented in the MATCOM program **CONGRAD**.

12.3. Krylov Subspace Methods for Linear Systems

Krylov subspace properties of the CG iteration. From the description of the CG algorithm and the discussions preceding it, the following can be proved (Exercise 12.10):

- *(Conjugate iterates span the Krylov subspace.)*

$$\begin{aligned}\mathcal{K}_k &= \text{span}\,\{x_1, x_2, \ldots, x_k\} \\ &= \text{span}\,\{p_0, p_1, \ldots, p_{k-1}\} = \text{span}\,\{r_0, r_1, \ldots, r_{k-1}\} \\ &= \text{span}\,\{b, Ab, \ldots, A^{k-1}b\}.\end{aligned} \qquad (12.43)$$

Convergence of the CG Method

In the absence of round-off errors the CG gradient method should converge in no more than n iterations as the following theorem shows.

Theorem 12.15. *The CG algorithm converges in no more than n steps.*

Proof. We know that r_n is orthogonal to $r_0, r_1, \ldots, r_{n-1}$. Again, from the above Krylov subspace identities (12.43), we have that $r_0, \ldots, r_{n-1}$ form a basis of $\mathbb{R}^n$. Since r_n is orthogonal to this entire basis, we conclude that $r_n = 0$, that is, $e_n = 0$, which means that $x_n = x$. □

Minimizing of the A-Norm Error

Minimizing $\phi(z)$ is equivalent to minimizing the A-norm of the error, as shown below.
  Define the function $\|\cdot\|_A$ by

$$\|x\|_A = \sqrt{x^T A x}. \qquad (12.44)$$

Then it can be verified (Exercise 12.4) that this function is a norm on $\mathbb{R}^n$ and is called the **A-norm**.
  Define the error $e = z - x$. Then from (12.37) we have

$$\phi(z) = \frac{1}{2}\|e\|_A^2 - \frac{1}{2}\|x\|_A^2.$$

Since $\frac{1}{2}\|x\|_A^2$ is constant, *minimizing $\phi(z)$ is equivalent to minimizing $\|e\|_A$.*

Rate of Convergence of the CG Method

The rate of convergence of the CG method is determined by the distribution of the eigenvalues of A. The following is an important result in this context (for a proof, see Greenbaum (1997, pp. 50–51)).

Theorem 12.16 (error bound for CG). *The error $e_k = x - x_k$ at the kth iteration is related to the initial error $e_0 = x - x_0$ as*

$$\|e_k\|_A / \|e_0\|_A \leq \min_{p_k} \max_{i=1,\ldots,n} |p_k(\lambda_i)|,$$

where $\lambda_1, \ldots, \lambda_n$ are the eigenvalues of A and the minimum is taken over all polynomials $p_k(x)$ of degree less than or equal to k with $p_k(0) = 1$.

As a corollary of Theorem 12.16, we also obtain the following.

Corollary 12.17. *If A has k distinct eigenvalues, then the CG method converges in at most k steps.*

Proof. Let $p_k(x) = \prod_{i=1}^{k} \left(1 - \frac{x}{\lambda_i}\right)$. Then $p_k(x) = 0$ at $x = \lambda_i, i = 1, \ldots, k$. □

The following well-known result shows how the *ratio of the largest eigenvalue to the smallest influences the rate of convergence* when nothing is known about the clustering. A proof of Theorem 12.18 can be found in Greenbaum (1997, pp. 51–52).

Theorem 12.18.
$$\|x_k - x\|_A \leq 2\alpha^k \|x_0 - x\|_A$$

or

$$\frac{\|e_k\|_A}{\|e_0\|_A} \leq 2\alpha^k,$$

where

$$\alpha = (\sqrt{\kappa} - 1)/(\sqrt{\kappa} + 1) \quad \text{and} \quad \kappa = \text{Cond}_2(A) = \|A\|_2 \|A^{-1}\|_2 = \lambda_n/\lambda_1.$$

Here λ_n and λ_1 are the largest and smallest eigenvalues of the symmetric positive definite matrix A (note that the eigenvalues of A are all positive).

Note: $\alpha = 0$ when $\text{Cond}(A) = 1$. When $\alpha \to 1$, $\text{Cond}(A) \to \infty$. Thus, *the larger $\text{Cond}(A)$ is, the slower the rate of convergence.*

12.3.7 Solving Indefinite Symmetric Systems Using CG-type Methods: MINRES and SYMMLQ.

The CG method was derived under the assumption that matrix A is symmetric and positive definite. In fact, the *positive definiteness of A ensures the minimization property of the CG.* In case A is symmetric indefinite, the minimization property can no longer be ensured. In such a case, two well-known alternatives, due to Paige and Saunders (1975), are the **MINRES** and **SYMMLQ** methods.

MINRES: MINRES (**minimum residual method**) aims at minimizing $\|Ax_k - b\|_2$ by extracting information from the symmetric Lanczos algorithm. It can be shown that

$$\|Ax_k - b\|_2 = \|D_{k+1}\tilde{T}_k y_k - \|r_0\|_2 e_1\|_2,$$

where $D_{k+1} = \text{diag}(\|r_0\|_2, \|r_1\|_2, \ldots, \|r_k\|_2)$, $\tilde{T}_k$ is the $(k+1) \times k$ tridiagonal matrix obtained after k steps of the symmetric Lanczos method, and y_k is the solution of the projected problem obtained after k steps of the Lanczos method that is, y_k satisfies

$$T_k y_k = \|r_0\|_2 e_1.$$

The above then is a minimum-norm least-squares problem and can be solved using Givens rotations, as in the case of the GMRES method.

12.3. Krylov Subspace Methods for Linear Systems

Theorem 12.19 (error bound for MINRES).

$$\|r_k\|/\|r_0\| \leq \min_{p_k} \max_{i=1,\ldots,n} \|p_k(\lambda_i)\|,$$

where p_k and λ are the same as in Theorem 12.16.

For a proof, see Greenbaum (1997, pp. 50–51).

Implications of Theorems 12.16 and 12.19 for the CG and MINRES methods. From the above theorems it follows that a favorable distribution of eigenvalues is one in which the polynomials p_k are small. This will happen, for example, if the *eigenvalues are tightly clustered around a single point c away from the origin*. This is because in this case the kth degree polynomial $p_k(z) = (1 - \frac{z}{c})^k$ at points close to c is small in magnitude and $p_k(0) = 1$. Similarly, an example of an *unfavorable distribution is one in which the eigenvalues are well separated, especially if they lie on both sides of the origin, because, in this case, it is difficult to find a low-order polynomial with its value 1 at the origin and which is small at a large number of points* (see Greenbaum (1997)). See the results of numerical experiments below for illustrations (Figures 12.2 and 12.3).

Comparison of CG and MINRES with two examples, one with favorable and another with not-so-favorable eigenvalue distributions.

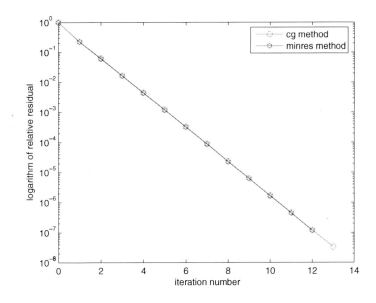

Figure 12.2. *Comparison of CG and MINRES on a diagonal matrix of order 1000×1000 with eigenvalues distributed in* $[0.5, 1.5]$.

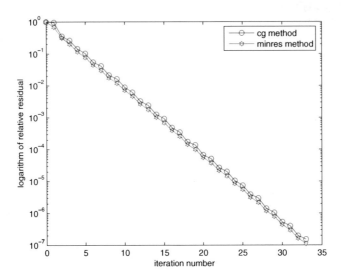

Figure 12.3. *Comparison of CG and MINRES on a* 1000×1000 *diagonal matrix with eigenvalues distributed in* $[-2.5, -1.5]$ *and* $[0.5, 1.5]$.

SYMMLQ: The SYMMLQ (**symmetric LQ method**) is based upon solving the symmetric tridiagonal system T_k by using an LQ decomposition. For details, see Paige and Saunders (1975).

12.3.8 The Nonsymmetric Lanczos Method

The nonsymmetric Lanczos algorithm (also known as the *two-sided Lanczos* algorithm) aims at transforming A into a nonsymmetric tridiagonal matrix T_m rather than a Hessenberg matrix, as is done by the Arnoldi method. However, having insisted on obtaining a tridiagonal matrix, we must give up the orthogonality of the vectors $\{v_i\}$. Instead, one computes two sets of **biorthogonal vectors** $\{v_1, \ldots, v_{m+1}\}$, $\{w_1, \ldots, w_{m+1}\}$ (that is, $v_i^T w_j = 0$, $i \neq j$, and $v_i^T w_i = 1$) by using a **three-term recurrence** in place of $(k+1)$th-term recurrence, as needed for the Arnoldi method. If the tridiagonal matrix T_m is given by

$$T_m = \begin{pmatrix} \alpha_1 & \beta_2 & & 0 \\ \delta_2 & \ddots & \ddots & \\ & \ddots & \ddots & \beta_m \\ 0 & & \delta_m & \alpha_m \end{pmatrix} \tag{12.45}$$

and δ_{m+1} and β_{m+1} are two scalars, then the *three-term recurrences* for generating $\{v_i\}_{i=1}^{m+1}$, $\{w_i\}_{i=1}^{m+1}$ (satisfying the biorthogonality relations) are

$$\begin{aligned} Av_k &= \alpha_k v_k + \beta_k v_{k-1} + \delta_{k+1} v_{k+1}, \\ A^T w_k &= \alpha_k w_k + \delta_k w_{k-1} + \beta_{k+1}, w_{k+1}. \end{aligned} \tag{12.46}$$

The reader is invited to fill in the details.

12.3. Krylov Subspace Methods for Linear Systems

ALGORITHM 12.6. The Nonsymmetric (Two-Sided) Lanczos Algorithm.

Inputs: $A \in \mathbb{R}^{n \times n}$, $v \in \mathbb{R}^{n \times 1}$, $w \in \mathbb{R}^{n \times 1}$, and m, a positive integer less than n.
Outputs: The set of vectors $\{v_1, \ldots, v_{m+1}\}$ and $\{w_1, \ldots, w_{m+1}\}$ such that $w_i^T v_j = 0$, $i \neq j$, $i \geq 1$, $j \leq m$, $w_i^T v_i = 1$. T_m is an $m \times m$ *nonsymmetric tridiagonal matrix* as defined above.

Step 0. Scale the vectors v and w to get the vectors v_1 and w_1 such that $w_1^T v_1 = 1$. Set $\beta_1 \equiv 0$, $\delta_1 \equiv 0$, $w_0 \equiv v_0 \equiv 0$.

Step 1. For $j = 1, 2, \ldots, m$ do

$$\alpha_j = w_j^T A v_j;$$
$$\hat{v}_{j+1} = A v_j - \alpha_j v_j - \beta_j v_{j-1};$$
$$\hat{w}_{j+1} = A^T w_j - \alpha_j w_j - \delta_j w_{j-1};$$
$$\delta_{j+1} = \sqrt{|\hat{w}_{j+1}^T \hat{v}_{j+1}|}; \quad \text{if } \delta_{j+1} = 0, \text{ stop};$$
$$\beta_{j+1} = \hat{w}_{j+1}^T \hat{v}_{j+1} / \delta_{j+1};$$
$$w_{j+1} = \hat{w}_{j+1} / \beta_{j+1};$$
$$v_{j+1} = \hat{v}_{j+1} / \delta_{j+1}.$$

End

The Nonsymmetric Lanczos Relations

- The vectors $\{v_i\}$ and $\{w_i\}$ form *biorthogonal bases* for the subspaces $\mathcal{K}_m(A, v_1)$ and $\mathcal{K}_m(A^T, w_1)$, respectively (provided that the algorithm does not break down before m steps).

- From the biorthogonality relations, it follows immediately that

$$V_m^T W_m = W_m^T V_m = I, \tag{12.47}$$

  where $V_m = (v_1, \ldots, v_m)$ and $W_m = (w_1, w_2, \ldots, w_m)$.

- Furthermore,

$$W_m^T A V_m = T_m \quad (\text{a } \textit{nonsymmetric tridiagonal matrix}) \tag{12.48}$$

  and

$$\left. \begin{array}{l} A V_m = V_m T_m + \delta_{m+1} v_{m+1} e_m^T, \\ A^T W_m = W_m T_m^T + \beta_{m+1} w_{m+1} e_m^T \end{array} \right\} \quad \begin{array}{l} \textit{(nonsymmetric Lanczos} \\ \textit{factorizations)}. \end{array} \tag{12.49}$$

For proofs of the above relations, see Saad (2003, pp. 218–229).

Breakdown of the Lanczos Method.

It is clear that Algorithm 12.6 will break down if $\hat{w}_{j+1}^T \hat{v}_{j+1} = 0$ for some j. In can happen either when (i) one of these vectors is zero or (ii) they are both nonzero, but their inner product is zero. In the first case, if $\hat{v}_{j+1} = 0$, then the approximate solution to the system $Ax = b$ is exact. If $\hat{w}_{j+1} = 0$, then we also have an invariant subspace with the vectors $\{w_k\}$, and the approximate solution is exact for the dual system; however, nothing can be said about the approximate solution of $Ax = b$.

In the second case, we have a "serious breakdown." A cure for this problem is to use the **look-ahead Lanczos method**. *The idea behind this is to continue to the next step even if there is a breakdown at the current step.*

We refer the reader to the papers of Parlett, Taylor, and Liu (1985), Freund, Gutknecht, and Nachtigal (1993), and Brezinski, Zagila, and Sadok (1991) for theory and implementations of the look-ahead Lanczos method. The look-ahead Lanczos algorithm is implemented in the software package QMRPACK (Lehoucq, Sorensen, and Yang (1998)).

12.3.9 Solving Linear System $Ax = b$ Using the Lanczos Algorithm

As in the Arnoldi method, the Lanczos vectors and the nonsymmetric tridiagonal matrix T_m can be used to solve $Ax = b$.

Step 1. Starting with x_0, $v_1 = r_0/\|r_0\|_2$, and an arbitrary vector w_1 such that $w_1^T v_1 = 1$, run m steps of the Lanczos algorithm (Algorithm 12.6) to generate the Lanczos vectors $\{v_1, \ldots, v_m, v_{m+1}\}$, $\{w_1, w_2, \ldots, w_m\}$, the tridiagonal matrix T_m, and δ_{m+1}.

Step 2. Solve the tridiagonal system

$$T_m y_m = \beta e_1, \quad \text{where } \beta = \|r_0\|_2. \tag{12.50}$$

Step 3. Compute the new approximation $x_m = x_0 + V_m y_m$, where $V_m = (v_1, v_2, \ldots, v_m)$.

Note that if one has to solve a dual system with A^T as well, this algorithm is quite suitable.

Restarting. As in the FOM case, the method can be restarted after every m steps, and the norm of the residual vector, needed for a convergence test, can be cheaply computed as (Saad (2003, p. 222))

$$\|b - Ax_j\|_2 = |\delta_{j+1} e_j^T y_j| \, \|v_{j+1}\|_2. \tag{12.51}$$

12.3.10 The Bi-conjugate Gradient Algorithm

For nonsymmetric systems, the CG method is not suitable. However, a CG-type method called the **bi-conjugate gradient method (Bi-CG)** can be developed for such systems, based on the nonsymmetric Lanczos algorithm by using the LU factorization of the tridiagonal matrix T_m (Exercise 12.20). The Bi-CG algorithm is a projection process onto the Krylov subspace $\mathcal{K}_m = \text{span } \{v_1, Av_1, A^2v_1, \ldots, A^{m-1}v_1\}$ orthogonal to $\mathcal{K}_m(A^T, w_1) = \text{span}\{w_1, A^T w_1, \ldots, (A^T)^{m-1}w_1\}$, taking $v_1 = r_0\|r_0\|_2$ and w_1 such that $w_1^T v_1 \neq 0$. We will, however, describe the algorithm here in a manner analogous to the CG method. The major difference between the two methods is that in the Bi-CG method, unlike the CG method, the two sets of residuals $\{r_i\}$ and $\{\tilde{r}_i\}$ are produced which are **biorthogonal.** The set $\{\tilde{r}_i\}$ is obtained by using A^T rather than A. Similarly, the two sets of direction vectors $\{p_i\}$ and $\{\tilde{p}_i\}$ are produced from the residuals, which are mutually A-conjugate.

12.3. Krylov Subspace Methods for Linear Systems

These two sets of the residuals and the directions are, respectively, given by

$$r_{k+1} = r_k - \alpha_k A p_k, \quad \tilde{r}_{k+1} = \tilde{r}_k - \alpha_k A^T \tilde{p}_k \tag{12.52}$$

and

$$p_{k+1} = r_{k+1} + \beta_k p_k, \quad \tilde{p}_{k+1} = \tilde{r}_{k+1} + \beta_k \tilde{p}_k. \tag{12.53}$$

The choices

$$\alpha_k = \tilde{r}_k^T r_k / \tilde{p}_k^T A p_k; \quad \beta_k = \tilde{r}_{k+1}^T r_{k+1} / \tilde{r}_k^T r_k \tag{12.54}$$

will yield the **biorthogonality** and **conjugacy** relations

$$\tilde{r}_i^T r_j = \tilde{p}_i^T A p_j = 0, \quad i \neq j. \tag{12.55}$$

ALGORITHM 12.7. The Bi-CG Algorithm for $Ax = b$.

Input: $A \in \mathbb{R}^{n \times n}$, $b \in \mathbb{R}^{n \times 1}$, and x_0, an initial approximate solution.
Output: Approximations $\{x_k\}$ of the solution x.

Step 0. Compute $r_0 = b - Ax_0$.
Choose $\tilde{r}_0$ such that $\tilde{r}_0^T r_0 \neq 0$.

Step 1. Set $p_0 = r_0$, $\tilde{p}_0 = \tilde{r}_0$.

Step 2. For $j = 0, 1, \ldots$, do until convergence
  2.1. Compute the step length $\alpha_j = \tilde{r}_j^T r_j / \tilde{p}_j^T A p_j$.
  2.2. Update the iterates $x_{j+1} = x_j + \alpha_j p_j$.
  2.3. Update the residuals $r_{j+1} = r_j - \alpha_j A p_j$.
  2.4. Update the dual residuals $\tilde{r}_{j+1} = \tilde{r}_j - \alpha_j A^T \tilde{p}_j$.
  2.5. Compute $\beta_j = \tilde{r}_{j+1}^T r_{j+1} / \tilde{r}_j^T r_j$.
  2.6. Update the direction vectors $p_{j+1} = r_{j+1} + \beta_j p_j$.
  2.7. Update the dual direction vectors $\tilde{p}_{j+} = \tilde{r}_{j+} + \beta_j \tilde{p}_j$.
End

Orthogonal properties. As noted before, the residual and direction vectors produced by the Bi-CG algorithm have the following properties (Exercise 12.22): $\tilde{r}_i^T r_j = 0$, $i \neq j$; $\tilde{p}_i^T A p_j = 0$, $i \neq j$.

Notes: (i) *(Relationship between CG and Bi-CG methods.)* The Bi-CG method produces the same iterates as CG if applied to the symmetric positive definite matrices.

(ii) *(Variants of Bi-CG.)* There now exist several variants of the Bi-CG method. These include the **Bi-CGSTAB (bi-conjugate stabilized)** (van der Vorst (1992, 1996, 2003)) and **CGS (conjugate gradient squared)** (Sonneveld (1989)) methods. *These two methods* *avoid computations with A^T, as required by the Bi-CG method.* The CGS method often converges much faster than the Bi-CG, but might have irregular convergence patterns. The Bi-CGSTAB method avoids this problem and at the same time maintains the same speed of convergence. Van der Vorst (1992) calls it "a more smoothly convergent variant of CGS." See Saad (2003) and van der Vorst (2003) and for details. Avoiding breakdown in the CGS method has been discussed in Brezinski and Sadok (1991).

(iii) *(GMRES versus Bi-CG.)* The GMRES method generates the smallest residual over the current search space, while the Bi-CG method does not minimize the residual in any suitable norm. The GMRES method, however, does so at the cost of longer recurrences than the Bi-CG method.

12.3.11 The QMR Algorithm

It is natural to wonder if there exists a GMRES analogue based on the nonsymmetric tridiagonal Lanczos reduction. Indeed, the **quasi-minimal residual** (QMR) is such a method. The method was developed by Freund and Nachtigal in an award-winning paper in 1991 (Freund and Nachtigal (1991)). The idea here is to minimize the norm of the residual $\|b - Ax\|_2$ using the Lanczos reduction. In this case, it can be shown (Exercise 12.17) that after m steps of the nonsymmetric Lanczos method, the residual is given by

$$r_m = b - Ax_m = V_{m+1}(\beta e_1 - \tilde{T}_m y_m), \qquad (12.56)$$

where

$$\tilde{T}_m = \begin{pmatrix} T_m \\ \delta_{m+1} e_m^T \end{pmatrix} \quad \text{and} \quad \beta = \|r_0\|_2.$$

Thus,

$$\|b - Ax_m\|_2 = \|V_{m+1}(\beta e_1 - \tilde{T}_m y_m)\|_2 \qquad (12.57)$$

Since the columns of V_{m+1} are not orthonormal in the Lanczos algorithm, *we do not have* $\|b - Ax\|_2 = \|\beta e_1 - \tilde{T}_m y\|_2$. However, the *QMR algorithm is obtained by minimizing* $\|\beta e_1 - \tilde{T}_m y\|_2$ *over y* anyway, and then the new approximation is obtained as $x_m = x_0 + V_m y_m$, where y_m is the solution of the above least-squares problem. This is why the method is referred to as the quasi-minimal residual method. The reader is invited to write the QMR algorithm in algorithmic form, as was done for GMRES (Exercise 12.21).

Remarks. (i) In actual implementations, the QMR method is implemented using the *lookahead variant of the Lanczos method* to deal with the breakdown. See Freund and Nachtigal (1991) for details.

(ii) *(Residual norm.)* The residual norm of the approximate solution x_m satisfies

$$\|b - Ax_m\|_2 \leq \|V_{m+1}\| \, |\gamma_{m+1}|,$$

where γ_{m+1}, as in the case of GMRES, is the last component of the vector $Q_m(\beta e_1)$; Q_m^T is the Q matrix of the QR factorization of $\tilde{T}_m$.

(iii) *(Relationship between QMR and GMRES residuals.)* Let r_m^Q and r_m^G be the residual norms after m steps of QMR and GMRES, respectively; then

$$\|r_m^Q\| \leq \text{Cond}_2(V_{m+1})\|r_m^G\|_2.$$

For proofs, see Saad (2003, p. 226).

(iv) *(Relationship between QMR and Bi-CG methods.)* The QMR method avoids the breakdown of the Bi-CG method using the look-ahead Lanczos idea. Thus, when the Bi-CG method temporarily stagnates, the QMR may still work (though slowly).

(v) A transpose-free QMR algorithm, called **TFQMR**, was developed by Freund (1993) from the CGS algorithm. For details, see Freund (1993) and Saad (2003, pp. 234–239).

(vi) *(Mixed Bi-CGSTAB–CGS method.)* A new Krylov subspace method, called the mixed the *Bi-CGSTAB–CGS method*, was developed by Chan and Ye (1997). The idea is to combine these two methods by switching from one to the other at each iteration step so that the stability of the CGS can be improved.

12.4 Preconditioners

For making the GMRES-type methods practically viable, it is almost mandatory to use a preconditioner so that the preconditioned system has a better spectral property.

For the idea to be effective, the preconditioner M should be such that

- it is a reasonably good approximation to the original matrix,

- its construction is not too expensive, and

- the preconditioned system is easier to solve than the original system.

If the preconditioner M is used to solve the preconditioned system

$$M^{-1}Ax = M^{-1}b,$$

then M is called a **left preconditioner.**

The Krylov subspace method in this case will construct an orthonormal basis of the Krylov subspace $\mathcal{K}(M^{-1}A, r_0) = \text{span}\{r_0, M^{-1}Ar_0, \ldots, (M^{-1}A)^{m-1}r_0\}$, where $r_0 = M^{-1}(b - Ax_0)$ and $v_1 = r_0/\|r_0\|_2$.

The computed residuals and their norms will be those of the preconditioned system rather than those of the original system.

M is called a **right preconditioner** if M is such that it solves

$$AM^{-1}z = b, \ z = Mx.$$

A compromise between a left and right preconditioner is to have a preconditioner M that can be factorized into $M = M_1 M_2$, resulting in the two successive systems:

$$M_1^{-1}AM_2^{-1}z = M_1^{-1}b \quad \text{and} \quad x = M_2^{-1}z.$$

In this case the matrix $\tilde{A} = M_1^{-1}AM_2^{-1}$ *should be as close as possible to the identity matrix.* Such a preconditioner is called a **two-sided preconditioner.**

There is now a wide range of preconditioners available in the literature. Some of the most common ones are stated below.

12.4.1 Classical Iterative Methods as Preconditioners

Here the idea is to use the iteration matrices that we encountered in developing Jacobi, Gauss–Seidel, and SOR methods as preconditioners. Recall that these methods are based on the iteration of the form $x_{k+1} = Bx_k + d$.

Suppose A has the following splitting: $A = M + N$. Then the left preconditioned system $M^{-1}Ax = M^{-1}b$ can be written as

$$x = -Bx + d, \quad \text{where } B = -M^{-1}N = I - M^{-1}A, \ d = M^{-1}b.$$

Jacobi and block Jacobi preconditioner. Recall that

$$\begin{aligned} B_J &= -D^{-1}(L+U) = -D^{-1}(A-D) \\ &= I - D^{-1}A \ (\text{since } A = L + D + U). \end{aligned}$$

Thus, we can choose D as the preconditioner M, denoted by M_{Jacob}:

$$M_{\text{Jacob}} = D = \text{diag}(a_{11}, a_{22}, \ldots, a_{nn}), \quad \text{provided that } a_{ii} \neq 0.$$

This is the simplest possible preconditioner and is known as the **Jacobi preconditioner.** If A is a **block matrix** $A = (A_{ij})_{k \times k}$, where A_{ii} are square, then the block matrix

$$M = \text{diag}(A_{11}, A_{22}, \ldots, A_{kk})$$

can be taken as a preconditioner, known as the **block Jacobi preconditioner.** The *block Jacobi preconditioners are suitable for structured linear systems arising from solutions of partial differential equations on regular grids.*

M_{SGS} and SSOR preconditioner. Since the splitting matrix M for SOR is not symmetric, a better preconditioner can be derived for a *symmetric matrix A* from SSOR iteration as

$$M_{\text{SSOR}} = \frac{1}{\omega(2-\omega)}(D + \omega L)D^{-1}(D + \omega L^T),$$

which is known as the **SSOR preconditioner.** The **symmetric Gauss–Seidel preconditioner** is then obtained as (taking $\omega = 1$)

$$M_{\text{SGS}} = (D+L)D^{-1}(D+L^T).$$

12.4.2 Polynomial Preconditioners

The idea here is to find a low-degree *polynomial matrix* of order, say, p, $p_p(A)$ (with better properties), so that the iterative method can be applied to $p_p(A)Ax = p_p(A)b$.

If such a polynomial can be found, then the preconditioner M can be defined by $M^{-1} = p_p(A)$.

One way to find a polynomial preconditioner is to use the low-order terms of the Neuman series of $(I-B)^{-1}$ if A is written as $A = I - B$, and when the series converges; that is, when $\rho(B)$, the spectral radius of B, is less than 1.

12.4. Preconditioners

More general polynomial preconditioners, involving the **shifted Chebyshev polynomials**, have been developed and are currently being used. For details, see Saad (2003, Chapter 10).

12.4.3 Incomplete LU (ILU) Factorization as a Preconditioner

The basic idea here is to compute the LU factorization of a sparse matrix, allowing fill-in only in certain positions, for example, computing the corresponding entries of the L and U matrices only when $a_{ij} \neq 0$ and leaving the zero entries of A in their places. We shall discuss this type of preconditioner in case A is symmetric positive definite in the context of the CG Method. Details can be found in Saad (2003, pp. 287–320). The stability of incomplete LU (ILU) factorization has been analyzed by Elman (1986). Bank and Wagner (1999) have discussed multilevel ILU decomposition.

12.4.4 Preconditioning with Incomplete Cholesky Factorization

Let A be symmetric positive definite; then a preconditioner M of type $M = (\tilde{L}\tilde{L}^T)$ will be a suitable candidate for the CG method. A natural choice for such an $\tilde{L}$ will be the *incomplete Cholesky factor* (a special case of the ILU preconditioner discussed in the last section). *Mathematically, this factorization is equivalent* to $A = \tilde{L}\tilde{L}^T + R$, where $R \neq 0$.

To generate this $\tilde{L}$, we use the **Cholesky factorization** of $A = \tilde{L}\tilde{L}^T$, as follows: If $a_{ij} = 0$, set $l_{ij} = 0$; otherwise calculate the appropriate l_{ij}.

ALGORITHM 12.8. Incomplete Cholesky Factorization.

Input: $A \in \mathbb{R}^{n \times n}$, large and sparse, and symmetric positive definite.
Output: The incomplete Cholesky factor $\tilde{L} = (l_{ij})$ of A.

Set $\ell_{11} = \sqrt{a_{11}}$

For $i = 1, 2, \ldots, n$ do
    For $j = 1, 2, \ldots, i - 1$ do
        If $a_{ij} = 0$, then $l_{ij} = 0$ else
            $\ell_{ij} = \frac{1}{\ell_{jj}}(a_{ij} - \sum_{k=1}^{j-1} \ell_{ik}\ell_{jk})$
    End
$\ell_{ii} = \sqrt{(a_{ii} - \sum_{k=1}^{i-1} l_{ik}^2)}$
End

Remark. Algorithm 12.8 requires computation of square roots. However, one can obtain a no-fill incomplete LDL^T factorization of A that avoids square root computations (Exercise 12.14).

MATCOM Notes: Algorithm 12.8 has been implemented in the MATCOM program **ICHOLES**. The no-fill incomplete LDL^T factorization algorithm has been implemented in the MATCOM program **NICHOL**.

> **ALGORITHM 12.9. The Preconditioned Conjugate Gradient Method.**
>
> **Inputs:** Same as in Algorithm 12.8.
> **Output:** Same as in Algorithm 12.8.
>
> **Step 1.** Find a preconditioner M.
> **Step 2.** Choose x_0 and ϵ.
> **Step 3.** Set $r_0 = b - Ax_0$, $p_0 = y_0 = M^{-1}r_0$.
> **Step 4.** For $i = 0, 1, 2, 3, \ldots$ do
>   4.1. $w = Ap_i$.
>   4.2. $\alpha_i = y_i^T r_i / p_i^T w$.
>   4.3. $x_{i+1} = x_i + \alpha_i p_i$.
>   4.4. $r_{i+1} = r_i - \alpha_i w$.
>   4.5.   Test for convergence: If $\|r_{i+1}\|_2^2 \geq \epsilon$, continue.
>   4.6. $y_{i+1} = M^{-1} r_{i+1}$.
>   4.7. $\beta_i = y_{i+1}^T r_{i+1} / y_i^T r_i$.
>   4.8. $p_{i+1} = y_{i+1} + \beta_i p_i$.
> End

Note: *If $M = I$, then the preconditioned conjugate gradient method reduces to the basic conjugate gradient.*

Remarks. At every iteration step in the preconditioned conjugate gradient method, one symmetric positive definite system (Step 4.6) has to be solved. However, since matrix M is the same at each iteration, *the incomplete Cholesky factorization or no-fill incomplete LDL^T* (Exercise (12.14)) *has to be computed once for all.*

12.4.5 Numerical Experiments on Performance Comparison

In this section, we present results on our numerical experiments with different solvers: CG, MINRES, QMR, Bi-CG, and GMRES, on several matrices using different preconditioners.

Experiment 1. Here we compare CG and MINRES, and CG with incomplete Cholesky as a preconditioner on a two-dimensional Poisson matrix of order 1024 (Figure 12.4).

Experiment 2. Here we compare GMRES, Bi-CG, and QMR on a nonsymmetric matrix of order 1042 constructed from the MATLAB gallery matrix **Wathen**, with and without several preconditioners. The preconditioners used are Jacobi, Gauss–Seidel, and ILU (Figures 12.5–12.7).

12.5 Comparison of Krylov Subspace Methods for Linear Systems

What method to pick? It is a difficult question to answer. *The choice of a method is often problem-dependent.* Here are some rough guidelines (see Table 12.4).

12.5. Comparison of Krylov Subspace Methods for Linear Systems

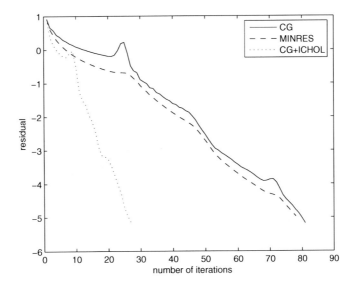

Figure 12.4. *Comparison of CG, MINRES, and CG with incomplete Cholesky preconditioner on two-dimensional Poisson matrix of order* 1024.

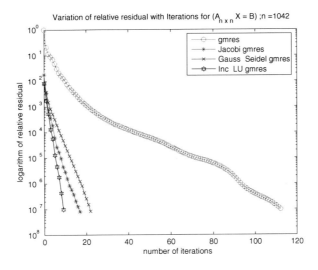

Figure 12.5. *Performance of GMRES with preconditioners.*

- **For symmetric definite systems:** An obvious choice is the CG method.

- **For symmetric indefinite systems:** The choices are between MINRES and SYMLQ. They suffer no breakdown. The MINRES method minimizes the residual and SYMLQ solves a projected problem.

472   Chapter 12. Iterative Methods: An Overview

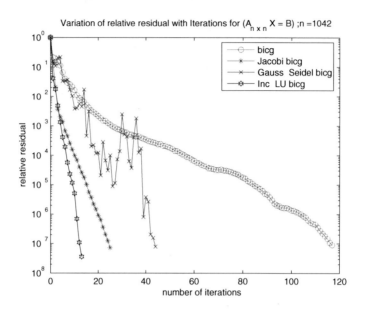

Figure 12.6. *Performance of Bi-CG with preconditioners.*

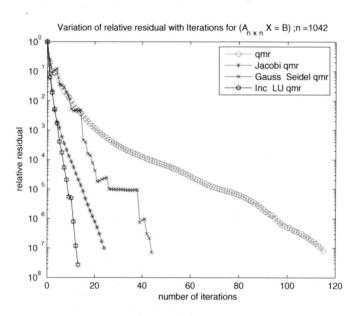

Figure 12.7. *Performance of QMR with preconditioners.*

- **For nonsymmetric systems:** The choices include GMRES, QMR, Bi-CG, CGS, Bi-CGSTAB, and some others (not described in this book; see Saad (2003) for these methods).

Table 12.4. *Comparison of some Krylov subspace methods for Ax = b.*

Method	Properties	Implementation
Full-orthogonalization and GMRES	Work for nonsymmetric matrices. The GMRES method is guaranteed to give the smallest residual. Full-orthogonalization solves a projected system. Both methods require restarting in practice.	Require only matrix-vector products, but both the storage requirement and work grow with m significantly.
CG	Applicable only to the symmetric positive definite systems. A-norm error is minimized at each iteration. The speed of convergence in general depends upon the distribution of the eigenvalues.	Requires only matrix-vector products with A.
Bi-CG	Applicable to nonsymmetric matrices. The convergence behavior might be quite irregular in some cases.	Requires matrix-vector products with both A and A^T.
QMR	Applicable to nonsymmetric matrices. Avoids breakdown of Bi-CG. May work when Bi-CG fails.	Requires the transpose matrix-vector product.
CGS and Bi-CGSTAB	CGS sometimes works well, but the round-off error in the method is a major concern. Bi-CGSTAB is the stabilized version of Bi-CG and convergence is often faster.	CGS does not require computation with A^T. Bi-CGS requires two matrix-vector products and four inner products.

Try GMRES first. If the matrix-vector multiplication is not too expensive and storage is an issue, then other methods can be tried. The QMR is generally recommended over Bi-CG. The choice between QMR, CGS, and Bi-CGSTAB is problem-dependent.

Conclusion: There is no clear winner among the Krylov subspace methods. A comparative study by Nachtigal, Reddy, and Trefethen (1992) shows that *while one method is the best for one specific class of problems, it may not work as well for other problems.* For more details, see also, for example, Saad (2003), Greenbaum (1997), van der Vorst (2003), and Barrett et al. (1994, pp. 33–35).

12.6 Eigenvalue Approximation Using Krylov Subspace Methods

Because of high storage and computational costs, it will be unrealistically ambitious to think of computing the whole spectrum (and the associated eigenvectors) using a Krylov subspace method. We will be lucky if approximations of only a few eigenvalues and eigenvectors are obtained in this way.

Here we will discuss

- eigenvalue approximations of a nonsymmetric matrix using the Arnoldi method;
- eigenvalue approximations of a symmetric matrix using the symmetric Lanczos method;
- Krylov subspace methods for the generalized eigenvalue problem $Ax = \lambda Bx$;
- Krylov subspace methods for the quadratic eigenvalue problem.

12.6.1 Eigenvalue Approximation Using the Arnoldi Method

If Algorithm 12.1 is run for $m = n$ steps, then we will obtain the $n \times n$ Hessenberg H_n such that
$$V_n^T A V_n = H_n,$$
where V_n is orthogonal. Thus, in this case the eigenvalues of A and those of H_n are the same. Thus, in theory if the Arnoldi method is carried out for n steps, all the eigenvalues of A are obtained by finding the eigenvalues of H_n. However, this is not practical for large n. The question, therefore, arises: When $m < n$, how well do the eigenvalues of H_m approximate those of A? To this end, note that the algorithm breaks down at step k when $\|\hat{v}\|_2$ is zero.

It can be shown that this happens if and only if the starting vector v_1 is a combination of the eigenvectors. Then the subspace $\mathcal{K}_k$ is an invariant subspace of A and the breakdown is a *happy breakdown*.

Definition 12.20. *An eigenvalue $\lambda_i^{(m)}$ of H_m is called a **Ritz value** and the vector $u_i^{(m)} = V_m y_i^{(m)}$, where $y_i^{(m)}$, is the eigenvector associated with the eigenvalue $\lambda_i^{(m)}$, is called the **Ritz eigenvector**. The pair $(\lambda_i^{(m)}, u_i^{(m)})$ is called a **Ritz pair**.*

A *small number of Ritz values typically constitute good approximations to the corresponding eigenvalues λ_i of A.*

However, in practice, it is desirable to compute one eigenpair at a time and then use *deflation* to compute the other pairs, as required. To obtain a good approximation of one pair, the Arnoldi method can be restarted explicitly with a fixed m as shown below.

Residual norm. Let $(\lambda_i^{(m)}, u_i^{(m)})$ be a Ritz pair. Then
$$\|r_i^{(m)}\|_2 = \|(A - \lambda_i^{(m)} I) u_i^{(m)}\|_2 = h_{m+1,m} |e_m^* y_i^{(m)}|,$$
where e_m is the mth unit vector.

Though this norm is not always a sure indicator of a good approximation, it can be used as a stopping criterion.

From experience it has been seen that the outermost eigenvalues are approximated first.

Choosing the starting vector. For eigenvalue computation, the starting vector for the Arnoldi method ideally should be chosen as the dominant one in the direction of eigenvectors. If such vectors are not known, a random vector is a reasonable choice.

12.6. Eigenvalue Approximation Using Krylov Subspace Methods

> **ALGORITHM 12.10. An Explicit Restarted Arnoldi Algorithm for Computing an Eigenpair.**
>
> **Step 1.** Run m steps of Algorithm 12.1.
>
> **Step 2.** Compute the rightmost eigenvalue $\lambda_1^{(m)}$ and the corresponding Ritz eigenvector $u_1^{(m)} = V_m y_1^{(m)}$.
>
> **Step 3.** If the residual norm $h_{m+1,m}|e_m^* y_1^{(m)}|$ is small, stop. Otherwise, return to Step 1 with $v_1 \equiv u_1^{(m)}$.

Computing several eigenpairs: Deflation. The above algorithm will approximate the rightmost eigenpair (λ_1, u_1) of A. If several pairs are needed, then the obvious thing to do is to use **deflation**. This deflation technique is as follows:

Suppose that an orthogonal basis $\{u_1, \ldots, u_{k-1}\}$ of the invariant subspace corresponding to the eigenvalues $\lambda_1, \ldots, \lambda_{k-1}$ has been computed. Set
$$U_{k-1} = (u_1, \ldots, u_{k-1}).$$
Then to compute the eigenvalue λ_k, form the matrix $\tilde{A} = A - U_{k-1} \sum U_{k-1}^*$, where $\sum = \text{diag}(\alpha_1, \alpha_2, \ldots, \alpha_{k-1})$ is the matrix of the shifts.

Then the eigenvalues of $\tilde{A}$ are $\{\lambda_1 - \alpha_1, \lambda_2 - \alpha_2, \ldots, \lambda_{k-1} - \alpha_{k-1}\} \cup \{\lambda_k, \ldots, \lambda_n\}$.

Choosing the shifts. The shifts can be chosen in the context of eigenvalues of interest. If the eigenvalues with the largest real parts are desired, then the shifts should be chosen so that λ_k becomes the next eigenvalue with the largest real part of A. Other types of deflations are also possible. See Bai et al. (2000).

12.6.2 Implicitly Restarted Arnoldi Method for the Nonsymmetric Eigenvalue Problem

The implicitly restarted Arnoldi method, developed in Lehoucq and Sorensen (1996) (see also (Lehoucq 1995)), is a method for extracting useful information (such as the *eigenvalues with largest (smallest) real parts or of largest (smallest)* magnitudes) from an m-step Krylov subspace by avoiding the storage and computational costs, using the standard QR iteration technique. For implementational aspects of the method, see the *ARPACK Users' Guide* by Lehoucq, Sorensen and Yang (1998). For details of this method, see Bai et al. (2000, pp. 169–175). The description here has been taken from that book.

Let $m = k + p$. Let $V_k = (v_1, v_2, \ldots, v_k)$.

The step-by-step process follows:

- Do m steps of the Arnoldi method to obtain V_m and H_m, yielding the Arnoldi factorization
$$AV_m = V_m H_m + f_m e_m^T.$$

- Compress this factorization to one of length k (containing the eigenformation of interest) by applying the QR iteration algorithm with p shifts, say, $\mu_1, \ldots, \mu_p$, resulting in
$$AV_m^+ = V_m^+ H_m^+ + f_m e_m^T Q,$$

where $V_m^+ = V_m Q$, $H_m^+ = Q^* H_m Q$, and $Q = Q_1 Q_2 \ldots Q_p$. Each Q_j is the orthogonal matrix associated with the shift μ_j used in the shifted QR iteration algorithm.

- Equating the first k columns on both sides of the above equation, compute V_k^+ and H_k^+ such that we have an updated k-step Arnoldi factorization

$$A V_k^+ = V_k^+ H_k^+ + f_k^+ e_k^T.$$

- Using the above as a starting point, now apply p additional steps of the Arnoldi method to obtain the new m-step Arnoldi factorization

$$A V_m = V_m H_m + f_m e_m^T.$$

ALGORITHM 12.11. Implicitly Restarted Arnoldi Method for Nonsymmetric Eigenvalue Problem.

Inputs: (i) A large and sparse matrix A, (ii) a starting vector v_1 of unit length, and (iii) positive integers k, p, and m such that $m = k + p$.
Output: k approximate eigenvalues of A.

Step 1. Run m steps of the Arnoldi method (Algorithm 12.1) to obtain the Arnoldi factorization (12.21): $A V_m = V_m H_m + f_m e_m^T$.

Step 2. Do until convergence:

    2.1. Compute the spectrum of H_m and choose p shifts $\mu_1, \ldots, \mu_p$.

    2.2. Initialize $Q = I_m$.

    2.3. For $j = 1, 2, \ldots, p$ do

    QR factorized: $H_m - \mu_j I = Q_j R_j$.

    Update: $H_m = Q_j^* H_m Q_j$, $Q = Q Q_j$.
End

    2.4. Compute $\beta_k = H_m(k+1, k)$; $\sigma_k = Q(m, k)$.

    2.5. Compute $f_k = v_{k+1} \beta_k + f_m \sigma_k$, where $f_m = h_{m+1,m} e_m^T$.

    2.6. Compute $V_k = V_m Q(:, 1:k)$ and $H_k = H_m(1:k, 1:k)$, where $Q(:, 1:k)$ denote the first k columns of Q and $H_m(1:k, k:k)$ denotes the $k \times k$ principal submatrix of H_m.
We have now the k-step Arnoldi factorization

$$A V_k = V_k H_k + f_k e_k^T.$$

    2.7. Now beginning with the above k-step factorization, apply p additional steps of the Arnoldi algorithm to obtain a new m-step factorization: $A V_m = V_m H_m + f_m e_m^T$. (Note that in this case the Arnoldi loop in Step 1 of Algorithm 12.1 runs from the index $(k+1)$ to m starting with the previous starting vector or the previous residual.)
End

12.6. Eigenvalue Approximation Using Krylov Subspace Methods

Remarks.

1. *(Choosing the shifts.)* The shifts should be chosen in the context of the eigenvalues of interest. If the eigenvalues of H_m are sorted into two groups, "wanted" and "unwanted," then the latter can be chosen as the shifts $\mu_1, \ldots, \mu_p$. With this strategy, the matrix H_k^+ will have the k "wanted" eigenvalues as its spectrum.

   Examples of the "wanted" set are (i) the k eigenvalues with largest real parts, (ii) the k eigenvalues with smallest real parts, and (iii) the k eigenvalues with largest (smallest) absolute values.

2. *Convergence.* Accept a Ritz pair (θ, u), where $u = V_m y$, as a convergent pair if the residual norm $\|f_k\| |e_k^T y|$ is less than $\|H_k\|\epsilon$, where θ is "wanted."

Deflation. Upon convergence, this pair should be deflated. There are two types of deflation, **locking** and **purging,** in the context of the implicitly restarted Arnoldi method, depending on whether a member of the wanted set of eigenvalues has converged or not. See Bai et al. (2000, pp. 176–177).

MATLAB Note: A slightly modified version of Algorithm 12.11 has been implemented in the MATLAB function **eigs.**

12.6.3 Computing the Eigenvalues of a Symmetric Matrix Using the Symmetric Lanczos Algorithm

In this section, we turn our attention to computing approximate eigenvalues of a **symmetric matrix** A using the symmetric Lanczos method (Algorithm 12.4).

We have remarked that, when a β_j is exactly equal to zero, we have an invariant subspace. This is indeed good news. Unfortunately, this happens very rarely in practice. *In practice, for large enough values of j, the eigenvalues of T_j provide very good approximations to the extremal eigenvalues of A.*

To answer the question of how well a Ritz pair of T_j approximates an eigenpair of A, we state the following result.

Theorem 12.21. *Let (θ_i, y_i) be a Ritz pair and let $R_i = Ay_i - y_i\theta_i, i = 1, \ldots, j$. Then in each interval $[\theta_i - \|R_i\|_2, \theta_i + \|R_i\|_2]$, there is an eigenvalue of A.*

Thus, it follows from the above theorem that $\|R_i\|_2$ is a good measure of how accurate the Ritz pair (θ_i, y_i) is. It also turns out that $\|R_i\|_2$ *can be computed cheaply from the Schur decomposition of T_j,* as shown by the following theorem.

Theorem 12.22 (residual theorem for symmetric Lanczos ritz pair). *Let T_j be the $j \times j$ symmetric tridiagonal matrix obtained after j steps of the symmetric Lanczos algorithm. Let $S_j^T T_j S_j$ denote the real Schur form of T_j:*

$$S_j^T T_j S_j = \mathrm{diag}(\theta_1, \ldots, \theta_j);$$

that is, $\theta_1, \ldots, \theta_j$ are the eigenvalues of T_j. Let

$$V_j S_j = Y_j = (y_1, y_2, \ldots, y_j).$$

Then for each i from 1 to j, we have

$$\|R_i\| = \|Ay_i - \theta_i y_i\| = |\beta_j| \, |s_{ji}|,$$

where

$$s_{ji} = e_j^T s_i; \quad s_i \text{ is the } i\text{th column of } S_j.$$

Proof.

$$\begin{aligned}
\|R_i\| &= \|Ay_i - y_i \theta_i\| = \|AV_j s_i - V_j s_i \theta_i\| \\
&= \|(AV_j - V_j T_j) s_i\| \quad \text{(because } T_j s_i = s_i \theta_i\text{)} \\
&= \|(\beta_j v_{j+1} e_j^T) s_i\| \quad \text{(note that } AV_j - V_j T_j = \beta_j v_{j+1} e_j^T\text{)} \\
&= |\beta_j| \, \|e_j^T s_i\| = |\beta_j| \, |s_{ji}| \quad \text{(note that } \|v_{j+1}\|_2 = 1\text{)}. \quad \square
\end{aligned}$$

A computable error bound. Combining Theorems 12.21 and 12.22, we obtain the following eigenvalue bounds for the eigenvalues λ's of A:

$$\min_\lambda |\theta_i - \lambda| \leq |\beta_k| \|s_{ki}|, \quad i = 1, 2, \ldots, k.$$

"Ghost" eigenvalues phenomenon. The loss of orthogonality of the Lanczos vectors due to round-off errors can have a severe effect on the eigenvalue approximations obtained by the Lanczos process. For example, the matrix T_k in exact arithmetic is an unreduced symmetric tridiagonal matrix and therefore should have all eigenvalues real and distinct. However, in practical computations, it might happen that the computed matrix $\tilde{T}_k$ has some multiple eigenvalues that correspond to simple eigenvalues of A. This is known as the "Ghost" eigenvalue phenomenon. For details and some cures of this problem, see Golub and Van Loan (1996, pp. 484–487).

12.7 The Bisection Method for the Tridiagonal Symmetric Positive Definite Generalized Eigenvalue Problem

As we have seen in several previous case studies, in many practical situations, matrices A and B are structured: *tridiagonal and banded cases are quite common*. Unfortunately, the *Cholesky QR algorithm for the symmetric definite pencil $A - \lambda B$ described in* Chapter 11 *(Algorithm 11.4), when applied to such structured problems, will very often destroy the sparsity. Even though A and B are banded, the matrix $C = L^{-1} A (L^T)^{-1}$ will in general be full.* Thus, the Cholesky QR algorithm is not practical for large and sparse matrices. The following is a straightforward generalization of the bisection method (Algorithm 10.1) for the single symmetric matrix A to the symmetric positive definite generalized pair (A, B). We assume that A and B are both symmetric tridiagonal and that B is positive definite. The method takes advantage of the tridiagonal forms of A and B.

12.7.1 The Bisection Method for Tridiagonal A and B

Let

$$A = \begin{pmatrix} \alpha_1 & \beta_1 & \cdots & 0 \\ \beta_1 & \ddots & \ddots & 0 \\ \vdots & \ddots & \ddots & \beta_{n-1} \\ 0 & \cdots & \beta_{n-1} & \alpha_n \end{pmatrix}, \quad B = \begin{pmatrix} \alpha'_1 & \beta'_1 & \cdots & 0 \\ \beta'_1 & \ddots & \ddots & \\ \vdots & \ddots & \ddots & \beta'_{n-1} \\ 0 & \cdots & \beta'_{n-1} & \alpha'_n \end{pmatrix}. \quad (12.58)$$

Define the sequence of polynomials $\{p_k(\lambda)\}$ given by

$$p_0(\lambda) = 1, \tag{12.59}$$
$$p_1(\lambda) = \alpha_1 - \lambda \alpha'_1, \tag{12.60}$$
$$p_r(\lambda) = (\alpha_r - \lambda \alpha'_r) p_{r-1}(\lambda) - (\beta_{r-1} - \lambda \beta'_{r-1})^2 p_{r-2}(\lambda), \tag{12.61}$$
$$r = 2, 3, \ldots, n.$$

Then it can be shown (Exercise 12.24) that these polynomials form a Sturm sequence.

The generalized eigenvalues of the pencil $(A - \lambda B)$ are then given by the zeros of $p_n(\lambda)$. The zeros can be found by **bisection** or any other suitable root-finding method for polynomials. For a proof of the algorithm, see Wilkinson (1965, pp. 340–341).

Computing a Generalized Eigenvector

Once a generalized eigenvalue is computed, *the corresponding eigenvector can be computed using the inverse iteration by taking full advantage of the tridiagonal forms of A and B as follows:*

Let y_0 be the initial approximation of an eigenvector corresponding to a computed generalized eigenvalue λ. Then this inverse iteration will be as follows:

For $i = 0, 1, 2, \ldots$, do until convergence occurs
  1. Solve for x_{i+1} : $(A - \lambda B) x_{i+1} = y_i$ by taking advantage of the tridiagonal structures of A and B.
  2. Form $y_{i+1} = B x_{i+1}$.
End

Remark. About two iterations per eigenvector are usually adequate.

MATCOM Note: The above process has been implemented in the MATCOM program **GENSTURM**.

12.8 Krylov Subspace Methods for Generalized Eigenvalue Problems

The Arnoldi and Lanczos methods described in Section 12.6 for the standard eigenvalue problem can be applied to the generalized eigenvalue problem (GEP): $Ax = \lambda B$ once the latter is reduced to a standard one. The followings are some of the possible cases. For details, see Bai et al. (2000).

- (*Invert B.*) If B is nonsingular, then the GEP is equivalent to the standard eigenvalue problem
$$(B^{-1}A)x = \lambda x.$$

To apply the Arnoldi or Lanczos method to $B^{-1}A$, one needs to evaluate the matrix-vector product of the form $r = (B^{-1}A)y$, which can be accomplished without inverting B as follows:

(a) Form $u = Ay$.

(b) Solve $Br = u$ for r.

One can take advantage of the sparsity of A and B in the above computations.

- (*Shift-and-invert.*) A and B are both singular or B is ill-conditioned. In this case, it is advisable to use the shift-and-invert technique as described below:

Let σ be a user-supplied shift such that $(A - \sigma B)$ is nonsingular. Then the GEP can be transformed to the standard eigenvalue problem

$$Cx = \mu x, \text{ where} \tag{12.62}$$

$$C = (A - \sigma B)^{-1} B \quad \text{and} \quad \mu = \frac{1}{\lambda - \sigma}. \tag{12.63}$$

To apply the Arnoldi or Lanczos method to C, the matrix-vector product of type $r = Cy$ can be computed as follows without explicitly computing the inverse:

(i) Find the sparse LU factorization: $A - \sigma B = LU$. (12.64)

(ii) Form $v = By$. (12.65)

(iii) Solve $Lw = v$ for w. (12.66)

(iv) Solve $Ur = w$ for r. (12.67)

- (*Symmetric positive definite generalized eigenvalue problem.*) Here A and B are both symmetric and, furthermore, B is positive definite. Assume that B is well-conditioned (which happens in some applications). Let $B = LL^T$ be the Cholesky factorization. Then, as we have seen before, the GEP: $Ax = \lambda Bx$ is transformed into the standard symmetric eigenvalue problem

$$(L^{-1}A(L^T)^{-1})\tilde{x} = \lambda \tilde{x}, \tag{12.68}$$

where $\tilde{x} = L^T x$. The symmetric Lanczos algorithm can now be applied to the symmetric matrix $L^{-1}A(L^T)^{-1}$. The matrix-vector product of the form

$$r = L^{-1}A(L^T)^{-1}y \tag{12.69}$$

needed in this implementation of the symmetric Lanczos algorithm can be computed without explicitly evaluating the matrix $L^{-1}A(L^T)^{-1}$, as follows:

(i) Solve $L^T u = y$ for u. (12.70)

(ii) Form $w = Au$. (12.71)

(iii) Solve $Lr = w$ for r. (12.72)

- (*Symmetric indefinite generalized eigenvalue problem.*) Here A and B are both symmetric, but neither A nor B nor a linear combination of them is positive definite. *This is the case arising mostly in structural dynamics.*

If B is nonsingular and one wants a few largest eigenvalues in magnitude, one may solve the standard eigenvalue problem $B^{-1}Ax = \lambda x$, which is symmetric with respect to A or B.

If one wants a few eigenvalues closest to a number σ, the *shift-and-invert* technique

$$((A - \sigma B)^{-1}B)x = (\lambda - \sigma)^{-1}x \tag{12.73}$$

can be used. Choose the shift $\sigma = 0$ if the smallest eigenvalues in magnitude are desired. For details of how to solve the above eigenvalue problem using the symmetric indefinite Lanczos method, see Bai et al. (2000, pp. 253–256).

Concluding remarks. *The shift-and-invert technique is a powerful tool for the largest and smallest eigenvalues of the pencil $A - \lambda B$, both in symmetric and general cases, provided a suitable sparse factorization technique of $(A - \sigma B)$ is available.* Otherwise, one can use the generalized version of the Jacobi–Davidson method. See Bai et al. (2000) for details.

12.9 Krylov Subspace Methods for the Quadratic Eigenvalue Problem

The quadratic eigenvalue problem (QEP) for large and sparse matrices M, D, and K can be solved by using the methods for the GEP described in the last section once the QEP is transformed into a linear problem as shown in Section 11.9.

Numerical methods discussed in Section 12.8 can be used to solve these linear formulations of the QEP. For example, in the MSC/NASTRAN software package, the later formulation is used and the "linear" problem is solved by using the block Lanczos method. When implementing a symmetric linearization, the symmetric nature of A and B can be exploited, thus reducing the cost to a significant amount. See Grimes, Lewis, and Simon (1994).

Spectral Transformations for the QEP

An iterative solver, such as the one described in Section 12.8, is suitable for solving the QEP via linearization to a GEP when only a few exterior eigenvalues and eigenvectors are desired. However, if one wants to compute some of the smallest (in magnitude) eigenvalues and eigenvectors, or the eigenvalues closest to a shift σ, some spectral transformation will be needed first.

- **Computing the Smallest Eigenvalues and Eigenvectors**
  In this case, a natural transformation is $\mu = \frac{1}{\lambda}$. This transforms the QEP $(\lambda^2 M + \lambda D + K)x = 0$ to the *inverted* QEP

$$(M + \mu D + \mu^2 K)x = 0. \tag{12.74}$$

Assuming that $\lambda \neq 0$, we will then have the GEP in terms of λ, rather then μ, as

$$Az = \frac{1}{\lambda}Bz, \tag{12.75}$$

where we may take

$$A = \begin{pmatrix} -D & -M \\ I & 0 \end{pmatrix}, \quad B = \begin{pmatrix} K & 0 \\ 0 & I \end{pmatrix}, \quad \text{and} \quad (12.76)$$

$$z = \begin{pmatrix} x \\ \lambda x \end{pmatrix}. \quad (12.77)$$

For *the case when* $M > 0$, $D = D^T$, *and* $K > 0$, we may formulate the GEP $Az = \frac{1}{\lambda} Bz$, with

$$A = \begin{pmatrix} D & M \\ M & 0 \end{pmatrix}, \quad B = \begin{pmatrix} -K & 0 \\ 0 & M \end{pmatrix}. \quad (12.78)$$

Note that A and B are both symmetric but indefinite.

- **Computing the Eigenvalues Closest to a Shift**

If one would like to approximate the eigenvalues of the QEP closest to the shift σ, then one may use the **shift-and-invert spectral transformation**, which transfers the QEP to $(\mu^2 \hat{M} + \mu \hat{D} + \hat{K})x = 0$, where $\mu = \frac{1}{\lambda - \sigma}$, and $\hat{M} = \sigma^2 M + \sigma D + K$, $\hat{D} = D + 2\sigma M$, and $\hat{K} = M$. The exterior eigenvalues μ approximate the eigenvalues λ of the original pencil closest to the shift σ.

The corresponding GEP to be solved (in terms of λ, rather than μ) is

$$\begin{pmatrix} -\hat{D} & -\hat{K} \\ \hat{I} & 0 \end{pmatrix} \begin{pmatrix} x \\ (\lambda - \sigma)x \end{pmatrix} = \frac{1}{\lambda - \sigma} \begin{pmatrix} \hat{M} & 0 \\ 0 & I \end{pmatrix} \begin{pmatrix} x \\ (\lambda - \sigma)x \end{pmatrix}, \quad (12.79)$$

or

$$\begin{pmatrix} \hat{D} & \hat{K} \\ \hat{K} & 0 \end{pmatrix} \begin{pmatrix} x \\ (\lambda - \sigma)x \end{pmatrix} = \frac{1}{\lambda - \sigma} \begin{pmatrix} -\hat{M} & 0 \\ 0 & \hat{K} \end{pmatrix} \begin{pmatrix} x \\ (\lambda - \sigma)x \end{pmatrix} \quad (12.80)$$

if the symmetry is to be preserved.

Sensitivity of the linearization. The accuracy of a computed eigenvalue λ of the quadratic pencil $P_2(\lambda)$ depends upon the type of linearization used. *It might happen that for the same eigenvalues λ the condition numbers are different for different linearizations, and thus the accuracy will be different.* See the results of a numerical experiment in support of this statement by Tisseur and Meerbergen (2001). Tisseur (2000) has shown that by knowing $\|M\|$, $\|K\|$, and $\|D\|$, and the structures of the left and right eigenvectors of $P_2(\lambda)$, it is possible to identify which formulations are preferred for the large and the small eigenvalues. For several recent results on these topics, see Higham, Mackey, and Tisseur (2006, 2009), Higham et al. (2008), and Higham, Li, and Tisseur (2007).

12.10 The Jacobi–Davidson Method for the Quadratic Eigenvalue Problem

The Jacobi–Davidson method belongs to a family of projection methods applied directly to the QEP. These methods build an orthonormal basis V_k for the subspace $\mathcal{K}_k$ and then solve a projected smaller problem: $V_k^* P_2(\lambda) V_k z = 0$. Here is the basic idea. For details see Sleijpen, Booten, Fokkema, and van der Vorst (1996), Bai et al. (2000), and Tisseur and Meerbergen (2001).

- Compute a Ritz pair $(\tilde{\lambda}, \tilde{x})$ corresponding to a pair (λ, x) of $(\lambda^2 M + \lambda D + K)x = 0$ with $x^*x = 1$, by finding an orthonormal basis $\{v_1, v_2, \ldots, v_k\}$ for $\mathcal{K}_k$.

- Compute the pair (v, η) by solving the correction equation

$$\begin{pmatrix} P_2(\tilde{\lambda}) & P_2'(\tilde{\lambda})\tilde{x} \\ 2\tilde{x} & 0 \end{pmatrix} \begin{pmatrix} v \\ \eta \end{pmatrix} = \begin{pmatrix} -r \\ 0 \end{pmatrix},$$

where $P_2(\tilde{\lambda}) = \tilde{\lambda}^2 M + \tilde{\lambda} D + K$, $P_2'(\tilde{\lambda}) = 2\tilde{\lambda} M + D$, and $r = P_2(\tilde{\lambda})\tilde{x}$ (*residual*).

- Obtain the new basis vector v_{k+1} by orthonormalizing v against the previous columns of the orthonormal basis matrix $V_k = (v_1, v_2, \ldots, v_k)$.

- Repeat the process until r is sufficiently small.

Notes: (i) The above linear system can be rewritten as

$$\left(I - \frac{P_2'(\tilde{\lambda})\tilde{x}\tilde{x}^*}{\tilde{x}^* P_2'(\tilde{\lambda})\tilde{x}}\right) P_2(\tilde{\lambda}) \left(I - \frac{\tilde{x}\tilde{x}^*}{\tilde{x}^*\tilde{x}}\right) v = -r,$$

which can be solved using an iterative solver such as GMRES.

(ii) The Jacobi–Davidson method targets one eigenvalue at a time, in contrast with the Krylov subspace methods, which can compute several eigenvalues simultaneously. This leads to fast local convergence but slow global convergence. See Sleijpen, van der Vorst, and van Gijzen (1996) and Sleijpen, Booten, et al. (1996) for details. For the Jacobi–Davidson method for linear problems, see Sleijpen and van der Vorst (1996).

12.11 Review and Summary

12.11.1 The Classical Iterative Methods

The Jacobi, Gauss–Seidel, and SOR methods have been discussed.

A generic formulation of these iterative methods is

$$x^{(k+1)} = Bx^{(k)} + d.$$

Different methods differ in the way B and d are chosen. Writing $A = L + D + U$, we have the following:

- For the Jacobi method,

$$B = B_J = -D^{-1}(L + U), \quad d = b_J = D^{-1}b.$$

- For the Gauss–Seidel method,

$$B = B_{\text{GS}} = -(D + L)^{-1}U, \quad d = b_{\text{GS}} = (D + L)^{-1}b.$$

- For the SOR method,

$$B = B_{\text{SOR}} = (D + \omega L)^{-1}[(1 - \omega)D - \omega U], \quad d = b_{\text{SOR}} = \omega(D + \omega L)^{-1}b$$

(ω is the relaxation parameter).

- The iteration
$$x^{(k+1)} = Bx^{(k)} + d$$
converges for any arbitrary choice of the initial approximation $x^{(1)}$ if and only if the spectral radius of B is less than 1 (Theorem 12.3).

  A sufficient condition for convergence is $\|B\| < 1$ (Theorem 12.3).

- Using this sufficient condition, it has been shown that both the Jacobi and Gauss–Seidel methods converge when A is a strictly row diagonally dominant matrix (Corollaries 12.4 and 12.5).

- The Gauss–Seidel method also converges when A is symmetric positive definite (Theorem 12.6).

- For the SOR iteration to converge for any arbitrary choice of the initial approximation, the relaxation parameter ω has to lie in $(0, 2)$ (Theorem 12.8).

- If the matrix A is symmetric positive definite, then the SOR iteration is guaranteed to converge for any arbitrary choice of ω in the interval $(0, 2)$ (Theorem 12.9).

- For a consistently ordered and 2-cyclic matrix A with nonzero diagonal entries, the optimal choice of ω, denoted by ω_{opt}, is given by
$$\omega_{\text{opt}} = \frac{2}{1 + \sqrt{1 - \rho(B_J)^2}},$$
assuming that the eigenvalues of B_J are real and $\rho(B_J) < 1$, where $\rho(A)$ stands for the spectral radius of A. For definitions of these matrices, see Varga (2000) and Young (1971).

12.11.2 Krylov Subspace Methods

A brief overview of Krylov subspace methods, both for linear systems and eigenvalue problems, has been given. First, two basic methods, *Arnoldi* (Algorithm 12.1) and *Lanczos* (Algorithms 12.4 and 12.6) (both symmetric and nonsymmetric) and their Krylov subspace properties, have been described. Then it was shown how these methods could be applied to solve linear systems and compute eigenvalues.

For linear systems, we described the following:

- Lanczos-based Conjugate Gradient (CG) method for symmetric positive definite systems (Algorithm 12.5).

- Arnoldi-based full-orthogonalization (Algorithm 12.2), GMRES (Algorithm 12.3), Bi-CG (Algorithm 12.7), and QMR methods for nonsymmetric systems. Other methods, such as MINRES, SYMMLQ for symmetric indefinite systems, and variants of Bi-CG, including CGS and Bi-CGSTAB, have been mentioned with proper references.

12.11.3 Large Eigenvalue Problem

A brief discussion of the eigenvalue problem includes

- an explicitly restarted Arnoldi algorithm (Algorithm 12.10);

- an implicitly restarted Arnoldi method for an eigenpair nonsymmetric eigenvalue problem (Algorithm 12.11);

- a Lanczos algorithm for the symmetric eigenvalue problem (Section 12.6.3);

- the bisection method for the symmetric definite generalized eigenvalue problem (Section 12.7);

- Krylov subspace methods for GEPs (Section 12.8);

- Krylov subspace methods for the QEP (Section 12.9);

- the Jacobi–Davidson method for the QEP (Section 12.10).

12.12 Suggestions for Further Reading

Some well-known books on the classical iterative methods include Varga (1992, 2000), Young (1971), Hageman and Young (1981), Ortega (1990), and Axelsson (1994). In recent years, several books on Krylov subspace methods have been published. These include Greenbaum (1997), Saad (2003), and van der Vorst (2003). Some of the seminal papers on Krylov subspace methods that made a profound impact on research in this area, such as Saad and Schultz (1986) on the GMRES method; Freund and Nachtigal (1991, 1994) and Freund, Gutknecht, and Nachtigal (1993) on the QMR method; Paige (1970, 1971, 1976, 1980) on the Lanczos method; and Parlett, Taylor, and Liu (1985) on the look-ahead Lanczos method are highly recommended for further readings on these methods. There are several interesting survey papers which contain a wealth of information, including Saad and van der Vorst (2000), van der Vorst and Chan (1997), Gutknecht (1992), Freund, Golub, and Nachtigal (1992), and Saad (1981). Most of the modern textbooks on matrix computations—Trefethen and Bau (1997), Demmel (1997), Golub and Van Loan (1996), and Watkins (2002)—also contain a fair amount of discussion on Krylov subspace methods. See also Hackbush (1994). A book devoted solely to the Lanczos method and its applications in industry is Komzsik (2003). For more on the Lanczos method and its applications and implementation see Parlett (1980, 1989, 1992) and Ye (1994, 1996).

The CG method was discovered independently by Lanczos (1952) and Hestenes and Stiefel (1952). An excellent overview of this history of the method and its development can be found in Golub and O'Leary (1989). See also Ashby, Manteuffel, and Saylor (1990), Shewchuk (1994), Greenbaum and Strakos (1992), and Golub and Ye (1999). See Elman, Saad, and Saylor (1986) for a hybrid Chebyshev–Krylov algorithm, and Ferng, Golub, and Plemmons (1991) for an adaptive Lanczos algorithm for recursive condition estimation. Some other earlier papers of interest related to the conjugate gradient method include Young et al. (1980, 1988). Bischof (1990) and Bischof et al. (1990) have discussed incremental condition estimation of sparse matrices.

A two-level preconditioned scheme for the CG method was developed by Pierce and Plemmons (1988). See Brezinski and Sadok (1991), Brezinski, Zagila, and Sadok (1991,

1992), and Ye (1994) for discussions on detailed breakdown in Lanczos-type algorithms. A concise and useful account of the iterative methods is given in Barrett et al. (1994). See Arioli, Demmel, and Duff (1989) for a discussion on solving sparse system with sparse backward error. For more on convergence of GMRES, see van der Vorst and Vuik (1993). For a hybrid GMRES algorithm, see Nachtigal, Reichel, and Trefethen (1992).

Saad's book (2003) contains a complete chapter on preconditioning. A fair amount of discussion on preconditioning also appears in Greenbaum (1997). See some of Saad's papers also in this context (Saad (1984, 1988, 1993)) and Axelsson (1985). See Tong and Ye (2000) on the analysis of the Bi-CG algorithm, Chan and Ye (1997) on the development of a hybrid Krylov subspace method combining CGS and Bi-CG, Golub and Ye (1999) on the inexact preconditioned conjugate gradient method, Reichel and Ye (2005) on the breakdown-free GMRES for singular systems, van de Vorst and Ye (2000) on residual replacement strategies for Krylov subspace methods, Bai, Hu, and Reichel (1994) for a Newton basis GMRES implementation, and Bank and Chan (1993) for analysis of the composite Bi-CG method. Saylor and Smolarki (1988) have described an optimal iterative method for solving any linear system with a square matrix.

Books exclusively devoted to large eigenvalue problems are Saad (1992), Bai et al. (2000), van der Vorst (2002), and Cullun and Willoughby (1995). Some interesting recent papers on sparse eigenvalue problems include Calvetti et al. (1994), Golub and van der Vorst (2000), Golub and Ye (2000), Li and Ye (2003), Bai, Day, and Ye (1999), Stewart (2001b), and Jia and Stewart (2000). Some earlier papers on this topic include Parlett and Reid (1981), Parlett, Simon, and Stringer (1982), Ericsson and Ruhe (1980), Jennings and Stewart (1975), Paige, Parlett and van der Vorst (1995), Stewart (1976a), and Ruhe (1994). See Money and Ye (2005) for a MATLAB program for the symmetric generalized eigenvalue problem.

For more recent developments on numerical methods for quadratic and higher-order eigenvalue problems, see Ye (2006), Hoffnug, Li, and Ye (2006), and several recent papers by Higham, Tisseur, and others mentioned in Chapter 11. A brief account of the recent developments on the quadratic inverse eigenvalue problem and its applications to finite element model updating can be found in the dissertation by Sokolov (2008).

For a collection of nonlinear eigenvalue problems, see Betke et al. (2008). Mehrmann and Voss (2004) have discussed some challenges of solving nonlinear eigenvalue problems.

Multigrid methods (not discussed in this book) are powerful tools for solving partial differential equations arising from discretization. They have superior performance compared to preconditioned Krylov subspace methods. See Briggs (1987), Demmel (1997), and Saad (2003) for details. A classical book on direct methods for sparse SPD systems (not discussed here) is by George and Liu (1981). A recent book is by Davis (2006).

The research in this area is indeed very dynamic and new papers are coming up all the time.

Exercises on Chapter 12

12.1 Construct an example to show that the convergence of the Jacobi method does not necessarily imply that the Gauss–Seidel method will converge.

12.2 Let the $n \times n$ matrix A be partitioned into the form $A = (A_{ij})$, where there are N diagonal blocks and each diagonal block A_{ii} is square and nonsingular.

(a) Write down the block Jacobi, block Gauss–Seidel, and block SOR iterations for the linear system $Ax = b$ (*Hint*: Write $A = L + D + U$, where $D = \text{diag}(A_{11}, \ldots, A_{NN})$, and L and U are strictly block lower and upper triangular matrices.)

(b) If A is symmetric positive definite, then show that $U = L^T$ and D is positive definite. In this case, from the corresponding results in the scalar cases, prove that, with an arbitrary choice of the initial approximation, block Gauss–Seidel always converges and block SOR converges if and only if $0 < \omega < 2$.

12.3 Show that the block Jacobi iteration for Poisson's equation (6.46) is given by

$$A_n x_i^{(k+1)} = x_{i+1}^{(k)} + x_{i-1}^{(k)} + b_i, \quad i = 1, \ldots, N.$$

Then write down the block Gauss–Seidel and block SOR iterations also for this system.

12.4 Show that the function $\|x\|_A = x^T A x$ is a norm in $\mathbb{R}^n$.

12.5 (a) Prove that $\sqrt{\lambda_1} \|x\|_2 \leq \|x\|_A \leq \sqrt{\lambda_n} \|x\|_2$, where A is a symmetric positive definite matrix with the eigenvalues $0 < \lambda_1 \leq \lambda_2 \leq \cdots \leq \lambda_n$.

(b) Using the result in (a), prove the A-norm error bound (Theorem 12.18) for the conjugate gradient (CG) method.

12.6 Compute $\rho(B_J)$ and $\rho(B_{GS})$ for the block diagonal system $Bx = d$, where $B = \text{diag}(A_5, \ldots, A_5)$, and A_5 has the same form as the matrix of the Poisson equation (6.46), and d is chosen so that $x = (1, \ldots, 1)^T$. Solve the system using 5 iterations of Gauss–Seidel and SOR with optimal value of ω. Compare the rates of convergence.

12.7 Prove that α_k given by (12.41) minimizes the function $\phi_\alpha(x_k + \alpha p_k)$.

12.8 Show that the eigenvectors of A are the direction vectors of the CG method.

12.9 (a) Apply the incomplete Cholesky factorization algorithm (Algorithm 12.8) to an unreduced tridiagonal matrix T and show that the result is the usual Cholesky factorization of T. Verify the above statement with a 5×5 matrix of the same form as in Poisson's equation.

(b) Apply the SOR iteration to the matrix T in (a) with $\omega = 1.5$ using $x^{(0)} = (0, 0, 0, 0, 0)^T$, and make a table of the results of the iterations.

12.10 Let $p_0, p_1, \ldots, p_{n-1}$ be the direction vectors generated by the classical CG algorithm. Let $r_k = b - Ax_k$, $k = 0, 1, \ldots, n-1$. Then prove that

(a) $r_k \in \text{span}(p_0, \ldots, p_k)$, $k = 0, 1, 2, \ldots, n-1$;

(b) $\text{span}(p_0, \ldots, p_i) = \text{span}(r_0, Ar_0, \ldots, A^i r_0)$, $i = 0, 1, \ldots, n-1$;

(c) $r_0, \ldots, r_{n-1}$ are mutually orthogonal.

12.11 (*Multisplitting.*) Consider solving $Ax = b$ by using the iteration $x^{(k+1)} = Bx^{(k)} + d$, where B and d are given by $B = \sum_{i=1}^{k} D_i B_i^{-1} C_i$, $d = (\sum_{i=1}^{k} D_i B_i^{-1})b$, and $A = B_i - C_i$, $i = 1, \ldots, k$, $\sum_{i=1}^{k} D_i = I$ ($D_i \geq 0$).

Develop the Jacobi, Gauss–Seidel, and SOR methods based on the multisplitting of A (O'Leary and White (1985), Neumann and Plemmons (1987)).

12.12 Develop the symmetric successive overrelaxation (SSOR) method in detail and do an illustrative example.

12.13 Apply the Jacobi, Gauss–Seidel, and SOR (with optimal relaxation factor) methods to the system in Example 12.10 and verify the statement made there about the number of iterations for different methods.

12.14 *(No fill-in incomplete Cholesky.)* Develop the preconditioned conjugate gradient algorithm with a *square root–free incomplete Cholesky factor* as a preconditioner of the form LDL^T.

12.15 Prove the residual expression (12.26) for the full-orthogonalization method method.

12.16 Prove formulas (12.31)–(12.32) for the GMRES method.

12.17 Deduce the relation (12.56) for the QMR method.

12.18 Prove the following formulas for the CG method: $p_k^T A p_j = 0$ and $r_k^T r_j = 0, j < k$.

12.19 Prove (a) the residuals $\{r_k\}$ in the CG method are orthogonal to $\mathcal{K}_k(A, r_0)$, $k = 1, 2, \ldots$; (b) $\|e_k\|_A = \min\{\|x - x^*\|_A \big| x^* \in x_0 + \mathcal{K}_k(A, r_0)\}$.

12.20 Develop the CG and Bi-CG methods from the symmetric and nonsymmetric Lanczos methods, respectively.

12.21 Develop the MINRES and QMR methods in detail.

12.22 Prove the biorthogonality and A-conjugacy relations for the Bi-CG method.

12.23 Develop the implicitly restarted Arnoldi algorithm for nonsymmetric eigenvalue problem (Algorithm 12.11) by incorporating the details of Step 2.7.

12.24 Prove that the polynomials (12.59)–(12.61) form a Sturm sequence.

12.25 (a) Develop a generalized Lanczos algorithm for the symmetric definite pencil $A - \lambda B$.

(b) Prove that the Lanczos vectors are *B-orthogonal*:
$$v_i^T B v_i = 1; \quad v_i^T B v_j = 0, \quad i \neq j.$$

(c) Find the generalized eigenvalues for the pair (A, B), where
$$A = \begin{pmatrix} 1 & 1 & 0 \\ 1 & 1 & 1 \\ 0 & 1 & 1 \end{pmatrix}, \quad B = \begin{pmatrix} 10 & 1 & 0 \\ 1 & 10 & 1 \\ 0 & 1 & 10 \end{pmatrix},$$
using this algorithm.

12.26 Develop an Arnoldi algorithm based on the shift-and-invert technique for the generalized eigenvalue problem $Ax = \lambda Bx$, using (12.62)–(12.67). Test your algorithm with an illustrative example.

12.27 Develop an Arnoldi algorithm for computing the smallest and largest eigenvalues of the symmetric indefinite generalized eigenvalue problem using the shift-and-invert technique (12.73). Test your algorithm with a 10×10 random example.

12.28 Develop an algorithm for computing the smallest eigenvalue of the quadratic pencil $P_2(\lambda)$ based on (12.74). Give an illustrative example.

12.29 Develop an algorithm for computing the eigenvalue closest to a shift of the pencil $P_2(\lambda)$ based on (12.79)–(12.80).

12.30 Construct an example to show that the different linearizations of the pencil $P_2(\lambda)$ can produce different sensitivities for the same eigenvalue λ.

MATLAB Programs and Problems on Chapter 12

Notes: (i) The functions **gmres, qmr, bicg, pcg,** and **minres** are available in MATLAB.
(ii) The functions **jacobi, gaused, sucov,** and **nichol** are available in MATCOM.

M12.1 Run the programs **jacobi, gaused**, and **sucov** (choosing ω as an optional relaxation parameter) from MATCOM on the 500 × 500 matrix A of the same type as that of Example 12.10 with the same starting vector $x^{(0)} = (0, 0, \ldots, 0)^T$. Find how many iterations each method will take to converge.

M12.2 (*Van der Vorst* (2003).) Run **gmres** from MATLAB with the following data: $A = SBS^{-1}$ of order $n = 200$, where

$$S = \begin{pmatrix} 1 & \beta & & 0 \\ & 1 & \ddots & \\ & & \ddots & \beta \\ 0 & & & 1 \end{pmatrix} \quad \text{and} \quad B = \begin{pmatrix} \lambda_1 & & 0 \\ & \ddots & \\ & & \ddots \\ 0 & & \lambda_n \end{pmatrix}.$$

M12.3 Run the program **nichol** from MATCOM implementing the "no-fill incomplete Cholesky factorization" on the nonsymmetric tridiagonal symmetric positive definite matrix T of order 200 arising in discretization of Poisson's equation. Compare your result with that obtained by running **chol**(T) on T.

M12.4 (a) Run MATLAB program **gmres** to implement the GMRES algorithm without a preconditioner ($M = I$) for solving $Ax = b$, with A taken as (i) an 500 × 500 diagonal matrix with eigenvalues clustered around 1, (ii) an 500 × 500 upper bidiagonal matrix with eigenvalues clustered around zero and 1's along the subdiagonal, (iii) the symmetric part of A in (ii), and (iv) **wilkinson** (500) from MATLAB.

(b) Plot relative residuals versus the number of iterations k for each of the systems in (a). Write down your conclusions.

M12.5 Repeat Problem M12.4 with MATLAB function **qmr** and compare the results with those of GMRES on the same data as in Problem M12.4.

M12.6 (*Comparison of GMRES, QMR, and Bi-CG.*) Run **gmres**, **qmr**, and **bicg** (without preconditioners) on the same 500 × 500 sparse matrix created by MATLAB function **spdiags** and plot the logarithms of the norms of the residuals versus number of iterations. Write your observations.

M12.7 (*Study of convergence of the CG method with varying eigenvalue distributions.*)

    (a) Run MATLAB program **pcg** to implement the CG method without any preconditioner ($M = I$) on the following systems: $Ax = b$, taking A as (i) the *symmetric positive definite matrix of the system* (6.46) arising in solution of *Poisson's equation* of order 500; (ii) the 500×500 *diagonal matrix with one isolated eigenvalue* 0.0001 *and the other* 499 *eigenvalues distributed equally over the interval* $[0.05, 1.5]$; (iii) a 500×500 diagonal matrix with *the first* 255 *eigenvalues set as* $1, 2, \ldots, 255$ *and the rest distributed equally over the interval* $[255, 500]$; (iv) a 500×500 *symmetric tridiagonal matrix* with $1, 2, \ldots, 500$ on the diagonal and 1's along the sub- and superdiagonals; and (v) a 500×500 *symmetric tridiagonal matrix with a parameter* τ such that as τ increases from zero, the matrix A becomes more and more ill-conditioned.

    (b) In one single graph, plot $\|r_k\|$ versus number of iterations k for all the above systems. Write your observations.

M12.8 For each system in Problem M12.7, present the following results in tabular form (in one single table): (i) number of iterations N_k for convergence; (ii) $\frac{\|e_{N_k}\|_A}{\|e_0\|_A}$; (iii) the condition number κ; (iv) $2(\frac{\sqrt{\kappa}-1}{\sqrt{\kappa}+1})^{N_k}$; and (v) actual residual norm $\|A - bx_{N_k}\|_2$.

M12.9 (*Comparison of the CG method with different preconditioners.*) Repeat Problem M12.7 with (i) *incomplete Cholesky preconditioner* (write a separate MATLAB program and then use it to feed into the program **pcg**), (ii) *Jacobi preconditioner*, and (iii) a *diagonal preconditioner having a few distinct eigenvalues* and the others closely clustered. Plot relative residuals versus number of iterations k in a separate graph for each of these preconditioners.

M12.10 Repeat Problem M12.8 with each of the preconditioners used in Problem M12.9. Make one separate table for each preconditioner.

M12.11 (*Study of convergence of the CG method with different matrices having the same condition number.*) Run MATLAB program **pcg** on five diagonal matrices with positive diagonal entries, each of order 500, having the same condition number (ratio of the largest to the smallest diagonal entries) and plot the relative residuals versus the number of iterations in each case.

M12.12 Run MATLAB program bicg on the matrices of Problem M12.11 with some diagonal entries negatives this time. Plot relative residuals versus the number of iterations in each case.

M12.13 (a) Run MATLAB program **minres** (without a preconditioner) on five different diagonal matrices of order 500 with eigenvalues contained in the intervals $[a, b] \cup [c, d]$, where $a < b < 0 < c < d$ and $b - a = d - c$, by choosing different values of $a, b, c,$ and d. Plot relative residuals versus the number of iterations.

    (b) Repeat part (a) in the special case when the two intervals are symmetrically placed; that is, $a = -d$ and $b = -c$.

M12.14 Repeat Problem M12.13 with the following preconditioners: Jacobi, SSOR, and ILU.

M12.15 Construct a parametric matrix A with a parameter τ as follows: $A = QDQ^T$, where Q is orthogonal, and $D = \text{diag}(\lambda_1, \ldots, \lambda_n)$, with $\lambda_i = \lambda_1 + \frac{i-1}{n-1}(\lambda_n - \lambda_1)\tau^{n-i}$, $i = 2, \ldots, n-1$. Run the unpreconditioned CG method on A with $n = 50$ and $\tau = 0.1, 0.5, 0.6, 0.7, 0.8, 0.9, 1.0$. Plot the A-norm error versus the number of iterations in a single graph. Tabulate (i) the condition number κ of A, and (ii) $2(\frac{\sqrt{\kappa}-1}{\sqrt{\kappa}+1})^n$ for each value of the parameter τ. Write your observations.

M12.16 Implement Algorithm 12.10 in MATLAB for finding the rightmost eigenvalue of A.

Test data: (i) A random matrix A of order 1000; (ii) $A = $ **numgrid** $('R', n)$ with $R = C, D$, and $n = 500$; (iii) $A = V^T B V$, where B is a diagonal matrix of order 500 with most eigenvalues chosen randomly from a normal Gaussian distribution and V is a 500 × 500 orthogonal matrix. For each A, plot the results of residual norms versus the number of iterations. Take $m = 20, 25, 50$, and 100. (*Note:* **numgrid** is a MATLAB function.)

M12.17 Find the 10 largest and smallest eigenvalues for each of the matrices in Problem M12.16 using MATLAB function **eigs**.

M12.18 Construct a diagonal matrix A of order 500 with only a few distinct eigenvalues and the rest clustered around the center of the spectrum, and then run the symmetric Lanczos method on A for $k = 5, 10, 15, 25$, and 50.

(a) Plot eigenvalues of T_k for each k and those of the original matrix A. Write your observations.
(b) Plot the relative errors $\frac{|\lambda_i(T_k) - \lambda_i(A)|}{|\lambda_i(A)|}$ for the first five and the last five eigenvalues.

M12.19 Repeat Problem M12.18 with a diagonal matrix of 1000 with eigenvalues chosen randomly from a normal Gaussian distribution.

M12.20 Illustrate by means of a plot the phenomenon of "ghost eigenvalues" by taking A as a 500 × 500 diagonal matrix whose eigenvalues are clustered in $(0, 1)$ and has two eigenvalues 2 and 3 lying outside the interval.

M12.21 Run each of the programs written for problems (12.27)–(12.29) with an appropriate problem of dimension 500.

Chapter 13
Some Key Terms in Numerical Linear Algebra

algorithm: An ordered set of operations, logical and/or arithmetic, which when applied to a computational problem defined by a given set of data, called the input data, produces a solution to the problem.

Arnoldi method: A Krylov subspace method that forms the basis of many modern large-scale algorithms for linear systems and eigenvalue computations.

back substitution: The process of solving an upper triangular system $Tx = b$. The entries of the column vector x are obtained one by one, starting from the bottom and working up through the elements of the vector.

backward stability: An algorithm is backward stable if the computed solution obtained by the algorithm is the exact solution of a nearby problem.

balancing: A process applied to A, before the start of the QR iteration algorithm, so that the entries of matrix A become somewhat uniform.

BLAS: Basic Linear Algebra Subroutines. There are three levels: Level 1 subroutines are for vector-vector operations, Level 2 are for matrix-vector operations, and Level 3 are for matrix-matrix operations.

catastrophic cancellation: A phenomenon that occurs when two numbers that are nearly equal are subtracted. Many significant digits are lost. It often indicates that errors were made in previous computations.

characteristic polynomial: For an $n \times n$ matrix A, the polynomial $p(\lambda) = \det(A - \lambda I)$ of degree n in λ.

Cholesky factorization: The factorization of a symmetric positive definite matrix A into HH^T, where H is a lower triangular matrix with positive diagonal entries.

companion matrix: An unreduced upper Hessenberg matrix with ones along the subdiagonal and possible nonzero entries on the last column and zeros everywhere else is an upper companion matrix. The coefficients of the characteristic polynomial of the upper companion matrix are given by the entries of the last column. The transpose of an upper companion matrix is a lower companion matrix.

complete pivoting: The form of pivoting where the search for the pivot entry at step k is made among the entries of the submatrix below the row $(k-1)$ during the Gaussian elimination process.

condition number: The number indicating the sensitivity of a problem. The condition number of the linear system problem $Ax = b$ is Cond $(A) = \|A\| \|A^{-1}\|$. It is also called the condition number of the matrix A.

condition number of eigenvalue: If λ is an eigenvalue of a nondefective matrix, then the number $\frac{1}{|y_i^T x_i|}$, where y_i and x_i are the left and right unit length eigenvectors of the matrix, respectively, is the condition number of λ.

deflation: A technique used in eigenvalue computations. Once an eigenvalue (or a pair of eigenvalues) is computed, an appropriate row and column (or pair of rows and columns) are deleted and the computations proceed with the remaining submatrix for the remaining eigenvalues.

dense matrix: A matrix in which most entries are nonzero. The zero entries are too few to devise any specialized algorithm.

direct method: A method that produces the solution of a problem in a finite number of steps. The Gaussian elimination, QR factorization, and Cholesky factorization methods are examples of direct methods for the linear system problem.

double-shift QR iteration: The shifted QR iteration where two single shifts are used successively.

efficiency of an algorithm: An algorithm involving matrices of order n is efficient if it does not require more than order n^3 floating point operations for implementation.

eigenvalue and eigenvector: A scalar λ is an eigenvalue of A if there exists a nonzero vector x such that $Ax = \lambda x$. x is called the eigenvector corresponding to λ.

elementary lower triangular matrix: A matrix of the form $I + m_k e_k^T$, where $m_k = (0, \ldots, 0, m_{k+1,k}, \ldots, m_{n,k})^T$. It is an identity matrix, except possibly for a few nonzero entries below the diagonal of a single column.

floating point number: A floating point number x has the form $x = \pm \gamma \beta^e$, where e is the exponent, γ is the significant, and β is the base of the number system. A floating point number x is denoted by $fl(x)$.

flop: Floating point operation.

forward elimination: The process of solving a lower triangular system $Ly = b$. The entries of the unknown vector x are obtained one by one, starting from the top and working down through the vector's elements.

forward stability: An algorithm is forward stable if the computed solution $\hat{x}$ by the algorithm is close to the exact solution x in some sense.

Frobenius norm: $\|A\|_F = \left[\sum_{j=1}^{n} \sum_{i=1}^{m} |a_{ij}|^2 \right]^{1/2}$, where $A = (a_{ij})$ is of order $m \times n$.

Gaussian elimination: An elimination process used to solve a linear system. It is named after the celebrated mathematician Karl Friedrich Gauss.

Gauss–Seidel method: An iterative method for solving $Ax = b$, named after the mathematicians Gauss and Seidel. The ith component of the $(k+1)$th iteration vector $x_i^{(k+1)}$ is computed using a combination of information from the kth step and that which is available at the current $(k+1)$th step:

$$x_i^{(k+1)} = \frac{1}{a_{ii}}\left(b_i - \sum_{j=1}^{i-1} a_{ij} x_j^{(k+1)} - \sum_{j=i+1}^{n} a_{ij} x_j^{(k)}\right).$$

generalized eigenvalue problem: The eigenvalue problem $Ax = \lambda Bx$ involving two matrices A and B.

Geršgorin disks: The disks in the complex plane $\{z : |z - a_{ii}| < r_i\}$, where $r_i = \sum_{j=1, j \neq i}^{n} |a_{ij}|$ associated with $A = (a_{ij})$. Each eigenvalue of A lies in at least one of the disks.

Givens matrix: A matrix $J(i, j, \theta)$ is a Givens matrix if it is an identity matrix except for the four entries $J_{ii} = c$, $J_{ji} = -s$, $J_{jj} = c$, where $c = \cos\theta$, $s = \sin\theta$. A Givens matrix is orthogonal.

GMRES method: The Generalized Minimal Residual Method. A Krylov subspace method for solving large and sparse linear systems.

Gram–Schmidt process: A process to generate an orthonormal basis of a subspace, starting from a given basis. If applied to the columns of a matrix A, it yields a QR factorization of A.

growth factor: The ratio of the largest element (in magnitude) of the matrices A, $A^{(1)}$, ..., $A^{(n-1)}$ (obtained during the Gaussian elimination process) to the largest element (in magnitude) of A. The growth factor is an indicator of stability or instability of the Gaussian elimination process.

guard digit: An extra digit in the lower end of the arithmetic register whose purpose is to catch the low-order digit that would otherwise be pushed out of existence when the decimal points are aligned.

Hilbert matrix: The matrix $H = (h_{ij})$, with $h_{ij} = \frac{1}{i+j-1}$, named after the celebrated mathematician David Hilbert. The higher-order Hilbert matrices are extremely ill-conditioned.

Householder matrix: A matrix of the form $I - 2\frac{uu^T}{u^T u}$, where u is a vector. A Householder matrix is symmetric and orthogonal. It is also known as a *projector*.

ill-conditioned problem: A problem is ill-conditioned if a small change in the input data can cause a significant change in the solution.

implicit QR iteration: A version of the QR iteration algorithm where one constructs the QR iteration step $A_{s+1} = R_s Q_s + \mu I$ implicitly without forming the matrix $A_s - \mu I$.

incomplete Cholesky factorization: The Cholesky factorization of a sparse symmetric positive definite matrix where calculations are made only with the nonzero entries, and the nonzero entries in the Cholesky factor are allowed only in those positions that have a nonzero in A.

inverse iteration: An iterative process for finding an eigenvector given an approximation to the eigenvalue for which the eigenvector is to be computed.

inverse power method: The same as *inverse iteration*.

iterative method: A method that produces the solution of a problem after a few number of iterations, starting from an initial approximation of the solution. Usually, the approximations become closer and closer to the solution as the iteration proceeds.

The Jacobi, Gauss–Seidel, and successive overrelaxation methods are examples of iterative methods for solving linear systems problems. The power method, the inverse power method, and the QR method are examples of iterative methods for solving eigenproblems.

iterative refinement: A procedure to iteratively improve a computed solution of a problem.

Jacobi method: An iterative method for solving the system $Ax = b$; named after the mathematician Karl Gustav Jacobi. The ith component of the $(k + 1)$th iteration vector $x_i^{(k+1)}$ is computed using only information from kth iteration step:

$$x_i^{(k+1)} = \frac{1}{a_{ii}} \left(b_i - \sum_{j=1, i \neq j}^{n} a_{ij} x_j^{(k)} \right).$$

Krylov subspace: The subspace spanned by the sequence of vectors $\{x, Ax, \ldots, A^{m-1}x\}$; m is the dimension of the subspace. The sequence $\{A^i x\}$ is called the Krylov sequence.

Krylov subspace method: A method based on generating an orthonormal basis of a Krylov subspace. The Lanczos, conjugate gradient, GMRES, and the Arnoldi methods are examples of such methods.

LU factorization: A factorization of A into LU, where L is lower triangular and U is upper triangular. Elementary lower triangular matrices are used to achieve an LU factorization.

LAPACK: A mathematical software package for linear algebra computations. The package was developed mainly for portable high-performance computing.

least-squares solution: A solution x to the linear system $Ax = b$ having the property that $\|Ax - b\|_2$ is minimized.

machine precision: The machine precision μ is the smallest positive floating point number in a computer such that $\mathrm{fl}(1 + \mu) > 1$.

MATCOM: A MATLAB-based software package implementing all the major algorithms of this book.

MATLAB: MATLAB stand for MATrix LABoratory. It is an interactive computing system designed for easy computations of various matrix-based scientific and engineering problems.

minimum-norm solution: Among all the least-squares solutions to the linear system $Ax = b$, the one having the minimum norm is the minimum-norm solution.

norm of a matrix: Let A be an $m \times n$ matrix. Then the norm of A, $\|A\|$, is a scalar such that $\|A\| \geq 0$, $\|\alpha A\| = |\alpha| \|A\|$, and $\|A + B\| \leq \|A\| + \|B\|$, where α is a scalar.

norm of a vector: Let x be a vector. Then the norm of x, $\|x\|$, is a scalar associated with x such that $\|x\| \geq 0$, $\|\alpha x\| = |\alpha| \|x\|$, and $\|x + y\| \leq \|x\| + \|y\|$, where α is a scalar.

normal equations: For the system $Ax = b$, the normal equations are the set of equations $A^T Ax = A^T b$.

orthogonal projection: P is the orthogonal projection onto a subspace S of $\mathbb{R}^n$ if range $(P) = S$, $P^2 = P$, and $P^T = P$.

overdetermined system: A linear system having more equations than unknowns.

overflow: A situation that results when an exponent of a computed quantity becomes too large to fit in a given computer.

partial pivoting: The form of pivoting where the search for the pivot entry at step k is made in the kth column below the row $(k-1)$ during the Gaussian elimination process.

perturbation analysis: Analysis pertaining to the effect on the solution of a problem by the perturbation of the data of the problem.

power method: An iterative method for computing the dominant eigenvalue and the corresponding eigenvector of a matrix A. The method is so called because it is based on implicitly computing the powers of A.

preconditioning: A process to improve the condition number of a matrix.

pseudoinverse: The matrix $A^\dagger$ defined by $A^\dagger = V\Sigma^\dagger U^T$, where $A = U\Sigma V^T$ is the SVD of A, $\Sigma^\dagger = \text{diag}(\frac{1}{\sigma_1}, \frac{1}{\sigma_2}, \ldots, \frac{1}{\sigma_r}, 0, \ldots, 0)$, and r is the rank of A.

pseudocode: Form of codes for describing algorithms which can be translated easily into computer codes.

QR iteration: An iterative process for finding the eigenvalues of a matrix A, based on repeated QR factorizations of matrices orthogonally similar to A.

QR factorization: A factorization of A into QR, where Q is orthogonal and R is upper triangular. Householder matrices, Givens matrices, and the Gram-Schmidt process are used to achieve a QR factorization.

QR factorization with column pivoting: The factorization of A in the form

$$Q^T A P = \begin{pmatrix} R_{11} & R_{12} \\ 0 & 0 \end{pmatrix},$$

where Q is orthogonal, P is a permutation matrix, and R_{11} is upper triangular and nonsingular.

quadratic eigenvalue problem: The eigenvalue problem of the form $(\lambda^2 A + \lambda B + C)x = 0$.

quasi-minimal residual method: A Krylov subspace method to solve a large and sparse linear system. It is popularly known as the QMR method.

QZ algorithm: An algorithm for finding the eigenvalues of the pencil $A - \lambda B$. The orthogonal matrices Q and Z are constructed so that $Q^T A Z$ is an upper quasi-triangular matrix T, and $Q^T B Z = S$ is upper triangular. The eigenvalues are then extracted from the eigenvalues of T and S.

rank-revealing QR: A QR factorization of the form

$$Q^T A P = \begin{pmatrix} R_{11} & R_{12} \\ 0 & 0 \end{pmatrix}$$

that reveals the rank of A in exact arithmetic. Rank of $A = \text{rank}(R_{11})$.

Rayleigh quotient: The quotient $\frac{x^T A x}{x^T x}$ is called the Rayleigh quotient of the vector x. If x is an approximation of an eigenvector, then the Rayleigh quotient is an approximation of the eigenvalue corresponding to which the approximate eigenvector is x.

real Schur form: A real quasi-triangular matrix whose diagonal entries are 1×1 or 2×2 matrices. Every real matrix A can be transformed to real Schur form by orthogonal similarity.

relative error: If $\hat{x}$ is an approximation to x, then the relative error is $\frac{\|\hat{x}-x\|}{\|x\|}$.

rounding error: The error made in rounding a computed quantity.

scaling (row): A process by which a diagonal matrix D is constructed so that the rows of $D^{-1}A$ have more or less equal infinity norms.

single-shift QR iteration: The shifted QR iteration where a single shift is used but the shift varies from iteration to iteration.

singular value decomposition: A decomposition of matrix A in the form $A = U \Sigma V^T$, where U and V are orthogonal and Σ is a "diagonal" matrix.

sparse matrix: A matrix with a large number of zero entries. Sparsity is an asset with a large problem. A sparse matrix may conveniently be stored in a computer and specialized algorithms can be devised.

special matrices: An $n \times n$ matrix $A = (a_{ij})$ is

> **diagonal** if $a_{ij} = 0, i \neq j$;
>
> **upper triangular** if $a_{ij} = 0, i > j$;
>
> **lower triangular** if $a_{ij} = 0, i < j$;
>
> **upper Hessenberg** if $a_{ij} = 0$ for $i > j + 1$;
>
> **lower Hessenberg** if $a_{ij} = 0$ for $j > i + 1$;
>
> **tridiagonal** if it is both lower and upper Hessenberg;
>
> **a permutation matrix** if there is exactly one nonzero entry in each row and column that is a 1 and the rest are zero;
>
> **unreduced upper Hessenberg** if A is upper Hessenberg and $a_{i,i-1} \neq 0, i = 2, \ldots, n$;
>
> **unreduced lower Hessenberg** if A is lower Hessenberg and $a_{i,i+1} \neq 0, i = 1, \ldots, n-1$;
>
> **(row) diagonally dominant** if $|a_{ii}| \geq \sum_{j \neq i} |a_{ij}|$ for all i;
>
> **symmetric (Hermitian)** if $A^T = A (A^* = A)$;
>
> **positive definite** if A is symmetric (Hermitian) and $x^T A x > 0 (x^* A x > 0)$ for every nonzero vector x;
>
> **a defective matrix** if it has fewer than n eigenvectors;
>
> **a convergent matrix** if $A^k \to 0$ as $k \to \infty$.

spectral norm: $\|A\|_2 = \sqrt{\text{maximum eigenvalue of } A^T A}$.

spectral radius: The spectral radius is max $|\lambda_i|, i = 1, \ldots, n$, where $\lambda_1, \ldots, \lambda_n$ are the eigenvalues of A.

spectrum: The set of all the eigenvalues of a matrix.

stopping criterion: A criterion that terminates an iterative method.

successive overrelaxation method: A modified Gauss–Seidel method. The ith component of the $(k+1)$th approximation is given by

$$x_i^{(k+1)} = \frac{\omega}{a_{ii}}\left(b_i - \sum_{j=1}^{i-1} a_{ij} x_j^{(k+1)} - \sum_{j=i+1}^{n} a_{ij} x_j^{(k)}\right) + (1-\omega) x_i^{(k)}.$$

ω is called the relaxation factor.

symmetric definite generalized eigenvalue problem: The eigenvalue problem $Ax = \lambda Bx$, where A and B are symmetric matrices and B is positive definite.

underdetermined system: A linear system having more unknowns than equations.

underflow: A situation that results when the exponent of a computed quantity becomes too small to fit in a given computer.

well-conditioned problem: A problem whose solution is not sensitive to small perturbations of the input data.

Wilkinson bidiagonal matrix: The bidiagonal matrix whose entries along the diagonal are $1, 2, \ldots, 20$, and those along the superdiagonal (subdiagonal) are $20, 19, \ldots, 1$. Certain eigenvalues of this matrix are extremely sensitive.

Wilkinson shift: A special shift, named after James H. Wilkinson, used in the symmetric QR iteration algorithm. If $\alpha_1, \ldots, \alpha_n$ and $\beta_1, \ldots, \beta_{n-1}$ are, respectively, the diagonal and off-diagonal entries of a symmetric tridiagonal matrix, then the Wilkinson shift is given by $\mu = \alpha_n + r - \text{sign}(r)\sqrt{r^2 + \beta_{n-1}^2}$, where $r = \frac{\alpha_{n-1}-\alpha_n}{2}$.

Bibliography

E. Anderson, Z. Bai, C. Bischof, S. Blackford, J. Demmel, J. Dongarra, J. Du Croz, A. Greenbaum, S. Hammarling, A. McKenney, and D. Sorensen (1999), *LAPACK Users' Guide*, 3rd ed., SIAM, Philadelphia.

H.C. Andrews and B.R. Hunt (1988), *Digital Image Processing*, Prentice–Hall, Englewood Cliffs, NJ.

A.C. Antoulas (2005), *Approximation of Large-Scale Dynamical Systems*, SIAM, Philadelphia.

M. Arioli, J.W. Demmel, and I.S. Duff (1989), "Solving Sparse Linear Systems with Sparse Backward Error," *SIAM J. Matrix Anal. Appl.*, 10, 165–190.

M. Arioli, I. Duff, and D. Ruiz (1992), "Stopping Criteria for Iterative Solvers," *SIAM J. Matrix Anal. Appl.*, 13, 138–144.

M. Arioli and A. Laratta (1985), "Error Analysis of an Algorithm for Solving an Underdetermined System," *Numer. Math.*, 46, 255–268.

M. Arnold (1992), *Algorithms and Conditioning for Eigenvalue Assignment,* Ph.D. Dissertation, Northern Illinois University, DeKalb, IL.

W. Arnoldi (1951), "The Principle of Minimized Iteration in the Solution of the Matrix Eigenvalue Problem," *Quart. Appl. Math*, 9, 17–29.

S.F. Ashby, T.A. Manteuffel, and P.F. Saylor (1990), "A Taxonomy for Conjugate Gradient Methods," *SIAM J. Numer. Anal.*, 27, 1542–1568.

K.E. Atkinson (1989), *An Introduction to Numerical Analysis*, 2nd ed., John Wiley and Sons, New York.

O. Axelsson (1985), "A Survey of Preconditioned Iterative Methods for Linear Systems of Equations," *BIT*, 25, 166–187.

O. Axelsson (1994), *Iterative Solution Methods*, Cambridge University Press, Cambridge, UK.

Z. Bai (1988), "Note on the Quadratic Convergence of Kogbetliantz's Algorithm for Computing the Singular Value Decomposition," *Lin. Alg. Appl.*, 104, 131–140.

Z.-J. Bai, B.N. Datta, and J. Wang (2009), "Robust and Minimum Norm Partial Quadratic Eigenvalue Assignment in Vibrating Systems," *Mechanical Systems and Signal Processing*, to appear.

Z. Bai, D. Day, and Q. Ye (1999), "ABLE: An Adaptive Block Lanczos Method for Non-Hermitian Eigenvalue Problems," *SIAM J. Matrix Anal. Appl.*, 20, 1060–1082.

Z. Bai and J.W. Demmel (1993a), "Computing the Generalized Singular Value Decomposition," *SIAM J. Sci. Comput.*, 14, 1464–1486.

Z. Bai and J. Demmel (1993b), "On Swapping Diagonal Blocks in Real Schur Form," *Lin. Alg. Appl.*, 186, 73–95.

Z. Bai, J. Demmel, J. Dongarra, A. Ruhe, and H. van der Vorst (2000), *Templates for the Solution of Algebraic Eigenvalue Problems: A Practical Guide*, SIAM, Philadelphia.

Z. Bai, J. Demmel, and A. McKenney (1993), "On Computing Condition Numbers for the Nonsymmetric Eigenproblems," *ACM Trans. Math. Software*, 19, 202–223.

Z. Bai, D. Hu, and L. Reichel (1994), "A Newton Basis GMRES Implementation," *IMA J. Num. Anal.*, 14, 563–581.

Z. Bai and G.W. Stewart (1998), "Algorithm 775: SRRIT—A FORTRAN Subroutine to Calculate the Dominant Invariant Subspace of a Nonsymmetric Matrix," *ACM Trans. Math. Software*, 23, 494–513.

R.E. Bank and T.F. Chan (1993), "An Analysis of the Composite Step Biconjugate Gradient Method," *Numer. Math.*, 66, 259–319.

R.E. Bank and C. Wagner (1999), "Multilevel ILU Decomposition," *Numer. Math.*, 82, 543–576.

J.L. Barlow (1993), "Error Analysis of Update Methods for the Symmetric Eigenvalue Problem," *SIAM J. Matrix Anal. Appl.* 14, 598–618.

R. Barrett, M.W. Berry, T.F. Chan, J. Demmel, J. Donato, J. Dongarra, V. Eijkhout, R. Pozo, C. Romine, and H. van der Vorst (1994), *Templates for the Solution of Linear Systems: Building Blocks for Iterative Methods*, SIAM, Philadelphia.

F.L. Bauer (1963), "Optimally Scaled Matrices," *Numer. Math.*, 5, 73–87.

F.L. Bauer (1965), "Elimination with Weighted Row Combinations for Solving Linear Equations and Least Squares Problems," *Numer. Math.*, 7, 338–352.

F.L. Bauer and C.T. Fike (1960), "Norms and Exclusion Theorems," *Numer. Math.*, 2, 137–141.

E.B. Becker, G.F. Carey, and J.T. Oden (1981), *Finite Element: An Introduction, Vol. 1*, Prentice–Hall, Englewood Cliffs, NJ.

A. Berman and R.J. Plemmons (1994), *Nonnegative Matrices in Mathematical Sciences*, 2nd ed., SIAM, Philadelphia.

M.W. Berry and M. Browne (2005), *Understanding Search Engines: Mathematical Modeling and Text Retrieval*, 2nd ed., SIAM, Philadelphia.

T. Betcke, N. Higham, V. Mehrmann, C. Schröder, and F. Tisseur (2008), "NLEVP: A Collection of Nonlinear Eigenvalue Problems", *MIMS Eprint* 2008.40, http://eprints.ma.man.ac.uk/1071/01/covered/MIMS_ep2008_40.pdf.

R. Bhatia (1996), *Matrix Analysis*, Springer-Verlag, New York.

R. Bhatia (2007), *Perturbation Bounds for Matrix Eigenvalues*, SIAM, Philadelphia.

A. Bhaya and E. Kaszkurewicz (2006), *Control Perspectives on Numerical Algorithms and Matrix Problems*, SIAM, Philadelphia.

C.H. Bischof (1990), "Incremental Condition Estimation," *SIAM J. Matrix Anal. Appl.*, 11, 312–322.

C.H. Bischof, J.G. Lewis, and D.J. Pierce (1990), "Incremental Condition Estimation for Sparse Matrices," *SIAM J. Matrix Anal. Appl.*, 11, 644–659.

A. Björck (1967), "Iterative Refinement of Linear Least Squares Solution I," *BIT*, 7, 257–278.

A. Björck (1968), "Iterative Refinement of Linear Least Squares Solution II," *BIT*, 8, 8–30.

A. Björck (1991a), "Component-wise Perturbation Analysis and Errors Bounds for Linear Least Squares Solutions," *BIT*, 31, 238–244.

A. Björck (1991b), "Error Analysis of Least Squares Algorithms," in *Numerical Linear Algebra, Digital Signal Processing and Parallel*, Springer-Verlag, Berlin, 41–73.

A. Björck (1994), "Numerics of Gram-Schmidt Orthogonalization," *Lin. Alg. Appl.*, 197/198, 297–316.

A. Björck (1996), *Numerical Methods for Least Squares Problems*, SIAM, Philadelphia.

A. Björck and G.H. Golub (1967), "Iterative Refinement of Linear Least Squares Solutions by Householder Transformation," *BIT*, 7, 322–337.

A. Björck and C.C. Paige (1992), "Loss and Recapture of Orthogonality in the Modified Gram–Schmidt Algorithms," *SIAM J. Matrix Anal. Appl.*, 13, 176–190.

Z. Bohte (1975), "Bounds for Rounding Errors in the Gaussian Elimination for Band Systems," *J. Inst. Math. Appl.*, 16, 133–142.

A.W. Bojanczyk (1995), *Linear Algebra for Signal Processing*, G. Cybenko, ed., Springer-Verlag, New York.

A.W. Bojanczyk, R.P. Brent, P. Van Dooren, and F.R. De Hoog (1987), "A Note on Downdating the Cholesky Factorization," *SIAM J. Sci. Statist. Comput.*, 8, 210–221.

D. Boley (1987), "Computing Rank-deficiency of Rectangular Matrix Pencils," *Systems Control Lett.*, 9, 207–214.

D. Boley (1990), "Estimating the Sensitivity of the Algebraic Structure of Pencils with Simple Eigenvalue Estimates," *SIAM J. Matrix Anal. Appl.*, 11, 632–643.

S. Brahma and B.N. Datta (2008), "An Optimization Approach for Minimum Norm and Robust Partial Quadratic Eigenvalue Assignment Problems for Vibrating Structures," *J. Sound and Vibration*, 324, 471–489.

M. Braun (1978), *Differential Equations and Their Applications*, Springer-Verlag, New York.

C. Brezinski and H. Sadok (1991), "Avoiding Breakdown in CGS Algorithm," *Numer. Alg.*, 199–206.

C. Brezinski, M. Zagila, and H. Sadok (1991), "Avoiding Breakdown and Near Breakdown in Lanczos Type Algorithms," *Numer. Alg.*, 1, 261–284.

C. Brezinski, M. Zagila, and H. Sadok (1992), "A Breakdown Free Lanczos-type Algorithm for Solving Linear Systems," *Numer. Math.*, 63, 29–38.

W. Briggs (1987), *A Multigrid Tutorial*, SIAM, Philadelphia.

R. Brualdi, D. Carlson, B.N. Datta, C.R. Johnson, and R. Plemmons, eds. (1985), *Linear Algebra and Its Role in Systems Theory*, Contemporary Mathematics 47, AMS, Providence, RI.

R.A. Brualdi and S. Mellendorf (1994), "Regions in the Complex Plane Containing the Eigenvalues of a Matrix," *Amer. Math. Monthly*, 101, 975–985.

J.R. Bunch (1971), "Analysis of the Diagonal Pivoting Method," *SIAM J. Numer. Anal.*, 8, 656–680.

J.R. Bunch (1987), "The Weak and Strong Stability of Algorithms in Numerical Linear Algebra," *Lin. Alg. Appl.*, 88/89, 49–66.

J.R. Bunch, J.W. Demmel, and C.F. Van Loan (1989), "The Strong Stability of Algorithms for Solving Symmetric Linear Systems," *SIAM J. Matrix Anal. Appl.*, 10, 494–499.

J.R. Bunch and L. Kaufman (1977), "Some Stable Methods for Calculating Inertia and Solving Symmetric Linear Systems," *Math. Comp.*, 31, 162–179.

J.R. Bunch, L. Kaufman, and B. Parlett (1976), "Decomposition of a Symmetric Matrix," *Numer. Math.*, 27, 95–109.

J.R. Bunch, C.P. Nielson, and D.C. Sorensen (1978), "Rank-One Modification of the Symmetric Eigenvalue Problem," *Numer. Math.*, 31, 31–48.

J.R. Bunch and B.N. Parlett (1971), "Direct Methods for Solving Symmetric Indefinite Systems of Linear Equations," *SIAM J. Numer. Anal.*, 8, 639–655.

R.L. Burden and J.D. Faires (2004), *Numerical Analysis*, 8th ed., Brooks/Cole, Pacific Grove, CA.

P.A. Businger (1968), "Matrices which Can Be Optimally Scaled," *Numer. Math.*, 12, 346–348.

P.A. Businger and G.H. Golub (1965), "Linear Least Squares Solutions by Householder Transformation," *Numer. Math*, 7, 269–276.

P.A. Businger and G.H. Golub (1969), "Algorithm 358: Singular Value Decomposition of a Complex Matrix," *Comm. ACM*, 12, 564–565.

R. Byers and B.N. Datta (2007), "Vector and Matrix Norms, Error Analysis, Efficiency and Stability," in *Handbook of Linear Algebra*, L. Hogben, ed., Chapman Hall/CRC Press, Boca Raton, FL, 37-1–37-22.

D. Callaerts, B. De Moor, J. Vandewalle, and W. Sansen (1990), "Comparison of SVD methods to Extract the Fetal Electrocardiogram from Cutaneous Electrode Signals," *Medical and Biological Engineering and Computing*, 217–224.

D. Calvetti, L. Reichel, and D. Sorensen (1994), "An Implicitly Restarted Lanczos Method for Large Symmetric Eigenvalue Problem," *Electron. Trans. Num. Anal.*, 2, 1–21.

S.L. Campbell (1985), "Rank Deficient Least Squares and the Numerical Solution of Linear Singular Systems of Differential Equations," in *Linear Algebra and Its Role in Systems Theory*, R. Brualdi et al., eds., *Contemporary Mathematics* 47, AMS, Providence, RI, 51–63.

S. Campbell and C.D. Meyer (1979), *Generalized Inverses of Linear Transformations*, Pitman, London.

D. Carlson and B.N. Datta (1979), "On the Effective Computation of the Inertia of a Non-Hermitian Matrix," *Numer. Math.*, 33, 315–322.

M.A. Celia and W.G. Gary (1992), *Numerical Methods for Differential Equations*, Prentice–Hall, Englewood Cliffs, NJ.

T.F. Chan (1982a), "An Improved Algorithm for Computing the Singular Value Decomposition," *ACM Trans. Math. Software*, 8, 72–83.

T.F. Chan (1982b), "Algorithm 581: An Improved Algorithm for Computing the Singular Value Decomposition," *ACM Trans. Math. Software*, 8, 84–88.

T.F. Chan (1987), "Rank-Revealing QR Factorizations," *Lin. Alg. Appl.*, 88/89, 67–82.

T.F. Chan and P.C. Hansen (1992), "Some Applications of the Rank Revealing QR Factorization," *SIAM J. Sci. Statist. Comput.*, 13, 727–741.

T. Chan and Q. Ye (1997), "A Mixed Product Krylov Subspace Method for Solving Nonsymmetric Linear Systems," *Asian J. Math.*, 1, 422–434.

S. Chandrasekaran and I.C.F. Ipsen (1994), "On Rank-Revealing Factorisations," *SIAM J. Matrix Anal. Appl.*, 15, 592–622.

S. Chapman (2009), *Essentials of MATLAB Programming*, 2nd ed., Cengage Learning, Toronto, Canada.

S.C. Chapra and R.P. Canale (2002), *Numerical Methods for Engineers*, McGraw–Hill, New York.

J.P. Charlier, M. Vanbegin, and P. Van Dooren (1988), "On Efficient Implementation of Kogbetliantz's Algorithm for Computing the Singular Value Decomposition," *Numer. Math.*, 52, 279–300.

M.T. Chu and G. Golub (2005), *Inverse Eigenvalue Problems: Theory, Algorithms, and Applications*, Oxford University Press, New York.

A.K. Cline, C.B. Moler, G.W. Stewart, and J.H. Wilkinson (1979), "An Estimate for the Condition Number of a Matrix," *SIAM J. Numer. Anal.*, 16, 368–375.

R.E. Cline and R.J. Plemmons (1976), "l_2-Solutions to Underdetermined Linear Systems," *SIAM Rev.*, 18, 92–106.

T.F. Coleman and C. Van Loan (1988), *Handbook for Matrix Computations*, SIAM, Philadelphia.

S.D. Conte and C. de Boor (1980), *Elementary Numerical Analysis: An Algorithmic Approach*, 3rd ed., McGraw–Hill, New York.

C.R. Crawford (1973), "Reduction of a Band Symmetric Generalized Eigenvalue Problem," *Comm. ACM*, 16, 41–44.

C.W. Cryer (1968), "Pivot Size in Gaussian Elimination," *Numer. Math.*, 12, 335–345.

J.K. Cullum and R.A. Willoughby (1985), *Lanczos Algorithms for Large Symmetric Eigenvalue Computations, Vol. 1: Theory* and *Vol. 2: Programs*, Birkhäuser, Basel.

J.J.M. Cuppen (1981), "A Divide and Conquer Method for the Symmetric Eigenproblem," *Numer. Math.*, 36, 177–195.

J. Daniel, W.B. Gragg, L. Kaufman, and G.W. Stewart (1976), "Reorthogonalization and Stable Algorithms for Updating the Gram-Schmidt QR Factorization," *Math. Comp.*, 30, 772–795.

B.N. Datta (1994), "Linear and Numerical Linear Algebra in Control Theory: Some Research Problems," *Lin. Alg. Appl.*, 197/198, 755–790.

B.N. Datta (2003), *Numerical Methods for Linear Control Systems Design and Analysis*, Elsevier, Boston.

B.N. Datta and K. Datta (1976), "An Algorithm for Computing the Powers of a Hessenberg Matrix and Applications," *Lin. Alg. Appl.*, 14, 273–284.

B.N. Datta, S. Deng, D. Sarkissian, and V. Sokolov (2009), "An Optimization Technique for Damped Model Updating with Measured Data Satisfying Quadratic Orthogonality Constraint," *Mechanical Systems and Signal Processing*, special issue on inverse problems, 1759–1772.

B.N. Datta, S. Elhay, and Y.M. Ram (1997), "Orthogonality and Partial Pole Assignment for the Symmetric Definite Quadratic Pencil," *Lin. Alg. Appl.*, 257, 29–48.

B.N. Datta, S. Elhay, Y.M. Ram, and D. Sarkissian (2000), "Partial Eigenstructure Assignment for the Quadratic Pencil," *J. Sound and Vibration*, 230, 101–110.

B.N. Datta, C.R. Johnson, M.A. Kaashoek, R. Plemmons, and E.D. Sontag, eds. (1988), *Linear Algebra in Signals, Systems, and Control*, SIAM, Philadelphia.

B.N. Datta, Y. Ram, and D. Sarkissian (2002), "Spectrum Modification in Gyroscopic Systems," *ZAMM*, 82, 191–200.

B.N. Datta and F. Rincón (1993), "Feedback Stabilization of a Second-Order System: A Nonmodel Approach," *Lin. Alg. Appl.*, 188/189, 135–161.

B.N. Datta and Y. Saad (1991), "Arnoldi Methods for Sylvester-Like Observer Matrix Equations and an Associated Algorithm for Partial Spectrum Assignment," *Lin. Alg. Appl.*, 154/156, 225–244.

B.N. Datta and D.R. Sarkissian (2001), "Theory and Computations of Some Inverse Eigenvalue Problems for the Quadratic Pencil," in *Structured Matrices in Operator Theory, Control, and Signal and Image Processing*, V. Olshevsky, ed., *Contemporary Mathematics* 280, AMS, Providence, RI, 221–240.

P.I. Davies, N.J. Higham, and F. Tisseur (2001), "Analysis of the Cholesky Method with Iterative Refinement for Solving the Symmetric Definite Generalized Eigenproblem," *SIAM J. Matrix Anal. Appl.*, 23, 472–493.

T.A. Davis (2006), *Direct Methods for Sparse Linear Systems*, SIAM, Philadelphia.

T. Davis and K. Sigmon (2005), *MATLAB Premier*, 7th ed., CRC Press, Boca Raton, FL.

G.J. Davis and C. Moler (1978), "Sensitivity of Matrix Eigenvalues," *Int. J. Numer. Meth. Eng.*, 12, 1367–1373.

J. Day and B. Peterson (1988), "Growth in Gaussian Elimination," *Amer. Math. Monthly*, 95, 489–513.

L.S. DeJong (1977), "Towards a Formal Definition of Numerical Stability," *Numer. Math.*, 28, 211–219.

J. Demmel (1983), *A Numerical Analyst's Jordan Canonical Form*, Ph.D. Thesis, University of California, Berkeley.

J. Demmel (1987a), "Three Methods for Refining Estimates of Invariant Subspaces," *Computing*, 38, 43–57.

J. Demmel (1987b), "On Condition Number and the Distance to the Nearest Ill-Posed Problem," *Numer. Math.*, 51, 251–259.

J. Demmel (1989), "On Floating Point Errors in Cholesky," *LAPACK Working Note* 14, Computer Science Department, University of Tennessee, Knoxville.

J. Demmel (1992), "The Componentwise Distance to the Nearest Singular Matrix," *SIAM J. Matrix Anal. Appl.*, 13, 10–19.

J.W. Demmel (1997), *Applied Numerical Linear Algebra*, SIAM, Philadelphia.

J. Demmel and A. Edelman (1995), "The Dimensions of Matrices (Matrix Pencils) with Given Jordan (Kronecker) Canonical Forms," *Lin. Alg. Appl.*, 230, 61–87.

J. Demmel, M. Gu, S. Eisenstat, I. Slapničar, K. Veselić, and Z. Drmač (1999), "Computing the Singular Value Decomposition with High Relative Accuracy," *Lin. Alg. Appl.*, 299, 21–80.

J. Demmel and B. Kagström (1993), "The Generalized Schur Decomposition of an Arbitrary Pencil, Part I: Theory and Algorithms," *ACM Trans. Math. Software*, 19, 160–174.

J. Demmel and W. Kahan (1990), "Accurate Singular Values of Bidiagonal Matrices," *SIAM J. Sci. Statist. Comput.*, 11, 873–912.

J. Demmel and K. Veselić (1992), "Jacobi's Method Is More Accurate than QR," *SIAM J. Matrix Anal. Appl.*, 13, 1204–1245.

B. De Moor (1992), "On the Structure and Geometry of the Product Singular Value Decompositions," *Lin. Alg. Appl.*, 168, 95–136.

B. De Moor (1991), "Generalizations of the Singular Value and QR Decompositions," *Signal Processing*, 25, 135–146.

B. De Moor and G.H. Golub (1991), "The Restricted Singular Value Decomposition: Properties and Applications," *SIAM J. Matrix Anal. Appl.*, 12, 401–425.

B. De Moor and P. Van Dooren (1992), "Generalizations of the Singular Value and QR Decompositions," *SIAM J. Matrix Anal. Appl.*, 13, 993–1014.

B. De Moor and H. Zha (1991), "A Tree of Generalizations of the Ordinary Singular Value Decomposition," *Lin. Alg. Appl.*, 147, 469–500, special issue on canonical forms of matrices.

E.F. Deprettere, ed. (1988), *SVD and Signal Processing: Algorithms, Applications and Architecture*, Elsevier (North–Holland), Amsterdam.

I.S. Dhilon (1998), "Current Inverse Iteration Software Can Fail," *BIT Numer. Math.*, 38, 685–704.

J.J. Dongarra, S. Hammarling, and J.H. Wilkinson (1992), "Numerical Considerations in Computing Invariant Subspaces," *SIAM J. Matrix Anal. Appl.*, 13, 145–161.

J.J. Dongarra, C.B. Moler, and J.H. Wilkinson (1983), "Improving the Accuracy of Computed Eigenvalues and Eigenvectors," *SIAM J. Numer. Anal.*, 20, 23–45.

J.J. Dongarra and D.C. Sorensen (1987), "A Fully Parallel Algorithm for the Symmetric Eigenvalue Problem," *SIAM J. Sci. Statist. Comput.*, 8, s139–s154.

Z. Drmač and K. Veselić (2008a), "New Fast and Accurate Jacobi SVD Algorithm. I," *SIAM J. Matrix Anal. Appl.*, 29, 1322–1342.

Z. Drmač and K. Veselić (2008b), "New Fast and Accurate Jacobi SVD Algorithm. II," *SIAM J. Matrix Anal. Appl.*, 29, 1343–1362.

I.S. Duff, R.G. Grimes, and J.G. Lewis (1989), "Sparse Matrix Test Problems," *ACM Trans. Math Software*, 15, 1–14.

I.S. Duff, R.G. Grimes, and J.G. Lewis (1992), "Users' Guide for the Harwell-Boeing Sparse Matrix Collection (Release 1)," *Report RAL*-92-086, Rutherford Appleton Laboratory, Oxon, UK, 84.

C. Eckart and G. Young (1939), "A Principal Axis Transformation for Non-Hermitian Matrices," *Bull. Amer. Math. Soc.*, 45, 118–121.

A. Edelman (1988), "Eigenvalues and Condition Numbers of Random Matrices." *SIAM J. Matrix Anal. Appl.*, 9, 543–560.

A. Edelman (1992a), "On the Distribution of a Scaled Condition Number," *Math Comp.*, 58, 185–190.

A. Edelman (1992b), "The Complete Pivoting Conjecture for Gaussian Elimination Is False," *Mathematica J.*, 2, 58–61.

A. Edelman and D. Friedman (1998), "A Counterexample to a Hadamard Matrix Pivot Conjecture," *Lin. Multilin. Alg.*, 273, 45–63.

A. Edelman and G.W. Stewart (1993), "Scaling for Orthogonality," *IEEE Trans. Signal Process.*, 41, 1676–1677.

B. Efron (1988), "Computer-Intensive Methods in Statistical Regression," *SIAM Rev.*, 30, 421–449.

L. Eldén (2007), *Matrix Methods in Data Mining and Pattern Recognition*, SIAM, Philadelphia.

L. Eldén and H. Park (1994), "Perturbation Analysis for Block Downdating of a Cholesky Decomposition," *Numer. Math.*, 68, 457–467.

H. Elman (1986), "A Stability Analysis of Incomplete LU Factorization," *Math. Comp.*, 47, 191–218.

H.C. Elman, Y. Saad, and P.E. Saylor (1986), "A Hybrid Chebyshev Krylov Subspace Algorithm for Solving Nonsymmetric Systems of Linear Equations," *SIAM J. Sci. Statist. Comput.*, 3, 840–855.

T. Ericsson and A. Ruhe (1980), "The Spectral Transformation Lanczos Method for the Numerical Solution of Large Sparse Generalized Symmetric Eigenvalue Problems," *Math. Comp.*, 35, 1251–1268.

J. Erxiong (1990), "An Algorithm for Finding Generalized Eigenpairs of a Symmetric Definite Matrix Pencil," *Lin. Alg. Appl.*, 132, 65–91.

K.V. Fernando and B.N. Parlett (1994), "Accurate Singular Values and Differential qd Algorithm," *Numer. Math.*, 67, 191–229.

W.R. Ferng, G.H. Golub, and R.J. Plemmons (1991), "Adaptive Lanczos Algorithms for Recursive Condition Estimation," *Numer. Alg.*, 1, 1–20.

J. Fish and T. Belytschko (2007), *A First Course in Finite Elements*, John Wiley and Sons, Chichester, UK.

G.E. Forsythe, M.A. Malcolm, and C.B. Moler (1977), *Computer Methods for Mathematical Computations*, Prentice–Hall, Englewood Cliffs, NJ.

G.E. Forsythe and C.B. Moler (1967), *Computer Solutions of Linear Algebraic Systems*, Prentice–Hall, Englewood Cliffs, NJ.

L.V. Foster (1986), "Rank and Nullspace Calculations Using Matrix Decompositions without Column Interchanges," *Lin. Alg. Appl.*, 74, 47–71.

L. Foster (1991), "Modifications of the Normal Equations Method that Are Numerically Stable," in *Numerical Linear Algebra, Digital Signal Processing and Parallel Algorithms*, G.H. Golub and P.M. Van Dooren, eds., *NATO ASI Series*, Springer-Verlag, Berlin, 501–512.

L.V. Foster (1994), "Gaussian Elimination with Partial Pivoting Can Fail in Practice," *SIAM J. Matrix Anal. Appl.*, 15, 1354–1362.

J.G.F. Francis (1961), "The QR Transformation: A Unitary Analogue to the LR Transformation, Parts I and II," *Comp. J.*, 4, 265–272, 332–345.

J.N. Franklin (1968), *Matrix Theory*, Prentice–Hall, Englewood Cliffs, NJ.

R.W. Freund (1993), "A Transpose-free Quasi-minimum Residual Algorithm for Non-Hermitian Linear Systems," *SIAM J. Sci. Comput.*, 14, 470–482.

R. Freund, G.H. Golub, and N. Nachtigal (1992), "Iterative Solution of Linear Systems," *Acta Numerica*, Vol. 1, 57–100.

R.W. Freund, M.H. Gutknecht, and N.M. Nachtigal (1993), "An Implementation of the Look-Ahead Lanczos Algorithm for Non-Hermitian Matrices," *SIAM J. Sci. Comput.*, 14, 137–158.

R. Freund and N. Nachtigal (1991), "QMR: A Quasi-Minimal Residual Method for Non-Hermitian Linear Systems," *Numer. Math.*, 60, 315–339.

R.W. Freund and N.M. Nachtigal (1994), "An Implementation of the QMR Method Based on Coupled Two-Term Recurrences," *SIAM J. Sci. Comput.*, 15, 313–337.

M.I. Friswell and J.E. Mottershead (1995), *Finite Element Model Updating in Structural Dynamics*, Kluwer Academic, London.

F.R. Gantmacher (1959), *The Theory of Matrices, Vols. 1 and 2*, Chelsea, New York.

J.E. Gentle (1998), *Numerical Linear Algebra for Applications in Statistics*, Springer-Verlag, New York.

W.M. Gentleman (1973), "Least Squares Computations by Givens Transformations without Square Roots," *J. Inst. Math. Appl.*, 12, 329–336.

W.M. Gentleman (1975), "Error Analysis of QR Decompositions by Givens Transformations," *Lin. Alg. Appl.*, 10, 189–197.

J.A. George and J. Liu (1981), *Computer Solution of Large Sparse Positive Definite Systems*, Prentice–Hall, Englewood Cliffs, NJ.

S. Geršgorin (1931), "Über die Abgrenzung der Eigenwerte Einer Matrix," *IZV. Akad. Nauk. USSR otd. Fiz.-Mat. Nauk*, 7, 749–754.

J.R. Gilbert, C. Moler, and R. Schrieber (1992), "Sparse Matrices in MATLAB: Design and Implementation," *SIAM J. Matrix Anal. Appl.*, 13, 333–356.

P.E. Gill, G.H. Golub, W. Murray, and M.A. Saunders (1974), "Methods for Modifying Matrix Factorizations," *Math. Comp.*, 28, 505–535.

P.E. Gill, W. Murray, and M. Wright (1991), *Numerical Linear Algebra and Optimization*, Addison–Wesley, Redwood City, CA.

W. Givens (1958), "Computation of Plane Unitary Rotations Transforming a General Matrix to Triangular Form," *SIAM J. Appl. Math.*, 6, 26–50.

G.M.L. Gladwell (2004), *Inverse Problems in Vibration*, 2nd ed., Springer- Verlag, Berlin.

I. Gohberg, P. Lancaster, and L. Rodman (1982), *Matrix Polynomials*, Academic Press, New York.

G.H. Golub (1965), "Numerical Methods for Solving Linear Least Squares Problems," *Numer. Math.*, 7, 206–216.

G.H. Golub (1969), "Matrix Decompositions and Statistical Computation," in *Statistical Computation*, R.C. Milton and J.A. Nelder, eds., Academic Press, New York, 365–397.

G.H. Golub and W. Kahan (1965), "Calculating the Singular Values and Pseudo-inverse of a Matrix," *SIAM J. Numer. Anal.*, 2, 205–224.

G.H. Golub and D.P. O'Leary (1989), "Some History of the Conjugate Gradient and Lanczos Algorithms: 1948–1976," *SIAM Rev.*, 31, 50–102.

G.H. Golub and C. Reinsch (1970), "Singular Value Decomposition and Least Squares Solutions," *Numer. Math.*, 14, 403–420.

G.H. Golub and H.A. van der Vorst (2000), "Eigenvalue Computation in the 20th Century," *J. Comp. Appl. Math.*, 123, 35–65.

G.H. Golub and C.F. Van Loan (1996), *Matrix Computations*, 3rd ed., Johns Hopkins University Press, Baltimore, MD.

G.H. Golub and J.H. Wilkinson (1966), "Note on the Iterative Refinement of Least Squares Solution," *Numer. Math.*, 9, 139–148.

G.H. Golub and J.H. Wilkinson (1976), "Ill-Conditioned Eigensystems and the Computation of the Jordan Canonical Form," *SIAM Rev.*, 18, 578–619.

G.H. Golub and Q. Ye (1999), "Inexact Preconditioned Conjugate Gradient Method with Inner-Outer Iteration," *SIAM J. Sci. Comput.*, 21, 1305–1320.

G.H. Golub and Q. Ye (2000), "Inexact Inverse Iterations for the Generalized Eigenvalue Problems," *BIT Numer. Math.*, 40, 672–684.

N. Gould (1991), "On Growth in Gaussian Elimination with Complete Pivoting," *SIAM J. Matrix Anal. Appl.*, 12, 354–361.

F.A. Graybill (1983), *Matrices with Applications in Statistics*, Wadsworth, Belmont, CA.

A. Greenbaum (1997), *Iterative Methods for Solving Linear Systems*, SIAM, Philadelphia.

A. Greenbaum and Z. Strakos (1992), "Predicting the Behavior of Finite Precision Lanczos and Conjugate Gradient Computations," *SIAM J. Matrix Anal. Appl.*, 13, 121–137.

R.G. Grimes, J.G. Lewis, and H.D. Simon (1994), "A Shifted Block Lanczos Algorithm for Solving Sparse Symmetric Generalized Eigenproblems," *SIAM J. Matrix Anal. Appl.*, 15, 228–272.

I. Griva, S. Nash, and A. Sofer (2009), *Linear and Nonlinear Optimization*, 2nd ed., Cambridge University Press, Cambridge, UK.

M. Gu and S.C. Eisenstat (1995a), "A Divide-and-Conquer Algorithm for the Symmetric Tridiagonal Eigenproblem," *SIAM J. Matrix Anal. Appl.*, 16, 172–191.

M. Gu and S.C. Eisenstat (1995b), "A Divide-and-Conquer Algorithm for the Bidiagonal SVD," *SIAM J. Matrix Anal. Appl.*, 16, 79–92.

W. Guorong, W. Yimin, and Q. Sanzheng (2004), *Generalized Inverses: Theory and Computations*, Science Press, Beijing, New York.

M.H. Gutknecht (1992), "A Completed Theory of the Unsymmetric Lanczos Process and Related Algorithms, Part I," *SIAM J. Matrix Anal. Appl.*, 13, 594–639.

W. Hackbush (1994), *Iterative Solution of Large Sparse Systems*, Springer-Verlag, Berlin.

L.A. Hageman and D.M. Young (1981), *Applied Iterative Methods*, Academic Press, New York.

W.W. Hager (1984), "Condition Estimates," *SIAM J. Sci. Statist. Comput.*, 5, 311–316.

W.W. Hager (1988), *Applied Numerical Linear Algebra*, Prentice–Hall, Englewood Cliffs, NJ.

S. Hammarling (1985), "The Singular Value Decomposition in Multivariate Statistics," *ACM SIGNUM Newsletter*, 20, 2–25.

P.C. Hansen, J.G. Nagy, and D.P. O'Leary (2006), *Deblurring Images: Matrices, Spectra, and Filtering*, SIAM, Philadelphia.

M. Heath (2002), *Scientific Computing*, 2nd ed., McGraw–Hill, New York.

M.E. Hestenes and E. Stiefel (1952), "Methods of Conjugate Gradients for Solving Linear Systems," *J. Res. Nat. Bur. Stand.*, 49, 409–436.

D.J. Higham and N.J. Higham (1998), "Structured Backward Error and Condition of Generalized Eigenvalue Problems," *SIAM J. Matrix Anal. Appl.*, 20, 493–512.

D.J. Higham and N.J. Higham (2005), *MATLAB Guide*, 2nd ed., SIAM, Philadelphia.

N.J. Higham (1987), "A Survey of Condition Number Estimation for Triangular Matrices," *SIAM Rev.*, 29, 575–596.

N.J. Higham (1990), "Experience with a Matrix Norm Estimator," *SIAM J. Sci. Statist. Comput.*, 11, 804–809.

N.J. Higham (1991), "Iterative Refinement Enhances the Stability of QR Decomposition Methods for Solving Linear Equations," *BIT*, 31, 447–468.

N.J. Higham (2002), *Accuracy and Stability of Numerical Algorithms*, 2nd ed., SIAM, Philadelphia (1st ed. published in 1996).

N.J. Higham (2008), "An Interview with Gene Golub," *MIMS Eprint 2008.8*, http://eprints.ma.man.ac.uk/1024/01/covered/MIMS_ep2008_8.pdf.

N.J. Higham and D.J. Higham (1989), "Large Growth Factors in Gaussian Elimination With Pivoting," *SIAM J. Matrix Anal. Appl.*, 10, 155–164.

N.J. Higham, R.-C. Li, and F. Tisseur (2007), "Backward Error of Polynomial Eigenproblems Solved by Linearization," *SIAM J. Matrix Anal. Appl.*, 29, 1218–1241.

N.J. Higham, D.S. Mackey, and F. Tisseur (2006), "The Conditioning of Linearizations of Matrix Polynomials," *SIAM J. Matrix Anal. Appl.*, 28, 1005–1028.

N.J. Higham, D.S. Mackey, and F. Tisseur (2009), "Definite Matrix Polynomials and Their Linearization by Definite Pencils," *SIAM J. Matrix Anal. Appl.*, 31, 478–502.

N.J. Higham, D.S. Mackey, F. Tisseur, and S. Garvey (2008), "Scaling, Sensitivity and Stability in the Numerical Solution of Quadratic Eigenvalue Problems," *Internat. J. Numer. Methods Eng.*, 73, 344–360.

N.J. Higham and F. Tisseur (2000), "A Block Algorithm for Matrix 1-Norm Estimation, with an Application to 1-Norm Pseudospectra," *SIAM J. Matrix Anal. Appl.*, 21, 1185–1201.

N.J. Higham and F. Tisseur (2003), "Bounds for Eigenvalues of Matrix Polynomials," *Lin. Alg. Appl.*, 358, 5–22.

R.O. Hill, Jr. (1991), *Elementary Linear Algebra with Applications*, Harcourt-Brace-Jovanovich.

R.O. Hill, Jr., and B.N. Parlett (1992), "Refined Interlacing Properties," *SIAM J. Matrix Anal. Appl.*, 13, 239–247.

L. Hoffnung, R.-C. Li, and Q. Ye (2006), "Krylov Type Subspace Methods for Matrix Polynomials," *Lin. Alg. Appl.*, 415, 52–81.

L. Hogben (2007), *Handbook of Linear Algebra*, Chapman Hall/CRC Press, Boca Raton, FL.

H. Hong and C.-T. Pan (1992), "The Rank-revealing QR Decomposition and SVD," *Math. Comp.*, 58, 575–596.

R.A. Horn and C.R. Johnson (1985), *Matrix Analysis*, Cambridge University Press, New York.

R.A. Horn and C.R. Johnson (1991), *Topics in Matrix Analysis*, Cambridge University Press, New York.

A.S. Householder (1958), "Unitary Triangularization of a Nonsymmetric Matrix," *J. ACM*, 5, 339–342.

A.S. Householder (1964), *The Theory of Matrices in Numerical Analysis*, Dover, New York.

T.-M. Hwang, W.-W. Lin, and V. Mehrmann (2003), "Numerical Solution of Quadratic Eigenvalue Problems with Structure Preserving Methods," *SIAM J. Sci. Comput.*, 24, 1283–1302.

IEEE Computer Society (1985), "IEEE Standard for Binary Floating Point Arithmetic," IEEE Computer Society Press, New York.

IEEE Computer Society (1987), "IEEE Standard for Radix-Independent Floating Point Arithmetic," IEEE Computer Society Press, New York.

D.J. Inman (2006), *Vibration with Control Measurement and Stability*, Prentice–Hall, Englewood Cliffs, NJ (1st ed. published in 1989).

D.J. Inman (2007), *Engineering Vibration*, 3rd ed., Prentice–Hall, Englewood Cliffs, NJ

A.K. Jain (1989), *Fundamentals of Digital Image Processing*, Prentice–Hall, Englewood Cliffs, NJ.

M. Jankowski and M. Wozniakowski (1977), "Iterative Refinement Implies Numerical Stability," *BIT*, 17, 303–311.

A. Jennings and G.T. Stewart (1975), "Simultaneous Iteration for the Partial Eigensolution of Real Matrices," *J. Inst. Math Appl.*, 15, 351–362.

E.R. Jessup and D.C. Sorensen (1994), "A Parallel Algorithm for Computing the Singular Value Decomposition of a Matrix," *SIAM J. Matrix. Anal. Appl.*, 15, 530–548.

Z. Jia and G.W. Stewart (2000), "An Analysis of the Rayleigh-Ritz Method for Approximating Eigenspaces," *Math. Comp.*, 70, 637–647.

Y. Jiang, W.W. Hager, and J. Li (2005), "The Geometric Mean Decomposition," *Lin. Alg. Appl.*, 396, 373–384.

Y. Jiang, W.W. Hager, and J. Li (2008), "The Generalized Triangular Decomposition," *Math. Comp.*, 77, 1037–1056.

R. Johnson and D.K. Wichern (1992), *Applied Multivariate Statistical Analysis*, Prentice–Hall, Englewood Cliff, NJ.

R.L. Johnston (1971), "Geršgorin Theorems for Partitioned Matrices," *Lin. Alg. Appl.*, 4, 205–220.

B. Kagström (1985), "The Generalized Singular Value Decomposition and the General $A - \lambda B$ Problem," *BIT*, 24, 568–583.

B. Kagström and A. Ruhe (1980a), "An Algorithm for Numerical Computation of the Jordan Normal Form of a Complex Matrix," *ACM Trans. Math. Soft.*, 6, 398–419.

B. Kagström and A. Ruhe (1980b), "Algorithm 560 JNF: An Algorithm for Numerical Computation of the Jordan Normal Form of a Complex Matrix," *ACM Trans. Math. Software*, 6, 437–443.

W. Kahan (1958), *Gauss-Seidel Method for Solving Large Systems of Linear Equations*, Ph.D. Thesis, University of Toronto.

W. Kahan (1966), "Numerical Linear Algebra," *Canadian Math. Bull.*, 9, 757–801.

D. Kahaner, C.B. Moler, and S. Nash (1988), *Numerical Methods and Software*, Prentice–Hall, Englewood Cliffs, NJ.

L. Kaufman (1993), "An Algorithm for the Banded Symmetric Generalized Matrix Eigenvalue Problem," *SIAM J. Matrix Anal. Appl.*, 14, 372–389.

D. Kincaid and W. Cheney (2002), *Numerical Analysis: Mathematics of Scientific Computing*, 3rd ed., Brooks/Cole, Pacific Grove, CA.

L. Komzsik (2003), *The Lanczos Method: Evolution and Application*, SIAM, Philadelphia.

C. Koukouvinos, E. Lappas, M. Mitrouli, and J. Seberrey (2001), "On the Complete Pivoting Conjecture for Hadamard Matrices of Small Orders," *J. Res. Practice Inform. Technol.*, 33, 298–302.

C. Koukouvinos, M. Mitrouli, and J. Seberry (2000), "Growth in Gaussian Elimination for Weighing Matrices $W(n, n - 1)$," *Lin. Alg. Appl.*, 306, 189–202.

C. Koukouvinos, M. Mitrouli, and J. Seberry (2001), "An Algorithm to Find Formulae and Values of Minors of Hadamard Matrices," *Lin. Alg. Appl.*, 330, 129–147.

C. Kravvaritis and M. Mitrouli (2009), "The Growth Factor of a Hadamard Matrix of Order 16 is 16," *Numer. Lin. Alg. Appl.*, 16, 715–743.

V.N. Kublanovskaya (1961), "On Some Algorithms for the Solution of the Complete Eigenvalue Problem," *USSR Comp. Math. Phys.*, 3, 637–657.

P. Lancaster (1966), *Lamda Matrices and Vibrating Systems*, Pergamon Press, Oxford, UK.

P. Lancaster (1969), *The Theory of Matrices*, Academic Press, New York.

P. Lancaster and M. Tismenetsky (1985), *The Theory of Matrices with Applications*, 2nd ed., Academic Press, New York.

C. Lanczos (1950), "An Iteration Method for the Solution of the Eigenvalue Problem of Linear Differential and Integral Operators," *J. Res. Nat. Bur. Stand.*, 45, 255–282.

C. Lanczos (1952), "Solution of Systems of Linear Equations by Minimized Iteration," *J. Res. Nat. Bur. Stand.*, 49, 33–53.

H. Langer, B. Najman, and K. Veselić (1992), "Perturbation of the Eigenvalues of Quadratic Matrix Polynomials," *SIAM J. Matrix Anal. Appl.*, 13, 474–489.

C.L. Lawson and R.J. Hanson (1995), *Solving Least Squares Problems*, SIAM, Philadelphia (revised version of 1974 Prentice–Hall edition).

D. Lay (2003), *Linear Algebra and Its Applications*, 3rd ed., Addison–Wesley, New York.

A.C. Lazer and P.J. McKenna (1990), "Large-Amplitude Periodic Oscillations in Suspension Bridges: Some New Connections with Nonlinear Analysis," *SIAM Rev.*, 32, 537–578.

T.-L. Lee, T.-Y. Li, and Z. Zeng (2009), "A Rank-Revealing Method with Updating, Downdating, and Applications. Part II," *SIAM J. Matrix Anal. Appl.*, 31, 503–525

R.B. Lehoucq (1995), *Analysis and Implementations of an Implicitly Restarted Arnoldi Iteration*, Ph.D. Thesis, Rice University, Houston, TX.

R.B. Lehoucq and D.C. Sorensen (1996), "Deflation Techniques with an Implicitly Restarted Arnoldi Iteration," *SIAM J. Matrix Anal. Appl.*, 17, 789–821.

R.B. Lehoucq, D.C. Sorensen, and C. Yang (1998), *ARPACK Users' Guide: Solution of Large-Scale Eigenvalue Problems with Implicitly Restarted Arnoldi Methods*, SIAM, Philadelphia.

S. Leon (2005), *Linear Algebra with Applications*, 7th ed., Pearson, New York.

R.-C. Li and Q. Ye (2003), "A Krylov Subspace Method for Quadratic Matrix Polynomials with Application to Constrained Least Squares Problems," *SIAM J. Matrix Anal. Appl.*, 25, 405–428.

T.Y. Li, K. Li, and Z. Zeng (1994), "An Algorithm for the Generalized Symmetric Tridiagonal Eigenproblem," *Numer. Alg.*, 8, 269–291.

T.Y. Li and Z. Zeng (2005), "A Rank-Revealing Method with Updating, Downdating and Applications," *SIAM J. Matrix. Anal. Appl.*, 26, 918–946.

D. Luenberger (1973), *Introduction to Linear and Nonlinear Programming*, Addison–Wesley, New York.

D. Luenberger (1979), *Introduction to Dynamic Systems: Theory, Models, and Applications*, John Wiley and Sons, New York.

D. S. Mackey, N. Mackey, C. Mehl, and V. Mehrmann (2006), "Structured Polynomial Eigenvalue Problems: Good Vibrations from Good Linearizations," *SIAM J. Matrix Anal. Appl.*, 28, 1029–1051.

D.J. Major and R.D. Sidman (1991), "A New Use of Singular Value Decomposition in Bioelectric Imaging of the Brain," in *SVD and Signal Processing* II: *Algorithms, Analysis, and Applications*, R.J. Vaccaro, ed., Elsevier, Amsterdam, 497–512.

P. Majumdar (2005), *Computational Methods for Heat and Mass Transfer*, Taylor & Francis, New York.

C. McCarthy and G. Strang (1973), "Optimal Conditioning of Matrices," *SIAM J. Numer. Anal.*, 10, 370–388.

V. Mehrmann and H. Voss (2004), "Nonlinear Eigenvalue Problems: A Challenge for Modern Eigenvalue Methods," *GAMM Mitt. Ges. Angew. Math. Mech.*, 27, 121–152.

C.D. Meyer (2000), *Matrix Analysis and Applied Linear Algebra*, SIAM, Philadelphia.

C.B. Moler (2004), *Numerical Computing with MATLAB*, SIAM, Philadelphia.

C.B. Moler and G.W. Stewart (1973), "An Algorithm for Generalized Matrix Eigenvalue Problems," *SIAM J. Numer. Anal.*, 10, 241–256.

P.C. Muller and W.O. Schiehlen (1985), *Linear Vibrations*, Martinus Nijhoff, Dordrecht, The Netherlands.

J. Money and Q. Ye (2005), "Algorithm 845: EIGIFP: A MATLAB Program for Solving Large Symmetric Generalized Eigenvalue Problems," *ACM Trans. Math. Software*, 31, 270–279.

N.M. Nachtigal, S.C. Reddy, and L.N. Trefethen (1992), "How Fast Are Nonsymmetric Matrix Iterations?," *SIAM J. Matrix Anal. Appl.*, 13, 778–795.

N.M. Nachtigal, L. Reichel, and L.N. Trefethen (1992), "A Hybrid GMRES Algorithm for Nonsymmetric Linear Systems," *SIAM J. Matrix Anal. Appl.*, 13, 796–825.

J.L. Nazareth (1989), "Updating the Triangular Factorization of a Matrix," *SIAM J. Matrix Anal. Appl.*, 10, 424–428.

J. Neter, W. Wasserman, and M. Kutner (1983), *Applied Linear Regression Models*, Richard D. Irwin Inc., Homewood, IL.

M. Neumann and R. Plemmons (1987), "Convergence of Parallel Multisplitting Iterative Methods for M-matrices," *Lin. Alg. Appl.*, 88, 559–575.

B. Noble and J.W. Daniel (1988), *Applied Linear Algebra*, Prentice–Hall, Englewood Cliffs, NJ.

J. Nocedal and S.J. Wright (2006), *Numerical Optimization*, 2nd ed., Springer-Verlag, Berlin.

W. Oettli and W. Prager (1964), "Compatibility of Approximate Solutions of Linear Equations with Given Error Bounds for Coefficients and Right Hand Sides," *Numer. Math.*, 6, 405–409.

S. Okamoto (1984), *Introduction to Earthquake Engineering*, 2nd ed., University of Tokyo Press.

D.P. O'Leary (1980), "Estimating Matrix Condition Numbers," *SIAM J. Sci. Statist. Comput.*, 1, 205–209.

D.P. O'Leary (2009), *Scientific Computing with Case Studies*, SIAM, Philadelphia.

D. O'Leary and R. White (1985), "Multisplitting of Matrices and Parallel Solution of Linear Systems," *SIAM J. Alg. Discrete Methods*, 6, 630–640.

P. O'Neil (1991), *Advanced Engineering Mathematics*, Wadsworth, Belmont, CA.

J. Ortega (1987), *Matrix Theory: A Second Course*, Plenum Press, New York.

J. Ortega (1990), *Numerical Analysis: A Second Course*, SIAM, Philadelphia.

J. Ortega and W.G. Poole, Jr. (1981), *An Introduction to Numerical Methods for Differential Equations*, Pitman, Boston, MA.

M. Overton (1988), "On Minimizing the Maximum Eigenvalue of a Symmetric Matrix," in *Linear Algebra in Signals, Systems, and Control*, R. Brualdi et al., eds., SIAM, Philadelphia, 150–169.

M. Overton (2001), *Numerical Computing with IEEE Floating Point Arithmetic*, SIAM, Philadelphia.

C.C. Paige (1970), "Practical Use of the Symmetric Lanczos Process with Reorthogonalization," *BIT*, 10, 183–195.

C.C. Paige (1971), *The Computation of Eigenvalues and Eigenvectors of Very Large Sparse Matrices*, Ph.D. Thesis, London University, London.

C.C. Paige (1976), "Error Analysis of the Lanczos Algorithm for Tridiagonalizing a Symmetric Matrix," *J. Inst. Math. Appl.*, 18, 341–349.

C.C. Paige (1980), "Accuracy and Effectiveness of the Lanczos Algorithm for the Symmetric Eigenproblem," *Lin. Alg. Appl.*, 34, 235–258.

C.C. Paige (1986), "Computing the Generalized Singular Value Decomposition," *SIAM J. Sci. Statist. Comput.*, 7, 1126–1146.

C.C. Paige, B. Parlett, and H.A. van der Vorst (1995), "Approximate Solutions and Eigenvalue Bounds from Krylov Subspaces," *Num. Lin. Alg. Appl.*, 2, 115–134.

C.C. Paige and M.A. Saunders (1975), "Solutions of Sparse Indefinite Systems of Linear Equations," *SIAM J. Numer. Anal.*, 12, 617–629.

C.C. Paige and M.A. Saunders (1981), "Towards a Generalized Singular Value Decomposition," *SIAM J. Numer. Anal.*, 18, 398–405.

C.C. Paige and P. Van Dooren (1986), "On the Quadratic Convergence of Kogbetliantz's Algorithm for Computing the Singular Value Decomposition," *Lin. Alg. Appl.*, 77, 301–313.

C.-T. Pan and K. Sigmon (1994), "A Bottom-Up Inductive Proof of the Singular Value Decomposition," *SIAM J. Matrix Anal. Appl.*, 15, 59–61.

B.N. Parlett (1965), "Convergence of the QR Algorithm," *Numer. Math.*, 7, 187–193; correction in *Numer. Math.*, 10, 163–164.

B.N. Parlett (1966), "Singular and Invariant Matrices under the QR Algorithm," *Math. Comp.*, 20, 611–615.

B.N. Parlett (1968), "Global Convergence of the Basic QR Algorithm on Hessenberg Matrices," *Math. Comp.*, 22, 803–817.

B.N. Parlett (1974), "The Rayleigh Quotient Iteration and Some Generalizations for Nonnormal Matrices," *Math. Comp.*, 28, 679–693.

B.N. Parlett (1980), "A New Look at the Lanczos Algorithm for Solving Symmetric Systems of Linear Equations," *Lin. Alg. Appl.*, 29, 323–346.

B.N. Parlett (1989), "Towards a Black Box Lanczos Program," *Comp. Phys. Comm.*, 53, 169–179.

B.N. Parlett (1991), " Symmetric Matrix Pencils," *J. Comp. Appl. Math.*, 38, 373–385.

B.N. Parlett (1992), "The Rewards for Maintaining Semi-orthogonality among Lanczos Vectors," *J. Num. Lin. Alg. Appl.*, 1, 243–267.

B.N. Parlett (1995), "The New QD Algorithms," *Acta Numerica*, 459–491.

B.N. Parlett (1998), *The Symmetric Eigenvalue Problem*, SIAM, Philadelphia.

B.N. Parlett and Chen (1990), "Use of Indefinite Pencils for Computing Damped Natural Modes," *Lin. Alg. Appl.*, 140, 53–88.

B.N. Parlett and I. Dhillon (1997), "Fernado's Solution to Wilkinson's Problem: An Application of Double Factorization," *Lin. Alg. Appl.*, 267, 247–279.

B.N. Parlett and J.K. Reid (1970), "On the Solution of a System of Linear Equations Whose Matrix Is Symmetric but Not Definite," *BIT*, 10, 386–397.

B.N. Parlett and J.K. Reid (1981), "Tracking the Progress of the Lanczos Algorithm for Large Symmetric Eigenproblems," *IMA J. Num. Anal.*, 1, 135–155.

B.N. Parlett and Reinsch (1969), "Balancing a Matrix for Calculation of Eigenvalues and Eigenvectors," *Numer. Math.*, 13, 292–304.

B.N. Parlett and D.S. Scott (1979), "The Lanczos Algorithm with Selective Orthogonalization," *Math. Comp.*, 33, 217–238.

B.N. Parlett, H. Simon, and L.M. Stringer (1982), "On Estimating the Largest Eigenvalue with the Lanczos Algorithm," *Math. Comp.*, 38, 153–166.

B.N. Parlett, D.R. Taylor, and Z.A. Liu (1985), "A Look-Ahead Lanczos Algorithm for Unsymmetric Matrices," *Math. Comp.*, 44, 105–124.

R.V. Patel, A.J. Laub, and P.M. Van Dooren, eds. (1994), *Numerical Linear Algebra Techniques for Systems and Control*, IEEE Computer Society Press, Piscataway, NJ.

G. Peters and J.H. Wilkinson (1979), "Inverse Iteration, Ill-Conditioned Equations and Newton's Method," *SIAM Rev.*, 21, 339–360.

P.H. Petkov, N.D. Christov, and M.M. Konstantinov (1991), *Computational Methods for Linear Control Systems*, Prentice–Hall International, Hemel Hempstead, UK.

D. Pierce and R.J. Plemmons (1988), "A Two-Level Preconditioned Conjugate Gradient Scheme," in *Linear Algebra in Signals, Systems, and Control*, R. Brualdi et al., eds., SIAM, Philadelphia, 170–185.

D.J. Pierce and R.J. Plemmons (1992), "Fast Adaptive Condition Estimation," *SIAM J. Matrix Anal. Appl.*, 13, 274–291.

R.J. Plemmons (1974), "Linear Least Squares by Elimination and MGS," *J. ACM*, 21, 581–585.

C.R. Rao and S.K. Mitra (1971), *Generalized Inverse of Matrices and Its Applications*, John Wiley and Sons, New York.

C.R. Rao and M.B. Rao (1998), *Matrix Algebra and Its Applications to Statistics and Economics*, World Scientific, River Edge, NJ.

J.N. Reddy (1993), *An Introduction to the Finite Element Method*, McGraw-Hill, New York.

L. Reichel and Q. Ye (2005), "Breakdown-free GMRES for Singular Systems," *SIAM J. Matrix Anal. Appl.*, 26, 1001–1021.

J.R. Rice (1966), "A Theory of Condition," *SIAM J. Numer. Anal.*, 3, 287–310.

J.R. Rice (1981), *Matrix Computations and Mathematical Software*, McGraw–Hill, New York.

A. Ruhe (1994), "Rational Krylov Algorithms for Nonsymmetric Eigenvalue Problems. II. Matrix Pairs," *Lin. Alg. Appl.,* 197/198, 283–295.

H. Rutishauser (1958), "Solution of Eigenvalue Problems with the LR Transformation," *Nat. Bar. Stand. Appl. Math. Ser.*, 49, 47–81.

Y. Saad (1981), "Krylov Subspace Methods for Solving Large Unsymmetric Linear Systems," *Math. Comp.*, 48, 651–662.

Y. Saad (1984), "Practical Use of Polynomial Preconditionings for the Conjugate Gradient Method," *SIAM J. Sci. Statist. Comput.*, 6, 865–881.

Y. Saad (1987), "On the Lanczos Method for Solving Symmetric Systems with Several Right Hand Sides," *Math. Comp.*, 48, 651–662.

Y. Saad (1988), "Preconditioning Techniques for Indefinite and Nonsymmetric Linear Systems," *J. Comp. Appl. Math.*, 24, 89–105.

Y. Saad (1992), *Numerical Methods for Large Eigenvalue Problem: Theory and Algorithms*, John Wiley and Sons, New York.

Y. Saad (1993), "A Flexible Inner-Outer Preconditioned GMRES Algorithm," *SIAM J. Sci. Comput.*, 14, 461–469.

Y. Saad (2003), *Iterative Methods for Sparse Linear Systems*, 2nd ed., SIAM, Philadelphia.

Y. Saad and M.H. Schultz (1986), "GMRES: A Generalized Minimal Residual Algorithm for Solving Nonsymmetric Linear Systems," *SIAM J. Sci. Statist. Comput.*, 7, 856–869.

Y. Saad and H.A. van der Vorst (2000), "Iterative Solution of Linear Systems in the 20th Century," *J. Comp. Appl. Math.*, 123, 1–33.

P. Saylor and D.C. Smolarski (1988), "An Optimum Iterative Method for Solving Any Linear System with a Square Matrix," *BIT*, 28, 163–178.

H. Schneider and G. P. Barker (1989), *Matrices and Linear Algebra*, Dover, New York.

J.R. Shewchuk (1994), "An Introduction to the Conjugate Gradient Method without the Agonizing Pain," *Tech. Report, CMU-CS* 94-125, Carnegie Mellon University, Pittsburgh, PA.

H. Simon (1984), "Analysis of the Symmetric Lanczos Algorithm with Reorthogonalization Methods," *Lin. Alg. Appl.*, 61, 101–132.

R. Skeel (1979), "Scaling for Numerical Stability in Gaussian Elimination," *J. ACM*, 26, 494–526.

R. Skeel (1980), "Iterative Refinement Implies Numerical Stability for Gaussian Elimination, *Math. Comp.*, 35, 817–832.

R.D. Skeel (1981), "Effect of Equilibration on Residual Size for Partial Pivoting," *SIAM J. Numer. Anal.*, 18, 449–454.

G.L.G. Sleijpen and H.A. van der Vorst (1996), "A Jacobi–Davidson Iteration Method for Linear Eigenvalue Problems," *SIAM J. Matrix Anal. Appl.*, 17, 401–425.

G.L.G. Sleijpen, A.G.L. Booten, D.R. Fokkema, and H.A. van der Vorst (1996), "Jacobi–Davidson Type Methods for Generalized Eigenproblems and Polynomial Eigenproblems," *BIT*, 36, 595–633.

G.L.G. Sleijpen, H.A. van der Vorst, and M. van Gijzen (1996), "Quadratic Eigenproblems Are No Problem," *SIAM News*, 29 (7), 8–9.

V. Sokolov (2008), *The Quadratic Inverse Eigenvalue Problems: Theory, Methods, and Applications*, Ph.D. Dissertation, Northern Illinois University, DeKalb, IL.

P. Sonneveld (1989), "CGS, a Fast Lanczos-Type Solver for Nonsymmetric Linear Systems," *SIAM J. Sci. Statist. Comput.*, 10, 36–52.

D.C. Sorensen (1992), "Implicit Application of Polynomial Filters in a k-step Arnoldi Method," *SIAM J. Matrix Anal. Appl.*, 13, 357–385.

D.C. Sorensen and P.T.P. Tang (1991), "On the Orthogonality of Eigenvectors Computed by Divide-and-Conquer Techniques," *SIAM J. Numer. Anal.*, 28, 1752–1775.

R.F. Steidel (1979), *An Introduction to Mechanical Vibration*, 2nd ed., John Wiley and Sons, New York.

G.W. Stewart (1973), *Introduction to Matrix Computations*, Academic Press, New York.

G.W. Stewart (1975), "Geršgorin Theory for the Generalized Eigenvalue Problem $Ax = \lambda Bx$," *Math. Comp.*, 29, 600–606.

G.W. Stewart (1976a), "A Bibliographical Tour of the Large Sparse Generalized Eigenvalue Problem," in *Sparse Matrix Computations*, J.R. Banch and D.J. Rose, eds., Academic Press, New York, 113–130.

G.W. Stewart (1976b), "Simultaneous Iteration for Computing Invariant Subspaces of Non-Hermitian Matrices," *Numer. Math.*, 25, 123–136.

G.W. Stewart (1977a), "On the Perturbation of Pseudo-inverses, Projections and Linear Least Squares Problems," *SIAM Rev.*, 19, 634–662.

G.W. Stewart (1977b), "Perturbation Bounds for the QR Factorization of a Matrix," *SIAM J. Numer. Anal.*, 14, 509–518.

G.W. Stewart (1978), "Perturbation Theory for the Generalized Eigenvalue Problem," in *Recent Advances in Numerical Analysis*, C. De Boor and G. H. Golub, eds., Academic Press, New York, 193–206.

G.W. Stewart (1979), "Perturbation Bounds for the Definite Generalized Eigenvalue Problem," *Lin. Alg. Appl.*, 23, 69–86.

G.W. Stewart (1983), "A Method for Computing the Generalized Singular Value Decomposition," in *Matrix Pencils*, B. Kagström and A. Ruhe, eds., Springer-Verlag, New York, 207–220.

G.W. Stewart (1984), "Rank Degeneracy", *SIAM J. Sci. Statist. Comput.*, 5, 403–413.

G.W. Stewart (1987), "Collinearity and Least Squares Regression," *Statist. Sci.*, 2, 68–100.

G.W. Stewart (1991), "Perturbation Theory for the Singular Value Decomposition," in *SVD and Signal Processing* II: *Algorithms, Analysis, and Applications*, R.J. Vaccaro, ed., Elsevier, Amsterdam, 99–109.

G.W. Stewart (1993a), "On the Perturbation of LU, Cholesky, and QR Factorizations," *SIAM J. Matrix Anal. Appl.*, 14, 1141–1145.

G.W. Stewart (1993b), "On the Early History of the Singular Value Decomposition," *SIAM Rev.*, 35, 551–566.

G.W. Stewart (1998a), *Afternotes Goes to Graduate School*, SIAM, Philadelphia.

G.W. Stewart (1998b), *Matrix Algorithms, Volume I: Basic Decomposition*, SIAM, Philadelphia.

G.W. Stewart (2001a), *Matrix Algorithms, Volume II: Eigensystems*, SIAM, Philadelphia.

G.W. Stewart (2001b), "A Krylov–Schur Algorithm for Large Eigenproblems," *SIAM J. Matrix Anal. Appl.*, 23, 601–614.

G.W. Stewart and J.G. Sun (1990), *Matrix Perturbation Theory*, Academic Press, San Diego.

G. Strang (1986), *Introduction to Applied Mathematics*, Wellesley-Cambridge Press, Wellesley, MA.

G. Strang (2003), *Introduction to Linear Algebra*, 3rd ed., Wellesley-Cambridge Press, Wellesley, MA.

G. Strang (2006), *Linear Algebra and Its Applications*, 4th ed., Cengage, 2006.

G. Strang and G. Fix (1973), *An Analysis of the Finite Element Method*, Prentice–Hall, Englewood Cliffs, NJ.

J.-G. Sun (1983), "Perturbation Analysis for the Generalized Singular Value Problem," *SIAM J. Numer. Anal.*, 20, 611–625.

J.G. Sun (1991), "Perturbation Bounds for the Cholesky and QR Factorizations," *BIT*, 31, 341–352.

R.A. Thisted (1988), *Elements of Statistical Computing*, Chapman and Hall, New York.

W.T. Thomson (1992), *Theory of Vibrations with Applications*, 4th ed., Prentice–Hall, Englewood Cliffs, NJ.

F. Tisseur (2000), "Backward Error and Condition of Polynomial Eigenvalue Problems," *Lin. Alg. Appl.*, 309, 339–361.

F. Tisseur and K. Meerbergen (2001), "The Quadratic Eigenvalue Problem," *SIAM Rev.*, 43, 235–286.

C. Tong and Q. Ye (2000), "Analysis of Finite Precision Bi-conjugate Gradient Algorithm for Nonsymmetric Linear Systems," *Math. Comp.*, 69, 1559–1575.

L. Trefethen (2007), "Obituary: Gene H. Golub (1932–2007)," *Nature*, 450, 962.

L.N. Trefethen and D. Bau, III (1997), *Numerical Linear Algebra*, SIAM, Philadelphia.

L.N. Trefethen and R.S. Schreiber (1990), "Average-Case Stability of Gaussian Elimination," *SIAM J. Matrix Anal. Appl.*, 11, 335–360.

R.J. Vaccaro, ed. (1991), *SVD and Signal Processing* II: *Algorithms, Analysis, and Applications*, Elsevier, Amsterdam.

A. van der Sluis (1969), "Condition Numbers and Equilibration of Matrices," *Numer. Math.*, 14, 14–23.

H.A. van der Vorst (1992), "Bi-CGSTAB: A Fast and Smoothly Converging Variant of Bi-CG for the Solution of Nonsymmetric Linear Systems," *SIAM J. Sci. Statist. Comput.*, 13, 631–644.

H.A. van der Vorst (1996), "Modern Methods for the Iterative Solution of Large Systems of Linear Equations," *Nieuw Archief voor Wiskunde, Vierde Serie Deel*, 14, 127–143.

H.A. van der Vorst (2002), "Computational Methods for Large Eigenvalue Problems," in *Handbook of Numerical Analysis, Vol. 8*, North–Holland (Elsevier), Amsterdam, 3–179.

H.A. van der Vorst (2003), *Iterative Krylov Methods for Large Linear Systems*, Cambridge University Press, Cambridge, UK.

H.A. van der Vorst and T.C. Chan (1997), "Linear System Solvers: Sparse Iterative Solvers," in *Parallel Numerical Algorithms, ICASE/LaRC Interdisciplinary Series in Science and Engineering* 4, Kluwer Academic, Dordrecht, The Netherlands, 91–118.

H.A. van der Vorst and C. Vuik (1993), "The Superlinear Convergence Behaviour of GMRES," *J. Comp. Appl. Math.*, 48, 327–341.

H.A. van der Vorst and Q. Ye (2000), "Residual Replacement Strategies for Krylov Subspace Iterative Methods for the Convergence of True Residuals," *SIAM J. Sci. Comput.*, 22, 835–852.

J. Vandewalle and B. De Moor (1988), "Variety of Applications of Singular Value Decomposition in Identification and Signal Processing," in *SVD and Signal Processing: Algorithms, Applications and Architectures*, E.F. Deprettere, ed., Elsevier (North–Holland), Amsterdam, 43–89.

P. Van Dooren (1981a), "A Generalized Eigenvalue Approach for Solving Riccati Equations," *SIAM J. Sci. Statist. Comput.*, 2, 121–135.

P. Van Dooren (1981b), "The Generalized Eigenstructure Problem in Linear System Theory," *IEEE Trans. Automat. Control*, 26, 111–128.

P. Van Dooren (1991), "Structured Linear Algebra Problems in Digital Signal Processing," in *Numerical Linear Algebra, Digital Signal Processing, and Parallel Algorithms*, NATO Adv. Sci. Inst. Ser. F Comput. Systems Sci. 70, Springer-Verlag, Berlin, 361–384.

C. Van Loan (1976), "Generalizing the Singular Value Decomposition," *SIAM J. Numer. Anal.*, 13, 76–83.

C. Van Loan (1987), "On Estimating the Condition of Eigenvalues and Eigenvectors," *Lin. Alg. Appl.*, 88/89, 715–732.

C. Van Loan (2000), *An Introduction to Scientific Computing: A Matrix-Vector Approach Using MATLAB*, 2nd ed., Prentice–Hall, Upper Saddle River, NJ.

J.M. Varah (1968a), "The Calculation of the Eigenvectors of a General Complex Matrix by Inverse Iteration," *Math. Comp.*, 22, 785–791.

J.M. Varah (1968b), "Rigorous Machines Bounds for the Eigensystem of a General Complex Matrix," *Math. Comp.*, 22, 793–801.

J.M. Varah (1970), "Computing Invariant Subspaces of a General Matrix When the Eigensystem Is Poorly Determined," *Math. Comp.*, 24, 137–149.

R.S. Varga (2000), *Matrix Iterative Analysis*, Springer-Verlag, Berlin, 2000 (1st ed. published by Prentice–Hall in 1962).

R.S. Varga (2004), *Geršgorin and His Circles*, Springer-Verlag, Berlin.

H.F. Walker (1988), "Implementation of the GMRES Method Using Householder Transformations," *SIAM J. Sci. Statist.*, 9, 152–163.

S. Wang and S. Zhao (1991), "An Algorithm for $Ax = \lambda Bx$ with Symmetric and Positive-Definite A and B, *SIAM J. Matrix Anal. Appl.*, 12, 654–660.

R.C. Ward (1981), "Balancing the Generalized Eigenvalue Problem," *SIAM J. Sci Statist. Comput.*, 2, 141–152.

D.S. Watkins (2002), *Fundamentals of Matrix Computations*, 2nd ed., John Wiley and Sons, New York.

P.A. Wedin (1973), "Perturbation Theory for Pseudo-inverses," *BIT*, 13, 217–232.

J.H. Wilkinson (1963), *Rounding Errors in Algebraic Process*, Prentice–Hall, Englewood Cliff, NJ.

J.H. Wilkinson (1965), *The Algebraic Eigenvalue Problem*, Clarendon Press, Oxford, UK.

T. Williams and A. Laub (1992), "Orthogonal Canonical Forms for Second-Order Systems," *IEEE Trans. Automat. Control*, 37, 1050–1052.

S.J. Wright (1993), "A Collection of Problems for which Gaussian Elimination with Partial Pivoting Is Unstable," *SIAM J. Sci. Comput.*, 14, 231–238.

Q. Ye (1994), "A Breakdown-Free Variation of the Nonsymmetric Lanczos Algorithms," *Math. Comp.*, 62, 179–207.

Q. Ye (1996), "An Adaptive Block Lanczos Algorithm," *Numer. Alg.*, 12, 97–110.

Q. Ye (2006), "An Iterated Shift-and-Invert Arnoldi Algorithm for Quadratic Matrix Eigenvalue Problems," *Appl. Math. Comp.*, 172, 818–827.

D.M. Young (1970), "Convergence Properties of the Symmetric and Unsymmetric Over-Relaxation Methods," *Math. Comp.*, 24, 793–807.

D.M. Young (1971), *Iterative Solution of Large Linear Systems*, Academic Press, New York.

D.M. Young (1972), "Generalization of Property A and Consistent Ordering," *SIAM J. Numer. Anal.*, 9, 454–463.

D.M. Young and K.C. Jea (1980), "Generalized Conjugate Gradient Acceleration of Nonsummetrizable Iterative Methods," *Lin. Alg. Appl.*, 34, 159–194.

D.M. Young, K.C. Jea, and T.-Z. Mai (1988), "Preconditioned Conjugate Gradient Algorithms and Software for Solving Large Sparse Linear Systems," in *Linear Algebra in Signals, Systems, and Control*, B.N. Datta et al. eds., SIAM, Philadelphia, 260–283.

Z. Zeng and T. Li (2008), "Numerical Computation of the Jordan Canonical Form," unpublished manuscript.

H. Zha (1993), "A Componentwise Perturbation Analysis of the QR Decomposition," *SIAM J. Matrix Anal. Appl.*, 14, 1124–1131.

Index

(Page numbers in *italic* type indicate online content.)

absolute error, 31
algorithm, 49
    for Arnoldi method, 447
    for Arnoldi method (implicit) for eigenpair, 476
    for Arnoldi method for eigenpair, 475
    for Arnoldi method for linear system, 450
    for back substitution, 51
    for bi-conjugate gradient, 465
    for bisection method, 356
    for block back substitution, 165
    for block forward elimination, 165
    for block LU factorization, 164
    for Cholesky factorization, 156
    for Cholesky factorization (incomplete), 469
    for condition estimation, 142
    for conjugate gradient (preconditioned), 470
    for conjugate gradient method, 458
    for diagonalization of symmetric definite pencil, 402
    for divide-and-conquer method, 362
    for forward elimination, 52
    for generalized eigenvector, 399
    for Givens matrix product, 196
    for GMRES method, 453
    for Hessenberg inverse iteration, 336
    for Householder Hessenberg reduction, 310
    for inner product, 50
    for inverse iteration, 299
    for iterative refinement (least-squares), 267
    for iterative refinement (linear system), 144
    for Lanczos method (nonsymmetric), 463
    for least-squares solution (rank-deficient) using SVD, 261
    for least-squares solution using MGS, 252
    for least-squares solution using normal equations, 252
    for least-squares solution using QR factorization (Householder), 256
    for least-squares solution using SVD, 259
    for least-squares solution via reduced SVD, 259
    for linear system with multiple right-hand sides, 134
    for linear system with partial pivoting, 130
    for linear system without explicit factorization, 132
    for LU factorization of tridiagonal matrix, 162
    for LU factorization with complete pivoting, 105
    for LU factorization with partial pivoting, 102
    for LU factorization without pivoting, 92
    for matrix inverse, 138
    for positive definite linear system (Cholesky), 157
    for power method, 296
    for pseudoinverse using SVD, 369
    for QR factorization (CGS), 204
    for QR factorization (Givens), 200
    for QR factorization (Householder), 191

for QR factorization (MGS), 204
for QR factorization of Hessenberg matrix, 202
for QR iteration (basic), 322
for QR iteration (implicit double-shift), 330
for QR iteration (symmetric), 358
for QR updating, *536*
for QZ iteration (complete), 398
for QZ iteration (one-step), 397
for QZ iteration for symmetric definite pencil, 401
for Rayleigh quotient iteration, 302
for Rayleigh quotient iteration (generalized), 404
for symmetric Lanczos factorization, 455
for underdetermined least-squares solution (full rank), 264
for underdetermined least-squares solution (minimum-norm) using QR, 264
for vector norm, 50
for zeroing in vector (Givens), 198
for zeroing in vector (Householder), 187
amplitude, 406
applications
   giving rise to block tridiagonal linear systems, 149
   giving rise to eigenvalue problem, 284, 285, 288, 290, 292
   giving rise to generalized eigenvalue problem, 405, 406, 409
   giving rise to linear systems, 119, 120, 122, 123, 125, 128
   giving rise to positive definite and tridiagonal linear systems, 151
   giving rise to quadratic eigenvalue problem, 421
   giving rise to special linear systems, 147, 150
   of least-squares problem, 242
   of SVD, 221
Arnoldi method, 446
   for eigenvalue approximation, 474
   for linear system solution, 448

Arnoldi, Walter Edwin, 447
asymptotically stable solution, 282

backward stability, 54
balancing, 334
Bauer–Fike theorem, 315
bi-conjugate gradient method, 464
bidiagonal reduction, 371
bisection method
   for symmetric eigenvalue problem, 354
   for symmetric positive definite eigenvalue problem, 478
block cyclic reduction, 167
bulge-chasing phenomenon, 328

catastrophic cancellation, 41
Cauchy–Schwarz inequality, 18
Cholesky algorithm, 156
Cholesky, Andre-Louis, 156
Cholesky factorization, 112, 154
companion matrix, 313
condition number, 57
   of matrix-vector product, 58
   and accuracy of linear system solution, 139
   and pivoting, 140
   and singular values, 216
   and singularity, 69
   estimation by Hager's algorithm, 142
   estimation of, 141
   of a function, 58
   of a rectangular matrix, 246
   of an eigenvalue, 317
   properties of, 67
   using pseudoinverse, 246
conjugate gradient method, 456
   convergence of, 459
   error bound for, 459
convergence
   of conjugate gradient method, 459
   of Gauss–Seidel method, 439, 440
   of iterative methods, 438
   of Jacobi method, 439
convergent matrix, 438
Courant, Richard, 352
Courant–Fischer theorem, 352
Cramer's rule, 117

Index

damping
    modal, 412
    nonproportional, 413
    proportional, 412
    Rayleigh, 412
deflation, 332
determinant, 11
    computation of, 139
diagonally dominant matrix, 160
divide-and-conquer method, 359
Doolittle reduction, 112

eigenvalue
    algebraic multiplicity of, 305
    and asymptotic stability, 283
    and principal component analysis, 290
    applications in European arms race, 284
    approximation using Arnoldi method, 474
    bounds using matrix norms, 295
    computation using QR iteration (implicit double-shift), 327
    computational difficulties using characteristic polynomial, 313
    condition number, 317
    defective, 305
    sensitivity, 315
    stock market example of, 292
eigenvector
    sensitivity, 319
elementary matrix
    lower triangular, 82
elementary reflector, 183
equilibrium solution, 283
equivalent matrices, 389
error
    backward, 53
    forward, 53
European arms race, 284

finite difference scheme, 125
floating point arithmetic, 29
    addition in, 36
    inner product in, 38
    multiplication in, 37
flop, 52

flop-count
    for bisection algorithm, 357
    for Hessenberg linear system, 158
    for Hessenberg triangular reduction, 392
    for Householder matrix product, 185
    for Householder QR factorization, 193
    for inverse computation, 138
    for least-squares solution using MGS process, 258
    for least-squares solution using SVD, 262, 379
    for linear systems solution using partial pivoting, 130
    for LU factorization with complete pivoting, 106
    for LU factorization with partial pivoting, 102
    for LU factorization without pivoting, 93
    for normal equations method, 252
    for one-step QZ iteration, 397
    for QR factorization using Givens matrices, 200
    for QR factorization using Householder matrices, 193
    for QR factorization with Gram–Schmidt process, 205
    for QR iteration, 333
    for QZ iteration (complete), 398
    for SVD, 377
    for symmetric QR iteration, 358
    for underdetermined least-squares solution using QR, 265
forward elimination, 51
forward error bound
    for inner product, 53
    for matrix multiplication, 54

Galerkin method, 449
Gauss, Karl Friedrich, 84
Gauss–Jordan matrix, 112
Gauss–Jordan reduction, 113
Gauss–Seidel iteration, 437
Gauss–Seidel method, 436

Gaussian elimination, 82
    with complete pivoting, 104
    with partial pivoting, 97, 100
    without pivoting, 89
generalized eigenvalue problem, 387
    symmetric definite, 388
generalized real Schur decomposition, 389
generalized SVD, 379
generalized triangular decomposition, 226
geometric interpretation
    of Householder transformation, 183
    of least-squares problem, 238
    of singular values and singular vectors, 214
geometric mean decomposition, 226
Geršgorin disk theorem, 293
Geršgorin disks, 293
Geršgorin, Semyon Aranovich, 293
Givens matrix, 194
Givens QR factorization, 199, 200
Givens rotation, 195
Givens, Wallace, 194
Givens–Hessenberg QR factorization, 202
GMRES method, 449, 451
Golub, Gene H., 256
Google matrix, 296
Gram, Jorgen Pedersen, 202
Gram–Schmidt process, 202
growth factor, 107
    for a banded matrix, 166
    for diagonally dominant system, 161
    for Gaussian elimination with complete pivoting, 109
    for Gaussian elimination with partial pivoting, 108
    for Hessenberg system, 158
    for positive definite system, 154
    for tridiagonal system, 162
guard digit, 35

Hadamard matrix, 109
Hager's norm-1 condition estimator, 142
heat diffusion equation, 124
heat distribution in a medium, 127
Hessenberg
    QR iteration, 324
    reduction using Givens matrices, 311
    reduction using Householder matrices, 307
Hessenberg matrix, 16
Hessenberg QR factorization, 202
Hessenberg reduction, 307, 311
    uniqueness of, 312
Hessenberg system, 158
Hessenberg, Karl, 307
Hessenberg triangular form, 391
Hilbert system, 128
Householder matrix
    properties, 183
Householder QR factorization, 188, 191
Householder transformation, 183
Householder vector, 183
Householder, Alston, 183

IEEE standard, 29
ill-conditioned eigenvalue problems, 70
ill-conditioned problem, 57
ill-conditioned system, 68
image compression
    by SVD, 222
image restoration
    using SVD, 222
implicit Q theorem, 312
inner product, 8
interconnected reactors, 120
interlacing property, 355
invariant subspace
    from real Schur form, 334
inverse computation using partial pivoting, 138
inverse iteration, 299
    for Hessenberg matrices, 335
inverse of a matrix , 13
inverse power method, 299
iterative refinement, 144
    and accuracy, 145
    for least-squares problem, 266
    for linear system, 144

Jacobi iteration, 436
Jacobi method
    for linear system solution, 436
    for symmetric eigenvalue problem, 363
Jacobi, Carl Jakob, 363

Jacobi–Davidson method
 for quadratic eigenvalue problem, 482
Jordan canonical theorem, 306
Jordan, Camile, 306

Kahan matrix, 217
Kirchhoff's law, 120
Krylov matrix, 445
Krylov methods
 for generalized eigenvalue problems, 479
 for linear systems, 445
 for quadratic eigenvalue problem, 481
Krylov subspace, 445
Krylov, Alexei Nikolaevich, 445

Lanczos method, 455
 for nonsymmetric matrix, 462
 loss of orthogonality, 456
Lanczos, Cornelius, 455
Laplace's equation, 124
leading principal minor, 91
least-squares problem, 238
 and generized inverse, 245
 applications of, 242
 existence and uniqueness, 240
 geometric interpretation of, 238
 sensitivity of, 246
 solution using MGS process, 257
 solution using normal equations, 252
 solution using projection, 241
 solution using QR factorization, 255
 solution using SVD, 259
linear system solution
 existence and uniqueness of, 119
 using Gaussian elimination, 129
 using Gaussian elimination with partial pivoting, 130
 using nonsymmetric Lanczos, 464
 using QR factorization, 208
 using SVD, 263
 with multiple right-hand sides, 134
 without explicit factorization, 132
linear systems
 arising from ordinary and partial differential equations, 122
 arising from partial differential equations, 123

LU factorization, 89, 102
 existence and uniqueness of, 91
 of a tridiagonal matrix, 162
 of block tridiagonal matrix, 164
 with complete pivoting, 105
 with partial pivoting, 100
 without pivoting, 92

mass matrix, 406
MATCOM, *567*
 chapterwise listing of programs, *568*
MATLAB, *552*
 books, 78, *555*
 frequently used operations, *558*
 program writing and examples, *562*
 tutorial, 78
matrix, 9
 block, 10
 block diagonal, 15
 companion, 313
 convergent, 438
 Givens, 194
 Hermitian, 16
 Hessenberg, 16
 Hilbert, 68
 Householder, 183
 Jordan, 306
 Kahan, 217
 Krylov, 445
 nonderogatory, 313
 normal, 318
 orthogonal, 15
 Pei, 69
 pencil, 387
 positive definite, 153
 sparse, 435
 symmetric, 16
 triangular, 15
 tridiagonal, 16
 unitary, 15
 Vandermonde, 69, 243
 variance-covariance, 367
 Wilkinson, 70
matrix pencil, 387
Millennium Bridge, 287

minimum-norm solution
    for rank-deficient least-squares problem, 262
    for underdetermined problem, 265
MINRES method, 460
    error bound for, 461
modal matrix, 410
mode participation factor, 414
model reduction, 415
Modified Gram–Schmidt QR factorization, 204

natural frequency, 406
norm, 17, 22
    spectral, 19
    consistent, 20
    defined by singular values, 216
    equivalent property, 18, 20
    Frobenius, 19
    Hölder, 17
    Hölder inequality, 17
    matrix, 18
    subordinate, 18
    triangle inequality, 17
    vector, 17
normal equations, 239
    difficulties with, 253
null space of a matrix, 12
numerical rank, 219

orthogonal matrix, 15, 181
    norm property of, 22
orthogonal projection
    definition of, 209
    and orthogonal bases, 209
    using QR factorization, 210
    using SVD, 220
orthonormal bases
    using QR factorizations, 210
    using SVD, 220
Ostrowski, Alexander, 443
Ostrowski–Reich theorem, 443
overdetermined system, 239
overflow, 30

PageRank, 296
permutation matrix, 96
    properties of, 96

perturbation analysis, 57
    of least-squares problem, 247, 248, 250
    of linear system, 61, 64, 65
    of singular values, 215
pivots, 89
Poisson's equation, 124, 149
polynomial preconditioner, 468
positive definite matrix
    definition and properties of, 153
power method, 296
    convergence of, 298
precision, 29
preconditioner
    using incomplete Cholesky factorization, 469
preconditioners, 467
    with classical iterative methods, 468
pseudoinverse, 246
    computation with SVD, 368

QMR algorithm, 466
QR factorization
    of Hessenberg matrix, 202
    of a complex matrix, 194
    of a matrix, 181
    rank-revealing, *535*
    reduced, 181
    relationship between Householder and Givens QR factorizations, 201
    uniqueness, 208
    updating, *535*
    using classical Gram–Schmidt process, 203
    using Givens rotations, 199, 200
    using Householder matrices, 188, 191
    using modified Gram–Schmidt process, 204
    with column pivoting, 211, *531*, *532*
QR iteration
    basic, 322
    double-shift (complex), 326
    double-shift (implicit), 327
    double-shift (real), 327
    for symmetric matrices, 357
    Hessenberg, 324

quadratic eigenvalue problem, 418
    linearization, 423
    numerical methods for, 422
    orthogonality of, 419
QZ iteration, 390, 393, 394

range of a matrix, 12
rank, 13
    properties, 13
rank-revealing QR factorization, *535*
Rayleigh quotient, 301
Rayleigh quotient (generalized), 404
Rayleigh, Baron (John William Strutt), 301
real Schur form, 321
    and invariant subspace, 334
    from QR iteration, 331
    generalized, 389
relative error, 31
resonance, 285
Ritz eigenvector, 474
Ritz value, 474
round-off error, 32
    for back substitution, *539*
    for Cholesky algorithm, 156
    for forward elimination, *537*
    for Gaussian elimination, 106, *540*, *542*
    for Gaussian elimination with pivoting, 107
    for Givens QR factorization, 200
    for Householder Hessenberg reduction, 311
    for Householder QR factorization, 193
    for least-squares solution using Modified Gram–Schmidt process, 258
    for least-squares solution using QR factorization, 257
    for linear system solution using LU factorization, *544*
    for linear system solution with partial pivoting, 131, *546*
    for matrix operations, 39
    for orthogonal matrix multiplication, 40
    for QR iteration, 333
    for QZ iteration, 398
    for SVD computation, 377

scaling, 135
scaling and conditioning, 136
Schmidt, Erhard, 202
Schur triangularization theorem, 321
Schur, Issai, 320
secular equation, 361
sensitivity
    of eigenvalues, 315, 317
    of eigenvalues of normal matrix, 318
    of eigenvectors, 319
    of individual eigenvalues, 317
    of least-squares problem, 246, 249, 251
    of linear systems solutions, 61
    of singular values, 215, 367
Sherman–Morrison formula, 137
significant digits, 32
similar matrices, 14
singular values, 212
    relationship with eigenvalues, 366
singular vectors, 212
Skeel's condition number, 143
smallest eigenvalue
    computation of, 303
software for matrix computations, *551*
SOR (successive overrelaxation method), 436
    iteration, 437
sparse matrix, 435
special linear systems
    arising from finite difference methods, 147
    arising from finite element methods, 150
spectral decomposition, 16
spectral radius, 12
spring-mass problem, 122
stability
    conditioning and accuracy, 60
    of arithmetic operations, 54
    of back substitution, 56
    of bisection algorithm, 357
    of block LU factorization, 164
    of Cholesky algorithm, 156

of diagonally dominant system, 161
　　　of Gaussian elimination, 106
　　　of Gaussian elimination for tridiagonal system, 162
　　　of Gram–Schmidt QR factorization, 205
　　　of Householder Hessenberg reduction, 311
　　　of Householder QR factorization, 193
　　　of inner and outer products, 55
　　　of inverse iteration, 300
　　　of least-squares QR factorization method, 257
　　　of QR iteration, 333
　　　of QZ iteration, 398
　　　of the Cholesky QR algorithm, 401
stiffness matrix, 406
SVD (singular value decomposition)
　　　and pseudoinverse, 368
　　　and rank deficiency, 217
　　　applications of, 221
　　　computation using Golub–Kahan–Reinsch algorithm, 370
　　　computation using MATLAB, 213
　　　definition of, 182
　　　existence theorem, 212, 365
　　　generalized, 379
　　　geometric interpretation, 214
　　　of a bidiagonal matrix, 374
　　　of a complex matrix, 213
　　　reduced, 214
　　　underdetermined least-squares solution using, 265
symmetric eigenvalue problem
　　　bisection method, 354
　　　comparison of methods, 364
　　　divide-and-conquer method, 359
　　　Jacobi method, 363
　　　QR iteration method, 357
　　　special properties of, 352

symmetric Lanczos algorithm
　　　for eigenvalues, 477
SYMMLQ method, 460

Table
　　　comparison of Krylov subspace methods, 473
　　　comparison of least-squares methods, 266
　　　comparison of LU factorization methods, 110
　　　comparison of methods for special systems, 167
　　　comparison of QR factorization methods, 207
Tacoma Bridge, 287
time response, 411
trace, 20
transient current for electric circuit, 288
triangle inequality, 17
triangular SVD, 378

underdetermined system, 263, 265
underflow, 30
unitary matrix, 15
　　　norm property, 22

Vandermonde matrix, 243
variance-covariance matrix, 367
vector, 7
　　　outer product, 9
　　　inner product, 9
　　　orthogonality, 8
vibration of a building, 405
　　　due to earthquake, 415

well-conditioned problem, 57
Wilkinson polynomial, 59
Wilkinson shift, 357
Wilkinson, James H., 54